# LAND PLANNER'S
# ENVIRONMENTAL HANDBOOK

# LAND PLANNER'S ENVIRONMENTAL HANDBOOK

by

**William B. Honachefsky**

**NOYES PUBLICATIONS**
Park Ridge, New Jersey, U.S.A.

Library of Congress Catalog Card Number: 90-23204
ISBN: 0-8155-1267-8
Printed in the United States

Published in the United States of America by
Noyes Publications
Mill Road, Park Ridge, New Jersey 07656

10 9 8 7 6 5 4 3 2 1

Library of Congress Cataloging-in-Publication Data

Honachefsky, William B.
    Land planner's environmental handbook / by William B. Honachefsky.
        p.    cm.
    Includes bibliographical references and index.
    ISBN 0-8155-1267-8 :
    1. Environmental protection. 2. Land use--Environmental aspects.
    I. Title.
    TD170.2.H63    1991
    363.7--dc20                                90-23204
                                                CIP

We have not only left the land for the cities, but worse, we have left behind any kind of feelings for that land. And it is only a question of time before the silent earth takes its vengeance against a people that denies the deepest sign of life—their own ability to feel.

Colman McCarthy, 1976
Newark Star Ledger

# Preface

The final decade of the twentieth century will bring dramatic changes in the attitudes of Americans, and the human population in general, towards the earth's environment. We simply have no other choice. Lingering and continuing damages to the environment, in combination with an unprecedented and growing human population, could exhaust or irreversibly damage some of the earth's major ecosystems by the middle of the 21st century—an alarming prospect that has prompted a worldwide resurgence of citizen activism. In America, citizen action groups are forming with increasing regularity, both as a result of this impending global threat and the public's general exacerbation with the nation's inability to decisively halt it's own environmental deterioration.

Much of the American public's ire is presently being directed at the nation's local land planners and the massive state and federal bureaucracies who have, for the last two decades, been entrusted with the responsibility to protect the national environment. The nation's land planners are being faulted primarily for their persistent and damaging piecemeal dissection of natural ecosystems. This in spite of an abundance of municipal land use plans, ordinances and regulations promulgated specifically to protect the environment. The environmental protection agencies on the other hand are being accused of having lost their sense of purpose; of lacking the necessary aggressiveness to get the job done and of being more preoccupied with self-preservation than environmental preservation.

In their defense, the environmental bureaucracies claim the public does not fully understand the complexity of the nation's environmental problems—a claim that has some validity because too few citizens have made it a point to keep up to date on contemporary environmental issues. Land planners, many of them untrained volunteers, simply shrug and say they are doing the best they can. Although, by their own admission, many have not kept fully abreast of the ever-changing agenda of contemporary environmental problems. As a consequence they often rely heavily, perhaps too heavily, on the recommendations of hired "experts"—too few of whom are environmental planners.

vii

Could we have done more in the past two decades? The answer is an unequivocal yes. The reasons we did not, however, are more complex than the public will admit, but less complex than the bureaucrats and planners would have us believe.

Our mission for the remainder of the twentieth century, however, is not to assign blame. There is little time for that. Rather we must focus our efforts on loftier goals, those of restoring the momentum of the American environmental movement; of enhancing the technical knowledge of the American public on environmental issues; of interjecting new standards into national land use-land planning schemes to give absolute priority to the preservation of complete ecosystems during land development and of inaugurating a National Environmental Master Plan.

Those who have traditionally measured the nation's well being solely by economic indices may find these new initiatives unsettling. However, Paul Ehrlich, professor of population studies at Stanford University, correctly points out that, "in rich nations economic growth is the disease, not the cure." We have been, as Ehrlich says, "living on our capital—a one time bonanza of fossil fuels, Pleistocene ground water, high-grade mineral ores, topsoil and biodiversity" ("The Population Explosion," Simon and Schuster, New York, 1990). Nature now demands our interest.

Local land planners and citizen activists, therefore, will welcome the arrival of the *Land Planner's Environmental Handbook.* It will greatly simplify what heretofore has been an intimidating task—that of procuring technical information on the environment from thousands, if not millions, of pages of published texts, reports and manuals. In addition to its convenience as a one-stop technical reference manual, the author has also included hindsight dramatization of many of the negative impacts of the nation's past and present land use planning efforts. Some of them will surprise you. Nearly all of the nation's hottest environmental issues are covered including: landfilling, incineration, composting, leaking underground tanks, AIDS, radon, Lyme disease, medical waste, ocean pollution and wildlife protection, to name but a few.

The book offers hope for the future with recommendations that some, not too many years ago, would have described as "tough" but now, in retrospect, are the nation's only hope to keep its quality of life.

March 1991                                    William B. Honachefsky

# Acknowledgments

The author wishes to thank the following groups, organizations and individuals who contributed various reference documents and photographs to this text:

The Association of New Jersey Environmental Commissions
The Township of Clinton, Hunterdon County, New Jersey
Mr. John Mihatov
Mr. Edward Markowski
Mr. B. Budd Chavooshian
The South Branch Watershed Association
Pollution Engineering Magazine
PURAC Engineering Inc.
Public Works Magazine
BioCycle Magazine
West Publishing Co.
CW Powers Co., Inc.
Prof. Marco de Bertoldi
Van Nostrand Reinhold Co.
Wisconsin Geological and Natural History Survey
Dr. Stanley Erlandsen

# About the Author

William Honachefsky, an environmental scientist, is also a licensed professional land surveyor, a professional land planner and a licensed health officer in the State of New Jersey. For the past 20 years, he has specialized in the field of environmental protection, both in private enterprise and in state and federal government. Most recently, he designed and coordinated a massive cleanup of New Jersey's coastal shorelines utilizing prison inmate labor.

A staunch consumer advocate, his prior book, "Danger—Real Estate for Sale," has helped innumerable consumer-buyers avoid the pitfalls of buying environmentally constrained real estate. He also lectures frequently on environmental protection.

## NOTICE

To the best of the Publisher's knowledge the information contained in this publication is accurate; however, the Publisher assumes no responsibility nor liability for errors or any consequences arising from the use of the information contained herein. Final determination of the suitability of any information, procedure, or product for use contemplated by any user, and the manner of that use, is the sole responsibility of the user.

The book is intended for informational purposes only. The reader is warned that caution must always be exercised when dealing with materials or procedures which might be considered hazardous. Advice from responsible experts should be obtained at all times when implementation is being considered.

Mention of trade names or commercial products does not constitute endorsement or recommendation for use by the Publisher.

# Contents

# 1

# The Environment in Perspective

## A BRIEF HISTORY

In order to place the current state of the nation's environment in proper perspective, I think it is necessary to review, if only briefly, the historical development of the American environmental ethic [1].

The current American philosophy of environmental protection which is now an important part of our national agenda, was not easily assimilated into American society. In fact, the process took place over a period of nearly two hundred years. Why it took so long for Americans to become serious about their roles as stewards of the nation's natural resources has been examined at great length by previous authors and historians. There is no need to repeat those detailed examinations in this text.

The development of the American environmental protection (formerly conservation) ethic occurred generally in three phases. The first, covering the time period from the end of the American revolutionary war to the beginning of the 20th century can be described as the Formulation Phase. Abundance was the key word of this period--an abundance of natural resources and human greed. In spite of this, it was during this time period that a small but diverse group of citizens, including philosophers, teachers, historians, public servants, and politicians began to catalog the nation's natural resources and to observe and critique the performance of their fellow Americans and the human race generally, towards their natural environment. It was these early naturalists who would formulate the relatively inviolate values of the American conservation ethic.

1

These values were best summed up by George Perkins Marsh who is often considered the nation's first ecologist. In his 1864 book entitled, "Man and Nature", he states:

> "The ravages committed by man subvert the relations and destroy the balance which nature has established...; and she avenges herself upon the intruder by letting loose her destructive energies...When the forest is gone, the great reservoir of moisture stored up in its vegetable mould is evapo- rated...The well-wooded and humid hills are turned to ridges of dry rock... and...the whole earth, unless rescued by human art from the physical degradation to which it tends, becomes an assemblage of bald mountains of barren, turfless hills, and of swampy and malarious plains. There are parts of Asia Minor, of Northern Africa, of Greece, and even Alpine Europe, where the operation of causes set in action by man, has brought the face of the earth to a desolation almost as complete as that of the moon...The earth is fast becoming an unfit home for its noblest in- habitant, and another era of equal human crime and human improvidence...would re- duce it to such a condition of impover- ished productiveness, of shattered sur- face, of climate excess, as to threaten the depravation, barbarism, and perhaps even extinction of the species."

It is not surprising that his concerns and ob- servations differ very little from those offered by contemporary American environmentalists.

The second phase, called the Period of Height- ened Concern lasted from around 1900 through 1960. This period is marked by the continued subservience of the nation's natural resources to the personal empires of a wealthy few. The conservation effort peaked and waned during this phase, interrupted by an economic depression and two world wars. Two notable peaks were the presidencies of Theodore Roosevelt, (1903-1909), and Franklin D. Roosevelt, (1933-1945). The pugna- cious "T. R." will be remembered mostly for his at- tempts to focus the attention of the nation on conservation, for retrieving over 150 million acres of American forests from the clutches of corrupt, vested interests and for commencing an inventory of the nation's natural resources. At the end of his presidency, however, Americans still remained prodigal with their resources. The black period just prior to Franklin Roosevelt's presidency is described in Stuart Udall's book, "The Quiet Crisis":

> "The economic bankruptcy that gnawed at our country's vitals after 1929 was closely related to a bankruptcy of land stewardship. The buzzards of the raiders had, at last, come back to roost, and for each bank failure, there were land failures by the hundreds. In a sense, the Great Depression, was a bill collector sent by nature, and the dark tidings were borne on every silt laden stream and every dust cloud that darkened the horizon."

Franklin Roosevelt made conservation, in this case land restoration, an integral part of his war against the depression. His greatest contributions were the creation of the Civilian Conservation Corps (C.C.C.), the Tennessee Valley Authority (T.V.A.), and the Soil Conservation Service (S.C.S.). The active incorporation of the general populace in his conservation programs established a new sense of personal responsibility to the environment for millions of Americans. Unfortunately, the 15 years that followed his presidency and which ended the second phase were relatively uneventful for the conservation movement.

The third phase of the environmental movement is best described as the Period of Panic, and commenced around 1960 and continues to the present.

Aldo Leopold, one of the nation's premier conservationists, had complained in 1933 (Ref. 1.3) that:

> "There is as yet no other dealing with man's relation to land and to the animals and plants which grow upon it. The lands relation is still strictly economic, entailing privileges, but not obligations... Obligations have no meaning without conscience, and the problem we face is the extension of the social conscience from people to land."

It was in 1961 that the extension of the national conscience to the land that Leopold spoke of would begin in earnest. Why at this period of history, one may ask? Perhaps it was the election of a vibrant and environmentally sensitive president, or the worsening condition of the nation's natural resources or a healthy economy, which now allowed citizens to earnestly focus on issues other than an empty stomach or an empty wallet. In all likelihood, it was a combination of all three. But, there was also another milestone in the conservation movement about to occur. An adroit, highly principled scientist laboring away in Maryland was already concluding what would probably be her magnum opus on environmental degradation. Armed with new technologies and a communication network previously unavailable to earlier naturalists and scientists, Rachel Carson jolted the national conscience with her 1962 book entitled, "Silent Spring". Her book's success was due largely to her ability to distill the incomprehensible into a few hundred pages of easily understood prose. Citizens who may have suspected, but had difficulty comprehending the magnitude of the nation's misuse of pesticides and chemicals, found Carson's book on the topic frighteningly clear. Equally as important was the fact that "Silent Spring" also stimulated a personal and national inquiry into the ramifications of man's other transgressions upon the environment.

The old values formulated by the likes of Marsh, Thoreau, Emerson, Muir, Powell and Leopold were suddenly resurrected by new legions of advocates. By the latter part of the sixties, environmental protection had risen to the top of the national agenda culminating in the passage of the National Environmental Policy Act in 1969. One year later, a nation filled with remorse would celebrate the world's first Earth Day.

In the decade that followed, the federal government produced environmental protection laws, rules and regulations, literally by the ton. States and municipal governments responded in a similar fashion, compelled either by federal mandates or citizen initiatives. Being anti-environment nearly equated to being anti-American, so deep was the guilt and remorse of the nation.

Despite the widespread popularity of environmental protection and anti-pollution legislation, little of it passed through Congress or state legislatures with ease. Political action committees, many newly formed and representing a broad cross-section of corporate America, lobbied very effectively for key modifications and changes. Confrontations were common and often bitter. Compromise would become a common, but distasteful word in the vocabulary of modern American conservationists, now generally referred to as environmentalists. This large scale production of environmental legislation had negative side effects other than the most obvious one of widening the schism between corporate America and American environmentalists. It also promoted the false theory of the "legislative cure," the hypothesis of which is that the mere passage of legislation will somehow cure the problem. The most significant negative impact of this mass production effort, however, was the fact that it promoted the diminution of individual participation in the environmental movement. Abetted by burn out from over a decade of concerted effort to produce legislation that often had unpalatable compromises, many Americans found it easy to assume that the massive bureaucracy they had set in place would diligently continue the battle for them. It was a false presumption.

## THE PROBLEMS CONTINUE

It is a rare day, indeed, in contemporary America when some incident of environmental degradation is not reported by the news media. With such negative reports appearing daily, it is difficult for Americans to maintain an optimistic outlook on the state of the national environment. Some claim we are on the brink of an environmental Armageddon while others scoff at such predictions as preposterous. While there is little doubt that we have made considerable improvements in our attitudes about the environment and in the environment itself, since the turn of the 20th century it is difficult to ignore the continued documentation of symptoms of an ailing environment. It is not possible to list the total sum of the nation's current environmental ailments, here. How-

ever, it is important that we mention  at least some
of them.

## Estuaries and Coastal Water Pollution

A significant portion of the nation's 32,000
square miles of estuaries, show either increasing
symptoms of stress or persistent degradation due
largely to accelerated development in these areas.
There was a 69% increase in coastal populations between 1950 and 1980 and the influx of persons into
these areas shows no signs of decreasing.  In fact,
the U.S. Census Bureau predicts that by the end of
1990, 75% of the nation's population will live within
50 miles of a coastline.  Whether it is Coos Bay,
Oregon or Buzzards Bay, Massachusetts, the causes of
environmental degradation in these estuarine areas are
strikingly similar and include discharges of raw and
partially treated domestic and industrial wastewater
as well as runoff of excess pesticides, fertilizers
and silt.  The question of whether this degradation is
gradually spreading to the earth's ultimate sink, the
oceans, was raised in the summer of 1987 along the
nation's eastern seaboard.  Some 740 bottle-nosed
dolphins succumbed to red algae-induced[2] food chain
toxicity along the Atlantic coastline from New Jersey
to Florida.  Accompanying these deaths were widespread
algae blooms and anoxic conditions along an extensive
part of the east coast from Long Island Sound in New
York to the Pamlico estuary in North Carolina.

## Ground Water Deterioration

More than 50% of the U.S. population depends on
ground water for a supply of drinking water.  Yet,
ground water in nearly all of the United States is
contaminated to some degree with a variety of materials ranging from nitrates to synthetic organic chemicals.  We have only just begun to assess how exten-
sively our ground water has been damaged, but the initial data is not encouraging at all.  While some of
our ground water problems are due to naturally occurring substances (usually inorganic) the U.S. EPA in
its 1986 report to Congress (Ref. 1.4) identified 16
major sources of ground water contamination in the

Table 1.1:  Major Ground Water Contaminants Reported by States

| Contaminant | Reported as a Major Contaminant | |
| --- | --- | --- |
| | No. of States | % of States |
| Sewage | 46 | 89% |
| Inorganic Chemicals: | | |
|   Nitrates | 42 | 75% |
|   Brine/Salinity | 36 | 69% |
|   Arsenic | 19 | 37% |
|   Fluorides | 18 | 35% |
|   Sulfur Compounds | 7 | 14% |
| Organic Chemicals: | | |
|   Synthetic | 37 | 71% |
|   Volatile | 36 | 69% |
| Metals | 34 | 65% |
| Pesticides | 31 | 60% |
| Petroleum | 21 | 40% |
| Radioactive Materials | 12 | 23% |

*Based on a total of 52 States and territories which cited ground-water contaminants in their 305(b) submittals.

Source:  Ref. 1.4

Figure 1.1: State investigators contemplate the remediation of a protected wetland that has been illegally covered with fill.

United States.   All of them are related to the activities of man.   (See Chapter 2, Table 2.4)

## Loss of Wetlands

Wetlands are areas of transition between terrestrial and aquatic systems where the water table is at or near the ground surface or the land is covered by shallow water.[3]   Often referred to as marshes, swamps, or bogs, they can be both freshwater and saltwater.   Approximately 2/3 of the major U.S. commercial fish species rely on estuaries and salt marsh wetlands for nursery or spawning grounds.   They also temporarily store floodwaters, reduce shoreline erosion, improve water quality, and provide habitat for large numbers of wildlife.   Yet, they have long been considered useless tracts of land to be converted, covered over, or drained for other uses.   As a consequence, 50% of the original 215 million acres of wetlands in the conterminous U.S. have been lost and we continue to lose 350,000 to 500,000 more acres annually (Ref. 1.5).   Some states like Iowa, Nebraska and California have lost over 90% of their original wetlands.

## Acid Deposition

Acid deposition, which includes both wet and dry acidic substances, has impacted significant areas of the United States and eastern Canada.   Caused largely by man-made emissions of sulfur dioxide and nitrogen oxides from automobile exhausts and the burning of fossil fuels (coal, oil and gas) these materials have severely affected our water, air and vegetative resources.

In 1980, the United States was responsible for the release of 26 million tons of sulfur dioxide and 22 million tons of nitrogen oxides into the air (Ref 1.6) where it combined with atmospheric moisture to produce what is now the most common symptom--lowered pH rainfall and snow.   Connecticut and New Jersey for example, have reported their rainfall has an average pH of 4.3 (compared to a 5.6 pH for uncontaminated rainfall).   Individual storms have deposited precipitation with a pH of 2.7 (equivalent to vinegar) in Kane, Pennsylvania and a pH of 1.5 on Wheeling, West Virginia.

The consequences of this acidic precipitation have been the deterioration of the exterior of many big city buildings, the acidification of numerous inland lakes with the subsequent death of all fish life and the loss of certain tree species at higher elevations. In addition, the dry portion of these pollutants, usually sulfates, has been linked to such human diseases as chronic bronchitis, asthma and emphysema.

## Infrastructure Deterioration

It is not only our natural environment that shows signs of stress. Our man-made service systems labeled generically as "infrastructure" are also in serious jeopardy. Large numbers of American sewage treatment plants, potable water supply systems, roadways, storm sewer systems and sanitary sewer systems are in drastic need of rehabilitation. It is estimated that it would cost between $260 billion and 4 trillion dollars over ten years to repair these systems.

Globally, the environment is exhibiting similar symptoms. The Worldwatch Institute in Washington, D.C. in its annual "State of the World 1988" (Ref. 1.7), indicated that the physical condition of the earth is continuing to deteriorate. They blame this deterioration of the earth's condition on the "profound misallocation" of capital for global military expenditures and the "unmanageable" debt of the Third World nations, both of which indirectly boost birth rates. As a result, populations continue to grow, destroying the environmental support systems on which future economic progress depends.

Since 1960, 70 million more persons have been added to the population of the United States alone, bringing the total here to 250 million. This unprecedented level of global human population (about 5.3 billion) is a major element contributing to the deterioration of the American and world environments.

The truth of the matter is that man has not yet learned to live in harmony with his natural environment despite a documented history of dire consequences for failing to do so. While the Third World countries may receive some leniency for such an attitude, the United States, as an originator of extraordinary advances in science and technology, should not.

Perhaps, the last comment seems unduly harsh. Certainly, the United States has made significant strides in environmental protection when compared to other parts of the world. If measured solely by the amount of environmental legislation it has produced, it would be the world leader.

As one who has examined nearly all of the environmental legislation, rules and regulations produced by our 50 states, I can attest to the fact that the amount very nearly defies comprehension. The mere production of legislation, however, has not guaranteed a solution. The state of New Jersey, the nation's most densely populated state, (nearly 1,000 persons per square mile), is a premier example of this national penchant to produce prodigious amounts of environmental protection legislation. There is not one form of recognized environmental degradation, that is not addressed by some state statute, rule or regulation in New Jersey. To their credit, however, some of this legislation has been very innovative. Yet, in spite of the presence of a plethora of stiff regulations, a massive environmental protection bureaucracy, and provisions for substantial civil and criminal penalties and sanctions, environmental degradation continues to exhaust the natural resources of the state and the patience of its residents.

In 1988, one public interest group, of which there is a growing number in New Jersey, labeled the state a "polluters playground" after documenting over 3,000 water pollution violations in two years under the state run NPDES (National Pollutant Discharge Elimination System) program. The shutdown of several thousand of the state's potable water supply wells due to ground water contamination and the imposition of sewer connection bans (see Chapter 2) in 167 of the state's 567 municipalities for failure to meet federal Clean Water Act deadlines serve as further reminders that environmental degradation cannot be simply legislated out of existence.

One of New Jersey's renowned environmental journalists, Gordon Bishop, commented in 1987 that:

"New Jersey is conspicuously perched at the tip of the population needle. More people, cars, buildings, roads, pipes and wires are crammed per square mile within this confined space than in any state in

the nation.  Wherever America is going in
the 21st century, New Jersey will get
there first on those indices that define
the quality of life."

Other states and even some other nations, much
less densely populated, have looked to New Jersey for
guidance and perhaps for a glimpse of their own future.  If Mr. Bishop's prediction, that as New Jersey
goes, so goes the nation, proves accurate, then we all
have significant reason to be alarmed over our future.

What is happening in New Jersey typifies what is
happening in varying degrees nationally with our environmental protection agenda.  Consequently, it seems
prudent to identify at least some of the factors
limiting the success of this state's efforts to clean
up and protect its environment.

1.  **In spite of strong endorsement of stringent
environmental protection programs, not nearly enough
of the state's residents have made it a truly personal
concern until very recently.**  Citizen involvement has
been accomplished largely by proxy through various
non-profit environmental groups.  This dearth of personal involvement has had a substantial impact on the
effectiveness of the state's environmental protection
programs.  For example, strong environmental bills are
often emasculated by special interest lobbying groups
prior to passage for lack of an overwhelming popular
lobbying effort.  As a result, legislation is produced
that often prolongs rather than truncates, environmental degradation.  An additional consequence is
the isolation of the public from the regulatory agencies which makes it difficult if not impossible to
pinpoint accountability when programs fail to meet
their objectives.

2.  **Despite the presence of innumerable land
use-land planning documents containing substantial
provisions for protection of the environment, development in the state continues in a manner that promotes
the degradation of the state's environment.**  New Jersey has 21 county governments and 567 municipal governments; each of which has its own master plan for
land development.  In addition, there are a dozen
Areawide Waste Treatment Management Plans and 125
Wastewater Treatment Facility Plans required under
Section 208 and 201, respectively, of the federal
Clean Water Act.  If this is not confusing enough,

there are even more special land use plans for the state's National Pinelands Reserve, the Hackensack Meadowlands area, the coastal wetlands area, and the Delaware River Basin. It would certainly appear that what the state does not need is another land use plan. Yet, that is exactly what is desperately needed. In spite of this glut of planning documents, there is no single state plan to unify this morass of planning documents into some logical pattern that will protect the state's environment. An attempt in this regard was made in 1978 with the production of a State Development Guide Plan. However, it lacked the provisions of authority to overrule local planning (often referred to as "home rule"). With the consequences of such disorganized planning now threatening the state's "quality of life," the concept of a state development guide plan has been resurrected. A new plan called the State Development and Redevelopment Plan has been drafted and is at present undergoing a lengthy review and acceptance process by all of the state's county and municipal governments. This plan proposes a 7 tier growth management system which is explained more fully in Appendix 1, Part A. Although scheduled to be finalized in 1990 it is receiving some withering criticism from politicians, builders, farmers and other special interest groups who foresee this document as a threat either to their personal wealth or the state's present economic prosperity. If these special interest groups prevail the plan's implementation could be delayed for years or worse yet emasculated of meaningful provisions.

3.  **The political influences on the state's Department of Environmental Protection (DEP) detrimentally affects its mission and productivity.** New Jersey's DEP is headed by a Commissioner, who is the Department's chief policy maker. The Commissioner is appointed by the Governor with approval of the state legislature. As a consequence, Commissioners come and go as frequently as the governorship changes (usually every 4 years). In the last 19 years, New Jersey's DEP has had ten Commissioners. Unfortunately, there has been little continuity between Commissioners regarding Departmental priorities. In addition, each new Commissioner tends to restructure the upper management of the Department; promoting and demoting

until he or she is comfortably surrounded by loyal
supporters.

Priorities in some regimes have been as superfi-
cial as making certain all correspondence from the
public is answered within seven days to assuring that
everyone in the Department wears a dress shirt and tie
in order to present a good public image. Unfortunately, there have been too few priorities with real
long term substance. After several changes of Commissioners, Departmental staff are left confused over
what the objectives of the Department really are and
ultimately their confusion deteriorates to indiffer-
ence.

The influx of substantial amounts of federal
grant monies from the U.S. EPA also dictates some of
the priorities that the DEP must pursue. These priorities can change yearly, usually reflecting the criti-
cisms EPA has received on its own programs. The EPA
priorities are often incompatible with state priorities.

Lastly, influential politicians and corporate
executives with the proper political contacts can and
do influence the decision making process, particularly
regarding the issuance of permits.

4.  **Enforcement of the state's environmental
protection laws is erratic and well below the level of
aggressiveness envisioned by the state's residents--
mirroring perhaps, the federal EPA's performance in
this regard.**

Aside from the confusion generated by the continuous change in priorities mentioned previously,
other factors contribute to this disappointing enforcement effort. First, the level of documentation
needed to successfully prosecute environmental cases
through the judicial process has increased dramatically in the past decade. Sample collection and sample analysis methodologies alone, have become extremely sophisticated in order to withstand the withering cross-examination of highly specialized corporate environmental attorneys--a number of whom, I
might add, were former state deputy attorney generals,
who previously represented the public interest in environmental cases. Consequently, case documentation
becomes extremely lengthy and costly. Some cases have
been in the courts for a decade or more while the appeals process is exhausted. Although satisfying, if

won, strong enforcement actions can consume substantial amounts of manpower and money. Spending $200,000 to $300,000 to prosecute a single case is not unheard of. As a result, New Jersey has resorted to negotiating settlements. Unfortunately, such bargaining often requires the state to make concessions to the violator which can dilute rigorous enforcement actions.

5. **The overall "assimilative capacity" of the states natural systems is inadequately evaluated when permits are issued to discharge wastes into those systems.**

Many of the state's residents falsely equate a DEP permitted facility with zero discharge; that is to say, they assume that the DEP issues permits allowing only minute traces of waste to be discharged. This, of course, is a completely erroneous assumption. For instance, the state run NPDES (National Pollutant Discharge Elimination System) permit system which regulates wastewater discharges to surface waters allows municipal sewage treatment plants to discharge between 30 mg/l and 45 mg/l of $BOD_5$[4], (these are seven day and 30 day average levels, respectively, and consequently the average daily levels can be higher or lower), which seem rather small amounts at first glance. When converted to tons per year, however, this amounts to between 45 tons and 68 tons of $BOD_5$ per year for a single municipal treatment plant processing a million gallons of wastewater per day (roughly the amount of wastewater generated by 10,000 persons). Industrial permittees are allowed to discharge even larger amounts of such wastes along with tons of other, far more toxic materials. With some 1600 plus permitted surface water dischargers in the state, the annual deposit of wastes in the state's waterways is enormous. Yet, the state has little knowledge of the maximum "assimilative capacity" of its surface water system.

6. **Some of the state's discharge permitting programs are seriously flawed.** The NPDES program (NJPDES in New Jersey) that we briefly described in the prior item, for example, has other glaring deficiencies. First, the state's Department of Environmental Protection conducts permit compliance sampling at less than 5% of the total permitted facilities each year. As a result the state has to rely heavily on permittee self-monitoring to determine

compliance. This might be acceptable except for the fact that many of the sampling locations used by the permittees do not provide an accurate sample of the wastewater being discharged. Some of the reasons for this are included in the following:

(a)     Many of the NJPDES permittees sampled do not have an accurate flow measuring device. Where they do exist some have been out of service or uncalibrated for a decade or more. Some never had any flow measuring devices whatever in spite of the fact that their permit limits have been in pounds or kilograms per day since 1972.

(b)     Many dischargers to tidally influenced waters have designated sampling points where effluent is routinely mixed and subsequently diluted with incoming brackish water thereby falsely skewing the samples.

(c)     Some industrial dischargers who withdraw raw water from surface water bodies for use as process water, do so at a point downstream of their permitted discharge points, thereby, falsely skewing the background parameters so that treatment efficiencies are less than would be normally required.

(d)     There is a widespread manipulation of flows at domestic sewage treatment plants whereby supernatant is returned to the head of the treatment plant at the designated influent sampling point thereby again falsely increasing influent values so that percent removal efficiencies will be less than required.

7.     **In spite of a staff of over 3,900 persons (up from 900 in 1970) many of the state Department of Environmental Protection's programs are understaffed.** This is the result of both poor management and an underestimation of the magnitude of some of the state's environmental problems. A classic example of the latter is the state's Environmental Cleanup and Responsibility Act (ECRA) administered by the DEP which became law in January, 1984 (see Appendix 1, Part A). This statute requires a pollution assessment and a cleanup, if necessary, of any New Jersey manufacturing site before it can be sold. When the state legislature

established ECRA it anticipated the program would process about 100 applications a year. As a consequence it began with a budget of $400,000 and a staff of 10 including secretarial personnel. In the first year they received 450 applications. The second year 850 and the third year 1,200. Lacking adequate staff, delays in processing applications became commonplace much to the irritation of applicants. The legislature had completely underestimated the task of cleaning up a hundred years of industrial pollution at the state's 15,000 manufacturing facilities. With a present staff of 104 and a budget of $5 million the processing of applications moves much quicker now, but not quick enough for the state's industrial-commercial interests. They continue to lobby for major reforms in the program in spite of the fact that it has saved taxpayers over $100 million in cleanup costs. Unfortunately, their allegations that excessive delays in processing applications are threatening the revitalization of the state's distressed urban areas has captured the interest of some key lawmakers.

      8. **Municipal planning boards who have histori-cally controlled most of the state's development** often **lack the expertise to fully understand the environmen-tal implications of their decisions.** Instead they rely heavily on the opinions of one or more of the following commonly hired experts: the professional engineer, the professional planner and the attorney--none of whom may be degreed in the environmental sciences. While local environmental commissions and the State Association of N.J. Environmental Commissions have provided some expertise the local planning boards need to hire one additional expert--The Environmental Planner.

**TOUGHER ENFORCEMENT AND PREVENTATIVE PLANNING--THE TWO KEYS**

      In spite of this somewhat bleak assessment, I am confident that the nation's seemingly overwhelming environmental afflictions can be remedied, but not without a dramatic change in the way we currently do business. Let there be no doubt, however, that the situation is very precarious. We still have not yet uncovered the full legacy of environmental damage left behind by our indifferent and unwitting predecessors.

As a consequence, any further contributions we may make to this existing damage could complicate our recovery. Consequently, we need to immediately adopt a more tenacious approach to enforcement of our current environmental laws. We also need to set priorities for remediation. The present "shotgun" approach to solve every problem simultaneously severely dilutes available resources and encourages the public perception that the existing environmental protection agencies are ineffective or even worse, incompetent. In addition, we need to encourage the revival of citizen involvement and to seek their active support. Likewise, we need to provide the public with candid information and reports, acknowledging that regulatory agencies make mistakes and that they will not always have immediate answers or solutions.

Lastly, and perhaps, most importantly we need to encourage preventative planning and to replace the theory of environmental subjugation with the theory of environmental sovereignty. In the remaining chapters of this book, we will emphasize the preventative planning concept presenting techniques and technologies that will be useful to anyone desiring to plan land use in harmony with the natural environment.

**FOOTNOTES**

**CHAPTER 1**

(1)    A more detailed examination of the development of the national environmental movement may be found in treatises by Udall and Petulla (Ref. 1.1, 1.2).

(2)    A team of scientists working under the auspices of the National Oceanic and Atmospheric Administration completed an 18-month long study of the bottle-nosed dolphin deaths in February, 1989. They concluded that an unusually long bloom of "red tide" algae along the West Coast of Florida had contaminated menhaden and Spanish mackerel, common prey of the dolphins with a powerful poison call brevetoxin. The red tide algae, menhaden and mackerel were all carried northward along the eastern seaboard by the Gulf Stream current and consequently supplied the dolphins with a steady diet of tainted prey during their migration. Some scientists, however, questioned the findings stating that red tide blooms are common in the Gulf of Mexico which serves as the home for thousands of bottle-nosed dolphins. Yet, there has been no previous record of dolphins being affected by this food chain toxicity. PCB's and DDT were found in the dolphins flesh but not at levels considered to be dangerous or lethal.

(3)    Wetlands as defined by the U.S. Fish and Wildlife Service must have one or more of the following three attributes: 1) at least periodically, the land supports predominantly hydrophytes; 2) the substrata is predominantly undrained hydric soil; and 3) the substrata is non-soil and is saturated with water or covered by shallow water at some time during the growing season of each year. (Ref. 1.8)

(4)    See Glossary for description of BOD. The term $BOD_5$ used here represents the measurement of B.O.D. taken at the end of a 5 day incubation period. Similarly if the measurement was made at the end of a 20 day period it would be

reported as $BOD_{20}$. The $BOD_5$ form, however, is the most commonly used measurement for reporting BOD.

**REFERENCES**

**CHAPTER 1**

1.1    Udall, Stewart L., "The Quiet Crisis", Holt, Rhinehart and Winston, New York, 1963.

1.2    Petula, Joseph M., "Environmental Protection in the United States", San Francisco Study Center, San Francisco, CA, 1987.

1.3    Leopold, Aldo, "The Conservation Ethic", Journal of Forestry, October, 1933.

1.4    United States Environmental Protection Agency, "National Water Quality Inventory--1986 Report to Congress", EPA 440/4-87-008, November, 1987.

1.5    United States Department of the Interior--Fish and Wildlife Service, "Wetlands of the United States: Current Status and Recent Trends", March, 1984.

1.6    La Bastille, Ann, "Acid Rain--How Great A Menace?", National Geographic, Vol. 160, No. 5, November, 1981.

1.7    Brown, Lester R. and William U. Chandler, Alan Durning, Christopher Flavin, Lori Heise, Jodi Jacobson, Sandra Postel, Cynthia Pollock Shea, Linda Starke, Edward C. Wolf, "State of the World--1988", Worldwatch Institute, W.W. Norton Co. Inc., New York, 1988.

1.8    Cowardin, Lewis M., Virginia Carter, Francis C. Golet and Edward T. LaRoe, "Classification of Wetlands and Deepwater Habitats of the United States", U.S. Dept. of the Interior, Washington D.C., 1979.

# 2

# Water Resources

## INTRODUCTION

Every day 4.2 trillion gallons of water (in the form of precipitation) falls on the United States (Ref. 2.1). This precipitation is part of a complex process of recirculation that has existed for millions of years. The molecules of water that fall to the earth's surface today are very likely the same ones, or a portion at least, of the ones that fell to earth millions of years ago. This recycling process has been termed the Hydrologic Cycle and is illustrated in Figure 2.1. There is no replenishment of the world's water supply from beyond the earth's atmosphere despite perceptions to the contrary.

Seventy-one percent of the earth's surface is covered by water and this apparent superabundance has fostered an illusion of inexhaustibility. Ninety-seven percent (97%) of the total world supply of water, however, is in the world's oceans and is generally unutilized due to its salinity. The remaining 3%, the freshwater, non-saline portion, is stored temporarily in the polar ice and glaciers, lakes, rivers, soil, ground water, hydrated earth minerals, plants, animals and of course, the atmosphere, Three-fourths (3/4) of this fresh water is locked up as ice and is likewise unavailable. The remaining one-quarter (1/4) fraction, which is largely ground water, has been the major source of water for human needs. Thus, despite the appearance of a copious supply of water, human life on earth is precariously linked to a very tiny fraction of the earth's total water supply.

Climatic conditions often produce extreme variations in the abundance of our water resources. In one instance we can have too much (i.e. flooding) and in

Figure 2.1:  The Hydrologic Cycle.

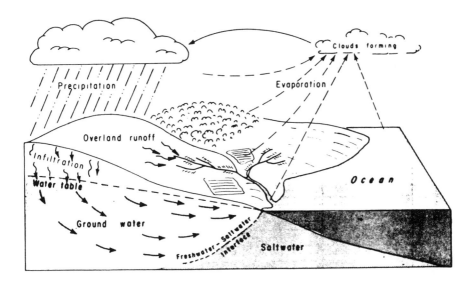

Source:  Ref. 2.5

Figure 2.2:  Percentage of total offstream withdrawals and freshwater consumptive use, by categories of use, 1980.

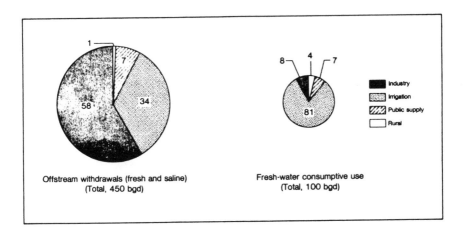

Source:  Ref. 2.2

another instance too little (i.e. drought). Both extremes can occur simultaneously as close as several hundred miles apart. Human cultural influences can exasperate these extremes. In 1980 for example, Americans were withdrawing 450 billion gallons of water a day from the nation's ground and surface water sources for a multitude of uses (Ref. 2.2, Table 2.1, 2.2 and Figure 2.2). As a result of such persistent withdrawals ground water levels have fallen to unprecedented lows and both surface and ground waters have been left with indelible traces of human contamination that threaten our health and that of generations to come. Likewise, our proclivity to reshape the earth's physical features to suit our needs has wreaked havoc with the earth's natural water drainage systems such that damages from flooding now cost the nation billions of dollars a year. [2]

It is expected that the nation's population, as well as the world's, will continue to expand, reaching unprecedented levels, thereby placing an even greater demand upon available water resources. Whether there will be adequate water resources available for these new consumers (competitors really) and further damages from flooding depends largely upon the strategies implemented by the nation's land planners in the next decade and a half. It is for certain, however, that land planning strategies which perpetuate the present haphazard dissection of the nation's watersheds and destructive alteration of natural drainage systems are an assured blueprint for failure.

## WATER SUPPLY

All of earth's life forms are dependent in some way upon water. Most, however, are passively liked to its cyclical abundance or absence. Not so with the human population which has actively exploited and manipulated the world's water resources for a multitude of uses including crop irrigation, wastewater treatment, cooling water and industrial processing. No use, however, is as essential as clean water for drinking. Wars have been fought for it and empires and cities have fallen because of the lack of it. Humans may go without food for 30 days or more but 3 to 7 days without a drink of water is almost always fatal. The human body itself is mostly water ranging

Table 2.1:  Summary of Estimated Water Use in the United States, in Billion Gallons per Day, at 5-Year Intervals, 1950-80

[The data generally are rounded to two significant figures; however, the percentage changes are calculated from unrounded numbers]

| | Estimated water use in billion gallons per day | | | | | | | Percentage increase (+) or decrease (−) | |
| --- | --- | --- | --- | --- | --- | --- | --- | --- | --- |
| | 1950[1] | 1955[1] | 1960[2] | 1965[2] | 1970[3] | 1975[4] | 1980[4] | 1970-75 | 1975-80 |
| Population, in millions | 150.7 | 164.0 | 179.3 | 193.8 | 205.9 | [5]216.4 | 229.6 | +5 | +6 |
| Offstream use: | | | | | | | | | |
| Total withdrawals | [5]180 | 240 | 270 | 310 | 370 | 420 | 450 | +12 | +8 |
| Public supply | 14 | 17 | 21 | 24 | 27 | 29 | 34 | +8 | +15 |
| Rural domestic and livestock | 3.6 | 3.6 | 3.6 | 4.0 | 4.5 | 4.9 | 5.6 | +10 | +14 |
| Irrigation | [5]89 | 110 | 110 | 120 | 130 | 140 | 150 | +11 | +7 |
| Self-supplied industrial: | | | | | | | | | |
| Thermoelectric power use | 40 | 72 | 100 | 130 | 170 | 200 | 210 | +18 | +9 |
| Other industrial uses | 37 | 39 | 38 | 46 | 47 | 45 | 45 | −6 | +1 |
| Source of withdrawals: | | | | | | | | | |
| Ground water: | | | | | | | | | |
| Fresh | 34 | 47 | 50 | 60 | 68 | 82 | 88 | +22 | +7 |
| Saline | (6) | .6 | .4 | .5 | 1 | 1 | .9 | −6 | −5 |
| Surface water: | | | | | | | | | |
| Fresh | [5]140 | 180 | 190 | 210 | 250 | 260 | 290 | +5 | +10 |
| Saline | 10 | 18 | 31 | 43 | 53 | 69 | 71 | +31 | +2 |
| Reclaimed sewage | (6) | .2 | [5].6 | .7 | .5 | .5 | .5 | +2 | −11 |
| Consumptive use | (6) | (6) | 61 | 77 | [7]87 | [7]96 | [7]100 | +10 | +7 |
| Instream use: | | | | | | | | | |
| Hydroelectric power | 1,100 | 1,500 | 2,000 | 2,300 | 2,800 | 3,300 | 3,300 | +21 | −2 |

[1] 48 States and District of Columbia.
[2] 50 States and District of Columbia.
[3] 50 States, District of Columbia, and Puerto Rico.
[4] 50 States, District of Columbia, Puerto Rico, and Virgin Islands.
[5] Corrected from published report.
[6] Data not available.
[7] Fresh water only.

Source:  Ref. 2.2

## Table 2.2:  Rural Freshwater Use, by State, in Million Gallons per Day, 1980

[Data generally are rounded to two significant figures; figures may not add to totals because of independent rounding]

| STATE | DOMESTIC USE | | | | LIVESTOCK USE | | | | TOTAL DOMESTIC AND LIVESTOCK USE | | | |
|---|---|---|---|---|---|---|---|---|---|---|---|---|
| | Withdrawals | | | Consumptive use | Withdrawals | | | Consumptive use | Withdrawals | | | Consumptive use |
| | By source | | Total | | By source | | Total | | By source | | Total | |
| | Ground water | Surface water | | | Ground water | Surface water | | | Ground water | Surface water | | |
| Alabama | 100 | 0 | 100 | 100 | 25 | 63 | 88 | 88 | 130 | 63 | 190 | 190 |
| Alaska | 11 | 0.1 | 11 | 0.1 | 0 | 0.1 | 0.2 | 0.2 | 11 | 0.3 | 11 | 0.3 |
| Arizona | 32 | 0 | 32 | 24 | 9.8 | 1.8 | 12 | 8.1 | 42 | 1.8 | 43 | 32 |
| Arkansas | 57 | 0 | 57 | 51 | 22 | 39 | 61 | 61 | 78 | 39 | 120 | 110 |
| California | 130 | 9.5 | 140 | 82 | 36 | 51 | 87 | 46 | 160 | 60 | 220 | 130 |
| Colorado | 35 | 62 | 98 | 24 | 19 | 86 | 110 | 35 | 54 | 150 | 200 | 59 |
| Connecticut | 53 | 0 | 53 | 32 | 0.4 | 1.8 | 2.2 | 2.2 | 54 | 1.8 | 56 | 34 |
| Delaware | 25 | 0 | 25 | 0 | 2.0 | 0 | 2.0 | 2.0 | 27 | 0 | 27 | 2.0 |
| D.C. | 0 | 0 | 0 | 0 | 0 | 0 | 0 | 0 | 0 | 0 | 0 | 0 |
| Florida | 250 | 0.1 | 250 | 42 | 39 | 20 | 59 | 59 | 290 | 20 | 310 | 100 |
| Georgia | 140 | 0 | 140 | 85 | 17 | 11 | 28 | 28 | 150 | 11 | 160 | 110 |
| Hawaii | 3.5 | 0.4 | 3.9 | 3.4 | 5.3 | 0.2 | 5.5 | 4.8 | 8.8 | 0.6 | 9.4 | 8.2 |
| Idaho | 44 | 2.0 | 46 | 11 | 9.3 | 13 | 22 | 19 | 53 | 15 | 68 | 30 |
| Illinois | 79 | 3.6 | 82 | 58 | 49 | 16 | 65 | 65 | 130 | 20 | 150 | 120 |
| Indiana | 110 | 5.6 | 120 | 120 | 24 | 19 | 42 | 42 | 130 | 24 | 160 | 160 |
| Iowa | 55 | 0.2 | 55 | 22 | 100 | 25 | 130 | 130 | 160 | 25 | 180 | 150 |
| Kansas | 58 | 4.3 | 63 | 59 | 35 | 46 | 81 | 79 | 93 | 50 | 140 | 140 |
| Kentucky | 54 | 6.3 | 61 | 48 | 1.9 | 37 | 39 | 39 | 56 | 43 | 99 | 87 |
| Louisiana | 54 | 0 | 54 | 39 | 12 | 5.2 | 18 | 18 | 67 | 5.2 | 72 | 57 |
| Maine | 26 | 0.5 | 26 | 26 | 1.0 | 0.7 | 1.7 | 1.7 | 27 | 1.2 | 28 | 28 |
| Maryland | 49 | 0 | 49 | 32 | 10 | 0.5 | 11 | 11 | 59 | 0.5 | 60 | 43 |
| Massachusetts | 32 | 0 | 32 | 3.9 | 0.7 | 0.5 | 1.2 | 1.2 | 32 | 0.5 | 33 | 5.1 |
| Michigan | 160 | 0 | 160 | 27 | 17 | 5.0 | 22 | 19 | 180 | 5.0 | 180 | 46 |
| Minnesota | 120 | 0 | 120 | 120 | 58 | 10 | 68 | 68 | 180 | 10 | 190 | 190 |
| Mississippi | 27 | 0 | 27 | 24 | 9.7 | 12 | 21 | 21 | 37 | 12 | 49 | 45 |
| Missouri | 68 | 24 | 92 | 39 | 17 | 48 | 65 | 58 | 85 | 72 | 160 | 98 |
| Montana | 60 | 0 | 60 | 60 | 14 | 14 | 28 | 28 | 74 | 14 | 88 | 88 |
| Nebraska | 49 | 0 | 49 | 49 | 93 | 23 | 120 | 110 | 140 | 23 | 170 | 160 |
| Nevada | 11 | 0.7 | 11 | 6.6 | 3.7 | 8.5 | 12 | 8.9 | 14 | 9.2 | 24 | 15 |
| New Hampshire | 9.1 | 0.2 | 9.3 | 0.5 | 0.2 | 0.5 | 0.8 | 0.7 | 9.3 | 0.8 | 10 | 1.2 |
| New Jersey | 75 | 0 | 75 | 15 | 2.0 | 1.0 | 3.0 | 2.5 | 77 | 1.0 | 78 | 17 |
| New Mexico | 32 | 1.1 | 33 | 15 | 9.6 | 9.6 | 19 | 9.6 | 42 | 11 | 52 | 25 |
| New York | 130 | 0 | 130 | 13 | 37 | 20 | 58 | 52 | 170 | 20 | 190 | 65 |
| North Carolina | 140 | 0 | 140 | 140 | 33 | 5.6 | 39 | 39 | 170 | 20 | 190 | 65 |
| North Dakota | 11 | 0.2 | 11 | 11 | 13 | 8.2 | 21 | 21 | 24 | 8.4 | 32 | 32 |
| Ohio | 80 | 8.8 | 89 | 62 | 24 | 16 | 40 | 36 | 100 | 25 | 130 | 98 |
| Oklahoma | 29 | 5.2 | 35 | 31 | 8.2 | 50 | 58 | 58 | 38 | 55 | 93 | 89 |
| Oregon | 130 | 19 | 150 | 150 | 7.1 | 19 | 26 | 26 | 140 | 38 | 170 | 170 |
| Pennsylvania | 150 | 0 | 150 | 15 | 54 | 7.0 | 61 | 41 | 200 | 7.0 | 210 | 56 |
| Rhode Island | 4.9 | 0 | 4.9 | 0.8 | 0.1 | 0.1 | 0.2 | 0.2 | 5.0 | 0.1 | 5.1 | 1.0 |
| South Carolina | 65 | 0.2 | 65 | 65 | 12 | 10 | 22 | 22 | 77 | 10 | 87 | 87 |
| South Dakota | 21 | 1.4 | 22 | 16 | 81 | 11 | 92 | 85 | 100 | 12 | 110 | 100 |
| Tennessee | 43 | 0 | 43 | 12 | 7.0 | 35 | 42 | 42 | 50 | 35 | 85 | 54 |
| Texas | 130 | 0 | 130 | 130 | 120 | 150 | 270 | 270 | 250 | 150 | 400 | 400 |
| Utah | 26 | 3.3 | 29 | 10 | 31 | 9.0 | 40 | 11 | 57 | 12 | 69 | 21 |
| Vermont | 17 | 2.6 | 20 | 1.0 | 5.7 | 3.5 | 9.2 | 9.2 | 23 | 6.1 | 29 | 10 |
| Virginia | 150 | 0.1 | 150 | 74 | 2.3 | 26 | 28 | 17 | 150 | 26 | 180 | 91 |
| Washington | 40 | 11 | 52 | 18 | 4.1 | 2.0 | 6.1 | 3.0 | 44 | 13 | 58 | 21 |
| West Virginia | 18 | 1.3 | 19 | 0.2 | 1.0 | 6.6 | 7.6 | 6.7 | 19 | 7.9 | 27 | 6.9 |
| Wisconsin | 72 | 0 | 72 | 7.0 | 72 | 3.0 | 75 | 75 | 140 | 3.0 | 150 | 82 |
| Wyoming | 8.8 | 0.8 | 9.6 | 6.7 | 3.1 | 12 | 15 | 15 | 12 | 13 | 25 | 21 |
| Puerto Rico | 3.0 | 3.0 | 6.0 | 1.0 | 15 | 15 | 30 | 7.0 | 18 | 18 | 36 | 8.0 |
| Virgin Islands | 2.0 | 0.1 | 2.1 | 1.0 | 0 | 0.1 | 0.1 | 0.1 | 2.0 | 0.2 | 2.2 | 1.1 |
| Total | 3,300 | 180 | 3,400 | 2,000 | 1,200 | 980 | 2,200 | 1,900 | 4,400 | 1,200 | 5,600 | 3,900 |

Source:  Ref. 2.2

from 55 to 65 percent water for women and 65 to 75 percent for men.

Water suitable for human consumption is termed potable water, and in the United States it comes generally from two major sources; surface waters such as springs, rivers, lakes and reservoirs or water captured in below ground rock or sand aquifers and retrieved by wells and pumps. Potable water is delivered to American homes either through a public water supply system or through a private well or spring (private supply). There are now 223,500 public water supply systems in the United States. The earliest U.S. public water supply system began operating in Boston in 1652. Approximately 40 percent of all public water supply systems operating in the United States today, depend on ground water as their source of supply (Table 2.3).

**Water Doctrines--East vs. West**

Laws regulating water use vary from state to state, but generally the nation is split by two specific surface water doctrines. In that portion of the nation east of the Mississippi River the Common Law Doctrine of Riparian Rights dominates, while in the western half of the United States, the Statutory or Appropriation Right Doctrine prevails (Figure 2.3).

The Common Law Doctrine of Riparian Rights is founded on English common law which gave certain water use rights to landowners whose property was adjacent to a watercourse. These original riparian landowners were entitled to water for domestic purposes that was unpolluted and undiminished in quantity except for the quantity used for domestic purposes by upstream riparian owners. The key element of the riparian doctrine is direct accessibility. If the land-water connection is lawfully severed, the riparian rights are lost. Theoretically, a riparian landowner could exhaust a stream to satisfy his domestic needs. When the supply of water is sufficient, there is generally no problem with satisfying needs. When water shortages occur, the riparian doctrine provides no guide as to how the remaining water will be apportioned.

The Statutory or Appropriation Right doctrine has a much different developmental history. This doctrine, believed to have been first applied in Califor-

Table 2.3:  Population Reliance on Ground Water for Drinking Water, by State

| State | Percent |
|-------|---------|
| Northern Marianas | 99 |
| Nevada | 95 |
| Hawaii | 95 |
| Mississippi | 93 |
| Florida | 90 |
| Idaho | 88 |
| New Mexico | 87 |
| Nebraska | 82 |
| Iowa | 82 |
| American Samoa | 80 |
| Virginia | 80 |
| South Dakota | 77 |
| Colorado | 75 |
| Washington | 71 |
| Guam | 70 |
| Louisiana | 69 |
| Wisconsin | 67 |
| Delaware | 67 |
| Utah | 66 |
| Wyoming | 65 |
| Arizona | 65 |
| North Dakota | 62 |
| Minnesota | 62 |
| Oregon | 60 |
| New Hampshire | 60 |
| Indiana | 59 |
| Maine | 57 |
| Vermont | 55 |

| State | Percent |
|-------|---------|
| Montana | 55 |
| Alabama | 54 |
| West Virginia | 53 |
| Tennessee | 51 |
| Michigan | 50 |
| Alaska | 50 |
| New Jersey | 50 |
| Arkansas | 50 |
| North Carolina | 50 |
| California | 50 |
| Illinois | 49 |
| Kansas | 49 |
| Georgia | 48 |
| Texas | 47 |
| Pennsylvania | 44 |
| Ohio | 42 |
| Virgin Islands | 42 |
| South Carolina | 42 |
| Oklahoma | 41 |
| New York | 35 |
| Missouri | 34 |
| Massachusetts | 33 |
| Puerto Rico | 33 |
| Connecticut | 33 |
| Kentucky | 31 |
| Rhode Island | 24 |
| Maryland | 15 |
| District of Columbia | 0 |

Source. 1986 State 305(b) Reports sup-
plemented by U.S. Geological
Survey's *National Water
Summary 1984*

Source:  Ref. 2.24

Figure 2.3

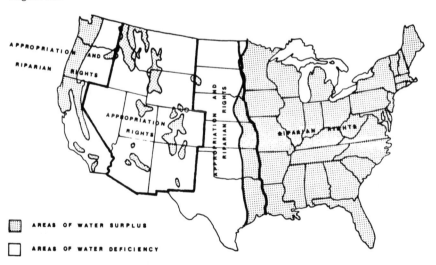

AREAS OF WATER SURPLUS

AREAS OF WATER DEFICIENCY

Source:  Ref. 2.23

nia by gold miners, permits water to be diverted, for
beneficial uses, from surface watercourses to non-
riparian and riparian lands alike regardless of any
diminution in flow. Those first in time are first in
right. Generally, we find this doctrine applied in
areas of the United States where there is a water budget deficit.

It is noteworthy that the thrust of these two
major doctrines is surface water. The rights to
ground water have not generally been as clearly
defined. The common law regarding ground water is
based on the theory that a landowner has absolute
right to everything above and below his land. If
applied strictly, there would be no limitation on the
nature or extent of use of the ground water by a
landowner--a position some states have supported even
as recently as 1987. We know, however, that ground
water, similar to surface water, is mobile (although
at a much slower rate) and will flow downgradient
irrespective of surficial, cadastral boundaries. Any
contamination that might be introduced will likewise
migrate with the flow. Conversely, ground water may
be withdrawn by an owner to such an extent that water
will be withdrawn from adjacent properties. Either
situation would be unacceptable to adjoining or
surrounding property owners. As a consequence, the
strict application of the common law ground water doctrine has been modified by various state legislatures
and the court systems--to protect the general public
health and safety as well as that of downgradient and
adjoining property owners.

## Current National Problems

The two major threats to the nation's potable
water supplies are contamination (Table 2.4) and over-
drafting (excessive withdrawals).

In regard to the former, the list of culpable
materials is as variable as it is lengthy. For Des
Moines, Iowa, it is farm pesticides. In Bald-
winsville, New York, it is the naturally occurring
organism Giardia and in Old Bridge, New Jersey,
Casper, Wyoming, and Dade County, Florida, it is industrial solvents and other toxics.

Excessive withdrawals from both ground and surface supplies is approaching critical proportions.

Table 2.4:  Major Sources of Ground Water Contamination Reported by States

| Source | No. of States Reporting Source[†] | % of States Reporting Source | No. of States Reporting as Primary Source[**] |
|---|---|---|---|
| Septic Tanks | 46 | 89% | 9 |
| Underground Storage Tanks | 43 | 83% | 13 |
| Agricultural Activities | 41 | 79% | 6 |
| On-site Landfills | 34 | 65% | 5 |
| Surface Impoundments | 33 | 64% | 2 |
| Municipal Landfills | 32 | 62% | 1 |
| Abandoned Waste Sites | 29 | 56% | 3 |
| Oil and Gas Brine Pits | 22 | 42% | 2 |
| Saltwater Intrusion | 19 | 37% | 4 |
| Other Landfills | 18 | 35% | 0 |
| Road Salting | 16 | 31% | 1 |
| Land Application of Sludge | 12 | 23% | 0 |
| Regulated Waste Sites | 12 | 23% | 1 |
| Mining Activities | 11 | 21% | 1 |
| Underground Injection Wells | 9 | 17% | 0 |
| Construction Activities | 2 | 4% | 0 |

[†]Based on a total of 52 States and territories which reported ground-water contamination sources in their 305(b) submittals.
[**]Some States did not indicate a primary source.

Source:  Ref. 2.24

The water table of the Ogallala aquifer which runs under six midwestern states (over an area of 60,000 square miles) has dropped between 10 and 15 feet.

The city of Tucson, Arizona, the largest city in the United States completely dependent on an underground water supply, has experienced a 150 foot drop in the water table of its aquifer. There, only about one-third of the water pumped out yearly is being recharged. Further south, its neighbor Texas, has had a 390 foot drop in water tables in the past 25 years in the Dallas/Fort Worth metropolitan area. To the east, ground water levels in parts of eastern North Carolina have dropped as much as 100 feet since 1900.

I think it is clear that no part of the nation has been exempted from threats to its water resources. While we labor to correct these past transgressions, we must also begin to incorporate innovative techniques into our land planning practices to prevent these or similar problems from reoccurring. As we legislate stricter standards for our drinking water, the cost of water treatment will continue to rise. It is better, therefore, to prevent despoilation of our water supplies so that additional treatment will not be necessary. We will now explore some of these preventative planning techniques.

## Planning for Future Supplies

### Surface Sources

It is appropriate that we begin our discussion with surface sources since these were, with few exceptions, the initial sources of potable water for early America. Even today the utilization of surface water for a potable water supply is still attractive--primarily for two reasons. First, it is usually easier to retrieve (unlike ground water, there are no expensive wells to drill and pumps to maintain). Secondly, pollution of a surface source is more likely to be immediately noticeable and easier to clean up, at least for some materials. The major disadvantage of using surface sources is an increased exposure to both natural and man-made pollutants.

It was this latter problem, the gross pollution of the nation's surface water bodies, that eventually drove Americans to seek, what seemed to be at least,

other less contaminated sources (ground water) for
their supply of water. Fewer Americans today rely on
surface water supplies than did in 1900, but where
they do, that reliance may be nearly 90 to 100
percent.

The key to successfully utilizing a surface
water source is watershed management. And, this, as
we shall see in the following section, applies to
ground water sources as well. "As the watershed goes,
so goes the water supply," is an axiom certainly worth
remembering. Stated in more businesslike terms, water
is the product of its watershed. Disrupt, degrade or
deplete the watershed and your product will be similarly affected.

Public or community water suppliers presently
utilizing or proposing to utilize a surface water
source, must begin to think in terms of preventative
planning for both the source and its concomitant watershed. How often have we seen potable water supply
reservoir sites selected and plans developed for their
construction without any attention being paid whatever
to developing companion watershed management or watershed protection plans? The commonly held opinion has
been, "Why worry, we can handle it at the treatment
plant." Unfortunately, not only have the standards
for potability become more stringent but new and more
complex contaminants have been discovered which require more complex and expensive treatment methodologies. Suddenly, the concept of preventative planning
has become more appealing and perhaps, is now considered more cost-effective.

Water suppliers who own an entire watershed are
rare but fortunate. Unlike partial owners they have
nearly total control over their water source. Partial
owners, on the other hand, must work harder developing
an intimate knowledge of current land uses, of surficial and sub-surface improvements and current land use
zoning in their contributing watershed. Unlike their
fee simple counterparts, the fate of their water
source(s) lies largely in the hands of others unless
they take the initiative. Taking the initiative, in
this instance, means more than acquiring the land-use
information mentioned above. It also means obtaining
complete information on all dischargers of wastewater
and other waste material into the watershed as well as
participating in the formulation of local land use

master plans and voicing opinions regarding any proposed development.  Regarding the former, the federal
Environmental Protection Agency and now some states
have made this data acquisition on wastewater dischargers somewhat easier.  Under the federal National
Pollutant Discharge Elimination System (NPDES) permit
program, all dischargers of wastewater to the surface
waters of the nation must have a permit to do so.
There are approximately 47,700 industrial and 15,300
municipal NPDES permittees nationwide.

These permits, sometimes derogatorily called
"licenses to pollute", identify and dictate the amount
and type of wastes that can be discharged to a surface
water body on a daily, weekly or monthly basis.  It is
assumed, albeit sometimes erroneously, that at the
levels permitted, these wastes can be easily assimilated by the receiving water body.  Assimilation in
some instances may mean simply dilution.

Permittees out of compliance with their permits
overburden the water resource, increase downstream water treatment costs and threaten the public health.
Consequently, water suppliers (purveyors) should know
at least on a monthly basis how well the dischargers
in their watershed are doing and be prepared to participate in the NPDES permit renewal process, which
usually occurs every 5 years.

Water quality monitoring throughout the watershed should likewise become a routine practice.  State
or federal agencies may already have such a monitoring
program in place and that data should be acquired either through the federal EPA STORET program or the
U.S. Geological Survey.

Sources of contamination without an easily
recognizable or continuous point source discharge
(e.g. runoff from sod farms, feedlots, livestock pastures and construction sites) will be somewhat harder
to monitor and control.  To combat such sources, a
number of regulations and ordinances have already been
adopted by various states, counties, and municipalities and include the following:

1.  Soil Erosion and Sediment Control
    Regulations

These ordinances deal primarily with the installation of structures to control erosion and the
runoff of sediment from construction sites.  We will
discuss these structures in more detail later in this

chapter.   It is important to note here that land disturbances associated with agricultural operations are
often excluded from such regulation but that their input can be just as detrimental to the water resource.
Historically, agriculturally related land disturbance
was controlled by voluntary participation in the soil
and water conservation programs offered by the U.S.
Department of Agriculture, Soil Conservation Service.
Unfortunately, in many states farmland has been sold
to a new breed of landlord--the land speculator.   New
Jersey, for example, has a farmland assessment act
which allows a substantial reduction in real estate
taxes for land being actively farmed.   Many speculators, after purchasing farmland, lease out the farming
rights to preserve this tax benefit until the land is
subdivided.   Lessee farmers with no particularly
strong ties to the land can often be seen tearing up
old soil erosion control swales, ditches and berms in
order to maximize profits on land that will probably
not be available in just a few years.

    2.   Vegetation Removal Ordinances

    Such ordinances usually regulate the amount
or aereal extent of trees that can be removed at any
one time and further provide for the preservation of
individual tree specimens with a unique or historical
value.

    3.   Pesticide Control Regulations

    Two cases immediately come to mind and amply
demonstrate why such regulations are an integral part
of a watershed protection program.   In the first instance, this writer was called upon to trace the
source of a massive fish kill in a water supply canal
traversing a large agricultural community.

    A day of calculated sleuthing and sampling re-
vealed the source of the kill to be two discarded pesticide containers that had been carelessly dumped next
to a small tributary of the water supply canal.   Several inquisitive school children had rolled the discarded drums into the stream where their residue was
dissolved in the stream water and carried into the
canal.

    In the second instance, an enthusiastic farmhand
began to apply what he thought was the proper mixture
of pesticide to an aphid infested alfalfa field.   A
mother in a nearby dwelling, downwind of the field,
looked out her rear window a short time after the

spraying had begun and saw her two young children lying unconscious in the rear yard. The undereducated farmhand had miscalculated the dilution rate and applied a dose three times the recommended level. Fortunately, the two children recovered.

Consequently, it is easy to see why these regulations now cover the training and licensing of pesticide applicators as well as the proper storage and disposal of pesticide containers.

4.  Steep Slope Ordinances

These ordinances are generally municipal in origin and are designed to protect areas of steep incline (i.e. greater than 15%) where vegetative cover is sparse or bedrock is close to the surface. Disturbing such fragile areas can wreak havoc with a natural equilibrium that may have taken hundreds or thousands of years to develop.

5.  Regulation of Salt Storage Piles

The application of rock salt to melt snow and ice on the nations roadways requires that substantial quantities of this material be stockpiled until used. If left uncovered and exposed to rainfall or other precipitation, these salt piles exude a strong brine that can flow into adjacent streams to kill fish or seep into the ground water thereby raising the level of dissolved solids and sodium. These regulations generally require that such salt piles be housed in a roofed structure to exclude rainfall (see also Chapter 12 section on Roadway Salting).

6.  Wetlands Regulations

No non-point source regulation program or watershed protection program would be complete without incorporating wetlands protection. Wetlands help maintain good water quality. They do so by removing excess nutrients, breaking down chemical and organic wastes and acting as filters [2] by intercepting pollutant laden runoff from land. The U.S. Fish and Wildlife Service has generally been regarded as the champion of the wetlands protection program, however, it was not until the passage of Section 404 of the federal Clean Water Act that some statutory protection was accorded these unique transition zones between land and water. While many states have passed coastal wetlands protection legislation, very few have acted to similarly protect interior wetland areas.

7.   Storm Water Management Regulations

Despite the fact that storm water conveyance systems, in most instances, discharge through pipes they are considered to be non-point sources. The effects of such discharges can be ameliorated to some extent by the use of various structures which we will discuss in more detail under Water Drainage. The primary emphasis of most storm water management regulations and ordinances (and our sample ordinance is no exception) is the minimization of changes in the quantity of water flowing from a watershed. They rarely specify water quality protection criteria--the assumption being that water quality improvement will be a secondary benefit.

Retrofitting existing storm drains with collection pits (or requiring their inclusion on new systems) so that the dirtiest part of the storm water runoff (the first flush) can be temporarily stored and later treated to remove impurities is a viable alternative and a suggested addition to storm water management regulations. The major obstacle presently preventing the inclusion of such systems nationally is cost. However, other problems, many of them logistical in nature, also need to be resolved on a case by case basis. These include:

(a)   The size of the collection tanks to be installed and their location particularly on existing systems in heavily developed areas.

(b)   Where will the collected water be treated (i.e. with a new treatment plant(s) or at a nearby existing sanitary sewage or industrial treatment plant).

(c)   The level of treatment to be required.

In the interim watershed managers and land planners can, as a minimum, encourage certain routine maintenance programs in developed areas that include weekly street and parking lot sweeping and the monthly removal of debris and silt from storm water catch basins and pipes.

Examples of most of the above regulations can be found in Appendix 1, Part B.

Special Considerations

1.    Springs and Seeps
     Land planners and watershed managers need to pay particular attention to the protection of springs and seeps commonly found in the upper reaches (headwaters) of the nation's watersheds. These miniscule capillaries are the origin of nearly every larger surface watercourse. Yet they are often regarded with nearly total disdain during land development--being regarded more as a nuisance than the asset they really are. As a result they have been covered over, routed through mazes of underground pipes and mixed with contaminated storm water runoff--all of which destroy their natural purity. This premature degradation in turn causes some rather severe problems for water users further downstream. Sewage treatment plants for instance often rely upon the purity of upstream water to treat and dilute the residual impurities being discharged in their treated wastewater. With stream water arriving in an already degraded condition these treatment plants may be forced to upgrade to a higher, more costly treatment level or worse yet may be prevented from further expansion altogether.
     Fortunately, new wetlands protection regulations will help to protect these valuable water sources. In the meantime, land planners are strongly encouraged to adopt regulations providing complete protection to all springs and seeps.
2.    Surface Impoundments--Loss of Capacity
     One of the major problems with utilizing on-stream surface impoundments as a source of water is the loss off capacity over time due to the entrapment of silt and sediment from the upstream watershed. Eventually all reservoirs, pond and lakes built directly on a stream will be filled in. This may take place in as little as 25 years or over several hundred years, depending upon the depth of the impoundment, its surface area and configuration and most importantly, the condition of the upstream watershed. Consequently, watershed management is a key element in the design of any on-stream surface water impoundment.
3.    Surface Impoundments--A New Biological Threat
     Surface water sources now face the threat of a relatively new contaminant; an organism called Giardia lamblia. This protozoan parasite produces a disease

in humans called giardiasis, which is characterized by weight loss, diarrhea, nausea, abdominal cramps and weakness.   It was previously referred to as the "backpackers disease" since its victims were primarily backpackers and other outdoors-persons, who drank from surface streams.   The microscopic Giardia organism exists as either an inert cyst (16,500 cysts can fit on the head of a pin) or a mobile form called a trophozoite (Figure 2.4).   The tough protected cyst form (the most prevalent source of infection) can survive in the stream environment for months even in the freezing cold.   Once ingested by man or other host, the cysts can transform themselves into the trophozoitic form, which quickly attach to the lining of the intestine.

The American beaver, Castor canadensis, has previously been implicated as the primary source of the Giardia organism in surface streams.   However, recent research by Dr. Stanley Erlandsen and Dr. William Bemrick of the University of Minnesota and others (Ref. 2.3, 2.4) has produced new evidence that the beaver may be receiving some undeserved blame.   They emphasize that:

> "The media-popularized term "beaver fever", used in regard to the role of these animals as a source of cysts in waterborne giardiasis, may actually be a misnomer, since human usage of the watersheds may have led to the contamination of the drinking water." (Ref. 2.4)

New regulations promulgated under the 1986 amendments to the federal Safe Drinking Water Act (Ref. 2.26) provide for both disinfection and filtration to combat Giardia lamblia in surface water supplies, (See synopsis of regulations in Appendix 1, Part B).   It is interesting to note that some of these new regulations require water purveyors to maintain a watershed control program to minimize the potential for contamination by human enteric viruses and Giardia lamblia cysts.


Ground Water Sources

It would be helpful, I believe, to briefly review the origin of ground water. A portion of the precipitation (rain, snow or ice) that falls to the

Figure 2.4: Giardia lamblia in the trophozoite stage shown here attached to the intestinal wall.

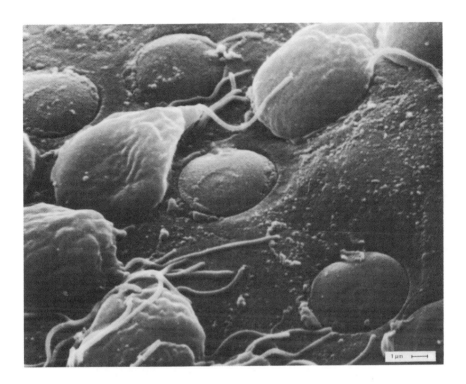

Source: Dr. Stanley Erlandsen, University of Minnesota.

Figure 2.5: The ground water system.

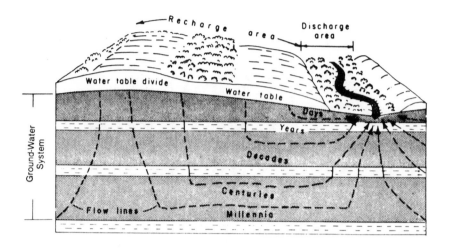

Source: Ref. 2.5

Figure 2.6: Underground water zones.

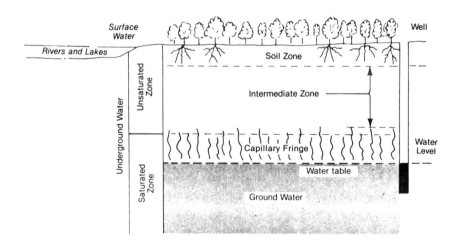

Source: Ref. 2.5

earth's surface percolates downward through the underlying soil and into crevices, cavities or fractures in
the underlying bedrock or into deep formations of
sand, gravel or other porous soils. This process is
called recharge (Figure 2.5). It will continue its
downward movement until a depth is reached where the
rock is no longer fractured (i.e. lacks storage
spaces) or reaches a soil strata that because of its
density prevents further downward penetration. Once
prevented from moving downward, the water begins to
accumulate, completely filling the voids or spaces
between the soil particles or filling the fractures
and cracks in the rock. The height to which this
accumulated water rises depends largely on how much
precipitation percolates into the ground. That
portion of the soil or rock where the spaces, cracks
or fractures are completely filled with water is known
as the "saturated zone". The top of this saturated
zone is called the ground water table. Figure 2.6
illustrates a typical cross-section of the ground
water system.

The ground water table is in most instances
sloped and generally follows surface contours, at
least in shallow aquifers. Thus, the water will flow
downgradient under the influence of gravity. The
speed or velocity at which this water moves is substantially slower than that of surface streams and can
range from several feet per day to several feet per
year. In such a conservative system, it is easy to
see why contamination can be difficult to remove once
it is introduced.

Hydrogeology--A Brief Overview

Land planners need not become experts in hydrology or geology, however, some understanding of basic
ground water hydrogeologic dynamics and terminology
will be beneficial if only to provide the planner with
some grasp of the complexity of the nation's ground
water systems. Four items worthwhile discussing include: hydraulic gradient, ground water velocity,
transmissivity and cone of depression. These are
described below.

1. Hydraulic Gradient
The rate of ground water movement
depends on the hydraulic gradient which is

defined as the change in head per unit of distance in a given direction. If the direction is not specified, it is understood to be in the direction in which the maximum rate of decrease in head occurs (Ref. 2.5). Using Figure 2.7 we can calculate the hydraulic gradient using the formula $h_L/L$, where $h_L$ is the head loss between wells 1 and 2 and L is the horizontal distance between them. Thus,

$$\frac{h_L}{L} = \frac{(100m-15m) - (98m-18m)}{780m} = \frac{85m-80m}{780m} = \frac{5m}{780m}$$

2. Ground Water Velocity

The ground water velocity is the rate of movement of the ground water. It is particularly important to planners who might want to calculate how quickly theoretical pollutants introduced into the ground water system might reach a potable water supply well. The general formula is:

$$V = \frac{Kdh}{ndl}$$

Where;

K = hydraulic conductivity
n = porosity (Table 2.5)
$\frac{dh}{dl}$ = hydraulic gradient

The following two calculations illustrate the wide variation in ground water velocities in the ground water system (Ref. 2.5).

a. Aquifer composed of coarse sand

K = 60 m/d
$\frac{dh}{dl}$ = 1m/1000m
n = 0.20

Figure 2.7: Gradient is determined by the difference in head (elevation) between two wells.

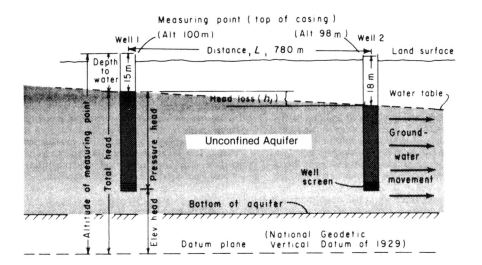

Source: Ref. 2.5

Table 2.5: Selected Values of Porosity, Specific Yield, and Specific Retention

[Values in percent by volume]

| Material | Porosity | Specific Yield | Specific Retention |
|---|---|---|---|
| Soil | 55 | 40 | 15 |
| Clay | 50 | 2 | 48 |
| Sand | 25 | 22 | 3 |
| Gravel | 20 | 19 | 1 |
| Limestone | 20 | 18 | 2 |
| Sandstone (semiconsolidated) | 11 | 6 | 5 |
| Granite | .1 | .09 | .01 |
| Basalt (young) | 11 | 8 | 3 |

Source: Ref. 2.5

$$V = \frac{K}{n} \times \frac{dh}{dl} = \frac{60m}{d} \times \frac{1}{0.20} \times \frac{1m}{1,000m}$$

$$= \frac{60m^2}{200md} = 0.3 \ md^{-1}$$

b.  Confining bed composed of clay

$$K = 0.0001 \ m/d$$
$$\frac{dh}{dl} = 1m/10m$$
$$n = 0.50$$
$$V = \frac{0.0001m}{d} \times \frac{1}{0.50} \times \frac{1m}{10m}$$

$$= \frac{0.0001m^2}{5md} = 0.00002md^{-1}$$

3.  Transmissivity

Transmissivity is defined as the capacity of an aquifer to transmit water of the prevailing kinematic viscosity. The general formula for calculating transmissivity is:

$$T = Kb$$

Where;

$K$ = hydraulic conductivity
$b$ = aquifer thickness

Using this formula we can calculate the transmissivity of the aquifer shown in Figure 2.8.

$$T = Kb = \frac{50m}{d} \times \frac{100m}{1} = 5,000m^2d^{-1}$$

4.  Cone of Depression

The cone of depression is particularly important to the land planner since, as we shall soon see, it can facilitate the delineation of those portions of a watershed that need to be protected.

When water is withdrawn from a well the water level inside the well begins to drop. As a result water begins to move from

Figure 2.8:  Water flow through an aquifer.

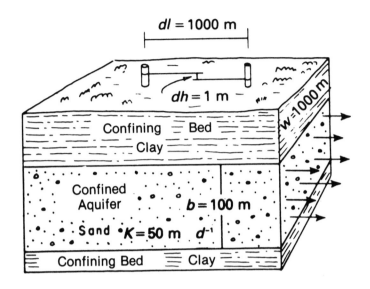

Source:  Ref. 2.5

Figure 2.9

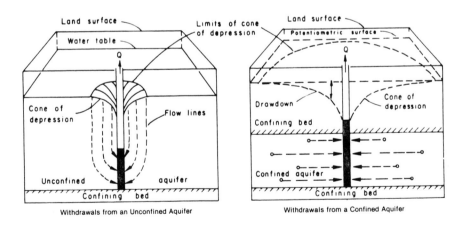

Source:  Ref. 2.5

the surrounding aquifer into the well. As pumping continues, the water level of the well continues to decline and the rate of flow into the well from the aquifer continues to increase until the rate of inflow equals the rate of withdrawal. This movement of water from the aquifer into the well results in the formation of a cone of depression. This is illustrated in Figure 2.9 which also demonstrates the differences in the size of the cones of depression between and unconfined aquifer and a confined aquifer. Cones of depression as large as 60 miles in diameter have been documented in some areas (Ref. 2.5).

Preventative Planning--The New Approach

Historically, the selection of a site for the installation of a public water supply well was generally based upon reviews of existing geological survey maps and prior studies, calculated yields and transmissivity rates, previously drilled wells and resultant yield data and sometimes preliminary test borings. Some sites were selected simply because they were on the only property that was owned. In the case of individual, residential supply wells, this latter reason was and is the only selection criteria. Not nearly enough attention was paid to the existing or potential surficial and subsurface improvements surrounding the proposed well location--an oversight we simply cannot afford to repeat. For the moment let us confine our discussion to public or community supply wells.

The Community Supply Well

Two crucial elements in the preventative planning process will be to establish the actual or potential "cone of depression" for each new well and an area ground water flow map (Figure 2.10). The cone of depression may be estimated mathematically prior to installation, particularly if the transmissivity of the aquifer is known and the aquifer is homogeneous. However, some engineers and planners prefer to install the well and then measure the actual cone of depres-

Figure 2.10: Typical base map for establishing a wellhead protection district.

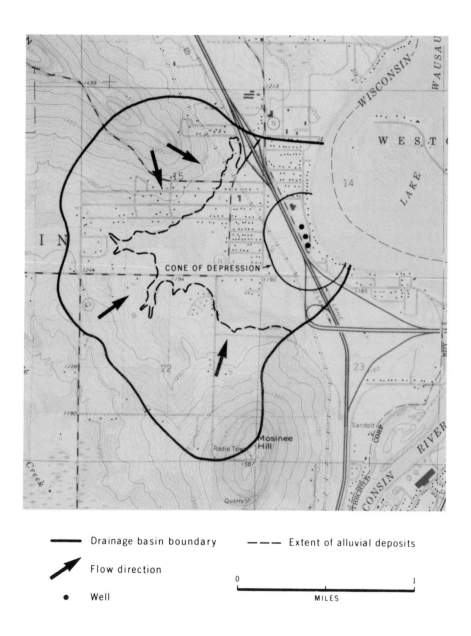

Drainage basin boundary        — — —  Extent of alluvial deposits

Flow direction

● Well

0 ———————————————— 1

MILES

Source:  Adapted from Ref. 2.25

sion under fully operating conditions. This is a lengthy procedure (usually 24 hours) and is accomplished by measuring the drawdown (drop in water level) in newly placed, small diameter monitoring wells or existing wells spaced at intervals away from the main well (Figure 2.11). Utilizing this method also allows the accurate measurement of ground water table surface elevations which are used to develop the ground water contour (flow) map. A further benefit is that these same small diameter monitoring wells may be left in place, and if regularly sampled for water quality, used as an early warning system to detect potential pollutants.

Either during or following the establishment of the periphery of the cone of depression a land use map and an inventory of all possible sources of pollution within this periphery should be made. This inventory may include such sources as dwellings on septic systems, gasoline service stations, industries, landfills, waste lagoons, mines, feedlots and agricultural areas. This land use-pollution source data should be overlain on the cone of depression and ground water contour maps and used to examine the potential for contamination from each of the identified sources. At the same time the density of future development within the delineated area and the future demands for water supply based on current zoning should be evaluated. Figures 2.12 and 2.13 demonstrate the variation in approaches used to map wellhead protection districts. In the former a 2 year travel time was used for delineating the wellhead protection district. In the latter much longer term travel times were used.

Compilation of this information may seem excessively burdensome at first glance, but it is only scientific information such as this that will withstand the scrutiny of judicial or regulatory agency review when land use zoning changes are suggested to protect the water resource. For instance, water resource protection zones which severely restrict land development within and near the cone of depression might be an appropriate zoning change if the data you have assembled indicates a high potential for ground water contamination.

The concept of establishing water resource protection zones around water supply wells, while new to the United States, has been commonly used in Europe

Figure 2.11

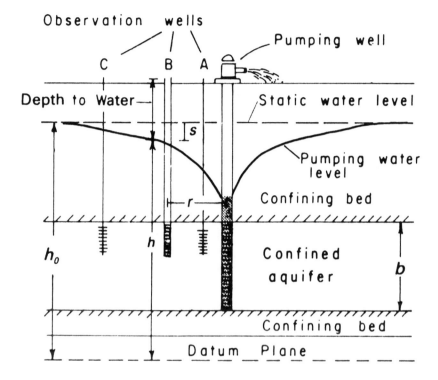

Source:  Ref. 2.5

Figure 2.12

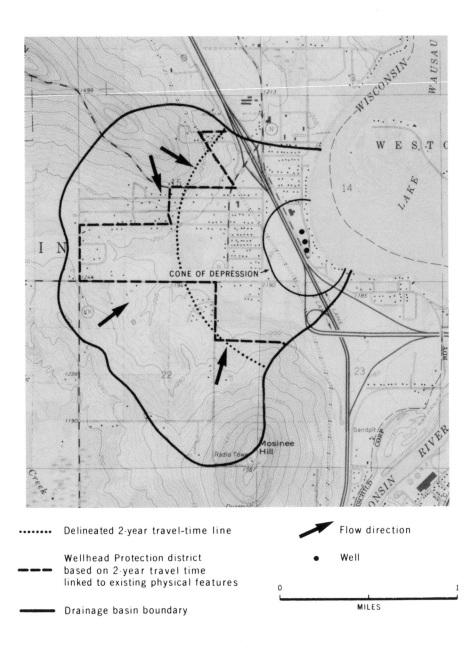

| ........ Delineated 2-year travel-time line | ➤ Flow direction |
| --- | --- |
| ▬▬▬ Wellhead Protection district based on 2-year travel time linked to existing physical features | • Well |
| ▬▬▬ Drainage basin boundary | 0 ————————————————— 1  MILES |

Source:  Adapted from Ref. 2.25

Figure 2.13

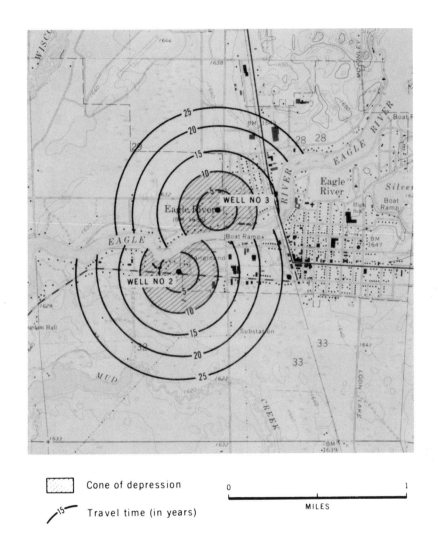

Cone of depression

Travel time (in years)

0 ⊢──────────────────┤ 1
MILES

Source:  Adapted from Ref. 2.25

for some time.   Eleven countries, including Germany and Switzerland, presently utilize this preventative planning concept.   The 1986 amendments to the federal Safe Drinking Water Act incorporated this concept for the first time.   It has been designated the Wellhead Protection Program.   The states will have the primary responsibility for managing this program, with the U.S. EPA providing both financial and technical assistance.   As land planners, we should be encouraging all state, county and municipal governments to participate.

Individual Supply Wells

Similar preventative planning concepts may also be applied to residential subdivisions where 50 to 100 or more individual wells may be installed on as many acres.   The threat here, however, may not necessarily be from external sources but from the individual dwelling units themselves and it may be twofold if the dwellings also utilize individual on-site septic systems.   Quantity wise we have had some major residential subdivisions where upstream (in relation to ground water flow) home owners have totally unsurped all available ground water and dried up the wells of their downstream neighbors.   This usually occurs in areas where ground water is stored in sparsely fractured rock or the aquifer is very shallow.   Quantity wise a concentration of individual septic systems can degrade the underlying ground water system with excessive levels of nitrates and fecal coliform bacteria.

In those areas of the nation where no community water supply or sanitary sewer is available to residential or commercial developments, land planners, municipal ones in particular, have begun to require more sophisticated documentation of the on-site ground water resources at the time of subdivision application.   Generally, this documentation must address the following three concerns:
1.   Is the ground water presently polluted?
2.   Is there sufficient ground water to dilute residual wastes (particularly nitrates) from on-site septic systems?
3.   Is there a sufficient quantity of ground water available to meet the proposed density of use?

Information to satisfy item 1 is generally collected by one of two different methods. The first method requires that the subdivider install one or more wells at various locations on the site prior to subdivision approval and to test the ground water for a variety of potential pollutants (Table 2.6). These analyses can cost from several hundred dollars to several thousand dollars or more depending on the number of pollutants checked. The second method requires the subdivider, after subdivision approval has been granted, but prior to the sale to third parties, to drill a well on each lot and to analyze the water from each well. The pollutants checked would generally be the same as those in the first method.

The same well(s) installed under either of the two previous methods may also be used to collect data for item 3. Now, however, the well(s) will be rigorously pumped to check their ability to provide adequate amounts of water. A recommended method for this procedure is the New Jersey modification (Ref. 2.6) of the "Connecticut minimum well formula" developed by the Connecticut Well Drillers Association (Ref. 2.7). New Jersey's procedure (which can be found in Appendix 1, Part B) consists of two parts--a peak demand test and a constant head test. The peak demand part of the two-part test procedure determines whether or not the volume of water stored in the well plus the volume which will flow from the aquifer to the well during peak time will be sufficient for peak needs. The constant head test, on the other hand, determines whether or not the aquifer contribution will meet long-term needs.

Applicants are not particularly happy about having to provide this added, rather expensive, documentation. They feel that some of this burden should be borne by the municipality. This complaint has some legitimacy. Some of this expense could be avoided if the municipality had designed its land use master plan and land use densities primarily on the carrying capacity of the available natural resources. By doing so it would have already made some evaluation of the ability of these resources to support human development. As you know, however, much of our land has already been zoned and densities assigned based on other than strictly environmental considerations. These new requirements are the first admissions that much of our

Table 2.6:  Recommended List of Parameters to be Measured to Determine
the Potability of a Domestic Water Supply

1.    Volatile organic scan

2.    Base neutral/acid extractable scan

3.    Total dissolved solids

4.    Heavy metal scan (As, Cd, Cu, Fe, Hg, Ni, Pb, Zn)

5.    Nitrates and Ammonia

6.    Fecal coliform and fecal strep bacteria

7.    Pesticides

8.    pH

Source:  Ref. 2.17

existing zoning is unrealistically related to the carrying capacity of the environment--in this instant case ground water resources.

How then do we make these pre-determinations which by the way is the preferred approach to land planning? Primarily with mathematical models. For ground water resources some competent models are readily available. Pizor and Nieswand, 1984 (Ref. 2.8) for instance have derived the following formula to pre-determine the number of dwelling units per acre that can be supported by various aquifer formations.

$$D_{ws} = \frac{Y}{640 Q_s P}$$

where $D_{ws}$ = development density in dwelling units per acre based on water supply;

$Y$ = ground water yield in gallons per square mile per day;

640 = conversion factor in acres per square mile;

$Q_s$ = water supply demand in gallons per capita per day;

$P$ = unit occupance in persons per dwelling unit.

In addition to this water supply model Pizor and Nieswand have also developed the following model to estimate the approximate lot size needed to dilute septic system effluent to any given level in the ground water.

$$D_{wq} = \frac{IC_1}{640 R C_e Q_e P}$$

where $D_{wq}$ = development density based on water quality for septic systems (in dwelling units per acre);

$I$ = infiltration to ground water recharge (in gallons per square mile per day)

$C_1$ = pollutant concentration limit (mg/l)

640 = conversion factor (acres per square mile)

$R$ = pollutant renovation factor (in decimal fraction) (3)

$C_e$ = pollutant concentration in septic systems effluent (mg/l)

$Q_e$ = septic system effluent generation (in gallons per capita per day)

P = unit occupance (persons per dwelling unit)

Other scientists (Trela and Douglas, Ref. 2.9) have also developed similar nitrate dilution models. Their work was the basis for one of the most recent nitrate dilution models to be developed.  This model is to be included as part of New Jersey's Final State Development and Redevelopment Plan (Ref. 2.10) and is shown below:

$$H = \frac{365 Q_p P (C_e - C_q)}{(27,154.29 R_i) C_q}$$

where H = the minimum lot size (in acres) that will avoid increasing the nitrate concentration above the selected standard;

365 = conversion factor to convert to an average annual rate of water usage;

$Q_p$ = the average daily water usage per person (in gallons per day);

P = residents per dwelling unit;

$C_e$ = nitrate concentration in effluent (in mg/l);

$C_q$ = the water quality standard needed to be met (in mg/l);

$R_i$ = the quantity of precipitation that infiltrates the shallow aquifer and is available to mix with ground water and septic effluent (in inches per year);

27,154.29 = conversion factor for $R_i$ to convert inches per year to gallons per acre per year. (1)

Additional, less sophisticated mathematical models to derive projected nitrate-nitrogen levels in ground water can be found in Appendix 1, Part B.

## WATER DRAINAGE

### Introduction

Be honest, now. When was the last time you seriously considered the flow of water across the earth's surface? The truth of the matter is that like many parts of our natural environment, we take that flow for granted, never contemplating its origins or concerning ourselves with its final destination.

It is only when these surface water pathways are filled with excess water or are obstructed in some way that we are forced to take notice. At these times, streams no longer flow quiescently in their normal channels, but rise up to inundate adjoining land, erode stream banks and channels, remove surface features and displace and re-deposit tons of silt and soil. It is a process as old as the earth itself, and a part of the perpetual geological cycle that continues to sculpt the features of the earth's surface. While these floods predate man's presence, they took on new significance with his arrival. Some of the surface features that these floods were now removing or inundating belonged to man. Even man himself became a victim of these excursions.

Man has been attracted to these surface water pathways (now commonly referred to as floodplains or flood hazard areas) for many reasons; to obtain drinking water, for commerce and transportation, to obtain water for irrigation of crops or use in industrial processing and for recreation. Less quantifiable reasons such as aesthetics or the serenity of moving water are equally as important. It was water's great utility that subdued man's common sense inclination to avoid such potentially lethal areas--a characteristic, I'm sorry to say, we seem to have inherited from our ancestors.

Man's encroachments into these flood hazard areas have been less than subtle, perhaps better characterized as brash and reckless. We have diverted, rerouted, filled in, dug out and covered over these pathways with impunity and then placed ourselves confidently in the niches we created. In the earth's "closed loop" system, however, there is always a price to pay. In the United States, property damage due to flooding amounts to billions of dollars

annually$^{(4)}$. We have paid a high price for our bold-
ness.

Even as late as the 1930's, we were still convinced that our structural methods could subdue the
nation's continuing flood problems. Under the federal
Flood Control Act of 1938, massive public work flood
control projects were undertaken across the United
States. It was not until 1960, however, that the con-
cession was finally made, thankfully, that the structural methods were not the complete answer. That year
a new federal Flood Control Act authorized the U.S.
Army Corps of Engineers to provide advice and flood
plain studies to local governments to assist them in
the land use planning process. It would be gratifying
to say that it has been a steadily improving situation
since then, but as most of you know, we cannot.

While more flood hazard area regulations are in
place today than in 1960, and we know a great deal
more about the characteristics of the nation's surface
water pathways we still have difficulty in removing
ourselves from and preventing further excursions into
these flood hazard areas. The fault, however, may be
with some of the newer regulations, themselves. For
example, the federal flood insurance program may actually be encouraging people to remain in flood hazard
zones by constantly paying to rebuild homes and other
structures that will, in all likelihood, be inundated
all over again. In addition, many state flood plain
regulations only cover a part of the complete flood
hazard area, principally the floodway (Figures 2.14,
2.15, 2.16). As a result the equally sensitive flood
fringe areas and other portions of the upper watershed
are left vulnerable to similar abuses. Enforcement
actions against those who continue to place fill or
other structures in stream corridors or divert or fill
in streams is virtually non-existent in many states.
New Jersey, for example, a state fraught with
innumerable such violations had until 1987 only one
person in its 3,900 plus employee Department of
Environmental Protection, to do follow up investigations of stream encroachment offenders.

It is highly unlikely that we will ever clear
our most impacted floodplain corridors of existing
structures and persons, but we need not exasperate the
situation by allowing the same type of poorly planned
development to continue to spread through the flood

Figure 2.14:  Flood hazard area—undeveloped valley.

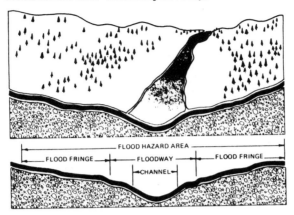

Figure 2.15:  Flood hazard area—developed valley.

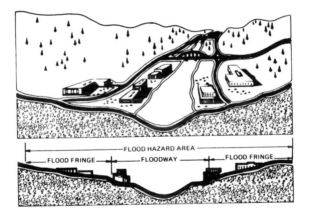

Figure 2.16:  Effects of improper development of a floodplain.

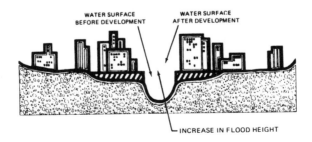

Source for Figures 2.14, 2.15 and 2.16:  Ref. 2.18

fringe areas and the remaining parts of our watersheds.  As with water supply, we again need to think
in terms of watershed management.

Municipal government will play a significant
role in watershed management since they have historically been granted authority by their respective
states to regulate land use through the use of land
zoning and land subdivision ordinances to protect the
public health and welfare.  These subdivision ordinances are the major tool being used today by municipalities to regulate the flow of storm water runoff
from developing areas into surface water courses.
Zoning has not yet been a major factor.  These municipal requirements have largely been of a structural
nature, some of which we will examine in this chapter.
State and federal agencies will also play an important
role by providing flood studies for stream corridors
that traverse multi-state or multi-municipal boundaries.

## Drainage Laws[5]

The drainage of surface water in the United
States generally falls under one of the following
rules.

1. Common Enemy Rule
   Under this rule, surface water runoff is
   considered a common enemy and each property
   owner may control or fight it off as he
   wishes by retention, diversion, altered
   transmission or repulsion.  No legal action
   can result from such interference even if
   some damage occurs.  A landowner, however,
   is prohibited from collecting a large or
   unusual quantity of surface water runoff on
   his land and then discharging it onto
   adjoining land to its injury.  Furthermore,
   most jurisdictions adhering to the common
   enemy rule prohibit a landowner from divert-
   ing surface water onto another's land where
   that water did not flow before.

2. The Natural Flow Rule
   This rule places an easement upon natural
   water courses on lower land.  The owner of
   such a lower property cannot obstruct this
   natural flow across his land to the detri-

ment of upstream property owners. Some jurisdictions where this rule applies have modified it to permit surface waters to be accelerated through natural surface watercourses. The landowner may not, however, overtax the capacity of a watercourse or divert runoff into it that would not have naturally drained there.

3.  The Reasonable Use Rule
    This rule provides that each property owner is permitted to make reasonable use of his land, even though by doing so, he alters or interferes with the flow of surface water and causes some harm to others. However, he incurs liability when that interference is unreasonable.

## Planning For Drainage

### Delineation of Flood Plains

Establishing the complete outside boundaries of the flood hazard areas along all stream corridors in the watershed is an important initial and recommended step in the preventative planning process (Figure 2.17). Historically, such delineations have been based on either a 100 year design storm or the use of flood elevations reached in a historical flood of record. A significant amount of 100 year floodway delineation has already been accomplished nationally thanks to the passage of the 1986 National Flood Insurance Act and the efforts of the Federal Insurance Administration (FIA) of the Federal Emergency Management Agency (FEMA). The FIA claims to have developed flood insurance rate maps (FIRMS) for approximately 99 percent of the flood prone communities in the United States. These maps can be obtained from the FEMA Flood Map Distribution Center, 6930 (A-F) San Tomas Road, Baltimore, Maryland, 21227-6227. Other sources of floodplain maps include: the U.S. Army Corps of Engineers, the U.S.D.A. Soil Conservation Service, the National Oceanic and Atmospheric Administration, the U.S. Geologic Survey, the Bureau of Land Management, the Bureau of Reclamation, the Tennessee Valley Authority and state and local flood control agencies.

Figure 2.17

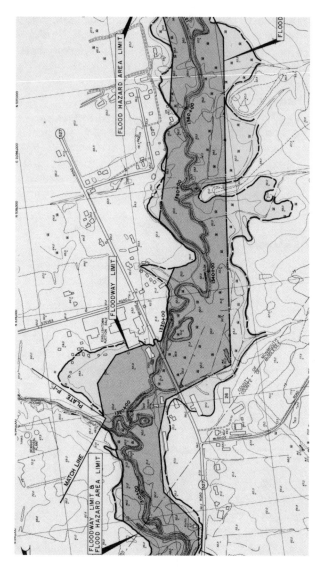

PLAN

Source:  Ref. 2.18

It is important to emphasize here, that such delineations should include all tributaries and extend into the uppermost reaches of the watershed as far as their points of origin (see discussion of Springs and Seeps under Special Considerations).

A less desirable method of delineating the outermost boundaries of these flood hazard areas is by the use of discontinuous stream encroachment lines (Figure 2.18). Used primarily during the development or subdivision of land such delineations are based on a significantly smaller engineering study. Such delineations may also cover less than a hundred lineal feet of a stream corridor or delineate only one side of the stream. In addition, such delineations may be accompanied by so called stream channel improvements (i.e. deepening, widening, straightening) that accelerate the flows for the design storm (usually a 100 year design) past the site exasperating downstream flooding. The philosophy of channel improvement to accelerate flows downstream should no longer be a recommended methodology. From the land developers point of view it is favored since it usually narrows the area needed for floodwater storage and passage thereby allowing more property to be developed. Always look to preserve the maximum width floodplain possible.

## Integrating the Delineated Floodplain With Other Environmental Buffer Zones

Some land developers believe that the corridors or buffer zones created by floodplain delineation represent the consummate effort at environmental protection in the watershed. They do not. They are, however, an important initial step in our enlightened approach to watershed management.

We spoke previously of the need to protect wetlands, springs and seeps and areas of steep topography (i.e. slopes of 15% or greater). These areas should be similarly guarded by either an established buffer zone or by regulations prohibiting their disturbance or a combination of both. It is only when all of these special areas and their concomitant buffer zones are linked together with the delineated floodplain corridors than we can claim to have provided meaningful protection to the watershed. In Figure 2.19 we

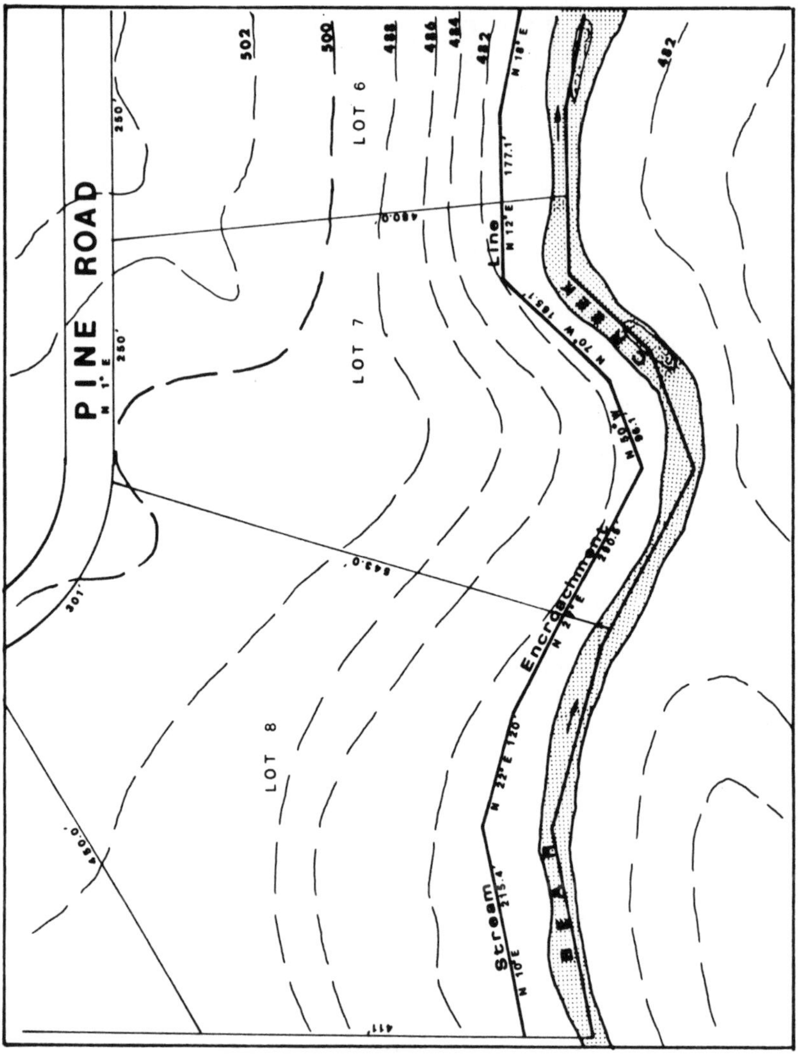

Figure 2.18:  Example localized stream encroachment lines.

Figure 2.19:  Composite floodplain and buffer map.

can see the resultant configuration of such integration. It is a pleasing blend that provides both environment protection and aesthetics.

Secondary benefits accruing from this combination of buffer zones include enhanced stream bank stabilization, runoff and sediment control, nutrient removal, stream temperature control and the provision of habitat for fish and other wildlife. The buffer zone widths provided by floodplain delineation and the inclusion of wetlands, springs and seeps and the steep slopes may not be sufficient to provide all of these secondary benefits. You should therefore compare your proposed buffer zone widths with those minimums recommended in Table 2.7 and if necessary adjust them accordingly.

## Storm Water Drainage Control

1. Retention and Detention Basins

The indispensable counterpart of flood hazard area delineation is the control of storm water at its source--primarily by municipal government through its land use ordinances. Converting woodland and heavily grassed or brushy areas to impervious pavement, homes and close-cropped lawns during the land improvement-subdivision process increases the amount of runoff that flows from the land during rainfall events. The old engineering philosophy of draining storm water away as quickly as possible is no longer applicable in the watershed management concept.

Municipal governments struggling to find satisfactory solutions seem to have found some solace in the use of storm water detention and retention basins. The use of such structures has become increasingly popular in the last decade and in some states they have become commonplace items in nearly every newly developed area. Their construction and function is relatively uncomplicated. Both types of basins are essentially designed as a dry pond. In the case of detention basins, they collect runoff during periods of rainfall, detain it, usually through the use of inlet and outlet pipe size differentials and release it at a slower rate to an adjacent waterway or storm drain system. This helps to stretch out

Table 2.7:  Recommended Buffer Widths in Order to Maintain Certain
Essential Stream Functions

| Function | Width (Feet) |
|---|---|
| Streambank stabilization | 25-50 |
| Sediment control* | 65-150 |
| Nutrient removal | 65-150 |
| Temperature control | 50-80 |
| Fish cover | 25-50 |
| Wildlife habitat | 100-330 |

*The protection of steeply sloped areas is critical to the
control of soil erosion and the resulting discharge of
sediment to surface streams.  Therefore it is recommended
that wherever streams pass through steeply sloped areas
(15% or greater) that the buffer zone boundary be expanded
to either the top of slope or to a point where topographic
grades become less than 15 percent.

Source:  Adapted from Ref. 2.19

or reduce the volume of peak flow in the stream. A retention basin, on the other hand, also collects rainfall, but retains it or stores it to be recharged into the soil. Such retention ponds usually only work where the soil is very porous.

While retention and detention basins are commendable and worthwhile structures, they are not as problem free as originally envisioned. Dampness and ponding water are often found in the bottom of these basins despite design features to eliminate it. This wetness encourages unsightly growths of weeds and grass which provide habitat for insect pests such as mosquitoes and gnats and even some undesirable wildlife. Often unfenced, they become an attractive nuisance utilized by both children and pets.

In July of 1989 two teenage boys in the community of East Brunswick, New Jersey were swept to their deaths after they were attracted to an unfenced storm water detention basin that had been filled to capacity by a brief torrential downpour. Sucked into the swirling vortex of the outlet their bodies were eventually found in downstream storm sewers located nearly 1-1/2 miles away.

The key to the successful operation of detention and retention basins is maintenance (Figure 2.20). Unfortunately, many municipalities have been slow to accept responsibility for the regular maintenance of these structures once they have been relinquished by the land developer. Often located within easements on private property, the responsibility for grooming them often falls upon the adjoining landowner(s). Some of these basins can range in size from an eighth of an acre up to several acres or more. Needless to say, the commitment in time and labor on the part of the landowner can be significant as well as short-lived. Consequently, it is this author's opinion that all such retention or detention basins be accepted as part of the municipality's regular storm water system to assure a regular program of maintenance.

Furthermore, it is important that all such basins be accessible to any equipment that might be necessary for their maintenance. I know of one recent case, in particular, where a developer in

Figure 2.20: These two photographs demonstrate the extreme variations in the type of maintenance provided to storm water retention basins.

placing a residential subdivision on the top of a steeply sloped (grades of 25% or more) mountaintop provided such a retention basin several thousand feet away at the bottom of the mountain. He provided a 25 feet wide easement down the mountain side between two lots but was not asked to provide a roadway for access. Subsequently, two years after the subdivision was completed and all bonds released, the basin malfunctioned due to a buildup of sediment. The municipality had no way to get the necessary heavy construction equipment down to the basin without constructing a new access roadway. This oversight eventually cost the town over $12,000 to provide the necessary access.

While we continue to encourage the use of these structural measures, not nearly enough municipalities are considering the collective impact of these detention basins scattered throughout the same watershed. For instance, are their releases timed so that their peak discharges reach the same downstream point simultaneously, thereby compounding an existing flooding problem? And, do these basins also extend the time period when stream flows are at full bank to bank capacity, thereby accelerating bank erosion?

2.   Retention Trenches

In some southeastern states several municipalities have taken the retention concept further and developed retention trenches (Figure 2.21). In particular, Titusville, Florida and North Myrtle Beach, South Carolina have used this concept. Perforated pipe is placed in below ground trenches surrounded by gravel and then wrapped in a fabric filter. Rainwater runoff is directed into these pipes where it is temporarily stored before being released slowly to the ground water. Such a system works best when the water table rises no higher than three feet from the ground surface.

**Erosion Control**

As discussed earlier, water flowing over the earth's surface is certainly not an unusual or recent phenomenon. Nor is the erosion and transport of soil, sediment and rock that occurs during the process.

Figure 2.21:  Cross-section of retention trench.

# PLAN

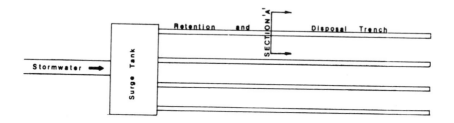

# SECTION 'A'

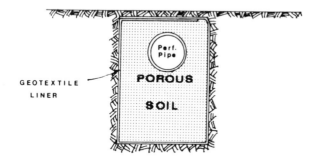

Man's activities, however, can significantly alter this natural process, accelerating erosion rates and stripping away layers of surface soils centuries ahead of their natural attrition rate.

The process of land development and subdivision construction frequently entails the removal of large quantities of soil-binding, protective vegetation ranging from grass to trees. Lacking vegetative protection, the exposed soils are highly susceptible to transport by rainfall runoff. Left unchecked, the soil is carried away to clog stream channels, smother aquatic insects, fish eggs and fish and add to the general deterioration of surface water quality.

Fortunately, we have acquired a substantial number of soil erosion control practices that we can employ during the land development-construction process to ameliorate this unnecessary erosion. A good deal of credit must go the United States Department of Agriculture's Soil Conservation Service whose research and assistance were invaluable in developing these practices. The SCS's Engineering Field Manual (Ref. 2.11) is an excellent technical reference for designing such erosion control measures. Some states have developed their own state soil erosion control manuals since soil types and erosion rates vary with geographic locale.

Lacking the space in this text to cover all of these practices, we will mention only a few. Before we do that, however, it is important to mention two salient observations. First, I have found that the expectations for these erosion control practices are sometimes too high. No method can guarantee a total prevention of all soil loss and sediment wash off. To expect to see totally clear water ensuing from a construction site where these practices are employed is unreasonable. Even one of the most effective practices, sediment basins, cannot remove some latent turbidity within their designed retention time without the addition of some type of chemical coagulant or flocculant.

Where these soil erosion and sediment control practices have failed completely or failed to achieve a constant maximum efficiency it is almost exclusively due to a lack of routine maintenance or any maintenance whatever. Sediment ponds need to be periodically cleaned out to remove accumulated silt. Hay bale barriers need to be replaced when rotted or broken and silt fences need to have the sediment they have trapped removed between rainstorms. The need to keep restoring the original capacity of these structures throughout the life of the construction project is often overlooked or ignored. As environmental planners, we have a responsibility to ensure that such maintenance programs are vigorously carried out on all construction sites.

## Some Important Erosion Control Structures

1. Sediment Basins

    There is actually little difference between a retention basin and a sediment basin. In fact, a single basin could ultimately serve in both capacities. A modification of the outlet of a retention basin by the placement of a perforated riser pipe (Figure 2.22) to facilitate the entrapment and accumulation of silt and soil is one of the major differences. It is an extremely effective sediment control structure and I recommend them highly. Even in some of our most steeply sloped mountains, these basins can be very effective, provided their size is reduced to minimize the depth of cut and accompanying vegetation and soil removal.

    Unfortunately, there seems to be a growing tendency to rely less on these types of basins and to substitute peripheral and less durable structures such as silt fences and hay bale filters which I feel are not nearly as effective. Perhaps they would be if they weren't just put in place and forgotten about.

    Planning for erosion control is not a "quick fix" science but requires careful pre-planning and a comprehensive look at staging or sequencing of construction. Further, every construction site needs a full time person whose sole responsibility is managing and maintaining these erosion control structures.

2. Gabions

    In spite of their fancy name, gabions are nothing more than stone filled wire baskets. But their simplicity belies the amazing utility of these structures. The concept of gabions is sev-

Figure 2.22: Sediment Basin. A well maintained and operated sediment basin. The depth of sediment however, indicates that the basin is nearing the end of its useful life and should be cleaned to restore maximum effectiveness.

eral thousand years old, having been utilized extensively throughout Asia and Europe long before they became popular in the United States. I can remember 25 years ago when gabion structures were a rarity in my home state of New Jersey. Not so today. Their effectiveness as stream bank protection devices is unequalled. The gabions shown in Figure 2.23 have just weathered a substantial hurricane and as you can see show no signs of damage whatever.

In addition to their use as stream bank protection devices, gabions have also proven to be effective and attractive steep slope protection structures. I encourage their application in this regard wherever possible. Their use in wooded mountainous terrain minimizes the aerial extent of slope cuts and excavation, which consequently reduces soil erosion potential, aids the preservation of existing vegetation and reduces overall excavation and restoration costs.

These same features also make them practical additions to storm water detention and retention basins where space is not available for extensive slope grading (Figure 2.24).

## Flood Prevention Practice

In spite of the increased public concern about flooding and stiffer governmental regulation of floodplains, it is apparent that residences and businesses are still being constructed in these hazardous areas. Consequently, I believe it is necessary to discuss some of the practices that can and have been used in floodplain areas to minimize the risks of flood damage. I re-emphasize that I am not encouraging development in floodplains in any form or manner, however, some of you may have need of such information. You should always consult the state, county or municipal environmental protection agencies in your area before constructing on any floodplain land.

One of the common methods of construction used in flood prone areas is to construct buildings (residence or business) on piers which raises the bottom floor level above the design storm (usually 100 year storm) flood elevation. During periods of flooding, however, access to the building may be unavail-

Figure 2.23: Gabions have proven to be one of the finest streambank erosion control structures yet devised.

Figure 2.24: Gabions in use as a retention basin.

able as the turbid waters cover access roads, sidewalks and stairways. You could be stranded in the
building at the height of the flood or denied access
until floodwaters recede. Emergency vehicles (fire,
police and first aid) would have similar access
problems in the event of a fire or other emergency
during flooding.

A second floodproofing practice has been to
place the building on a concrete slab (i.e. without a
foundation) and to raise the bottom floor elevation
one foot above the design storm flood levels. There
will be similar problems with access as in the first
floodproofing method.

A third practice, used less frequently, has been
to install flood gates, walls or dikes around the entire building or across doorways with a height sufficient to hold back the surrounding floodwaters. Your
building may remain dry but you will have the same
problem with ingress and egress as in the two previous
methods.

By now, some of you may have recognized the one
major flaw in all three methods. What happens if instead of a 100 year flood we get a 200 year or 500
year flood? The answer is obvious. Hence, my earlier, strong encouragement to you to stay out of
floodplain areas.

## SPECIAL CONSIDERATIONS--WATER RESOURCES

### 1. Fluming Existing Streams

Nothing has been as exasperating to this
writer as the difficulty in determining the
sources of pollution emanating from surface
streams that have been placed in underground
pipes, box culverts or other conduits. In almost
all instances the reason for enclosing these
streams has been to maximize the amount of developable area on a lot. However, another reason is
the desire to produce a level lot, free of encum-
bering stream banks, slopes and floodways. Unfortunately, in most commercial and industrial
districts, all sorts of floor drains eventually
end up connected to these subsurface stream conduits. Washwaters, including some hazardous organic chemicals from auto dealerships, milk
spilled in grocery stores, and oil from machine

shops have all been found flowing from these underground streams. Trying to track down the culpable parties is a nearly impossible task given the limited access available to environmental investigators who are not particularly fond of crawling up a quarter mile or more of underground pipe. Had the stream been left in its natural state, there would have been some deterrent to those who want to find a free way to dispose of their floor drain wastes. Furthermore, the darkened inert interiors of these subsurface conduits lack the biological, chemical and physical mitigation and removal systems usually found in a natural, undisturbed stream corridor. It is these systems that trap sediment, consume nutrients and provide habitat for a variety of ecologically significant macroinvertebrates--all of which help maintain good water quality.

My firm recommendation to planners today is not to allow any surface stream to be piped underground unless there is some overwhelming benefit to the public health and safety.

## FOOTNOTES

### CHAPTER 2

(1)    One acre foot equals approximately 325,928 gallons.

(2)    A mathematical model developed for Maryland's Coastal Zone program (Ref. 2.19, 2.21) indicates the following sediment trapping efficiencies of vegetation based on width:

| WIDTH OF VEGETATION (in feet) | SEDIMENT TRAPPING EFFICIENCY (Percent) |
|:---:|:---:|
| 200 | 90 |
| 100 | 82 |
| 50 | 75 |

(3)    Pizor and Nieswand assumed that 20% of the effluent is denitrified, leaving 80% of the pollution to be renovated (Ref. 2.8). This relationship is expressed as a pollutant renovation factor of 0.80.

(4)    In 1986 national flood damages were more than $6 billion and flood related fatalities totaled 208 (Ref. 2.12).

(5)    The surface water drainage laws applicable to your particular area should be verified through legal counsel.

**REFERENCES**

**CHAPTER 2**

2.1    United States Environmental Protection Agency, "You and Your Drinking Water", EPA Journal, Vol. 12, Number 7, September, 1986.

2.2    United States Geological Survey, "Estimated Use of Water In the United States In 1980", Circular 1001, 3rd Printing, Alexandria, Virginia, 1985.

2.3    Erlandsen, S.L. and W.J. Bemrick, "Waterborne giardiasis: sources of Giardia cysts and evidence pertaining to their implication in human infection", p. 227-236. In P.M. Wallis and B.R. Hammond (ed.), Advances in Giardia research. University of Calgary Press, Calgary, Alberta, Canada.

2.4    Erlandsen, S.L., Lee Ann Sherlock, William J. Bemrick, Helmy Ghobrial and Walter Jakubowski, "Prevalence of Giardia spp. in Beaver and Muskrat Populations in Northeastern States and Minnesota: Detection of Intestinal Trophozoites at Necropsy Provides Greater Sensitivity than Detection of Cysts in Fecal Samples", Applied and Environmental Microbiology, Vol. 56, No. 1, Jan. 1990, p. 31-36.

2.5    United States Environmental Protection Agency, "Protection of Public Water Supplies From Ground Water Contamination", EPA/625/4-85/016, September, 1985.

2.6    Hoffman, Jeffrey L., Robert Canace, "Two Part Pump Test for Evaluating the Water Supply Capabilities of Domestic Wells", New Jersey Geological Survey Ground Water Report Series No. 1, 1986.

2.7    Hunt, Joel, "How Much is Enough? A Minimum Well Formula", Water Well Journal, Vol. 32, No. 2, 1978, pg. 53-55.

2.8    Pizor, Peter J., George H. Nieswand, "A Quantitative Approach to Determining Land Use Densities From Water Supply and Quality", Journal of Environmental Management, Vol. 18, No. 1, January, 1984.

2.9   Trela, John J. and Lowell A. Douglas, "Soils, Septic Systems and Carrying Capacity in the Pine Barrens, N.J.", Agricultural Experiment Station, Rutgers University, New Brunswick, N.J., 1978.

2.10  New Jersey State Planning Commission, "Development of Nitrate Dilution Model-- Technical Reference Document", prepared by Rogers, Golden and Halpen, February, 1988.

2.11  U.S.D.A. Soil Conservation Service Engineering Field Manual, Vol. 1 and 2, PB 85-175164/AS, NTIS, July, 1984.

2.12  United States Geological Survey, "National Water Summary 1986", Water Supply Paper 2325, 1988.

2.13  United States Environmental Protection Agency, "Environmental Progress and Challenges--An EPA Perspective," June, 1984.

2.14  United States Environmental Protection Agency, "Wellhead Protection--A Decision-Maker's Guide", May, 1987.

2.15  Dorram, Peter B., "Water vs. Land Use--A New Concept in Zoning and Land Use Management Where Geologic Data Help Determine Densities", New Jersey Federation of Planning Officials, Vol. IX, No. 4, Winter, 1984.

2.16  Saunders, Wayne R., John J. Trela, Joseph A. Benintente, "Water Quality and Septic Systems, Ocean Acres Subdivision, Stratford and Barnegat Townships, Ocean County, N.J.", N.J. Department of Environmental Protection, April, 1979.

2.17  Honachefsky, William B., "Danger--Real Estate For Sale", Site Evaluations, Ringoes, New Jersey, 1985.

2.18  State of New Jersey, Flood Hazard Report No. 17, Matchaponix Brook System, N.J. Division of Water Resources, March, 1973, by Anderson-Nichols and Company Inc.

2.19  New Jersey State Planning Commission, "The New Jersey Freshwater Wetlands Protection Act as It Relates to Stream Corridor Buffer Considerations in the State Development and Redevelopment Plan", prepared by Rogers, Golden and Halpern, January 11, 1988.

2.20    Palfrey, Raymond and Earl Bradley, "The Buffer Area Study", Maryland Department of Natural Resources, Annapolis, MD, 1983.

2.21    Wong, Stanley L. and Richard H. McCuen, "Design of Vegetative Buffer Strips for Runoff and Sediment Control", (research paper) University of Maryland, Department of Civil Engineering, College Park, MD, 1981.

2.22    United States Environmental Protection Agency, "Controlling Thermal Pollution in Small Streams", EPA-R2-72-083, by George W. Brown and Jon R. Brazier, Washington, D.C., 1972.

2.23    United States Environmental Protection Agency, "Process Design Manual, Land Treatment of Wastewater", EPA 625/1-81-013, October, 1981.

2.24    United States Environmental Protection Agency, "National Water Quality Inventory, 1986 Report to Congress", EPA 440/4-87-008, November, 1987.

2.25    Born, S.M., D. A. Yanggen, A. R. Czecholinski, R. J. Tierney and R. G. Hennings, "Wellhead Protection Districts in Wisconsin: An Analysis and Test Applications, Wisconsin Geological and Natural History Survey Special Report 10, 75 pg., 1988.

2.26    Federal Register, 40 CFR Parts 141 and 142, Vol. 54, No. 124. Thursday, June 29, 1989.

# 3

# Geology and Topography

## THE HUMAN MANIPULATION

Man's manipulation of the earth's surface contours to more desirable configurations is a long
standing tradition in human culture. In the United
States, advanced earth moving technologies, developed
after World War II, greatly enhanced the national
ability to tear down, alter and even obliterate the
most durable features of the earth's topography. Land
heretofore considered marginal for building suddenly
became very viable. Unfortunately, the nation was not
as well equipped to deal with the consequences of
these often massive disturbances of the earth's contours. Nor were we particularly concerned about the
characteristics of the underlying geological formations, unless of course it was comprised of bedrock.
Then we were concerned only because rock was more
expensive to remove than soil. Perhaps we were naive
or indifferent or infatuated with our new found
prowess to bully the environment.

Nowhere was this attitude better demonstrated
than in Los Angeles, California in the late 1940's.
Enterprising builders and large developers, anxious to
satisfy a resurging demand for low cost housing, began
to acquire and develop land in the foothills and steep
mountain slopes surrounding the city's considerably
more expensive flatland areas. Unfortunately, little
attention was paid to the geological characteristics
of these areas, in particular, the colluvial soils.
These soils, or more correctly, sediments, lack internal structure and when stripped of surface cover and
saturated with water, become unstable and will flow
downhill as mud flows and land slides. Abetted by
several years of abnormally dry weather, it was not

until the winter of 1951-52 that Los Angeles realized the folly of this poorly regulated development. Heavy rains that particular winter washed out tons of soil, rock and other debris from the newly formed cut slopes and filled areas and carried them into the lower flatland areas. Storm drains were clogged, streets were filled and lawns and other property were damaged. By the time it was over $7.5 million worth of damage had been done. As a consequence of this disaster, the city of Los Angeles in 1952 became the first city in the United Stated to adopt a land grading ordinance. The passage of this ordinance did not end Los Angeles' dilemma nor prevent similar incidences from occurring elsewhere in California. Neither was the California experience an isolated incident. On the other side of the continent, the states of Maryland, Virginia and New Jersey were likewise experiencing the negative impacts of massive, poorly regulated land development. All of these states eventually adopted either a land grading code or soil erosion and sediment control standards. California's Chapter 70 land grading code finally adopted in 1970 is probably one of the most rigorous and comprehensive codes ever developed.

A prodigious amount of technical bulletins and recommended standards concerning grading and soil erosion and sediment control have subsequently been made available nationally. It is pleasing to note that the consideration and implementation of rudimentary soil erosion and sediment control practices have become a routine part of the land development process throughout much of the United States. While there are still problems with regular maintenance of these structures as discussed in Chapter 2, it is still a significant milestone for environmental planning. The detailed examination of subsurface geological features, however, has not become nearly as routine or as widespread, in spite of 40 years of documented, calamitous consequences for failing to do so. The inattentiveness to geological characteristics that plagued California and other states over four decades ago is still prevalent in many other parts of the nation today. It is this deficiency we intend to emphasize here.

I know there are contemporary land planners out there who are being severely criticized and accused of overzealousness for their attempts to incorporate complete geological evaluations into the land planning process. It is somewhat difficult to reconcile the fact that we can demand and be willing to fund, almost without question, expensive and comprehensive geological evaluations at contaminated Superfund sites, yet we are reluctant, almost recalcitrant, to require such detailed analyses as part of a preventative planning approach.

The need for thorough geological site evaluations can best be demonstrated, I believe, by case example. Consequently, I have selected the following two case histories to illustrate the type of problems we are perpetuating by our reluctance to demand comprehensive environmental data.

Case No. 1

This case involved a 40 lot residential subdivision in a sparsely settled rural township of about 8,000 persons in the northeastern part of the United States. The rectangular tract of about 100 acres was comprised largely of overgrown or fallow agricultural land on gently sloping grades with an overall drop in elevation from one end of the site to the other of about 50 feet. The underlying bedrock of the site was comprised of red shale and siltstone overlain by 4 to 6 feet of U.S.D.A. Soil Survey typed Bucks silt loam and Penn shaly silt loam soils.

Because of the rural nature of the township, there was no public sanitary sewer or city water available within the immediate vicinity and consequently each proposed two acre lot was to have an individual septic system and water supply well.

Due to the high clay content of the on-site soils, the local planning board and the local and county boards of health spent a great deal of time reviewing percolation tests and soil logs, even going so far as to request additional percolation tests where results were marginal. No attention, however, was directed to the capability of the underlying sparsely fractured bedrock to support the 40 home demand for ground water.

Preliminary approval was granted in 1985 and construction commenced a few months thereafter. As is typical in such subdivision construction the work was carried out in phases. Phase 1 consisted of 10 lots situated at the lowest (elevation-wise) end of the tract. When the individual wells were drilled for these ten homes, the results were not encouraging. The final developed flow rates averaged a sparse 3 to 5 gallons per minute with drawdowns of 50 to 60 feet.

Upon completion of the first phase in 1986, construction began on a group of 15 lots lying approximately 30 feet higher in elevation than the previous ten lots. The individual wells drilled in this second section had flow rates that were only slightly better than those in Phase 1, averaging 6 to 8 gallons per minute. This second section of homes was completed in early 1987 and the developer moved quickly to the final section of 15 lots.

With 25 homes now occupied, a curious thing began to happen. Home owners on the ten lots in Phase 1 began to complain of dirty water in their wells, then of an inability to fill their water tanks. By the end of October, 1987, eight of the wells on the original ten lots went dry. It was surmised and later proven that the uphill homes were intercepting and totally unsurping whatever sparse amount of ground water was available in the fractured bedrock.

Having learned a valuable lesson here, the municipality is in the process of adopting an amendment to its land use ordinance requiring the submission of complete geological data, including an evaluation of the capability of the underlying geology to supply the required amounts of ground water.

Case No. 2

In 1983, this municipal planning board was presented with a site plan for a proposed office building to be built on one of the towns four remaining commercial lots. Since no public sewer was available the applicant was proposing to install an 1800 gallon per day on-site septic system. The municipality expressed immediate concern because one if its two community water supply wells would be located about 100 feet from the edge of the proposed wastewater disposal area. The configuration and rather small size of the lot, however, allowed no room for maneuvering the proposed disposal system to a location that would increase the horizontal distance to the well.

Queries were made to the municipal engineer and the applicant's engineer as to the potential threat to the water supply well. Both engineers concurred that the threat would be minimal despite the fact that the area was underlain by cavernous limestone. They concluded that the 250 foot depth of the municipal well would be a sufficient buffer. Neither engineer, however, was a geologist or a geological engineer. Consequently, based upon the opinion of these "experts", no further geological evaluation was sought by the municipality. Approval was subsequently given and construction commenced in early 1984 and was completed in December of 1985.

Two years later, the routine bacterial sampling of the town's potable water system began to show elevated levels of fecal coliform bacteria where heretofore none existed. Following further intensive sampling and investigation, it was determined that the source of the fecal coliform bacteria was the municipal well located adjacent to the office building septic field. The 250 foot depth had proven to be a totally unreliable guarantee against pollution.

Additional geological and topographical concerns that land planners ought to be familiar with are discussed in the remaining portions of this chapter.

## EARTHQUAKES--NATURAL

Earthquakes are one of the earths most frightening and potentially catastrophic events. Anyone who has experienced a strong earthquake can attest to the unnerving sense of helplessness it produces. Let us for the moment briefly review the causes of such quakes.

The outer layer of the earth's lithosphere is broken into rock (tectonic) plates that constantly move and shift relative to each other. They may slide over or past one another thereby causing tremendous stress (also called elastic energy) to build up in the rock much like a steel spring that is being compressed. When this stress exceeds the strength of the rock, the rock breaks releasing the energy in the form of seismic waves which cause the earth to quake or undulate up and down much like waves on the ocean. Some of the waves travel within the earth while others travel along the surface. It is the surface waves that produce most of the damage in an earthquake.

In the United States when one mentions earthquakes, the state of California immediately comes to mind. The association is a valid one, for California lies within the Circum-Pacific seismic belt--an area in which 80% of all the earth's quakes originate. This seismic belt runs along the entire western coastline and highland region of Oregon and Washington and continues up to southern Alaska and over to Asia and beyond (Figure 3.1). California's predisposition to frequent earthquakes is further enhanced by the presence of the San Andreas fault formation--an 800 mile long junction point between the Pacific Plate and the North American Plate. This fault zone extends across California in a southeast-northwest direction. It was this fault that was responsible for the disastrous earthquakes of April 18, 1906 and October 17, 1989.

Those of us living on the eastern seaboard are not exempt from earthquakes.

The states of New Jersey, New York, Massachusetts, Connecticut and Pennsylvania have all experienced small earthquakes at one time or another. Even such an unlikely state as South Carolina has experienced an earthquake of significant magnitude. Fortunately, for the eastern United States, the closest major seismic belt, the Mid-Atlantic Ridge Belt lies well out in the Atlantic Ocean, thereby sparing them the frequency of earthquakes California and other western states must endure.

Quakes can sever or crack gas pipelines, water lines, sewer lines and electrical lines. They also topple houses, buildings, bridges and roadways and displace large quantities of rock and soil in sloped areas.

It would seem obvious to most us, therefore, that as land planners, we should be carefully considering the potential for earthquakes on any tract of land being considered for development, especially in high risk areas. Further, we should be discouraging development in those areas of extraordinary danger. That is to say, on or immediately adjacent to recognized fault zones. Yet, as obvious as these hazards are, planners have allowed large residential subdivisions and other structures to be built directly over California's San Andreas fault line.

Figure 3.1: Location of major earthquake belts near the North American continent.

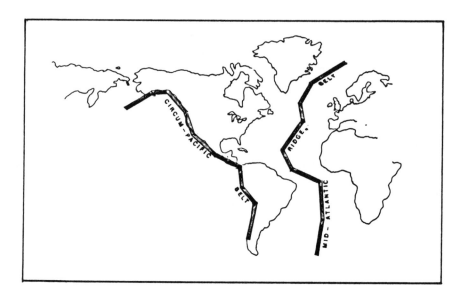

Figure 3.2: Establishing adequate buffer zones along known geologic fault lines is wise land use planning.

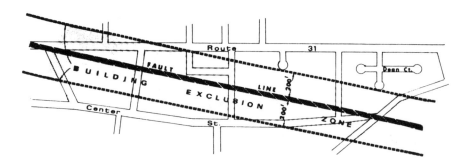

When confronted with plans for proposed development in a high risk earthquake zone, planners have several options available that they may recommend and implement. First, and as will be mentioned repeatedly throughout this text, we must begin with the preventative planning approach. All fault hazard areas should be identified and delineated. Following this delineation, minimum setbacks or buffer zones should be established along these fault lines (Figure 3.2). These buffer zones may vary in width depending upon the type of development and density proposed. However, this writer prefers a uniform width throughout.

Secondly, building on sanitary landfills or other unconsolidated fill materials should be prohibited. Such filled areas accentuate seismic waves during an earthquake, thereby magnifying the potential for damage. Whenever possible, any new buildings in these hazardous zones should be built on bedrock.

Thirdly, California's October 17, 1989 earthquake dramatized the need to also consider preventative planning measures for subsurface utility lines particularly gas and water. Gas fed fires from ruptured pipelines proved difficult to control with water lines that were similarly cracked and leaking. Consequently in high risk areas, such lines need to incorporate, as a minimum, flexible joints to accomodate movement of the earth during a quake. Additionally the 1989 California quake also revealed some serious flaws in California's highway bridge design--flaws which unfortunately proved fatal for a number of citizens who happened to be traveling on those structures when the quake struck. In the past so much emphasis has been placed on preventative planning for large office buildings that these structures were often overlooked or assigned a lesser importance.

Finally, all buildings regardless of their location in an earthquake hazard area should be designed to withstand a certain level of earthquake. California's Division of Mines and Geology, for example, has recommended both a "maximum credible earthquake" and a "maximum probable earthquake" likely to occur during a 100 year interval. See Appendix I, Part C for full details.

## EARTHQUAKES--MAN-MADE

Not all earthquakes are the result of natural occurrences. We have learned that man can induce earthquakes as well. In 1962 for instance, the Rocky Mountain Arsenal near Denver, Colorado decided to dispose of its wastewater through a 12,000 foot dry well beneath its site. The well bottom ended in a highly fractured metamorphic bedrock. The injection of this wastewater increased the fluid pressure in the bedrock and facilitated the slippage of existing fractures which produced a number of measurable earthquakes.

Even such projects as the building of dams and reservoirs can produce earthquakes. In the decade that followed the construction of Hoover Dam in Arizona for instance, more than 600 earth tremors were felt as a result of the increased load of water on the land's surface.

Finally, we know that our underground nuclear explosions can also trigger earthquakes and this has been observed numerous times at the Nevada test site. I have mentioned these man-made inducements of earthquakes primarily for two reasons: to further demonstrate that man is as capable as nature in producing environmental impacts of a significant magnitude and to alert you to the fact that deep well injection of wastewater, an increasingly favored technology, may produce totally unexpected results.

## TECTONIC CREEP

Just as there are sudden explosive shifts in the earth's tectonic plates that produce earthquakes, so too are there slower almost imperceptible movements that produce no earthquakes. This is termed tectonic creep (Figure 3.3). It can, however, do considerable damage to roads, sidewalks, buildings, sewer lines and other man-made structures. In some places in California, this movement has been as much as 1/2 inch per year.

As planners, therefore, we need to understand that in any earthquake prone area damage may occur to infrastructure and buildings regardless of whether an earthquake occurs or not.

Figure 3.3:  Typical curb and pavement displacement due to tectonic creep.

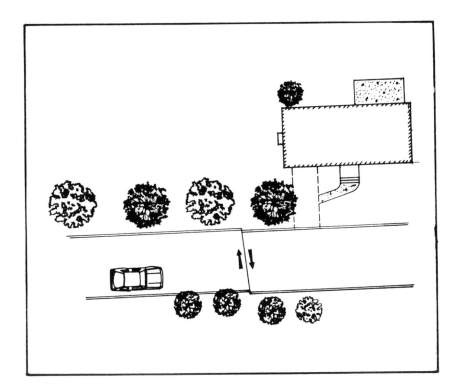

Figure 3.4:  Cross-section showing typical irregular surface profile of weathered carbonate bedrock.

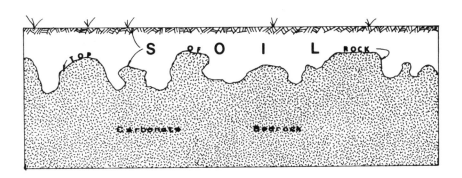

## CAVERNOUS LIMESTONE

You will recall in Chapter 2 that we discussed the excessive withdrawal of ground water and its effect of lowering the overall ground water table. Such drops in the ground water table can be particularly hazardous in areas underlain by limestone or carbonate rock[1] containing large solution cavities[2]. If these cavities occur below the water table surface they will undoubtedly be filled with water. Lowering the water table can empty these large voids and thereby remove any supportive buoyancy the water may have been providing to the overlying soil and rock. As a result, the roof of these cavities may collapse inward creating a sinkhole at the ground surface that can threaten buildings or structures lying overhead. The state of Florida is one of several states presently experiencing this phenomenon.

Even limestone cavities located above the water table can pose some hazards if similarly disturbed by man's activities. Here the cavities are often filled with insoluble clays--the residues of the limestone rock dissolved away by percolating ground water[3].

These filled cavities are very vulnerable to disintegration, particularly if part of the overlying soil has been removed or excess amounts of water or wastewater are discharged into or onto the overlying soil. We must not forget that undisturbed bedrock and soil are in some kind of equilibrium. The removal of even part of the overlying soil for land grading purposes allows faster infiltration of rainwater or wastewater thereby short-circuiting some of the buffering capacity of the soil.

Excessive discharges of wastewater, even without any soil being removed, can likewise accelerate the chemical weathering of the remaining limestone or wash away the clay contents of the existing cavities--the consequences of which are the acceleration of wastewater to the ground water system. With treatment plants rarely operating at maximum efficiency this acceleration of wastewater to the ground water system can represent a hazard to human health.

Land planners are therefore encouraged to incorporate into their land use ordinances and regulations some requirements for additional geohydrological investigations in areas underlain by carbonate bedrock.

An example of such an ordinance can be found in Appendix 1, Part C.

Finally, planners should be aware that the top of carbonate bedrock is typically very irregular as shown in Figure 3.4. The depth of the overlying soil is therefore equally as irregular. Consequently, determining an average soil depth from the excavation of a small number of widely spaced soil logs or test pits can be very inaccurate. Historically where information on subsurface geology has been sparsely collected a great deal of unanticipated blasting of bedrock has been necessary during site preparation. This is particularly true where uniform grades are needed for the proper installation of below ground storm or sanitary sewers or septic system absorption areas. Regarding the latter, it is recommended that at least one trench be excavated to the top of bedrock for the full length of any area in which a septic system absorption bed is to be installed. This should be done prior to design. Likewise for all proposed alignments of storm or sanitary sewer lines a series of closely spaced soil borings to determine the top of bedrock is recommended. Providing the planner with an accurate profile of the top of bedrock makes the job of judging the efficacy of any proposed construction or site improvements considerably easier.

**FOLIATED METAMORPHIC ROCKS**

Foliated or layered metamorphic rocks such as gneiss, slate and schist are frequently encountered during the land development process. The movement of water through such rock formations and its potential to slide vary with the orientation of the foliation planes. When roadways or other construction are planned where this type of rock is present, some determination should be made of the orientation of these foliations. As shown in Figure 3.5, the orientation of these foliation planes can affect slope stability and drainage, thereby affecting the ultimate costs of construction. Working with rather than against such geological formations can save substantial amounts of construction costs.

Figure 3.5: Construction that disturbs foliated bedrock formations must take into account the orientation of the planes of that rock. Failure to do so often produces chronic maintenance problems in the form of rock slides and soil slumps as well as bottom of slope drainage problems.

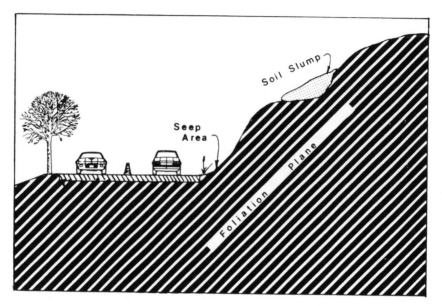

Figure 3.6: Gabion retaining walls have proven to be effective steep slope control structures.

## EXCESSIVELY STEEP SLOPES

The utilization of steeply sloped land (grades of 15% or larger) for land development purposes continues unabated across the United States.

In some cities and towns, such land may represent the only land still available for development. Planning boards and planners are frequently asked to approve grades for proposed roadways in such areas that are steeper than what is permitted under the local zoning ordinance on land use master plan, usually 12%. The developer argues that meeting mandated road grades increases the amount of slope area exposed, thereby producing more environmental impact. This argument is generally true, however, there are methods available to reduce the aereal extent of these slope cuts. Planners no longer need to feel so compelled to grant relief for fear of increasing environmental impact. An effective solution has been the use of gabion retaining walls (Figure 3.6).

### FOOTNOTES

### CHAPTER 3

(1)   These sedimentary rocks are comprised primarily of the minerals calcite and dolomite--two carbonate minerals that are nearly insoluble in pure water but are readily dissolvable by carbonic acid in ground water (See Footnote 3 also).

(2)   Where an excessive number of sinks or land subsidence pockets are visible at the surface, the area is often described as exhibiting karst topography--named after the Karst region in Yugoslavia where this condition is well developed. Approximately, 10% of the nation is under underlain by karstified carbonate rocks.

(3)   Rain falling to the earth's surface collects small amounts of carbon dioxide ($CO_2$) from the atmosphere as well as from the unsaturated soil zone. When combined with the water molecule, as shown in the following equation, carbonic acid is formed:

$$CO_2 \quad + \quad H_2O \quad ----> \quad H_2CO_3$$

(carbon          (water)          (carbonic
dioxide)                            acid)

which ionizes to $H^+$ + $(HCO_3)^-$

(hydrogen     (bicarbonate
ion)            ion)

The hydrogen ions then attack the calcite as shown below:

$$CaCO_3 \quad + \quad 2H^+ \quad ----> \quad H_2O \quad + \quad CO_2 \quad + \quad Ca^{2+}$$

(calcite)   (hydrogen   (water)   (carbon     (calcium
ion)                   ioxide)      ion)

## REFERENCES

## CHAPTER 3

3.1   Keller, Edward A., "Environmental Geology", Charles A. Merrill Publishing Company, 1976.

3.2   Longwell, Chester R.; Richard F. Flint; John E. Sanders, "Physical Geology", John Wiley and Sons Inc., 1969

3.3   Scullin, C. Michael, "Excavation and Grading Code Administration, Inspection and Enforcement", Prentice Hall, Inc., New Jersey, 1983.

# 4

# Soils

The definition of soil varies somewhat from profession to profession. For instance, geologists describe soil generally as "regolith" which is further defined as, "The noncemented rock fragments and mineral grains derived from rocks, which overlies bedrock in most cases." Soil scientists, on the other hand, have defined soil as, "Earth material modified by biological, chemical and physical processes such that the material will support rooted plants." Engineers may simply call it, "Earth material that can be removed without blasting."

It's not so important that there are differences in definition but that as land planners, we recognize that these differences do exist. Consequently, when we solicit or entertain expert opinions, we will be aware that each expert speaks with a bias toward his own definition and his own concerns.

## SOIL FORMATION--GENERALLY

The formation of soil usually begins with the physical and chemical weathering of rock, which is related directly to the interaction between rock and the hydrologic cycle. Subsequent biological, chemical and physical processes complete the transformation of rock to a mature soil. This soil may remain in place and therefore be classified as residual soil or may be transported from its source by other natural processes and labelled a transported soil. Transported soil may undergo further changes once removed thereby forming an entirely new soil.

Soil formation is an extremely slow process, often measured in terms of thousands of years. It is likewise a continuing process. Figure 4.1 illustrates

97

Figure 4.1:  Idealized soil profile.

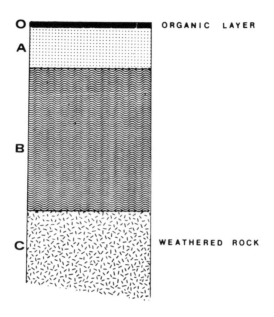

Figure 4.2:  General division of soils in the United States.

Source:  Adapted from Ref. 4.3.

a typical soil profile of a mature residual soil. Horizontal layers called "horizons" can be clearly identified in the profile based on their texture, color, chemical makeup and mineral constituents. For convenience, we have labelled these horizons as O, A, B, and C, starting from top to bottom. These horizons can be further subdivided into narrower bands to provide more detail about the profile.

The A horizon has lost part of its original substance through leaching of soluble materials by either carbonic or humic acid or through mechanical removal by percolating water. The B horizon, on the other hand, is a zone of accumulation, having gained part of the constituents that the A horizon has lost. Iron oxides and clay are common accumulations in the B horizon. The C horizon consists of the parent material--either rock or transported regolith.

While parent materials strongly influence the character of soils, the influence of climate can be an even stronger influence in determining the final soil type. Using this hypothesis, the soils of the United States can be divided into two general classes based on average annual rainfall (Figure 4.2). An artificial boundary line extending north from central Texas to western Minnesota divides the more humid eastern United States from the drier west. East of this boundary, the A horizons of the mature soils have lost large fractions of their calcium and magnesium carbonates due to leaching. Consequently, soils tend to be acidic. In addition, the B horizons have a high iron and clay content. These characteristics are a direct result of a good supply of rainfall. West of this boundary where rainfall is less abundant, carbonates are leached less. Furthermore, a higher evaporation rate brings up dissolved salts from below, consequently adding carbonates to the surface horizon. These soils tend to be alkaline. In large areas of western Texas and some adjoining states, these carbonates have built up a solid, almost impervious, white crust called "caliche", (Ref. 4.3). Soils with a calcium rich, upper horizon as found in the west, have been termed "pedocals". Soils with a high iron and clay content in the B horizon have been called "pedalfers". There are areas west of this artificial boundary, however, that have pedalfer soils-- particularly moist mountain area that received higher

than the average rainfall.  Where coniferous forests
exist, pedalfer soils have been intensely leached by
humic acids and consequently the upper part of the
soil profile is light grey or nearly white.  We refer
to these intensely leached soils as podzols.

As a general rule of thumb, we can state that
soils on convex topographic positions or on steep
slopes are usually well drained, well oxidized, thin-
ner and subject to erosion.  Soils on concave landscapes or on broad flat areas are more poorly drained,
less well oxidized and deeper.

## SOILS AND LAND PLANNING

A land planner's first association with soils
usually occurs during the preparation of a land use
master plan or land zoning ordinance.  It is at this
time that he must inventory a multitude of environmental characteristics, including soils, while simultaneously assessing their potential constraints on development.  Additional, site-specific evaluations of soil
conditions will occur as site plans and subdivision
plans are subsequently processed under the adopted
land use master plan and zoning ordinances.  The most
utilitarian documents presently available to planners
to assist them in evaluating soils are a series of
reports prepared under the National Cooperative Soil
Survey conducted by the United States Department of
Agriculture (U.S.D.A.), Soil Conservation Service
(SCS).  Over 30 years of effort have gone into this
identification and classification of the nation's
soils.  Figures 4.3 and 4.4 are examples of the types
of information available in these soil survey reports.
Of particular importance is the engineering and land
development limitation assessment prepared for each
soil type.  While I have found these soil survey maps
to be quite accurate, I do caution you that they are
not site specific.  Consequently, you should consider
any soils data collected on site as being the most accurate, providing it is collected by a qualified
individual.

The disposal of wastewater, the design of buildings, highways and retaining walls, the placement of
subsurface utility lines and the disposal of sludges
are just a few of the activities in the land development process for which land planners will find it nec-

**Figure 4.3**

## Neshaminy Series

The Neshaminy series consists of deep, gently sloping to steep, well-drained gravelly and stony soils. These soils are on the top and sides of the traprock ridges and on hills south of the terminal moraine of the Wisconsin glaciation. They formed in material weathered from the underlying basalt bedrock, and the content of coarse fragments generally increases with depth.

In a representative profile the surface layer is dark-brown gravelly silt loam about 8 inches thick. The upper part of the subsoil, to a depth of about 11 inches, is reddish-brown gravelly clay loam. The middle 12 inches of the subsoil is yellowish-red gravelly clay loam, and the lower 31 inches is yellowish-red cobbly clay loam. The substratum to a depth of 60 inches is yellowish-red sandy loam and weathered fragments of basalt.

Permeability and available water capacity are moderate. These soils are susceptible to frost heave and are difficult to work during cold months.

Most areas of the gently sloping soils were formerly cleared and farmed, but most areas have been abandoned, and the fields have reverted to trees. Yellow-poplar, oaks, hickory, and ash are the dominant trees. In places redcedar is abundant in the early stages of succession. In recent years developing communities have expanded onto the traprock ridges, and these soils are now highly regarded as residential areas. Neshaminy soils are slightly erodible to severely erodible, depending on the steepness of the slope and the amount of vegetative cover. Practices to help control runoff, erosion, and sediment are used on farms and in residential areas. The hard bedrock is difficult to excavate without blasting.

Representative profile of Neshaminy gravelly silt loam, 8 to 15 percent slopes, in the middle of a hayfield, 1,400 feet west of the junction of North Long Hill Road, Long Hill Road, and Mountain Avenue, 740 feet north of Long Hill Road, in Passaic Township:

Ap—0 to 8 inches, dark-brown (7.5YR 4/2) gravelly silt loam; moderate, medium, granular structure; friable; many, fine, fibrous roots; 15 percent angular basalt gravel; medium acid; abrupt, smooth boundary. 6 to 8 inches thick.

B1—8 to 11 inches, reddish-brown (5YR 4/4) gravelly clay loam; moderate, fine and medium, subangular blocky structure; friable; many fine fibrous roots; 15 percent angular gravel, many of which are strongly weathered; medium acid; gradual, wavy boundary. 3 to 7 inches thick.

B21t—11 to 23 inches, yellowish-red (5YR 4/6) gravelly clay loam; moderate, medium and coarse, subangular blocky structure; friable; common roots; 20 percent angular basalt cobbles and gravel; nearly continuous, thin to thick, waxy clay films on ped faces and nearly all sand grains are coated; medium acid; diffuse, wavy boundary. 3 to 12 inches thick.

B22t—23 to 39 inches, yellowish-red (5YR 4/6) cobbly clay loam; many, medium, distinct, black (5YR 2/1), semiglossy stains on ped faces and coarse fragments; weak, coarse, subangular blocky structure; firm when dry, slightly sticky and slightly plastic when wet; common roots increase in size and number with increasing depth; 30 percent coarse fragments, mostly angular basalt cobbles and gravel; slightly darker, waxylike, thin to thick, discontinuous clay films on ped faces; medium acid; gradual, wavy boundary. 10 to 25 inches thick.

B3—39 to 54 inches, yellowish-red (5YR 4/6) cobbly clay loam; massive; friable; 40 percent coarse fragments of angular cobbles and gravel composed of basalt; thick waxylike clay films in channels and voids and rarely as bridges between sand grains; medium acid; diffuse, wavy boundary. 10 to 15 inches thick.

C—54 to 60 inches, yellowish-red (5YR 4/6) sandy loam and weathered soft basalt gravel and cobble-size fragments; massive; friable; medium acid.

The solum ranges from 40 to 54 inches in thickness. Depth to bedrock ranges from 4 to 10 feet, but in less sloping areas it is more than 6 feet. Angular coarse fragments generally occur throughout the soil and range from 10 percent to 50 percent gravel, cobbles, and stones. In areas that are not limed, reaction is medium acid.

The A horizon has hue of 10YR or 7.5YR, value of 3 or 4, and chroma of 2 to 4.

The B horizon is 7.5YR to 5YR in hue, 4 or 5 in value, and 4 to 6 in chroma. This horizon has more clay than the A horizon.

The C horizon is commonly more than 50 percent coarse fragments, but in many places it consists of soft weathered saprolite and contains a few hard fragments of basalt.

Neshaminy soils are associated with Ellington variant, Klinesville, Penn, and Boonton soils. Neshaminy soils do not have the mottles that are common in Ellington and Boonton soils. They are deeper than Klinesville and Penn soils.

**Neshaminy gravelly silt loam, 3 to 8 percent slopes** (NeB).—This soil is in long narrow areas on the crests of the Watchung Mountains in the southern part of the county. Included in mapping are small areas of more stony Neshaminy soils, soils that are similar to this Neshaminy soil but have a mottled subsoil, the Ellington loamy subsoil variant, and Rock outcrop.

In places this soil contains a few cobbles, stones, and boulders and has moderate runoff and a moderate hazard of erosion.

This soil is well suited to farming. It has moderate limitations, and practices are needed to help control runoff, erosion, and sedimentation. Good management includes such practices as using crop rotation, cover crops, and contour cultivation. In places stripcropping is needed on long slopes. Most high-value crops are irrigated. This gently sloping soil is desirable for community development because it is deep and fertile and has good surface drainage and only a moderate hazard of erosion. Practices are needed to help control runoff, sedimentation, and erosion for this use. Among the suitable practices are early establishment of lawns, sequential development of large tracts so that only a small part of the soil is bare and exposed at any one time, and special seeding, fertilizing, and mulching in critical areas, such as waterways. Capability unit IIe-55.

**Neshaminy gravelly silt loam, 8 to 15 percent slopes** (NeC).—This soil has the profile described as representative of the series. Included in mapping are small areas of more stony and more sloping Neshaminy soils, the Ellington loamy subsoil variant, and soils that are similar to this Neshaminy soil but have gray mottles in the subsoil. Also included are small areas of bedrock outcrop.

This soil generally contains a few cobbles and in places stones and boulders. It has rapid runoff and a moderately severe hazard of erosion.

This soil is well suited to hay, pasture, and row crops but requires complex measures to control runoff, reduce

**Source:  Ref. 4.4**

**Figure 4.4:  Soil survey tables.**

SOIL  SURVEY

*Limitations of the soils for*

| Soil series and map symbols | Foundations for dwellings— | | Lawns, landscaping, and golf fairways | Septic tank absorption fields |
|---|---|---|---|---|
| | With basement | Without basement | | |
| Muck:  Ms, Mu | Severe:  frequent flooding; seasonal high water table at surface; low bearing strength; severe subsidence. | Severe:  frequent flooding; seasonal high water table at surface; low bearing strength; severe subsidence. | Severe:  frequent flooding; seasonal high water table at surface. | Severe:  frequent flooding; seasonal high water table at surface; low bearing strength. |
| Neshaminy: NeB | Slight:  hard bedrock at a depth of more than 6 feet in most places. | Slight | Moderate:  gravel | Slight:  bedrock at a depth of 6 feet or more in most places; stony in places. |
| NeC | Moderate:  slope; hard bedrock at a depth of 4 to 10 feet or more. | Moderate:  slope | Moderate:  gravel; slope; hazard of erosion. | Moderate:  hard bedrock at a depth of 4 to 10 feet or more; stony in places. |
| NfD | Severe:  steep; excessive stones. | Severe:  steep; excessive stones. | Severe:  steep; excessive stones. | Severe:  steep; excessive stones. |
| Netcong: NtB | Slight | Slight | Moderate:  gravel content. | Slight:  stony in places; hazard of ground-water pollution. |
| NtC | Moderate:  slope | Moderate:  slope | Moderate:  gravel content; slope; hazard of erosion. | Moderate:  slope makes special design and careful installation necessary. |
| Otisville: OtC | Moderate where slopes are 8 to 15 percent. Slight where slopes are 3 to 8 percent. | Moderate where slopes are 8 to 15 percent. Slight where slopes are 3 to 8 percent. | Severe:  coarse texture; low available water capacity; low fertility; low organic-matter content. | Slight where slopes are 3 to 8 percent. Moderate where slopes are 8 to 15 percent; hazard of ground-water pollution. |
| OtD | Severe:  steep | Severe:  steep | Severe:  coarse texture; low available water capacity; low fertility; low organic-matter content. | Severe:  steep slopes; hazard of ground-water pollution. |
| Parker: PuC | Moderate where bedrock is at a depth of 4 to 5 feet. Slight where bedrock is at a depth of 5 to 10 feet; stony in places. | Moderate where slopes are 8 to 15 percent. Slight where slopes are 3 to 8 percent. | Moderate:  gravelly and cobbly; stony in places. | Moderate where bedrock is at a depth of 4 to 10 feet. Slight where gently sloping; deep to bedrock; hazard of ground-water pollution. |
| PbD | Severe:  steep | Severe:  steep | Severe:  steep | Severe:  steep |
| PeC | Moderate:  moderate stone content; bedrock at a depth of 4 to 10 feet. | Moderate:  moderate stones. | Severe:  excessive stones, cobbles, and gravel. | Moderate:  moderate stones. |
| PeD | Severe:  steep; moderate stones. | Severe:  steep | Severe:  steep; excessive stones. | Severe:  steep; moderate stones. |

Figure 4.4:  (continued)

MORRIS COUNTY, NEW JERSEY

*town and country planning.*

| Local roads, streets, and parking lots | Athletic fields | Picnic and play areas | Campsites, trailers, and tents | Sanitary landfill[1] |
|---|---|---|---|---|
| Severe: frequent flooding; seasonal high water table at surface; low bearing strength; severe subsidence. | Severe: frequent flooding; seasonal high water table at surface; low bearing strength; severe subsidence. | Severe: water table above a depth of 20 inches or more for 1 month during season of use. | Severe: frequent flooding; water table above a depth of 20 inches during season of use. | Severe: frequent flooding; seasonal high water table at surface. |
| Moderate: moderate frost-action potential. | Severe: excessive gravel. | Moderate: gravel content. | Moderate: gravel content. | Slight to moderate: hard bedrock at a depth of 6 to 10 feet or more. |
| Moderate: slope; hazard of erosion; moderate frost-action potential; hard bedrock at a depth of 4 to 10 feet. | Severe: slope; excessive gravel. | Moderate: slope; gravel. | Moderate: slope; gravel. | Slight to severe: hard bedrock 4 to 6 feet or more. |
| Severe: steep; excessive stones. | Severe: steep | Severe: steep | Severe: steep | Severe: hard bedrock at less than 6 feet in most places. |
| Slight | Severe: excessive gravel; cobbles. | Moderate: gravel content. | Moderate: gravel content. | Severe: hazard of ground-water pollution. |
| Moderate: slope; hazard of erosion. | Severe: excessive gravel and cobbles; strong slopes. | Moderate: strong slopes; gravel content. | Moderate: strong slopes; gravel content. | Severe: hazard of ground-water pollution. |
| Slight where slopes are 3 to 8 percent. Moderate where slopes are 8 to 15 percent. | Severe: coarse texture; low available water capacity; low fertility. | Severe: coarse texture; poor trafficability. | Severe: loose sand; poor trafficability. | Severe: rapid permeability permits ground-water pollution. |
| Severe: steep | Severe: steep | Severe: steep | Severe: steep | Severe: rapid permeability permits ground-water pollution. |
| Moderate: strong slopes and stony in places. | Severe: excessive gravel, cobbles, and stones; slope. | Moderate: moderate gravel; strongly sloping to gently sloping. | Moderate: strongly sloping and gently sloping; moderate gravel and cobble. | Severe: insufficient filter material and rapid permeability in C horizon permit hazard of ground-water pollution. |
| Severe: steep; excessive coarse fragments in places. | Severe: steep | Severe: steep; excessive gravel, cobbles, and stones. | Severe: steep; excessive gravel, cobbles, and stones. | Severe: insufficient filter material and rapid permeability in C horizon permit hazard of ground-water pollution. |
| Moderate: moderate stones. | Severe: slope; excessive stones. | Moderate: moderate stones. | Moderate: moderate stones. | Severe: excessive stones. |
| Severe: steep; hazard of erosion; moderate stones. | Severe: steep | Severe: steep | Severe: steep | Severe: excessive stones. |

Source:  Ref. 4.4

essary to have accurate soils data available.  Some soil related problem areas have already been identified in Chapters 2 and 3.  The following are items of additional concern to today's land planners.

## Clay

Most of us are aware of the implications of clay in the soil profile.  After all, clay has been used for centuries to make pots for holding water, to build dams, and most recently to line the bottom of sanitary landfills.  You will recall that we spoke previously of clay accumulation in the B horizon.  Poor drainage is usually a synonymous characteristic of soils containing substantial amounts of clay and should serve as an immediate red flag to the planner.  Unfortunately, too many land planning decisions involving clay bearing soils have been based upon incomplete information.  It is important not only to recognize that clay exists in the soil profile but to further identify the type of clay, its thickness, depth or concentration, aereal extent, and physical, chemical and mineral characteristics.  Some clays, particularly bentonite, a derivative of altered volcanic ash has a high expansion and contraction coefficient when exposed to water.  Highway pavements have been cracked, building foundations skewed and retaining walls toppled where insufficient attention was paid to the presence of bentonite clay.  Bentonite clays in the Eagle Ford Formation of Austin, Texas have caused extensive damage to both property and infrastructure.

## Fragipans and Other Restrictive Soil Horizons

Less well known, but nonetheless important strata in soil are fragipans.  A fragipan is a dense, brittle, almost cement like layer of soil that is very low in organic matter and clay, but rich in silt and very fine sand.  Like clay, it is very slowly permeable to water.

Fragipan layers may range in thickness from several inches to several feet.  Thin fragipans are frequently overlooked and unrecognized by untrained professionals.  Yet, their impact on such things as septic system installation or wastewater treatment, can be severe.  Other restrictive soil strata that

react to water much like fragipans and clay may also
be found in some soil profiles. These are identified
below.

1.  Compaction Pan - A layer of fine grained
sediments that have been reduced in bulk volume or
thickness in response to the increasing weight of
overlying material that is continually being
deposited.

2.  Duripan - A soil layer that is characterized
by cementation of silica. Duripans occur mainly in
areas of volcanism that have arid or Mediterranean
climates.

3.  Petrocalcic Pan - A soil horizon that is
characterized by an induration with calcium carbonate
or sometimes with magnesium carbonate.

4.  Iron Pan - A soil horizon in which iron
oxides are the principal cementing agents. Iron pans
can be found in both wet and dry areas.

5.  Plow Pan - A layer of soil compacted by
repeated plowing to the same depth. Sometimes referred to as a plow sole.

## PERCHED WATER TABLE

The presence of any of the pan formations
enumerated above may contribute to a soil condition
known as a perched water table. It is not a part of
the true water table and in fact, is usually separated
from the true water table by a layer of unsaturated
soil (see Figure 4.5). The presence of a perched
water table can be confusing, leading to an erroneous
denial of land development projects based upon the
assumption that it is the true ground water table. In
areas where individual septic disposal systems are
utilized, a perched water table may not necessarily
preclude the use of these subsurface disposal systems.
Septic systems have been successfully installed on
such areas after a portion of the retarding pan layer
has been excavated and the overlying saturated soils
replaced with the required depth of select fill.

I would caution you, however, that such septic
system installations are not completely hazard free.
The possibility that breaching the pan will divert the
perched water into the newly installed disposal bed is
high. Whether this will occur depends upon the slope
and extent of the retarding pan, the size of the open-

ing through it and effectiveness of diversion structures placed around the perimeter of the disposal bed.

Figure 4.5: Diagram showing a perched water table.

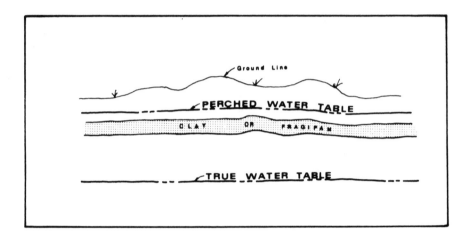

## REFERENCES

## CHAPTER 4

4.1   Flann, Peter T., "Environmental Geology",
      Harper's Geoscience Series, 1970.
4.2   Keller, Edward A., "Environmental Geology",
      Charles E. Merril Publishing Co., Columbus,
      Ohio, 1976.
4.3   Longwell, Chester R., Richard F. Flint, John E.
      Sanders, "Physical Geology", John Wiley and Sons
      Inc., 1969.
4.4   United States Department of Agriculture, Soil
      Conservation Service, "Soil Survey, Morris
      County, New Jersey", August, 1976.

# 5

# Wastewater Disposal

## HISTORIC PERSPECTIVE

While contemporary Americans may fret over the way the nation presently treats and disposes of its wastewaters, the current practices are a marked improvement over the conditions that existed in 19th century America. For instance, human waste as well as livestock and industrial wastes were frequently dumped into alleyways and city streets where it remained to putrefy until washed away by the next rainstorm. Some cities advanced as far as building cesspool pits or vaults beneath buildings into which human wastes were dumped. When these were filled to capacity, they were simply covered over and a new pit was built. The smell of putrefying waste must have been horrendous, especially when accompanied by the odors from huge stockpiles of horse manure that were amassed throughout the cities.

Where storm drainage systems had been built, they frequently served double duty, transporting both storm water and wastes directly to any number of nearby streams--streams I might add, that also served as sources of potable water for the very same cities. With such conditions, it was little wonder that outbreaks of waterborne diseases such as cholera, dysentery and typhoid were frequent occurrences.

From a land planners point of view, it is interesting to note that the technological advancements that eventually replaced the horse with streetcars, electric trolleys, cable cars, subways and railroads also changed urban land use patterns from what Petulla (Ref. 5.1) calls "walking cities" to sprawling, densely populated, congested metropolises. It was this surge in urban population that eventually drove

107

up water consumption by 50 to 100 percent. Consequently, the limited capacities of cesspools and privy vaults were no longer adequate nor was the curbside disposal of waste.

The first sewer system in America, designed to carry wastewater exclusively, was built in Lenox, Massachusetts in 1875-76. Its designer, Col. George E. Waring, was a friend of the renowned landscape architect, Frederick Law Olmstead. Waring's separate sewer system was not conceived as a method to provide treatment to wastewater at the end of the pipe, but rather to prevent the accumulation of putrescible fecal matter and subsequent generation of disease-inducing sewer gas. Waring still subscribed to the miasma (foul air) theory of disease. Even after the germ theory had largely displaced the miasma theory there was no great rush to provide end of pipe treatment. As America entered the 20th century, the nation's sanitary engineers continued to rely upon the British practices of using water as a carrier of wastes and streams for the dilution of wastewater. The consensus at the time was that it was cheaper to treat incoming river water than to treat outgoing raw sewage.

It is further disturbing to note how little, if any, attention was paid to the discharge of industrial wastes, even as late in American history as World War II. The reasons for this were twofold. First, there was the popular conception that industrial wastes, by their very nature, killed "germs" in the water. Unwittingly, they had accurately identified the toxicity problem of industrial wastes long before it became a buzz-word in contemporary society. Secondly, many American industries, as major employers in American cities, were able to dissuade municipalities from passing regulations, environmental or otherwise, that were unfavorable to industrial operations.

I have presented this brief overview of American wastewater treatment history, not to minimize the nation's current wastewater disposal problems, but to demonstrate that contrary to popular perception, we have made slow but significant advancements in the last 100 years in the handling of our wastewaters. Unfortunately, this rather slow pace of improvements has left us a legacy of damage that will not allow us

to proceed at such a leisurely pace over the next
hundred years.

## WASTEWATER--GENERALLY

Wastewater (literally used water) is approximately 99.9% water by weight with the remainder comprised of a wide variety of suspended or dissolved
materials. The high ratio of water to solids is the
result of our persistent use of water as a dilutor and
carrier of our liquid and solid waste materials. For
instance, the standard water closet uses 3 to 5 gallons of water to flush away as little as 4 to 6 ounces
of urine or feces. The average American family of
four discharges approximately 400 gallons per day of
wastewater. With the population of the United States
hovering around 250 million people, the amount of
wastewater that must be treated daily from this source
alone is enormous. When we add in the volume of industrial wastewater also generated daily, the job of
adequately cleaning the nation's wastewaters seems to
be a nearly impossible task.

As land planners, we will generally encounter
two basic types of wastewater treatment systems. The
first is the community-wide, municipal, subsurface
collection and pumping system usually found in densely
populated areas. These systems collect and transport
wastewater to some kind of treatment facility for
cleansing. Once treated, it is then returned to the
hydrological cycle via discharge to the surface or
ground water. The degree of treatment varies considerably at these plants, ranging from primary (the lowest level) through secondary to advanced or tertiary
treatment. Let's briefly discuss each.

The major goal of primary treatment is to remove
those pollutants that will either settle or float. A
good primary plant will typically remove 60% of the
suspended solids in raw sewage and 35% of the biochemical oxygen demand (BOD). Soluble pollutants are not
removed. Federal water pollution control regulations
now require a level of treatment higher than primary.
Consequently, secondary treatment has become the mandated level of treatment at municipal treatment plants
throughout the United States.

Secondary treatment is designed to remove the
soluble fraction of the waste that passes through pri-

mary treatment plus it also provides further removal
of suspended solids.  Secondary treatment is usually a
controlled biological process utilizing the same processes that would occur naturally in a receiving waterbody.  In addition, it speeds up those processes so
that the breakdown of organic pollutants can be
achieved in relatively short periods of time.  This
treatment relieves the burden that would be placed on
our streams to purify this wastewater.  Secondary
treatment is expected to removed more than 85% of the
suspended solids and BOD.  However, it does not remove
significant amounts of nitrogen, phosphorus or heavy
metals.  These materials must be removed by further
tertiary treatment methods.

Secondary treatment processes usually include
one of the following methods or a variation thereof:
1.   Trickling filters.
2.   Activated sludge.
3.   Oxidation ponds or lagoons.
4.   Rotating biological contactors.
5.   Activated biofilters.
6.   Physical--chemical treatment.
(Table 5.1).

Later in this chapter you will also be introduced to land application systems for wastewater
treatment that have increased in popularity and use in
the last two decades.

Industrial plants may have their own internal
collection system and treatment plants utilizing some
of the processes already discussed.  Because their
wastewaters may contain substantial toxic or hazardous
material and are highly variable as to content from
industry to industry, we cannot discuss them in detail
here.  As land planners, however, you are advised to
seek expert advice when confronted with the approval
of a proposed industrial wastewater treatment plant or
the introduction of industrial wastewaters into your
municipal sewerage system.  It might surprise you to
learn that 160,000 industries nationwide presently
discharge more than 3.2 billion gallons of wastewater
per day into publicly owned treatment plants.  Your
initial request for information should include, as a
minimum, a complete analysis of the anticipated
wastewater plus bench scale treatment tests on a
sample of the wastewater utilizing the proposed
method(s) of treatment.

Table 5.1:  Summary of Treatment Alternatives Available for
Pollutant Removal

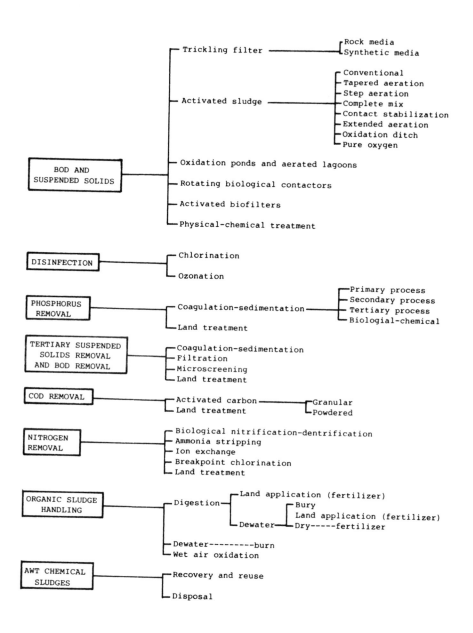

Source:  Ref. 5.5

The second basic form of wastewater treatment system is usually found in more rural and less densely populated areas.  In such areas, wastewaters are usually collected and treated by individual underground systems called septic systems (Figure 5.1).  As we shall soon see, there are even variations of such seemingly simple systems as these.

## SEWAGE TREATMENT PLANTS

### Problems in Operation

1.   Odors and Fugitive Emissions
Let no one kid you, cleaning wastewater is a smelly business, even in the most modern plants, in spite of claims to the contrary by designers and engineers.  No treatment plant operates at 100% efficiency one hundred percent of the time, and it is not unusual to see entire city blocks fumigated by nauseating odors or worse yet, by chemical vapors from industrial wastes, released during the normal agitation or aeration of the wastewater.  As land planners, we have an obligation to insure that new treatment plants have sufficient buffer zones established around them and that new residential developments are diverted away from existing sewage treatment plants.  A micro-climatological study would be helpful in establishing the appropriate widths for such buffer zones by determining the prevailing or predominant wind rose patterns.
2.   Insects
Some of the secondary treatment processes particularly trickling filters and oxidation ponds and lagoons can produce enormous numbers of nuisance insects.  The trickling filter fly, Psychoda alternata, is probably the most widely encountered.
The larvae of this species are valuable to the efficiency of trickling filters, but their adult forms can be extreme nuisances, both at the treatment plant and adjacent neighborhoods.  While they do not bite, they can carry pathogens on their bodies and can

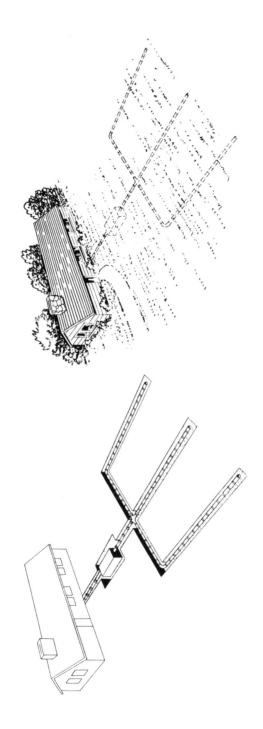

Figure 5.1: Typical septic tank-absorption field system.

Source: Ref. 5.15

invade the ears, eyes, mouth and nostrils.
They are not strong fliers, but can be wind-
borne up to a kilometer radius of the treatment plant. Similar problems with insects
can occur at treatment plants utilizing oxidation ponds and lagoons. Here, midge flies
Chironomus sp. and mosquitoes Culex sp. are
generally the major nuisance insects. Like
the trickling filter fly, they can swarm
into adjoining neighborhoods, with
mosquitoes carrying the further threat of
such diseases as encephalitis.

3.   Sludge Digestion and Disposal

The second largest number of complaints
regarding wastewater treatment plants are
related to either digestion or removal and
disposal of sludge. Large quantities of
sludge are generated by the primary and secondary treatment of wastewater. In fact,
for each million gallons of wastewater
treated an average of 945 cubic feet of wet
sludge is produced which must likewise be
treated. Since the organic content and con-
sequent oxygen demand are both extremely
high, it is often more economically feasible
to digest this sludge in closed tanks, utilizing anaerobic (i.e. no oxygen) conditions
and anaerobic bacteria. The process can
take from 15 days to 60 days to complete.
During the process, substantial volumes of
methane gas and carbon dioxide will be generated. This methane-carbon dioxide mixture
may be used as a supplemental fuel or wasted
to the atmosphere through burners that ignite the flammable gas as it passes through
the vents. Such releases may be accompanied
by strong, putrid odors caused by minor concentrations of hydrogen sulfide in the exit-
ing gas. If the burners fail to ignite or
the treatment plant operator does not
closely watch the digestion process, the
odors released to the atmosphere may increase to highly objectionable levels. The
subsequent removal of this digested sludge
from the tanks to the drying beds for de-

watering has also generated numerous complaints of foul odors.

As planners, we should be aware of
these potential problems and provide adequate buffer zones around such treatment
facilities.

4.   Bypasses Due to Hydraulic Overload

While the nation's design engineers can
estimate rather accurately the hydraulic
capability of a wastewater collection system
and its appurtenances, no design will ever
be fail safe.

Flows of wastewater may end up substantially larger than anticipated because land
development exceeded projections or subsurface sewerage lines and manholes were poorly
constructed and now have cracks and leaks
that allow storm water (inflow) or ground
water (infiltration) into the system.  In
addition there may have been widespread sur-
reptitious connections of home sump drains
and foundation drains to the sanitary sewer
system. [1]  When excess flows occur, threatening to overtop treatment units or collection tanks (wet wells) in pump stations,
bypass lines may be used to reduce the
amount of incoming wastewater.  Sewage is
then discharged, often without any treatment
whatever, except perhaps disinfection, to
the nearest waterbody.  The impact on the
receiving waterway can be substantial, resulting in large fish kills and the deposition of grease, tampons, feces and condoms
along the stream bottom.  At some treatment
plants, this is a regular occurrence.

In addition, blockages in poorly constructed or underdesigned sewer lines can
occur at any point in the collection system
resulting in overflowing manholes in
streets, backyards and basements or as in
several recent cases, into subsurface caverns in the underlying limestone bedrock.

Can land planners help to prevent such
occurrences?  Most definitely.  First, where
new sewer lines are proposed to connect to
existing ones, there must be a determination

made as to whether the existing downstream sewer lines can carry the increased amount of wastewater. Such evaluations must take into consideration the presence of any excessive infiltration or inflow and the history of any bypasses or overflows in these existing lines, particularly during wet weather. If such problems exist, I would urge you to proceed cautiously. You may be asked to approve the new flows based on dry weather conditions and there may be legitimate capacity available at these periods. If you feel obliged to recommend approval, I would urge you not to accept it without the condition that the applicant submit a remedial plan to remove two to three gallons of infiltration and inflow for every gallon of proposed wastewater flow. Further it must be clear that this remedial work is to be completed prior to the introduction of any new discharge. Such remedial work can include such things as internal manhole repair or replacement, removal of illegal household sump pit connections, raising the rims of manholes near waterways above flood levels, installing watertight manhole covers or televising the existing sewer lines to locate cracks and leaks, and then grouting or sealing them off.

Secondly, planners should discourage the installation of gravity sewer lines and manholes in areas of excessively steep topography, in bedrock or in areas of high water table. If such installations are unavoidable, then an extra-ordinary construction inspection program should be instituted. It is this stage that is the weakest link in even the most impeccable engineering design.

5.   Spills of Oil and Other Materials
     Into Treatment Plants

Nothing causes greater panic amongst wastewater treatment plant operators than the unexpected arrival of a large slug of oil, acid, industrial solvent or other similar material at the plant's incoming wet

well.  Often there is little that can be
done to stop the material from entering the
treatment units, and in biological secondary
treatment plants, the results can be devastating.  Multi-million gallon per day,
municipal treatment plants have lost nearly
all of their treatment ability because of
the death of their resident biological
organisms as a result of a spill.  Treatment
plants can be knocked out of operation for
several weeks.  In addition, explosions and
collapse of incoming sewer lines have also
occurred.  While treatment units are drained
and cleaned or sewer lines excavated and
repaired nearly    raw  sewage   must  be
bypassed.

Since many of these spills are from
industrial sources, planners may be able to
prevent some of these occurrences by either
of the following:

A.  Requiring all contributing industries to construct reinforced containment
structures around any indoor or outdoor liquid waste, liquid raw material or liquid by-
product storage tank that is located near
floor drains connected to any sanitary sewer
line.

B.  Adopting and enforcing a strict
industrial wastewater pre-treatment ordinance.

6.   AIDS Virus and Other Wastewater Viruses

Viruses are much smaller in size than
bacteria and can be found in significant
quantities in raw sewage.  Due to their
minute size (20 to 300 $\mu$m), they can easily
pass through filters that hold back most
bacteria.  Unlike bacteria, viruses serve no
role in the treatment or stabilization of
wastewater.  Their presence, however, is a
constant threat to treatment plant personnel
and to residents of neighborhoods in close
proximity to such facilities.  While the
wastewater treatment facility can reduce the
number of viruses, it cannot eliminate them
completely.  Any wastewater treatment plant
that produces large quantities of aerosol

sprays can also transport viruses over substantial distances. There are two main routes of viral infection for humans: by ingestion or direct inhalation of aerosols. Table 5.2 lists some viruses commonly found in wastewater.

The disinfection of the final effluent from all wastewater treatment plants is necessary because large numbers of bacteria and viruses will still be present. Chlorine is a commonly used disinfectant, however, this oxidizer may not kill all of the viruses present, since they react differently than bacteria to chlorine. Alternative disinfectants are available and they include bromine chloride, chlorine dioxide, ozone and ultraviolet light. These are less commonly used because of negative environmental side effects and higher cost.

A virus of particular concern today is the human immunodeficiency virus (HIV) which produces AIDS (acquired immunodeficiency syndrome). It has been found in the blood, saliva, semen, tears, urine and even breast milk of infected men and women. Since these fluids could enter the wastewater system through fecal waste, urine or other materials containing the virus, it is highly probable to expect the AIDS virus to be found in wastewater arriving at a sewage treatment plant. How much of a threat it poses to workers and residents surrounding the treatment plant is not completely known at this time. The Center for Disease Control in Atlanta, Georgia has so far found no cases of AIDS attributable to wastewater collection and treatment facilities.

A brief review of some of the characteristics of the AIDS virus may help you to formulate your own opinion as to the potential for being infected by the operations of a wastewater treatment facility (Ref. 5.2).

A.  The AIDS virus is a relatively weak virus that does not remain viable for long periods in a harsh environment such as wastewater.

Table 5.2:  Virus Groups of Concern in Wastewater

| Virus Group | Disease |
|---|---|
| Astrovirus | Mild gastroenteritis |
| Calcivirus | Mild gastroenteritis |
| Coronavirus | Upper respiratory tract infection |
| Enteroviruses | |
| Adenovirus | Acute respiratory disease; conjunctivitis; pharyngitis; pharynoconjunctival fever |
| Coxsackie A | Upper respiratory tract infection |
| Coxsackie B | Upper respiratory tract infection; myocarditis; aspectic meningitis; Bornholm's disease |
| Echo | Common cold; aspectic meningitis; conjunctivitis; gastroenteritis |
| Hepatitis A | Infectious hepatitis |
| Polio | Poliomyelitis |
| Reovirus | Upper respiratory tract infection |
| Norwalk | Acute gastroenteritis |
| Rotavirus | Acute gastroenteritis |

Source:  Adapted from Ref. 5.2

B.  The AIDS virus is quickly inacti-
vated by alcohol and chlorine and even exposure to air.

C.  The AIDS virus replicates in white
blood cells, not in the intestinal tract of
man or in wastewater.

D.  The AIDS virus must enter the blood
stream directly.  Therefore, AIDS contaminated wastewater would have to come into
direct contact with an open wound or skin
abrasion in order to be transmitted.

E.  The Center for Disease Control has
found no evidence that the AIDS virus can be
transmitted through water, air, food or
casual contact.

I hope this discussion reinforces my earlier
recommendations to you to require adequate buffer
zones around all wastewater treatment facilities and
to eliminate as many unnecessary bypasses of untreated
sewage as possible.

## THE NPDES PERMIT SYSTEM

The major program regulating wastewater treatment plant discharges to the nation's surface waters
is the National Pollutant Discharge Elimination System
(NPDES) authorized under Section 402 of the 1972
Federal Water Pollution Control Act (later amended as
the Clean Water Act).  While the U.S. Environmental
Protection Agency (U.S. EPA) had initial responsibility for all permit issuance that responsibility has
gradually been shifted to various states and U.S. territories with 35 states and territories assuming primacy for permit issuance and reissuance, as of
December 31, 1982.

The National Pollutant Discharge Elimination
System made it illegal to discharge pollutants into
the nation's navigable waters without obtaining an
NPDES permit.  These permits specify:  (a) the discharge limits for specific pollutants or substances,
(b) the schedules setting forth the types of actions
required and the time frames necessary to comply with
the discharge limits, (c) the requirements for self-
monitoring of the wastewater flows as well as the
periodic reporting of compliance.

These permits are to be renewed and upgraded if needed every five years. As of May 4, 1990 63,000 NPDES permits had been issued.[2]   Municipal wastewater dischargers were to meet secondary level treatment by July 1, 1977. Some municipal dischargers, however, were given extensions of this requirement until July 1, 1988.

Industrial dischargers were to install certain types of waste treatment technology with deadlines of July 1, 1977 and July 1, 1983.

As a planner, you may not expect to get directly involved in an enforcement action against either a municipal or industrial wastewater treatment plant. However, an enforcement action may very well involve a land use issue. For instance, a municipal sewage treatment plant may exceed its NPDES permitted flow limits, causing excursions or violations in other NPDES permitted parameters. Regulatory agencies, besides imposing fines, may also impose "sewer bans" (see discussion on Sewer Bans at the end of this chapter). In New Jersey, the imposition of such bans is a frequent and effective practice. The ban prohibits any further hookups and discharges to a community collection system until such time as the wastewater treatment plant no longer exceeds its permitted flows or other permit conditions. In effect, a sewer ban becomes a ban on building, an issue in which planners do have an interest. While sewer bans are not specifically addressed in the 1972 federal Clean Water Act and its subsequent amendments, some states have incorporated such authority in their own water pollution control acts and regulations when they assumed primacy for the NPDES permit program. In any event, the remedy for lifting a sewer ban can only be accomplished by achieving compliance with the limits of the facility's NPDES permit. Planners, therefore, will benefit from knowing something about the current effectiveness of the NPDES permit system.

In 1978, six years after the NPDES program began, the U.S. General Accounting Office (GAO) reported significant noncompliance with permit limits by major industrial facilities and in 1980, found a similar level of noncompliance at major municipal facilities (Ref. 5.4). A further evaluation of the NPDES system was again made by GAO in 1982 with a report released publicly in December 1983 (Ref. 5.7).

During this evaluation, the records of 531 major dischargers—274 municipal treatment plants and 257 industrial plants in six states were examined. These six states accounted for about 21 percent of the total 68,000 NPDES permits issued as of October 30, 1982, including 23 percent of all the major dischargers. Of the six states selected, Iowa, Missouri, New Jersey and New York were administering their own NPDES program while the EPA was administering it in Texas and Louisiana.

Their findings listed below sounded very similar to those reached in the prior 1978 and 1980 evaluations:

a.  Overall municipal noncompliance was 86% and industrial noncompliance was 79%.

b.  Forty-two of the 206 industrial dischargers that exceeded their permits were in significant noncompliance. Eighty-eight of the 238 municipal dischargers that exceeded their permit limits were in significant noncompliances (Tables 5.3 and 5.4).

c.  Both EPA and the states have a general policy that exempts municipal treatment plants from enforcement if they have applied for federal funds to upgrade existing facilities or build new ones. This policy allows noncompliance to continue as long as grant funds are pending or construction is underway.

d.  The EPA cannot assess fines directly for permit noncompliance, but must refer all such cases to the Department of Justice for litigation. This is a time consuming and expensive process and not as effective a deterrent as the ability to levy a fine.

e.  Thousands of dischargers have no permit or hold expired permits.

f.  The EPA and the states rely on the submission of self monitoring reports (SMR's or if you prefer, discharge monitoring reports DMR's) to detect permit noncompliance. However, many discharge monitoring reports were not submitted or were incomplete.

g.  The SMR data provided by a large percent of the laboratories was inaccurate[3].

Table 5.3:  Summary of Frequency of NPDES Permit Noncompliance by Selected Dischargers During the Period October 1, 1980 to March 31, 1982

| State | Discharger type | Sample size | 1 to 3 C[a] | 1 to 3 Q[b] | 4 to 6 C | 4 to 6 Q | 7 to 9 C | 7 to 9 Q | 10 to 12 C | 10 to 12 Q | Over 12 C | Over 12 Q | Total dischargers C | Total dischargers Q |
|---|---|---|---|---|---|---|---|---|---|---|---|---|---|---|
| Iowa | Municipal | 38 | 9 | 7 | 7 | 4 | 3 | 7 | 5 | 2 | 10 | 9 | 34 | 29 |
| | Industrial | 25 | 8 | 1 | 4 | 2 | 2 | 2 | 3 | 1 | 3 | 1 | 20 | 7 |
| | Total | 63 | 17 | 8 | 11 | 6 | 5 | 9 | 8 | 3 | 13 | 10 | 54 | 36 |
| Louisiana | Municipal | 38 | 3 | 3 | 5 | 5 | 4 | 5 | 8 | 2 | 13 | 9 | 33 | 24 |
| | Industrial | 50 | 13 | 15 | 9 | 4 | 5 | 5 | 4 | 1 | 2 | 5 | 33 | 30 |
| | Total | 88 | 16 | 18 | 14 | 9 | 9 | 10 | 12 | 3 | 15 | 14 | 66 | 54 |
| Missouri | Municipal | 43 | 11 | 0 | 7 | 0 | 8 | 0 | 2 | 0 | 11 | 0 | 39 | 0 |
| | Industrial | 37 | 11 | 8 | 4 | 1 | 5 | 2 | 4 | 1 | 6 | 1 | 30 | 13 |
| | Total | 80 | 22 | 8 | 11 | 1 | 13 | 2 | 6 | 1 | 17 | 1 | 69 | 13 |
| New Jersey | Municipal | 51 | 10 | 11 | 6 | 8 | 5 | 3 | 7 | 6 | 11 | 5 | 39 | 33 |
| | Industrial | 49 | 8 | 12 | 4 | 8 | 3 | 3 | 1 | 5 | 3 | 5 | 19 | 33 |
| | Total | 100 | 18 | 23 | 10 | 16 | 8 | 6 | 8 | 11 | 14 | 10 | 58 | 66 |
| New York | Municipal | 50 | 12 | 6 | 2 | 5 | 4 | 4 | 6 | 0 | 19 | 6 | 43 | 21 |
| | Industrial | 50 | 13 | 14 | 12 | 5 | 4 | 2 | 1 | 6 | 5 | 1 | 35 | 28 |
| | Total | 100 | 25 | 20 | 14 | 10 | 8 | 6 | 7 | 6 | 24 | 7 | 78 | 49 |
| Texas | Municipal | 54 | 14 | 17 | 8 | 7 | 6 | 1 | 7 | 3 | 6 | 4 | 41 | 32 |
| | Industrial | 46 | 19 | 13 | 9 | 4 | 3 | 5 | 0 | 1 | 1 | 1 | 32 | 24 |
| | Total | 100 | 33 | 30 | 17 | 11 | 9 | 6 | 7 | 4 | 7 | 5 | 73 | 56 |
| Total | Municipal | 274 | 59 | 44 | 35 | 29 | 30 | 20 | 35 | 13 | 70 | 33 | 229 | 139 |
| | Industrial | 257 | 72 | 63 | 42 | 24 | 22 | 19 | 13 | 15 | 20 | 14 | 169 | 135 |
| | Total | 531 | 131 | 107 | 77 | 53 | 52 | 39 | 48 | 28 | 90 | 47 | 398 | 274 |

[a]C=concentration.
[b]Q=quantity.

Table 5.4:  Summary of NPDES Permit Noncompliance by Selected Dischargers During the Period October 1, 1980 to March 31, 1982

| State | Discharger type | Sample size | At least one noncompliance instance, concentration or quantity | Percent of dischargers with at least one noncompliance instance | Significant non-compliers | Significant noncompliance as percent of total noncompliers |
|---|---|---|---|---|---|---|
| Iowa | Municipal | 38 | 36 | 95 | 13 | 36 |
| | Industrial | 25 | 20 | 80 | 4 | 20 |
| Louisiana | Municipal | 38 | 33 | 87 | 15 | 45 |
| | Industrial | 50 | 40 | 80 | 5 | 13 |
| Missouri | Municipal | 43 | 39 | 91 | 15 | 38 |
| | Industrial | 37 | 33 | 89 | 8 | 24 |
| New Jersey | Municipal | 51 | 44 | 86 | 17 | 39 |
| | Industrial | 49 | 37 | 76 | 12 | 32 |
| New York | Municipal | 50 | 43 | 86 | 20 | 47 |
| | Industrial | 50 | 38 | 76 | 9 | 24 |
| Texas | Municipal | 54 | 43 | 80 | 8 | 19 |
| | Industrial | 46 | 38 | 83 | 4 | 11 |

Source for Tables 5.3 and 5.4:  Ref. 5.7

   h.  Inspection by the EPA and the states to
       independently verify the accuracy of SMR
       data are being reduced or are at already low
       levels.
   i.  Many instances of noncompliance continued
       for extended periods before formal enforce-
       ment action was taken and in some cases con-
       tinued for years even after enforcement
       action had been taken.

The GAO further studied the causes of noncompliance and although they varied considerably, the most
prevalent causes were:

   1.   Operation and maintenance deficiencies
which included; limited staffing and lack of training
for plant operators, inadequate laboratory facilities
and lack of enforcement of sewer use ordinances.
   2.   Equipment deficiencies.
   3.   Treatment plant overloading.

These causes of noncompliance parallel the findings of a similar study which the Natural Resources
Defense Council (NRDC) also conducted in 1982. Here
the records of 40 industrial permittees in New York
and New Jersey, considered to be significant noncom-
pliers were examined. The primary causes of noncompliance identified here were:

   1.   Operation and maintenance problems.
   2.   Poorly designed plants.
   3.   Lack of expertise or negligence by company
engineers.
   4.   Improper equipment for the given treatment
process.

Conditions have not improved greatly since then.

## SEWAGE PUMP STATIONS

   Sewage pump stations are little more than temporary collection and transfer points for raw sewage.
They are used wherever gravity sewer lines will not
work (i.e. in depressions, valleys or extremely flat
areas). The two biggest problems at pump stations are
pump failure and power failure and the accompanying
bypasses. Lesser problems include odors and releases
of toxic vapors.

   Consequently, as a planner, you should require
the following as minimum standard equipment at all
pump stations:

1.    A standby, fuel operated generator in the event of a power failure.

2.    A duplicate, backup pump or pumps in the event one pump fails.

3.    Carbon absorption filters on any external vents if industrial wastes are included in the wastewater received at the pump station.

Further, be sure to keep the pump station out of floodways and at least several feet above the 100 year design storm elevation.

## ILLEGAL SEWER CONSTRUCTION

In spite of an abundance of strict state and municipal regulations prohibiting it, the illegal installation of sanitary sewer lines and their connection to existing sewerage systems remains a national problem. In some cases, such installations have even been sanctioned, although sometimes unwittingly by local planning boards and other officials.

There are two major problems with such illegally installed sewer lines. First, these illicit installations are noted for their shoddy workmanship. Common deficiencies include unchanneled manhole floors, insufficient slope to pipes, cracked and collapsed pipes, manhole frames not cemented to the tops of manholes, manholes without steel covers, illegal bypass pipes, uncemented manhole sidewalls and underdesigned pumps at pump stations. This poor workmanship abets the introduction of excess ground water and storm water into the overall sanitary sewerage systems. As a result the downstream sewage treatment plant receives excess quantities of water that cause plant treatment efficiency to drop. This in turn causes the plant to exceed its NPDES permit limitations. In addition, the episodes of plant bypassing become more frequent. The common regulatory agency response to such hydraulically overloaded sewage treatment plants has been the imposition of a "sewer ban". (See discussion of Sewer Bans at the end of this chapter.)

Secondly, those developers installing such illegal sewer lines gain a decided economic advantage over competitors who choose to follow the letter of the law. When they are able to avoid the permit process these criminal entrepreneurs do not have to pay out application fees or inspection fees, which can be sub-

stantial.   Nor do they have to delay construction,
sometimes for several months or more, while plans are
reviewed and permits are processed.   Lastly, with no
regulatory agency overseeing installation these developers are free to use less costly and frequently infe-
rior materials and workmanship without fear of being
discovered.   While they gain, their competitors and
society, in general, lose.

Consequently, I cannot emphasize strongly enough
that anyone found to have installed sanitary sewer
lines illegally be dealt with in the severest manner.
And here I do not mean merely penalties.   The developer must be required to show that his sewage system
can meet all prescribed standards.   This will require,
in almost all instances, the televising of the entire
underground sewer line to check for shoddy workmanship, leaks, cracks or collapsed sections of pipe.   If
such defects are found they must be repaired and then
the system retelevised to verify their correction.   In
addition, the municipality should hire an expert
professional engineer, at the developer's expense of
course, to inspect all manholes and any pump stations
to determine the adequacy of their design and installation.   He should further verify the adequacy of all
pipe slopes by checking elevations on all of the
inverts (bottoms) of the pipe lines.   If any of these
latter components are found to be inadequate they
should also be repaired or replaced.

How could such installations occur?   Easier it
seems than most planners would expect.   Putting aside
for a moment possible ineptness and collusion on the
part of the municipal planning board or other officials, the problem very often lies in poor communication between the regulating authorities which in most
instances is between local and state government.

In general, state agencies are responsible for
issuing permits to construct and operate sewage treatment systems.   A planning board may grant approval of
a subdivision or a site plan with the condition that
all state environmental agency permits be obtained.
Because sanitary and storm sewer lines are generally
the first improvements to be installed on any construction site following rough grading, a developer
may apply for but not wait to obtain, the necessary
state approval and permit before he installs the
underground lines.   With projects taking two to five

years to build out, the conditions of local approval can become clouded, municipal personnel can change and before very long no one is quite sure who approved or inspected what. Furthermore, the state agency frequently does not follow up on an unprocessed application. Other municipalities have simply assumed that since they have authority to allocate flows to their treatment plant they can likewise authorize extensions and hookups without any further state agency approval. An erroneous assumption in most instances.

Lastly, some municipalities may only operate a collection system. In this case their wastewater is simply transferred downstream to another municipally operated collection system for treatment. Lacking a specific regulatory permit of their own they may feel they have no obligation to limit additional connections or extensions--another erroneous assumption.

To avoid the problems associated with poorly constructed illegal sewer extensions, planners should not allow a developer or builder to install any sanitary sewer lines until the municipality has been provided with a copy of the finally approved state agency permit(s).

## LAND APPLICATION SYSTEMS FOR WASTEWATER TREATMENT

Within the last decade, there has been an increasing trend by environmental regulatory agencies to encourage treatment of sanitary wastewater and some industrial wastewater by applying them to the land and letting the natural vegetation, soil and soil microorganisms remove pollutants. This was certainly not a new concept, having been utilized in Europe for well over a hundred years, and in America primarily since World War II. Both industry and municipalities found this method to be a cheap way to get rid of wastewater. However, on American soil, the practice was essentially unregulated. As such, many abuses occurred, resulting in a cornucopia of toxic and hazardous materials leaching through the soil and into the ground water table. Despite these unfortunate occurrences there is more than sufficient data to indicate that a variety of wastewaters can be effectively treated by land application.

## Types of Systems Generally
## and Wastewater Constituents of Concern

Land planners will generally encounter three types of land treatment systems (Figure 5.2):
1.    Irrigation or slow-rate.
2.    Rapid infiltration or percolation.
3.    Overland flow.
In some instances, one or more of the above systems may be used jointly. See Table 5.5 for a comparison of the expected effectiveness of each method.

A comprehensive site evaluation is critical to the land planner's evaluation of a land application system regardless of type. Likewise so is an environmental impact statement and both documents should be prepared and submitted simultaneously by every applicant. Guidelines for the preparation of such documents can be found in Part D of Appendix 1. The composition of the wastewater itself deserves some attention as well. Some constituents of concern are listed in Table 5.6 and described in more detail below:

(a)    Total Dissolved Solids (Ref. 5.11)

The aggregate of the dissolved compounds in wastewater is the TDS (total dissolved solids). The TDS content which is related to the EC (electrical conductivity), is generally more important than the concentration of any specific ion. High TDS wastewater can cause a salinity hazard to crops, especially where annual evapotranspiration exceeds annual precipitation. A general classification as to salinity hazard by TDS content and electrical conductivity is given in Table 5.7. It should be noted that these values were developed primarily for the arid and semiarid parts of the country. High TDS wastewater may also create problems if allowed to percolate to the permanent ground water.

(b)    Suspended Solids

Suspended solids in applied effluents are important because they have a tendency to clog sprinkler nozzles and soil pores and to coat the land surface. A large percentage of the suspended solids can be removed easily by sedimentation. When applied to the land at acceptable

Figure 5.2: Methods of land application of wastewater.

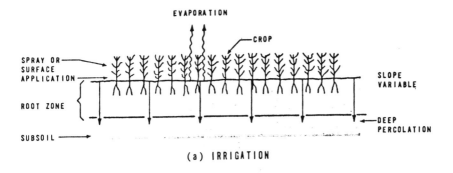

(a) IRRIGATION

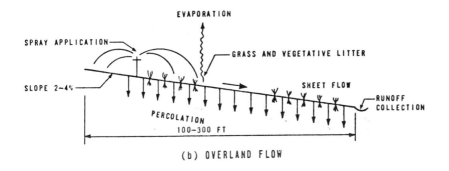

(b) OVERLAND FLOW

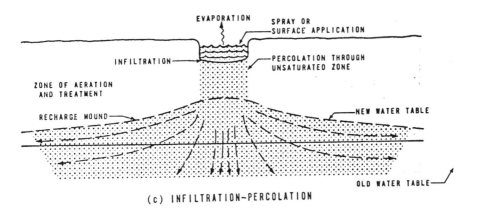

(c) INFILTRATION-PERCOLATION

Source: Ref. 5.11

Table 5.5:  Removal Efficiencies of Major Constituents for Municipal
Land Application Systems

| | Removal efficiency, % | | |
| | Application method | | |
| Constituent | Irrigation | Overland flow | Infiltration-percolation |
|---|---|---|---|
| BOD | 98+ | 92+ | 85-99 |
| COD | 95+ | 80+ | 50+ |
| Suspended solids | 98+ | 92+ | 98+ |
| Nitrogen (total as N) | 85+ | 70-90 | 0-50 |
| Phosphorus (total as P) | 80-99 | 40-80 | 60-95 |
| Metals | 95+ | 50+ | 50-95 |
| Microorganisms | 95+ | 98+ | 98+ |

Table 5.6:  Important Constituents in Typical Domestic Wastewater (in mg/l)

| | Type of wastewater | | |
| Constituent | Strong | Medium | Weak |
|---|---|---|---|
| BOD | 400 | 220 | 110 |
| Suspended solids | 350 | 220 | 100 |
| Nitrogen (total as N) | 85 | 40 | 20 |
|   Organic | 35 | 15 | 8 |
|   Ammonia | 50 | 25 | 12 |
|   Nitrate | 0 | 0 | 0 |
| Phosphorus (total as P) | 15 | 8 | 4 |
|   Organic | 5 | 3 | 1 |
|   Inorganic | 10 | 5 | 3 |
| Total organic carbon | 290 | 160 | 80 |

Source for Tables 5.5 and 5.6:  Ref. 5.8

## Table 5.7:  Water Quality Guidelines

| Problem and related constituent | No problem | Increasing problems | Severe |
|---|---|---|---|
| **Salinity**[a] | | | |
| EC of irrigation water, in millimhos/cm | <0.75 | 0.75 3.0 | >3.0 |
| **Permeability** | | | |
| EC of irrigation water, in mmho/cm | >0.5 | <0.5 | <0.2 |
| SAR (Sodium adsorption ratio) | <6.0 | 6.0-9.0 | >9.0 |
| **Specific ion toxicity**[b] | | | |
| From root absorption | | | |
| Sodium (evaluate by SAR) | <3 | 3.0-9.0 | >9.0 |
| Chloride, me/l | <4 | 4.0-10 | >10 |
| Chloride, mg/l | <142 | 142-355 | >355 |
| Boron, mg/l | <0.5 | 0.5-2.0 | 2.0-10.0 |
| From foliar absorption[c] (sprinklers) | | | |
| Sodium, me/l | <3.0 | >3.0 | -- |
| Sodium, mg/l | <69 | >69 | -- |
| Chloride, me/l | <3.0 | >3.0 | -- |
| Chloride, mg/l | <106 | >106 | -- |
| **Miscellaneous**[d] | | | |
| $NH_4$-N, $NO_3$-N  mg/l for sensitive crops | <5 | 5-30 | >30 |
| $HCO_3$, me/l  [only with overhead | <1.5 | 1.5-8.5 | >8.5 |
| $HCO_3$, mg/l  sprinklers ] | <90 | 90-520 | >520 |
| pH | Normal range = | 6.5-8.4 | -- |

Guideline values

a.  Assumes water for crop plus needed water for leaching requirement (LR) will be applied.  Crops vary in tolerance to salinity.  Refer to tables for crop tolerance and LR.   mmho/cm x 640 = approximate total dissolved solids (TDS) in mg/l or ppm; mmho x 1,000 = micromhos.

b.  Most tree crops and woody ornamentals are sensitive to sodium and chloride (use values shown). Most annual crops are not sensitive (use salinity tolerance tables).

c.  Leaf areas wet by sprinklers (rotating heads) may show a leaf burn due to sodium or chloride absorption under low-humidity, high-evaporation conditions.  (Evaporation increases ion concentration in water films on leaves between rotations of sprinkler heads.)

d.  Excess N may affect production or quality of certain crops, e.g., sugar beets, citrus, grapes, avocados, apricots, etc.  (1 mg/l $NO_3$-N = 2.72 lb N/acre-ft of applied water.)  $HCO_3$ with overhead sprinkler irrigation may cause a white carbonate deposit to form on fruit and leaves.

Note:  Interpretations are based on possible effects of constituents on crops and/or soils.  Guidelines are flexible and should be modified when warranted by local experience or special conditions of crop, soil, and method of irrigation.

Source:  Ref. 5.11

loading rates, almost complete removal can be expected from the percolate.

(c)  Organic Matter

Organic matter, as measured by BOD, COD and TOC, is present in the dissolved form as well as in the form of suspended and colloidal solids. Ordinarily, concentrations are low enough not to cause any short-term effects on the soil or vegetation. Organic compounds, such as phenols, surfactants and pesticides, are usually not a problem but in high concentrations they can be toxic to microorganisms.

BOD is removed from the wastewater very efficiently by each land-application method. The loading applied, however, will greatly influence the resting period for soil reaeration and may influence liquid loading rates.

(d)  Nitrogen Compounds

Nitrogen contained in wastewater may be present as: ammonium, organic, nitrate and nitrite; with ammonium and organic nitrogen usually being the principal forms. In a nitrified effluent, however, nitrate nitrogen will be the major form. Because nitrogen removal is sensitive to a variety of environmental conditions, monitoring of nitrogen concentrations is usually required. To avoid confusion, the concentrations of each form should be expressed as nitrogen.

Nitrogen is important because when it is converted to the nitrate form, it is mobile and can pass through the soil matrix with the percolate. In ground water, nitrates are limited to 10 mg/l.[4] In surface waters nitrates may aggravate problems of eutrophication.

Because of the mobility of nitrates, it is imperative that an annual accounting be made of applied nitrogen. This accounting should take into consideration crop uptake (Tables 5.8, 5.9 and 5.10) denitrification, volatilization, additions to the ground and surface waters and storage in the soil. The collection of samples from the soil, ground and surface water and plant tissue will be required. In addition, the total annual nitrogen loading must be calculated. This latter quantity may be calculated using either of the following formulas:

Table 5.8:  Nitrogen Uptake Rates for Selected Crops[a]

| Crop | Nitrogen uptake rate, kg/ha·yr |
|---|---|
| Forage | |
| Alfalfa | 300 |
| Bromegrass | 130 |
| Coastal bermudagrass | 400 |
| Kentucky bluegrass | 200 |
| Quackgrass | 240 |
| Reed canarygrass | 340 |
| Ryegrass | 200 |
| Sweet clover | 180 |
| Tall fescue | 160 |
| Field | |
| Barley | 70 |
| Corn | 180 |
| Cotton | 80 |
| Milomaize (sorghum) | 90 |
| Potatoes | 230 |
| Soybeans | 110 |
| Wheat | 60 |

a.  Values represent lower end of ranges

1 kg/ha·d = 0.893 lb/acre·d


Table 5.9:  Estimated Net Annual Nitrogen Uptake in the Overstory and Understory Vegetation of Fully Stocked and Vigorously Growing Forest Ecosystems in Selected Regions of the United States

| | Tree age, yr | Average annual nitrogen uptake, kg/ha·yr |
|---|---|---|
| Eastern forests | | |
| Mixed hardwoods | 40-60 | 220 |
| Red pine | 25 | 110 |
| Old field with white spruce plantation | 15 | 280 |
| Pioneer succession | 5-15 | 280 |
| Southern forests | | |
| Mixed hardwoods | 40-60 | 340 |
| Southern pine with no understory | 20 | 220[a] |
| Southern pine with understory | 20 | 320 |
| Lake states forests | | |
| Mixed hardwoods | 50 | 110 |
| Hybrid poplar[b] | 5 | 155 |
| Western forests | | |
| Hybrid poplar[b] | 4-5 | 300-400 |
| Douglas-fir plantation | 15-25 | 150-250 |

a.  Principal southern pine included in these estimates is loblolly pine.

b.  Short-term rotation with harvesting at 4-5 yr; represents first growth cycle from planted seedlings.

Source for Tables 5.8 and 5.9:  Ref. 5.8

Table 5.10:  Biomass and Nitrogen Distributions by Tree Component for Stands in Temperate Regions

Percent

| | Conifers | | Hardwoods | |
|---|---|---|---|---|
| Tree component | Biomass | Nitrogen | Biomass | Nitrogen |
| Roots | 10 | 17 | 12 | 18 |
| Stems | 80 | 50 | 65 | 32 |
| Branches | 8 | 12 | 22 | 42 |
| Leaves | 2 | 20 | 1 | 8 |

Source:  Ref. 5.8

Table 5.11:  Recommended Maximum Concentrations of Trace Elements in Irrigation Waters[a]

| Element | For waters used continuously on all soil, mg/l | For use up to 20 years on fine-textured soils of pH 6.0 to 8.5, mg/l |
|---|---|---|
| Aluminum | 5.0 | 20.0 |
| Arsenic | 0.10 | 2.0 |
| Beryllium | 0.10 | 0.50 |
| Boron | 0.75 | 2.0-10.0 |
| Cadmium | 0.010 | 0.050 |
| Chromium | 0.10 | 1.0 |
| Cobalt | 0.050 | 5.0 |
| Copper | 0.20 | 5.0 |
| Fluoride | 1.0 | 15.0 |
| Iron | 5.0 | 20.0 |
| Lead | 5.0 | 10.0 |
| Lithium | 2.5[b] | 2.5[b] |
| Manganese | 0.20 | 10.0 |
| Molybdenum | 0.010 | 0.050[c] |
| Nickel | 0.20 | 2.0 |
| Selenium | 0.020 | 0.020 |
| Zinc | 2.0 | 10.0 |

a.  These levels will normally not adversely affect plants or soils.  No data are available for mercury, silver, tin, titanium, tungsten.

b.  Recommended maximum concentration for irrigating citrus is 0.075 mg/l.

c.  For only acid fine-textured soils or acid soils with relatively high iron oxide contents.

Source:  Ref. 5.11

$$N = 2.7CL \quad (5)$$

where

   N = annual nitrogen loading, lb/acre/yr
   C = total nitrogen concentration, mg/l
   L = annual liquid loading, ft/yr

or:

$$N = 0.1CL \quad (5)$$

where

   N = annual nitrogen loading kg/ha/yr
   C = total nitrogen concentration, mg/l
   L = annual liquid loading, cm/yr

(e)  Phosphorus

Phosphorus contained in wastewater occurs mainly as inorganic compounds, primarily phosphates, and is normally expressed as total phosphorus. Phosphorus removal is accomplished through plant uptake and by fixation in the soil matrix. The long-term loadings of phosphorus are important because the fixation capability of some soils may be limited over the normal expected life span of the system. Phosphorus that reaches surface waters as a result of surface runoff or interception of ground water flow may aggravate problems of eutrophication.

(f)  Heavy Metals and Trace Elements

Although some heavy metals are essential in varying degrees for plant growth, most are toxic, at varying levels, to both plant life and microorganisms. The major risk to land treatment systems from heavy metals is in the long-term accumulation in the soil, because they are retained in the soil matrix by adsorption, chemical precipitation and ion exchange. Retention capabilities are generally good for most metals in most soils especially for pH values above 7.

Generally, zinc, copper and nickel make the largest contributions to the total heavy metal content. Zinc is used as a standard for plant toxicity, with copper being twice as toxic and nickel being eight times as toxic. A "zinc equivalent" can thus be determined for these two metals (Table 5.11).

(g)   Exchangeable Cations

The effect of concentrations of sodium, calcium and magnesium ions deserves special consideration. They are related by the sodium adsorption ratio (SAR), defined as:

$$SAR = \frac{Na}{\sqrt{\dfrac{Ca + Mg}{2}}}$$

where Na, Ca and Mg are the concentrations of the respective ions in milliequivalents per liter of water. High SAR (greater than 9) values may adversely affect the permeability of soils. Other exchangeable cations, such as ammonium and potassium, may also react with soils. High sodium concentrations in soils can also be toxic to plants, although the effects on permeability will generally occur first (Table 5.7).

(h)   Bacteria and Other Microorganisms

Microorganisms, primarily bacteria, are normally present in large quantities in wastewater. The bulk of these microorganisms can be removed by conventional treatment and the soil mantle is quite efficient in the removal of bacteria and probably viruses through the processess of filtration and adsorption. Problems may arise, however, in the actual application process, especially in spraying, where aerosols could present a health hazard. High degrees of preapplication treatment, including disinfection, may be necessary, particularly in cases in which public access to the application area is allowed. (See further discussion under Problem Areas.)

## Irrigation or Slow Rate Systems

Irrigation or slow rate systems are the most common method of land treatment. They are also capable of producing the highest degree of treatment of all land treatment systems (Table 5.12). Wastewater is applied either through controlled spraying or by surface spreading. Once applied the wastewater receives treatment as it flows through the plant-soil matrix. Generally irrigation systems are designed to

Table 5.12: Expected Quality of Treated Water from Land Treatment Processes[a]

mg/L Unless Otherwise Noted

| Constituent | Slow rate[b] | | Rapid infiltration[c] | | Overland flow[d] | |
|---|---|---|---|---|---|---|
| | Average | Upper range | Average | Upper range | Average | Upper range |
| BOD | <2 | <5 | 5 | <10 | 10 | <15 |
| Suspended solids | <1 | <5 | 2 | <5 | 10 | <20 |
| Ammonia nitrogen as N | <0.5 | <2 | 0.5 | <2 | <4 | <8 |
| Total nitrogen as N | 3[e] | <8[e] | 10 | <20 | 5[f] | <10[f] |
| Total phosphorus as P | <0.1 | <0.3 | 1 | <5 | 4 | <6 |
| Fecal coliforms, No./100 mL | 0 | <10 | 10 | <200 | 200 | <2,000 |

a.  Quality expected with loading rates at the mid to low end of the range

b.  Percolation of primary or secondary effluent through 1.5 m (5 ft) of unsaturated soil.

c.  Percolation of primary or secondary effluent through 4.5 m (15 ft) of unsaturated soil; phosphorus and fecal coliform removals increase with distance

d.  Treating comminuted, screened wastewater using a slope length of 30-36 m (100-120 ft).

e.  Concentration depends on loading rate and crop.

f.  Higher values expected when operating through a moderately cold winter or when using secondary effluent at high rates.

Source:  Ref. 5.8

avoid off-site surface runoff. Consequently, either
the rate of application is stringently controlled or
artificial methods are used to collect any excess
wastewater. See Figure 5.3 for the hydraulic pathways
of irrigation or slow rate systems.

Wastewater application rates range up to four
inches per week during the growing season, with a maximum yearly limit of eight feet per year. Three methods are generally used to apply the wastewater to the
land; spraying by the used of nozzles or sprinkler
heads; gravity flow through ridges and furrows or by
flooding the land with several inches of wastewater.
Preferred site characteristics for irrigation systems
include:

a. Climate - Warm to arid climates are preferred. Colder climates may be acceptable if adequate
storage is provided for freezing conditions or wet
weather.

b. Topography - Slopes up to 15 percent can be
used if runoff and erosion are controlled.

c. Soil Type - Loam soils are preferred.

d. Soil Drainage - Well drained soil is preferred.

e. Soil Depth - A uniform depth of 5 to 6 feet
is preferred.

f. Geological Formations - None that would
retard drainage or provide short circuits to the
ground water table.

g. Ground Water - A minimum depth to ground
water of five feet is preferred to maintain aerobic
conditions and prevent surface waterlogging.

## Rapid Infiltration or Percolation Systems

The rapid infiltration or percolation method of
land application of wastewaters is designed to use the
soil matrix almost exclusively for treatment. Unlike
irrigation or slow rate systems there is little or no
consumptive use of the wastewater by vegetation.

Wastewater is applied at high rates through the
use of ponds or lagoons or by spraying. The higher
application rates require less land area than would be
required for spray irrigation or overland flow. The
major portion of the wastewater percolates to the
ground water with the remainder lost through evapora-

Figure 5.3:  Slow rate hydraulic pathways.

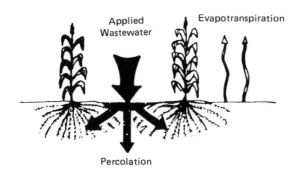

(a) Application Pathway

(b) Recovery Pathways

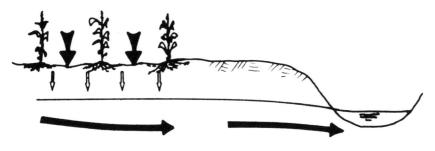

(c) Subsurface Pathway

Source:  Ref. 5.8

Figure 5.4:  Rapid infiltration hydraulic pathways.

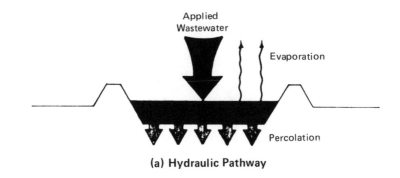

(a) **Hydraulic Pathway**

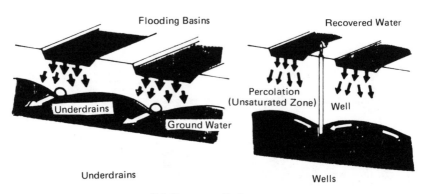

(b) **Recovery Pathways**

(c) **Natural Drainage Into Surface Waters**

Source:  Ref. 5.8

tion (Figure 5.4).   Preferred site characteristics include:

a. Soil Type - Sand, sandy loams, loamy sands and gravels.   Very coarse sand and gravel are not ideal because they will allow too rapid percolation in the first few feet of surface soil where the major chemical and biological treatment take place.

b. Soil Depth - A minimum depth of 15 feet is recommended.

c. Geological Formations - None which would retard drainage or allow water mounding closer than four feet to the surface.

d. Soil Drainage - Well-drained.

e. Ground Water - A minimum depth to ground water of 15 feet is recommended.

## Overland Flow Systems

The third method of land treatment of wastewater, overland flow, was utilized primarily by the nation's food-processing industries, that is until the mid 1970's.   Since then, this method has been used successfully on municipal wastewater as well.   Wastewater renovation occurs by physical, chemical and biological processes as the wastewater flows through vegetation on a sloped surface.   It has been used in both woodland areas and successional field areas.   The wastewater is generally sprayed on the uppermost portions of the site and is then collected at the bottom of the slope.   Some is lost by percolation and evapotranspiration (Figure 5.5).   Collected runoff is almost always discharged directly to surface waters. Consequently, water quality criteria may dictate a minimum of secondary treatment of the wastewater prior to application.

Preferred site characteristics include:

a. Climate - Warm, but cold climate sites have been successful.

b. Topography - Slopes of 2 to 4 percent are preferred.

c. Soil Type - Heavy clays or clay loams work best but must have a minimum of 6 to 8 inches of topsoil.

d. Geological Formations - Impermeable clay lenses of hardpans are acceptable.

e. Ground Water - Four feet preferred.

Figure 5.5:  Overland flow hydraulic pathways.

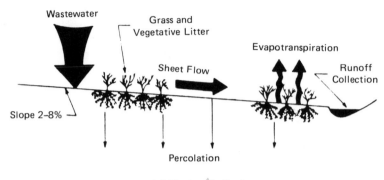

(a) Hydraulic Pathway

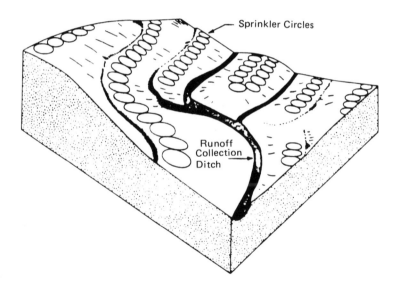

(b) Pictorial View of Sprinkler Application

(continued)

Figure 5.5: (continued)

(c) The owner of this land application (spray irrigation) site has applied more wastewater to the land than the soil can treat. As a result, vegetation has been killed and potable water supply wells in an adjacent residential area have been contaminated.

(d) Lacking storage ponds to hold his wastewater during inclement weather (in this case snow and ice) the owner of this land applied wastewater treatment system must discharge his wastewater year-round. Treatment is often ineffective and the arrival of warmer weather causes severe runoff and flooding on adjoining properties.

Source: Ref. 5.8

**Problem Areas**

The predominant problem associated with the use of these land treatment systems is hydraulic overloading which results in waterlogged soil, excessive ground water mounding, vegetation death or vegetative transition and ultimately the loss of treatment efficiency and contamination of ground and surface waters. The causes of hydraulic overloading are almost always related to either a failure to accurately predetermine the true capacity of a site or failure to adhere to the recommended flow limits.

A second common problem is wintertime operation, particularly in the northern latitudes of the United States. Many states in these areas recommend storage during periods of subfreezing weather. However, case studies of spray irrigation and overland flow systems in the northern latitudes of the United States indicate that wastewater can be sprayed year around with some minimal treatment continuing to occur even in near freezing temperatures. Problems associated with this wintertime application include: frozen spray heads, the buildup of ice and snow on trees and branches resulting in broken branches and split bark, the buildup of substantial mounds of ice and snow on the soil surface around each spray head resulting in substantial flooding and runoff during the spring thaw.

A third problem associated primarily with irrigation and overland flow systems is the drift of aerosol sprays onto adjoining properties. Aerosols are small airborne droplets of water that are less than 20 microns in diameter. Aerosols generated by sprinkler heads tend to be even smaller in size with two thirds or more in the 1 to 5 micron range--a size that makes them easily respirable by humans. Aerosols may carry bacteria and viruses whose concentration is a function of their concentration in the applied wastewater as well as the aerosolization efficiency of the spray heads. Undisinfected wastewater poses the greatest threat. Tables 5.13 and 5.14 illustrate actual field measurements of bacteria and virus concentrations in the air near various spray irrigation sites.

Even disinfected wastewater is no guarantee that the wastewater is totally free of all potential

Table 5.13:  Aerosol Enteroviruses at Land Treatment Sites

| Wastewater type | Location | Distance downwind from sprinkler, m | Wastewater enteroviruses, PFU/L | | Aerosol enteroviruses, PFU/m$^3$ | |
|---|---|---|---|---|---|---|
| | | | Range | Mean | Range | Mean |
| Nondisinfected secondary effluent | Pleasanton, California | 50 | 45-330 | 188 | 0.011-0.017 | 0.014 |
| Raw wastewater | Kibbutz Tzora, Israel | 36-42 | 0-650 | 125 | 0-0.82 | 0.015 |
| | | 50 | -- | 650 | -- | 0.14 |
| | | 70 | 170-13,000 | 6,585 | 0-0.026 | 0.013 |
| | | 100 | 0-82,000 | 16,466 | 0-0.10 | 0.038 |

Table 5.14:  Aerosol Bacteria at Land Treatment Sites

| Wastewater type | Location | Distance downwind from site, m | Bacteria | Density range[a], No./m$^3$ | |
|---|---|---|---|---|---|
| Raw or primary | Germany | 90-160[b] | Coliforms | -- | |
| | Germany | 63-400[b,c] | Coliforms | -- | |
| | California | 32[b] | Coliforms | -- | |
| | Kibbutz Tzora, Israel | 10 | Coliforms | 11-496 | |
| | | 10 | Fecal coliforms | 35-86 | |
| | | 20 | Coliforms | 0-480 | |
| | | 60 | Coliforms | 0-501 | |
| | | 70 | Salmonella | | |
| | | 100 | Coliforms | 30-102 | |
| | | 150 | Coliforms | 0-88 | |
| | | 200 | Coliforms | 4-32 | |
| | | 250 | Coliforms | 0-17 | |
| | | 300 | Coliforms | 0-21 | |
| | | 350 | Coliforms | 0-7 | |
| | | 400 | Coliforms | 0-4 | |
| Ponded, chlorinated | Deer Creek, Ohio | Control value | Standard plate count | 23-403 | (111) |
| | | 21-30 | Standard plate count | 46-1,582[d] | (485) |
| | | 41-50 | Standard plate count | 0-1,429[d] | (417) |
| | | 200 | Standard plate count | <0-223[d] | (37) |
| Secondary, nondisinfected | Ft. Huachuca, Arizona | Control value | Standard plate count | 12-170 | (28) |
| | | Control value | Coliforms | 0-58 | (2.4) |
| | | 45-49[c] | Standard plate count | 430-1,400 | (day) |
| | | | | 560-6,300 | (night) |
| | | | Klebsiella | 1-23 | |
| | | 120-152[c] | Standard plate count | 86-130 | (day) |
| | | | | 170-410 | (night) |
| | Pleasanton, California | Control value | Standard plate count | 300-805 | |
| | | 30-50 | Standard plate count | 450-1,560 | |
| | | | Total coliforms | 2.4-2.5 | |
| | | | Fecal coliforms | 0.4 | |
| | | | Fecal streptococci | 0.3-1.7 | |
| | | | Pseudomonas | 34 | |
| | | | Klebsiella | <5 | |
| | | | Clostridium perfringens | 0.9 | |
| | | | Mycobacterium | 0.8 | |
| | | 100-200 | Standard plate count | 330-880 | |
| | | | Total coliforms | 0.6-1.2 | |
| | | | Fecal coliforms | <0.3 | |
| | | | Fecal streptococci | 0.3-1.9 | |
| | | | Pseudomonas | 43 | |
| | | | Klebsiella | <5 | |
| | | | Clostridium perfringens | 1.1 | |
| | | | Mycobacterium | 0.8 | |

a.  Numbers in parentheses indicate mean values.

b.  Distance quoted is maximum distance at which coliforms were detected.

c.  Upper values occurred during night hours.

d.  Corrected for upwind background value.

Source for Tables 5.13 and 5.14:  Ref. 5.8

pathogens. Many states now require substantial buffer zones around such land treatment systems, depending on; topography, site microclimate, whether or not the wastewater is disinfected and the proximity of receptors (i.e. houses, persons, livestock).

Other mitigating measures include: use of sprinklers that spray laterally or downward with low nozzle pressure, rows of trees or shrubs around the periphery of the spray site and cessation of spraying or spraying only interior plots during high winds.

A fourth problem associated with land treatment systems is the presence of excess suspended solids in the wastewater that is being applied. Two New Jersey enforcement cases immediately come to mind in this regard. One involved a tomato processor who manufactured ketchup and the other was a paper mill. The former industry before initiating corrective pretreatment screening had tremendous amounts of seeds and tomato pulp suspended in its wastewater. This large amount of biodegradable material covered the soil surface with a thick slime, retarded infiltration of additional wastewater and created severe odors. In the second case, the paper mill wastewater contained large amounts of suspended pulp and paper fibers. Less than 5 months after spray irrigation began all vegetation within the perimeter of each spray head had been smothered by a mound of suspended solids sludge. As a result, further applications of wastewater were merely short-circuited to the nearest downstream ditch.

### Where Should Renovated Wastewater Meet Permit Limits?

The exact point at which land treated wastewater must meet its permit limits is often not as clear as with a direct discharge to surface waters. For overland flow systems which contribute comparatively little flow to the ground water, the permit conditions must generally be met at the bottom of the slope after a single pass through. For rapid infiltration-percolation and spray irrigation systems which utilize a larger part of the soil profile and contribute larger quantities of wastewater to the ground water the choice is less clear. Here, two schools of thought presently prevail. The first says that since the soil profile is the actual treatment system, the treated wastewater should meet its permit conditions the moment it reaches the top of the ground water table. The second says that further credit should be made for any additional treatment that occurs as the wastewater combines with the ground water and flows under and eventually off of the site. Advocates of this school of thought say that permit limits should be met 100 feet downgradient of the site. The author's personal preference is for the first option.

**Monitoring**

The correct placement of monitoring devices is critical to the ability of the regulatory agency and the permittees to accurately assess the performance of a land treatment system. For instance, if some pre-treatment of the wastewater is required prior to land application, a proper sampling chamber must be installed so that samples of the pretreated wastewater can be periodically collected to assure that concentrations of certain materials such as BOD, suspended solids or bacteria are not exceeded.

Generally, ground water monitoring wells will be required at all rapid infiltration-percolation and spray irrigation sites (See Chapter 11, Figs. 11.8 and 11.9, for specifications). These should be installed prior to the application of any wastewater. By doing so, an accurate water table contour map can be derived and the ambient or background ground water quality measured before it is influenced by the applied wastewater. At least one of these monitoring wells should be placed upstream (ground water-wise) of the site. Two or more wells should be placed downstream of the site to measure any effect that the land application system may have on the ground water system. Where soils are deep and homogeneous as in deep sandy soils, well installation and the collection of samples presents few problems. In areas where soil profiles are non-homogeneous with variable strata or where ground water is stored in bedrock, well clusters may be necessary (Figure 5.6). This is merely a group of wells installed at one location but screened in the ground water table at varying depths. Elevations should be set on the top of each well casing so that the depth to ground water can be measured each time samples are collected. This will help determine the fluctuation of the ground water table both seasonally

Figure 5.6: Typical well cluster configuration. Note that this method utilizes a single borehole for 3 wells. An alternate method would be the installation of each well within its own borehole.

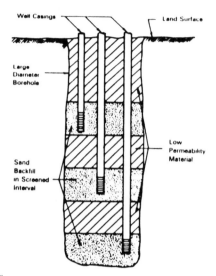

Source:  Ref. 11.15

Figure 5.7:  Components of a standard septic system.

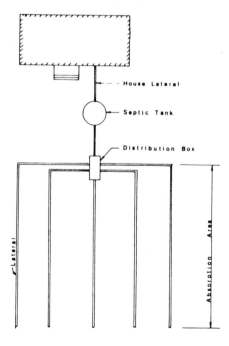

and yearly. Parameters to be measured in samples from
these monitoring wells will vary according to the type
of wastewater applied.  See Appendix 1, Part D, for
the recommended methods for evacuating and sampling
monitoring wells.

## SEPTIC TANK-ABSORPTION FIELD SYSTEMS

As mentioned at the beginning of this chapter,
the septic tank-absorption field system is the second
most common system used for treatment of wastewater.
Figure 5.7 details the basic components of a standard
septic system which generally includes a septic tank,
a distribution box and a subsurface absorption field.
While originally designed to treat domestic
wastewater from single family homes, the role of the
septic system has been expanded to include treatment
of wastewaters from commercial establishments, multi-
family developments and regrettably from some less
than desirable industries.  There are two units in the
system that provide the actual treatment.  The first
is the buried, watertight septic tank which separates
some solids from the liquid phase and provides limited
digestion of organic matter and stores solids and
floatable grease and scum.  The clarified wastewater
then flows to and through the distribution box to the
second treatment unit--the absorption field.  Here,
wastewater is discharged to the soil matrix which provides further physical, chemical and biological treatment.
Of particular importance at this second and
final stage of treatment is the removal of bacteria,
viruses and other pathogens which pass through the
septic tank relatively undiminished in concentration.
Standard septic tank systems are designed to operate
with little routine maintenance.  Of the two treatment
units in the standard system (the septic tank and
absorption field) the septic tank requires the least
amount of work to design and operate.  With most
septic tanks now constructed of durable material such
as concrete, fiberglass and plastic, they should last
up to fifty years.  A major cause of disruption of
treatment in septic tanks is the failure to periodically remove sludge solids from the bottom of the
tank.  Tanks should be inspected at intervals of about
3 years and the levels of scum and sludge measured

during the inspection. The tank should be cleaned if either of the following conditions are observed:

    (a)   the bottom of the scum layer is within 3 inches of the bottom of the outlet device (Figure 5.8)

    (b)   the top of the sludge layer is within 8 inches of the bottom of the outlet device

Sludge removal should occur somewhere between every 3 or 5 years. Special additives are not generally needed to improve or assist septic tank operation even after a routine clean out.

Aside from the routine removal of sludge, owners of septic tank systems can make a major contribution to maintaining the effectiveness of their septic tanks by keeping sanitary napkins, coffee grounds, cooking fats, bones, wet strength towels, disposable diapers and cigarette butts out of the septic tank. These materials will not be degraded and can clog inlets and outlets and add unnecessary bulk to bottom sludge.

The absorption field, unlike its counterpart, the septic tank, is not as easily designed despite claims to the contrary. If septic system failure occurs it will almost always be located at this final component of the septic system. Planners, therefore, have a major responsibility to assure that the site evaluation of a proposed absorption area is complete and accurate and that the design is compatible with existing site conditions. In addition, the subsequent installation must be closely monitored so that poor construction practices do not doom an absorption field to failure.

In Appendix 1, Part D you will find various tables which provide detailed information on the composition of residentially generated wastewater. This information will be useful to the land planner who wishes to verify the adequacy of proposed designs for septic systems.

## Progressive Clogging of Infiltrative Surfaces of Absorption Fields

Liquid flow by gravity is the most common method of distributing waste effluent over the infiltrative surface of the soil absorption field (Ref. 5.14). Wastewater is generally distributed by perforated, 4-inch, diameter pipes laid level or at a uniform slope of 2 to 4 inches per 100 feet, with the holes placed

Figure 5.8

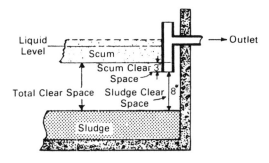

Source: Ref. 5.9

Figure 5.9: Progressive clogging of the infiltrative surfaces of subsurface absorption systems.

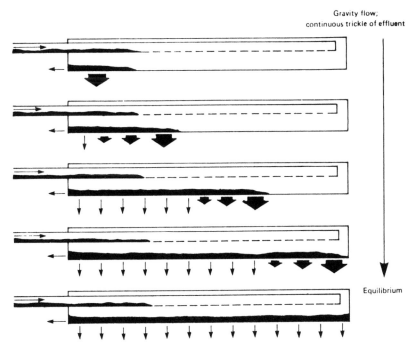

TRADITIONAL SUBSURFACE SEEPAGE BED

Source: Ref. 5.14

downward. Such a system does not provide uniform distribution. The liquid trickles out the holes nearest the absorption field inlet. This encourages clogging because the large holes permit too much liquid to be discharged near the inlet. As a result the underlying soil receives a continuous trickle of water and is constantly ponded. As clogging develops the liquid is forced to migrate further down the trench where the infiltrative surface of the soil is still fresh. This sequence continues until the entire bottom is clogged (Figure 5.9).

Such progressive clogging can be prevented or at least reduced by certain design and operational changes. For instance, the periodic dosing of large volumes of effluent into the field improves distribution and allows the soil to drain between applications. This drainage exposes the infiltrative surface to air thereby reducing clogging. A much better method is to use a pressure dosing system utilizing smaller diameter pipes with small holes. This allows the entire distribution system to fill before much liquid passes out of the holes thus achieving a more uniform distribution. Because the field is dosed intermittently an aerobic environment is maintained in the soil. This promotes the growth of microorganisms that destroy clogging materials.

## Restoring the Infiltration Capacity of a Clogged Absorption Field

One effective but simple method of restoring the infiltrative capacity of a clogged absorption field is resting the system. Resting allows the absorption field to gradually drain, exposing the clogged surface to air. After several months of rest, the clogging material(s) will be broken down by physical and biochemical processes. This method requires the use of a second bed to allow continued use of the disposal system while the failed bed rests.

The infiltrative surface can also be rejuvenated by the addition of an oxidizing agent to the absorption field. Oxidation serves the same function as resting, but the clogging zone is destroyed in several days rather than several months. Such a method does not necessitate taking the clogged bed out of service. The preferred oxidizer is hydrogen peroxide $(H_2O_2)$

usually in a 50 percent solution. The addition of such oxidizing agents should be undertaken only by experienced personnel. It is best to add the $H_2O_2$ when the system is not in use; for example, during a vacation. This will give the reagent time to work without being diluted by peak periods of effluent.

## Percolation Tests as Design Criteria

Two methods commonly used to evaluate a site's suitability for the installation of a septic system are the percolation test and the soil log. The "perc" test has been one of the most misunderstood and abused test procedures since it was first developed by Henry Ryon; an employee of the then N.Y. State Engineers office, in the mid-1920's.

The test, in reality, measures neither percolation nor infiltration but is merely an empirical test to observe the rate at which water seeps into a hole in the ground. Yet, it has been proven to be a useful tool.

Ryon's original procedure is described below:
"Soil Test--A practical but rather empirical test to determine the absorption quality of the soil which has proved very satisfactory may be made as follows: A hole about one foot square and 18 inches deep (the depth of the proposed trench) is dug at the site of the proposed field. The hole is filled to a depth of about 6 inches with water, taking care to wet the soil before pouring in the water for the test if it appears dry. (It is well to repeat the test to eliminate the effect of a very dry soil.) The time that it takes the surface of the water in the hole to fall one inch is then observed." (Ref. 5.12)

The United States Public Health Service (USPHS) later undertook a two decade effort, starting in the mid-1940's, to study septic tank practices in the United States. In the process, they evaluated Ryon's percolation test procedure and determined it was a valid methodology. However, when the Health Service produced their own Manual of Septic Tank Practice in 1967, they made some significant modifications to Ryon's procedures. Of primary importance was the clarification of the pre-soaking period described somewhat ambiguously in Ryon's procedure. The USPHS recommended that after the initial filling of the hole with 12 inches of clear water (Ryon had used 6 inches) that the level be continuously maintained by means of an automatic syphon, if necessary, for at least four hours or preferably overnight. Their reasoning was, "...to insure that the soil is given ample opportunity to swell and approach the condition it will be during the wettest season of the year." They further explained that, "It is important to distinguish between saturation and swelling. Saturation means that void spaces between soil particles are full of water. This can be accomplished in a short period of time. Swelling is caused by intrusion of water into the individual soil particle. This is a slow process, especially in a clay type soil and is the reason for requiring a prolonged soaking period."

The failure to adequately presoak a percolation test hole prior to the actual perc test is one of the major errors committed by contemporary percolation testers. The author consequently recommends a minimum four hour pre-saturation period for all percolation test holes except where sandy soil with little or no clay is present. During periods of extended drought conditions, the presaturation period should be extended to between 12 hours and 24 hours.

A second major source of error in the performance of a percolation test is the procedure used for measuring the drop in water level during the actual test. Some testers use a nail or twig stuck in the side of the perc hole as a measurement reference. Others measure from a board laid across the top of the hole, while some measure up from the bottom of the hole. There seem to be as many methods as there are technicians. Each technician has his personal visual limitations. Measurement errors as large as 1/8" or 1/4" are not uncommon. Therefore, it is recommended that a mechanical float gauge be used for all percolation tests. Most of these gauges can be accurately read to 0.02 inches (Figure 5.10). It is also strongly recommended that all percolation tests be witnessed by an agent of the appropriate regulatory

Figure 5.10: One variation of a percolation tester incorporating both a float (hidden in hole) and a graduated measuring rod.

authority. The minimum number of passing or acceptable[6] percolation tests for a single family absorption field should be two.   Larger absorption fields
for commercial establishments will generally require
more satisfactory tests.   Obviously both of these
passing tests should be within the area of the proposed bed.   Likewise, they should be within the same
soil strata (Figure 5.11).   In those areas of a lot
where failing percolation tests equal or outnumber
passing ones, the administrative authority is justi-
fied in disallowing the use of some or all of the
satisfactory percolation tests.   Some municipalities
for example do not allow the use of satisfactory percolation tests if they are located within 30 to 40
feet of failing tests and the failing tests are at the
same depth ($\pm$ 8") as the satisfactory ones (Figure
5.12).

The soil log (i.e. examination and interpretation of the soil profile) is likewise important in the
overall evaluation of any site to be used for the installation of a septic system.   Only trained individuals should perform such examinations.   A minimum of
one soil log is recommended for each proposed septic
system site, with the examination taking place during
the period when the seasonal high ground water table
is expected to be at its closest proximity to the
ground surface.   Significant errors which have occurred during the examination of the soil profile have
included the misidentification of geological mottling
for seasonal high water table mottling, the misidentification of soft or fractured bedrock (e.g. particularly soft red shale or horizontally fractured gneiss)
as soil and the failure to recognize and denote various pan layers.

## Alternatives to the Percolation Test

In many instances the percolation test cannot be
performed on sites where a restrictive soil horizon or
high water table exists.   Alternative test methods
have therefore been devised that allow some determination to be made of the soil's permeability in spite of
these limitations.   These tests include:
1. The Tube Permeameter Test
2. The Soil Permeability Class Rating Test
3. The Basin Flooding Test

Figure 5.11: Without a complete soils evaluation by means of soil borings or soil pits, percolation tests by themselves could lead to the installation of at least part of the absorption area in a zone of slow or no permeability. As shown here the use of the deepest passing percolation test led to the installation of a substantial part of the absorption area in a restrictive soil horizon.

SITE   EVALUATION   PHASE

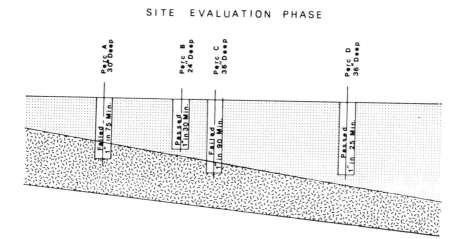

INSTALLATION   PHASE

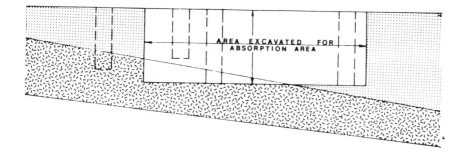

Figure 5.12: Although this lot met the minimum requirement for two passing percolation tests, the number and close proximity of the failed tests to passing ones deserves special attention by the land planner. Certainly more field investigation should be required before any approval is given.

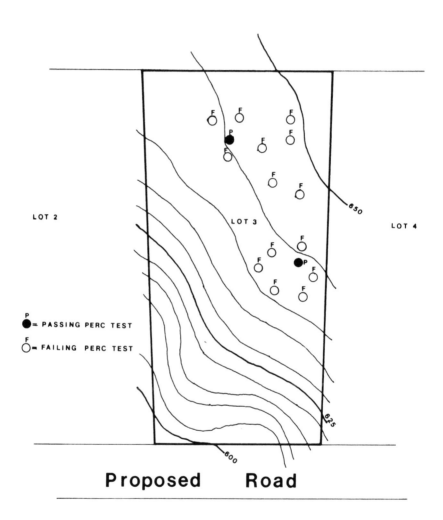

4.  The Pit-Bailing Test
5.  The Piezometer Test

These alternate procedures, however, have generated considerable controversy.  First, it is argued
that these alternatives are generally more involved
(See Appendix 1, Part D) and therefore even more vulnerable to human error than the regular percolation
test.  Secondly, some methods, in particular the permeability tests, may require the testing of field
collected samples at a laboratory far removed from the
actual site where quality assurance of lab procedures
is generally unregulated and unwitnessed by the administrative authority.  Lastly, but certainly not least
importantly is the fact that such methods have opened
up areas for construction (i.e. wetlands and transition zones, floodplain fringe areas and moderately
steep slopes) where it once would have been essentially prohibited due to the lack of successful percolation tests.  In effect builders argue that to deny
the use of such alternate tests in these areas is the
same as declaring a building moratorium.

Land planners, therefore, need to consider the
full implications of allowing or denying the use of
such alternate test methods.  However, contrary to
the opinions of some builders and developers, the
protection of environmentally sensitive areas must be
considered, at least initially, the highest and best
use until proven otherwise.

## Careful Installation - Critical to Successful Operation

Dr. Timothy Winneberger, a recognized expert in
the field of septic system design and installation and
author of "Septic Tank Systems, A Consultants
Toolkit", writes in the July 1980 issue of "Land and
Water":

> "Percolation tests do not predict a suc-
> cessful septic tank system in a soil.
> They only tell us that the soil can perco-
> late.  Construction practices in a soil
> can later ruin its permeability upon which
> proper function of a disposal field
> depends.  Destruction of permeability is
> done deliberately when building an earth
> dam, but sometimes it is done unintention-

ally when digging a trench for a septic system."

Consequently, it is imperative that careful construction techniques be employed during the installation of any septic system. Some of these techniques are listed below:

1. Excavation should proceed in clayey soils only when the moisture content is below the soil's plastic limit.

2. Front end loaders or bulldozer blades should not be used for excavation because the scraping action of the blade or bucket can smear the soil.

3. Excavation equipment must not be driven onto the bottom of the absorption area.

4. The bottom and sidewalls of the excavation should be left with a rough open surface. Any smeared or compacted surfaces should be removed.

5. The best guarantee that these preventative practices will be adhered to is to have the entire installation process witnessed, and if need be, directed by a representative of the regulatory agency.

## Select Fill

You will hear the term "select fill" used on more than an occasional basis when septic systems are being discussed. The fact is select fill is being used extensively in some areas to replace soils whose percolation rates are considered excessively fast—usually 1" in 4 minutes or less. The soil beneath an absorption bed is an integral part of the overall treatment system. Soil that allows wastewater to pass through too quickly will not effectively remove bacteria or viruses or perform other physical or chemical pollutant removal. Consequently, a man-made blend of soil called select fill has been used to slow down the rate of the seeping wastewater. Select fill is usually made up of both sand and clay. The following is a typical blend:

## Specification

| | |
|---|---|
| 1. Clay and Silt | Between 5% and 15% |
| 2. Sand | Between 85% and 95% |
| 3. Clay | Minimum 5% |
| 4. Coarse Fragments (1/4" or larger) | 15% maximum by weight |

Obviously, with such great reliance being placed upon this artificial soil it is imperative that some assurance be obtained that the material meets the required specifications. Some regulatory agencies only require a yearly quality control sample from the manufacturer of the select fill. This is not enough. At every site where select fill is being used, a sample is to be taken from the soil material delivered to the site. It will then be analyzed to assure that it meets the required specifications.[7]

## Curtain Drains and Other Subsurface Drains

It has been said that our engineers can find a solution for every problem. For instance, the diversion of high water tables, perched water tables and seasonal high water tables away from septic system absorption fields, by the use of subsurface drains is surely a classic example of such engineering acumen (Figures 5.13 and 5.14). Unfortunately, however, in this particular instance, while it may be a good engineering solution, it may not be the best environmental solution.

Curtain drains are merely trenches, at the bottom of which, perforated drain pipes have been placed. These drains are placed upslope of the absorption site to intercept subsurface water which then enters the drain pipe to be carried around and downstream of the absorption area. This diversion and acceleration of this subsurface water away from the absorption area can be very effective, too effective, some have argued. The consequences of such diversions, especially if the captured water is diverted to a storm drainage system, is the removal of water that could potentially recharge the ground water system

Figure 5.13:  Typical subsurface drains

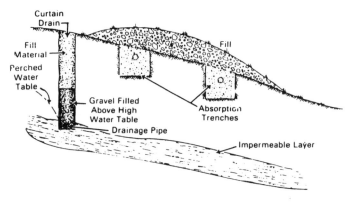

Figure 5.14:  Underdrains used to lower the water table.

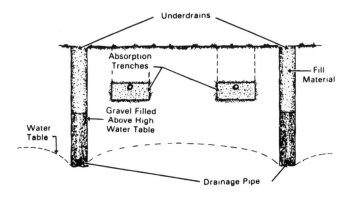

Source for Figures 5.13 and 5.14:  Ref. 5.9

and/or the accelerated removal of water directly from the ground water system itself.  This is a situation land planners ought not to be encouraging especially in areas where individual wells are the main source of potable water.  Yet, I have seen rural subdivisions with as many as 10 lots in a row with such curtain drains.

In addition, these curtain drains are impacting downstream property owners.  Where storm drain ditches used to flow only seasonally or intermittently, water is now present year around.  Maintenance and mowing, therefore, become real problems as the soggy ground encourages unsightly growth of water loving plants. Secondly, these diverted waters have caused severe problems in roadside ditches during the wintertime. The warmer ground water causes successive buildup of ice in the ditches which eventually overflows onto the adjoining roadways creating severe safety hazards due to sheet-ice formation.  Road departments are acquiring a strong dislike for these curtain drain diversions and their representatives are now asking to be included in the overall agency review process.

**Adaptability to Commercial Uses**

Septic systems have been used with great success for wastewater generated from commercial establishments as long as the waste strength and characteristics are comparable to that emanating from a single family dwelling unit.  Under no circumstances should laboratory or process wastewater of any kind be allowed into a septic system.  It is also recommended that all septic systems serving commercial buildings have a dual absorption area.  It has proven very difficult in some instances to correct a failing septic system at a commercial site when there are multiple tenants who may be unable to shut down operations or an absentee landlord who is living out of state. Without a dual absorption area, there is no place to divert ongoing flows while repairs are made to the old system.  In addition, many  commercial sites may be "maxxed out" with other improvements such as paved parking lots, buildings, drainage structures and roadways.  An acceptable compromise to the installation of a ready to go dual absorption area is the reservation of a portion of the site during site plan approval for the future installation of a dual absorption field. The area of course would have to have satisfactory percolation tests and soil logs done prior to approval.

## Alternate Designs for Absorption Areas

Two alternate designs frequently used for the soil absorption area are the mound-system (Figure 5.15) and the evapotranspiration bed (Figure 5.16).

The mound system as its name implies, is elevated above the existing ground surface by the use of select fill and is used to overcome such site restrictions as slowly permeable soils, shallow, permeable soils over creviced or porous bedrock, and permeable soils with a high water table. Treatment of the wastewater occurs largely as the wastewater flows through the select fill, however, some treatment also occurs in the existing unsaturated soil remaining below the fill material. Seepage has occurred at the bottom of these mounds where the select fill and existing soil interface. Therefore, it is imperative that at least 24 to 36 inches of existing unsaturated soil be present below the mound system to assure that the wastewater can flow away without surfacing. Where winters are severe and frost penetration is deep, these systems may malfunction,

The evapotranspiration (ET) bed is designed to be used where geological limitations preclude the use of any subsurface disposal. The ET system consists of a sand bed overlying an impermeable bottom liner plus the normal distribution piping. The bed functions by raising the wastewater to the top portion of the bed by capillary action where it is then evaporated to the atmosphere. Any vegetative cover on top of the mound also assists in removal of some of the wastewater by transpiring it to the atmosphere through its leaves. Large land areas are usually required for the installation of an ET system with approximately 4,000 to 6,000 square feet of bed required for an average single family home. The most significant constraint on the use of ET systems, and the one most often violated is climatic conditions. It is recommended that ET systems be used only in the arid and semi-arid portions of the western and southwestern United States where evaporation rates exceed rainfall.

Figure 5.15:  Typical mound systems.

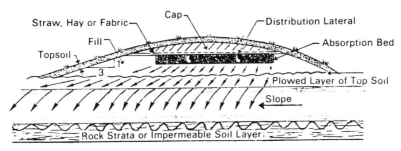

(a) Cross Section of a Mound System for Slowly Permeable
Soil on a Sloping Site.

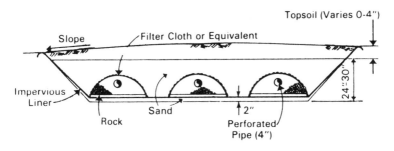

(b) Cross Section of a Mound System for a Permeable Soil,
with High Groundwater or Shallow Creviced Bedrock

Figure 5.16:  Cross-section of a typical ET bed.

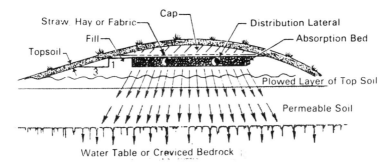

Source for Figures 5.15 and 5.16:  Ref. 5.9

A hybrid of the ET system used in other areas of the United States is the evapotranspiration-absorption (ETA) bed which unlike the ET system has no impermeable liner at the bottom of the sand fill.   Here a combination of both evapotranspiration and absorption are used to treat and remove the wastewater.

The reader is encouraged to procure a copy of the U.S.E.P.A. "Design Manual--Onsite Wastewater Treatment and Disposal Systems" (Ref. 5.9) for further details on the systems described above.

## Community Septic System

Development in America's rural areas is proceeding so rapidly that such standard infrastructure as municipally operated wastewater collection systems and sewage treatment plants cannot be built fast enough. Further, in many of these areas such large scale collection and treatment systems and concurrent surface water discharges, may be discouraged or prohibited entirely in order to maintain a high level of water quality in surface waters.   In addition, some lots on a major subdivision may not have soils suitable for the installation of an individual septic system.

As a result, engineers have offered developers and land planners the concept of the community septic system.

As its name implies, the community septic system is little more than a greatly enlarged septic system. Whereas a septic system serving a single family dwelling may receive 400 gallons per day, a community septic system may receive 8,000 to 10,000 gallons per day.   Each dwelling within the subdivision will still have its own septic tank, but all wastewater from these tanks will flow into a general collection system and from there to a community absorption field.   The wastewater may flow solely by gravity but most of these systems will have some kind of pump station. Unlike a single septic system, these community systems need a full time operator and full time maintenance which has led to one of the major operational problems of these systems.   Many municipalities and counties are reluctant to assume responsibility for their operation and have left the operation to uninformed and ill-prepared home owner associations.

As land planners, I encourage you not to approve such systems unless the municipality or county agrees beforehand to take over and operate them. They are simply too complex to be run by novice associations. Other concerns of the planner include assuring that flows are metered into the system and that an adequate buffer zone has been established around the absorption area and that it is completely fenced to keep out pedestrians and motor vehicles. Monitoring wells upstream and downstream of the absorption area will also be required as well as a schedule for their quarterly sampling. Finally, be sure to keep any pump stations a sufficient distance from individual dwellings to avoid future complaints about odors.

## HOLDING TANKS

Many an unwary home buyer has learned all too quickly about holding tank systems. Unlike the conventional septic system which has a septic tank and a soil absorption area, a holding tank has neither. Wastewater is merely collected in a water tight tank with no outlet until it is picked up by a waste scavenger or as they are affectionately known a "honey dipper".

Holding tanks are most often used in areas where sewer bans have been imposed or a conventional septic system has failed and there is little possibility of correction. They are generally regarded as a temporary measure until some long term alternative is available such as public sewers. Unfortunately, temporary has turned into years. It is surprising how quickly a family of four can fill a 1,000 or 2,000 gallon holding tank with wastewater even with water conservation methods in use. With collection of wastewater occurring as frequently as weekly or biweekly and each evacuation costing $150 or more, the negative aspects of utilizing a holding tank quickly becomes evident. Planners should discourage holding tank systems unless they are used in conjunction with water conservation plumbing.

## SEWER BANS

As mentioned earlier in this chapter, nothing brings development and construction to an abrupt halt faster than a good old fashioned sewer ban (a cessation of all hookups to a wastewater collection system). It is one of the best enforcement tools ever devised by regulatory agencies to motivate recalcitrant municipalities to bring their sewerage system operations into compliance with permit conditions. The consequences of a sewer ban are remarkably unambiguous--no sewer hookups, no development, no new ratables, no new tax revenues. I do not want to mislead you, the municipality is not the only one to suffer with the imposition of a ban. The regulatory agency itself, must likewise endure some hardship, most of which will be of a political nature. It will not be long after a sewer ban is imposed, that politically connected builders, corporations and others seek redress through their representatives in the state legislature. As a consequence, there will be a persistent flow of inquiries to the agency as to why such a ban was imposed and how it can best be lifted. If handling their repetitive inquiries is not enough, there must also be a review board set up to handle possible exemptions to the ban for reasons of hardship (not economic) or to protect the public health and welfare. Some states have even had to adopt comprehensive regulations to govern the exemption process to eliminate subjective or biased exemption decisions.

In spite of these difficulties, this writer still recommends the use of the sewer ban as a practical enforcement option.

## THE HOLDING TANK-DOSING SYSTEM

In some of the nation's communities, where sewer connection bans have been imposed, builders and developers are asking local planners to approve the use of holding tanks as part of a complex system for dosing sewage to normally overtaxed sewer lines during periods of reduced flow. These entrepreneurs and their engineers contend with some accuracy that there are periods during the daily 24 hour cycle when sewer lines are not operating at full capacity and therefore could accommodate their sewage with no resultant environmental damage. These time periods generally coincide with periods of reduced human activity (i.e. after midnight and up to 4 or 5 a.m.). The holding tank allows the retention of sewage until it can be released during these predetermined low flow periods. While the developers and their consultants celebrate their ingenuity, land planners are not nearly as enthusiastic.

The fact of the matter is that the holding tank-dosing system has very limited application. It should not, for example, be allowed:

1) In any sewerage system where excessive infiltration (from ground water) and inflow (from surface water) were the primary cause for the imposition of the sewerage connection ban;

2) In combined storm water and sanitary sewerage systems, especially in tidal areas where tidal gates are known to be broken, missing or stuck in the open position;

3) Where the treatment plant treatment level is less than secondary and where treatment plant malfunction was the primary cause for imposition of the ban; and

4) Where industrial wastewater is a major constituent (>80%) of the wastewater.

In the first two instances, any protracted period of rainfall will in all likelihood eliminate any period of low flow for up to several days or more as infiltrating ground and flood waters slowly recede from the system. This will leave holding tank users with no place to discharge. In the last instance, periods of low flow may be non-existent or unpredictable, especially where industries are operating 24 hours a day.

Only sanitary sewage collection and treatment systems having the following characteristics would be acceptable systems to apply the holding tank-dosing system:

1) The treatment plant provides a minimum of secondary treatment and does not bypass even when operating at maximum capacity;

2) The sewage collection system is structurally sound throughout and although frequently at maximum capacity does not overflow or bypass;

3) The treatment plant has no or minor NPDES permit limit violations;

4) The sewage collection system is not combined with any storm water sewer line.

When such systems are actually approved, the use of an automated timer to release the stored sewage is discouraged. Such timer systems are not responsive to the vagaries of weather or downstream sewer line blockages. Therefore daily inspection, maintenance and operation by a knowledgeable operator would seem to be a mandatory requirement.

## NITRATE POLLUTION FROM SEPTIC SYSTEMS

Please refer to the Chapter 2 discussion of this topic.

## FOOTNOTES

### CHAPTER 5

(1)   In some towns where a high water table was a chronic problem, these illegal connections of sump, floor and foundation drains were actually condoned by municipal authorities so that storm and ground water would not flow into the streets in winter and cause ice to buildup in roadways and sidewalks.

(2)   The actual number of NPDES permittees varies monthly. As of May, 1990 the number was estimated by EPA headquarters in Washington, D.C. to be 63,000 (15,300 municipal and 47,700 industrial. However, due to tracking problems the number of permits could be as high as 70,000.)

(3)   Subsequently, the EPA has initiated a laboratory quality assurance and quality control program.

(4)   Nitrates are limited to a maximum concentration of 10 mg/l in ground water since an excess of nitrates can cause methemoglobinemia in young infants.

(5)   Source: Ref. 5.8

(6)   The maximum passing or acceptable percolation rate varies regionally in the United States but generally ranges from 1" in 45 minutes to as high as 1" in 120 minutes.

(7)   Further assurances can be obtained by requiring the performance of an acceptable percolation test in the select fill after it has been put in place.

**REFERENCES**

**CHAPTER 5**

5.1     Petulla, Joseph M., "Environmental Protection in the United States", San Francisco Study Center, 1987.

5.2     Gerardi, Michael H., Adam P. Maczuga, Melvin C. Zimmerman, "An Operator's Guide to Wastewater Viruses", Public Works, April, 1988.

5.3     Gerardi, Michael H., James K. Grimm, "Insects Associated With Wastewater Treatment", Public Works, 1983.

5.4     United States General Accounting Office, "EPA and State Progress in Administering the National Pollutant Discharge Elimination System Permit Program", Statement of Hugh J. Wessinger, March 7, 1984.

5.5     United States Environmental Protection Agency, "Environmental Pollution Control Alternatives: Municipal Wastewater", EPA-625/5-76-012.

5.6     United States Department of Agriculture, Agricultural Research Service, "Factors Involved in Land Application of Agricultural and Municipal Wastes", October, 1975.

5.7     United States General Accounting Office, "Wastewater Dischargers Are Not Complying With EPA Pollution Control Permits", GAO/RCED-84-53, December 2, 1983.

5.8     United States Environmental Protection Agency, "Process Design Manual, Land Treatment of Municipal Wastewater", EPA 625/1-81-013, October, 1981.

5.9     United States Environmental Protection Agency, "Design Manual--Onsite Wastewater Treatment and Disposal Systems", EPA 625/1-80-012, October, 1981.

5.10    United States Environmental Protection Agency, "Process Design Manual For Land Treatment of Municipal Wastewater--Supplement on Rapid Infiltration and Overland Flow", EPA 625/1-81-013a, October, 1984.

5.11    United States Environmental Protection Agency, "Evaluation of Land Application Systems", EPA 430/9-75-001, March, 1975.

5.12    Winneberger, John H. Timothy, "Septic Tank Systems--A Consultant's Toolkit", Vol. I, Butterworth Publishers, 1984.

5.13    United States Environmental Protection Agency, "Septic Systems and Ground Water Protection, A Program Manager's Guide and Reference Book, GPO No. 055-000-00256-8, July, 1986.

5.14    United States Environmental Protection Agency, "Alternatives for Small Wastewater Treatment Systems", Vol. 1, EPA-625/4-77-011, October, 1977.

5.15    United States Department of Agriculture, Soil Conservation Service, "Soils and Septic Tanks", Bulletin 349, 1978.

5.16    Soil Conservation Society of America, "Land Application of Waste Materials", Ankeny, Iowa, 1976.

5.17    United States Environmental Protection Agency, "Handbook--Remedial Action at Waste Disposal Sites", EPA-625/6-82-006, June, 1982.

# 6

# Vegetation

I think it is fair to say that without the exploitation of the earth's vegetative resources, the human society we know today would probably be remarkably different. Perhaps, it might not exist at all.

When we were a society of nomadic hunters and gatherers, our prehistoric ancestors needed several square miles of territory for each individual to assure an adequate supply of food. When territory was unlimited (i.e. absent competition from other humans) the system worked well. When the population of man increased, however, individual territories began to overlap and huntable food species declined and the scattered communities of edible plants were overused. Fortunately for man, he acquired the knowledge to cultivate certain vegetation to mass produce edible grains. He likewise learned to domesticate various wild herbivores that could be herded and maintained by the controlled grazing of vegetation. Such cultivation and domestication encouraged an economy of space and the assimilation of small scattered groups into larger enclaves, which eventually became settlements and cities. The sedentary nature of cereal grain cultivation and animal husbandry released a significant number of individuals from the process of food procurement, thereby allowing more leisure time to develop further organization and even the arts.

On the American continent, in more modern times, it was the vegetative resources that made colonization possible. Without American forests supplying the first New England colonists with logs for cabins, lumber for furniture, stockades and fences, and fuel for heat, they simply could not have survived. At that

Figure 6.1:  Our attempts to save trees and other vegetation during land development are often haphazard. The developer of this subdivision made an effort to keep fill material away from the root zone of this ash tree, however, he did not consider the consequences of trapped stormwater. As a result this tree's roots were drowned and it died in less than a year.

time, almost one half of the American continent (some 1,065,000,000 acres) was covered by forest. Unfortunately, America's woodlands in spite of their enormous utility to these early Americans were often regarded as an impediment to "civilizing" the continent. As a result, these vegetative resources were severely exploited for the next three centuries. The enormous expanse of grasslands of the Great Plains met a similar fate when America's western frontier eventually pushed past the Mississippi River.

In spite of the fact that we sometimes perceive ourselves as an autonomous species, we remain inextricably bound to the earth's vegetative resources, from the cereal grain, fruits and vegetables we consume daily to the cotton fibers in our clothing, the lumber in our homes, and the wooden poles that carry our electrical and telephone lines. Even some of our most notable medicines have been derived from vegetative resources.

Curare, the powerful muscle relaxant, was first distilled from the Peruvian climbing liana vine Chondodendron tomentosum. The malaria combatant, quinine, came from the bark of the Cinchona ledgeriana tree in Ecuador.

Even more common plant species have yielded significant derivatives. For example, the common garden flower of Europe, foxglove, gave us the cardiostimulant, digitalis. The discovery and utilization of these naturally occurring alkaloids accelerated the development of modern medical techniques by decades if not centuries. Even the air we breathe is dependent in part upon the oxygen produced by the daytime respiration of plants.

In biological terms, the human relationship with the plant kingdom has been largely predatory. Our vegetative resources, however, are renewable thus capable of being used by successive generations of human consumers.

In some parts of the world, particularly in the tropics and in what we often refer to as third world countries, man's predation on native vegetation, largely forested land, has accelerated proportionally with population increases. Unfortunately, the conservation practices that would assure replenishment over time are rarely considered, let alone implemented. We spoke just previously of the benefits man has derived

from the plant kingdom and of the plant derived
medicines that accelerated the treatment of human diseases by several decades or more.  Many of those compounds came from the tropical forests that now face
wholesale removal.  We can only wonder what potential
medicinal discovery will slip from our grasp completely unnoticed.

Few of us will have the opportunity to actively
involve ourselves in the global protection of our vegetative resources.  We can, however, continue to make
significant contributions on a localized and national
scale.

## Impacts of Vegetative Removal

The most obvious and immediate impact of the removal of vegetation from the land surface is the
direct exposure of the top of the soil profile to the
effects of wind and rainfall.  In steeply sloped areas
the effects of rainfall on such exposed ground can be
particularly devastating, with soil particles dislodged, transported and deposited considerable distances from their point of origin--often to areas
(such as stream bottoms) where they can be extremely
detrimental.    Fortunately, we have developed a
national awareness of this potential hazard and have
formulated control strategies to combat it during land
development.  As mentioned in Chapter 4, however, we
need to remain constantly vigilant that such control
practices are properly installed and maintained.

Less frequently occurring, but no less significant impacts, include the destruction of ecologically
significant vegetative communities (e.g. wetlands),
the destruction of rare or endangered plant species
and the loss of individual tree species of some historical, aesthetic or scientific value.  How often
have we seen site and subdivision plans with the
generic notation "wooded" denoted with no further explanation given or requested as to the constituents of
that area.  A classic but sad example of our failure
to adequately address the last mentioned impacts was
recently relayed to me by a colleague of mine.  Those
of you familiar with the plight of the American Chestnut (Castanea dentata) can commiserate with me as I
relay this terrible tragedy.  The American Chestnut
was once the most valuable hardwood of the eastern

forests from Maine through the mid-Atlantic states down to Georgia and Alabama, providing large crops of nuts for both man and wildlife as well as tough durable lumber. Then, in the early 1900's, as a result of the importation of some chestnut seedlings from Japan, a devastating fungal blight was unleashed upon the vulnerable American Chestnut. By 1950, the chestnut had been all but extirpated throughout its natural range. In the years since then chestnut lovers and scientists alike have searched in vain for a cure that would restore the American chestnut to its former dominance in America's woodlands. Years of protracted research proved unsuccessful and most scientists finally conceded that if the chestnut was to survive, it must do it by itself. Perhaps, amongst the sprouts that now encircled the barren remains of the mature chestnut stumps would come a cure. As the decades passed, there was some optimism as the sprouts began to bear nuts which later fell to the ground and germinated to produce new seedlings. Each generation seemed to grow a bit stronger and a little taller before succumbing to the blight. It was just such a stand of American Chestnut saplings that is the focal point of this story. My colleague explained that this stand had been discovered accidentally by a hunter who frequently scouted a secluded mountainous region of northwestern New Jersey in search of white-tailed deer. As an amateur naturalist, he quickly recognized the value of his find and was particularly ecstatic to see how large some of the young chestnut saplings had grown. Some were nearly 8 inches in diameter. For nearly 6 years, he visited the small grove regularly, even when hunting seasons were not open, keeping a diary and collecting nuts in the fall. In all that time not once did he see any indication of the blight. Selfishly, he never divulged the location to anyone fearing perhaps others would seek to reap the annual bounty of nuts.

In 1986, he was devastated to find that this previously secluded region had become part of a large residential subdivision encompassing several hundred acres. The local planning board and other review agencies never once inquired or requested information about the composition of the woodland on the tract. The hunter, perhaps hoping that somehow the trees would be spared, kept silent. Even when the first paved roads passed within a hundred feet of the tiny grove, he never uttered a word of protest.  Much to his sorrow he returned in the spring of 1987 to find that his beloved grove, perhaps the progenitors of America's first blight resistant generation of chestnut, had been cut down and their stumps rooted from the earth to make way for the placement of a new dwelling.

We can only hope that somewhere in the northeast another group of chestnut saplings is similarly prospering and that our land planners will recognize their presence and their value and make great efforts to preserve them.

## CONVERSION OF WOODLAND TO GRASS AND LAWN

As planners, we need to encourage the preservation of existing vegetation wherever possible, particularly in woodland areas and areas of steep topography.  The best known and documented consequences of converting woodland to manicured lawns is the resulting increase in the rate of storm water runoff.  A lesser known and less frequently considered consequence is the introduction of fertilizers, pesticides and herbicides into portions of the watershed where none previously existed.  Poorly planned or excessive applications of these chemicals can wreak havoc on downstream surface waterbodies resulting in fish kills and unsightly algae blooms.

The lawn care industry in the United States has experienced unprecedented growth in the last decade doing an annual $1.5 billion worth of business.  It is estimated that as many as 11 percent of America's single family households now utilize the service of a commercial applicator (Ref. 6.5).  Sales of lawn care pesticides have increased to $700 million annually and by latest estimates result in about 67 million pounds of these materials being applied yearly for non-agricultural purposes.[1]

Table 6.1 is a list of 34 of the nation's major lawn care pesticides as compiled by the U.S. EPA.  Of these, two compounds are the most widely used; 2,4-D and Diazinon.  The former compound is a weed killer that has been in use for over 40 years.  Almost 4 million pounds are used annually on residential lawns. Based on evidence of increased cancer risk among

Table 6.1: List of 34 Major Lawn Care Pesticides

| Pesticide | Type |
|---|---|
| 2,4-D (2,4-dichlorophenoxyacetic acid) | Herbicide |
| Acephate | Insecticide |
| Atrazine | Herbicide |
| Balan | Herbicide |
| Bayleton | Fungicide |
| Bendiocarb | Insecticide |
| Benomyl | Fungicide |
| Betasan | Herbicide |
| Carbaryl | Insecticide |
| Chlorothalonil | Fungicide |
| Chlorpyrifos | Insecticide |
| DDVP (dichlorvos) | Insecticide |
| DSMA (disodium methanearsonate) | Herbicide |
| Dacthal | Herbicide |
| Diazinon | Insecticide |
| Dicamba | Herbicide |
| Diphenamid | Fungicide |
| Endothall | Herbicide |
| Glyphosate | Herbicide |
| Isoxaben | Herbicide |
| MCPA (2-methyl-4-chlorophenoxyacetic acid) | Herbicide |
| MCPP (potassium salt) | Herbicide |
| MSMA (monosodium methanearsonate) | Herbicide |
| Malathion | Insecticide |
| Maneb | Fungicide |
| Methoxychlor | Insecticide |
| Oftanol | Insecticide |
| PCNB (pentachloronitrobenzene) | Fungicide |
| Pronamide | Herbicide |
| Siduron | Herbicide |
| Sulfur | Fungicide |
| Trichlorfon | Insecticide |
| Triumph | Insecticide |
| Ziram | Fungicide |

Source: Ref. 6.5

farmers handling similar types of pesticides the U.S. EPA may eventually designate 2,4-D for Special Review as a human carcinogen thereby restricting its further use.   Diazinon is applied to home lawns in even greater quantities (an estimated 6 million pounds annually).   However, the environmental consequences for this compound are more well known.   In particular this chemical has caused the death of numerous waterfowl and other bird species not only on residential lawns but on golf courses, sod farms and corn and alfalfa fields as well.

Weaning Americans away from their addiction to manicured lawns and quick and easy lawn care chemical use will not be easy.   Some will remain unfazed by the enormity of the amount of chemical additives we are contributing to our watersheds.   Therefore land planners may find it necessary to recommend an outright ban on such chemicals in certain sensitive watersheds. However, it would be better if there were a voluntary reduction or elimination of such chemicals.   Consequently land planners also need to keep the public apprised of the present availability of alternative, non-chemical methods for maintaining a healthy lawn. These include practices such as:

1) Mowing higher to shade low growing weeds and dormant seeds.

2) Mowing more frequently to provide the lawn with a steady but light diet of grass clippings thereby negating the need for artificial fertilizers.

3) Removing weeds and their roots by hand.

4) Planting a more weed and insect resistant grass.

5) Applying diatomaceous earth, predatory nematodes and certain bacteria to discourage or kill insect pests (Ref. 6.6).

## FOOTNOTES

### CHAPTER 6

(1) This is about 8 percent of the 814 million pounds of such materials applied annually for agricultural purposes (Ref. 6.5).

# REFERENCES

## CHAPTER 6

6.1   Crawley, Michael J., "Plant Ecology", Blackwell Scientific Publications, Boston, Massachusetts, 1986.

6.2   Sears, Paul B., "Lands Beyond the Forest", Prentice-Hall Inc., Englewood Cliffs, New Jersey, 1969.

6.3   Schery, Robert W., "Plants For Man", Second Edition, Prentice-Hall Inc., Englewood Cliffs, New Jersey, 1972.

6.4   Swain, Tony, "Plants In the Development of Modern Medicine", Harvard University Press, Cambridge, Massachusetts, 1972.

6.5   United States General Accounting Office, "Lawn Care Pesticides, Risks Remain Uncertain While Prohibited Safety Claims Continue", GAO/RCED-90-134, March, 1990.

6.6   Jay Burnett, "Organic Lawn Care", Organic Gardening, May/June, 1990, pg. 70.

# 7

## Utilities

### RIGHT OF WAYS AND EASEMENTS--MORE DIFFICULTY ON THE WAY

Utility companies will begin to find it much more difficult to acquire new right of ways for the placement of their natural gas, electric, oil and gasoline transmission lines. Those lines that are transcontinental in nature will probably encounter the stiffest opposition. The reasons for this are twofold: a heightened awareness, sometimes bordering on paranoia, amongst the public concerning the perceived and potential hazards and environmental impact of such transmission lines; and the development of an environmental provincialism amongst the states and other levels of government. As a consequence, existing rights of way and easements will become more valuable to utility purveyors and the sharing of easements amongst utilities will become more common. Planners, therefore, may be asked to approve the location of multiple utilities or enlarged transmission structures in the same right of way. This practice, while acceptable in the respect that it alleviates the need to disturb other areas of the environment, may not be acceptable to landowners adjoining these easements and right-of-ways.

### HEALTH AND ENVIRONMENTAL IMPACTS

Local planning agencies will be called upon to address the issue of approving residential subdivisions whose lots may incorporate some portion of these utility easements. Of major concern will be the minimum distances, if any, that dwellings ought to be located, for health and safety reasons, from such utilities as high voltage electrical transmission lines and petroleum or natural gas pipelines. Few municipal land use ordinances incorporate such minimum distances today.

The concerns of planning officials and the public as well, relating to natural gas pipelines have usually been confined to the threat of explosions. A recent case of national notoriety, however, involving a transcontinental, natural gas, pipeline traversing 14 states raised the specter of additional problems-- in this instance, polychlorinated biphenyls (PCB's). The owner of the pipeline in this case acknowledged the fact that from approximately 1960 through 1977, it had used PCB laden lubricating oil in its operations. While it discontinued the use of the PCB oil in 1977, some residual oil remained within the pipeline and was periodically removed during routine pipeline cleaning. This flushed material was subsequently buried in earth pits at the company's pumping stations. Approximately 89 of these burial areas have been identified along the pipeline corridor and a full cleanup of all of them may take as long as a decade to complete. The additional concern that some of this PCB contaminated oil may have also been released through various vents along the pipeline, has not been completely ruled out.

**Electric Powerlines**

The concerns regarding electrical transmission lines are not quite as clear. There is a trend by electrical power suppliers to build larger, higher voltage, transmission lines for economic and engineering reasons. For instance, a single 765 kV line can carry as much power as thirty 138 kV lines at one tenth the construction cost (Ref. 7.1). The passage of electricity through these lines generates ions (charged atoms) that produce both low level electrical and magnetic fields. The electro-magnetic fields in turn influence surrounding electric fields and ion concentrations.

Such electrical and magnetic fields can have an effect on the biochemical activity of some body tissue. The primary interaction point between the electrical field and body tissue appears to be at the cell membrane which among other things controls the transport of ions into and out of the cell. Electrical fields can, therefore, interfere with normal ion transport which in turn affects the cell's biochemical activity.

Measuring the intensity of electric and magnetic fields near powerlines is fairly easy. Estimating the consequences of human exposure is not. People constantly change their proximity to the source of the field and even things like posture (i.e. upright or bending over) can affect exposure. A farmer plowing his fields, and moving back and forth under a high voltage line will receive an exposure much different than a housewife inside her home all day, but located 100 to 300 feet from the same high voltage line.

Powerlines produce a 60 Hz field frequency almost exclusively. The maximum unperturbed strength of an electric field directly under a 765 kV transmission line is about 10 kV per meter at the height of a man's head. At approximately 180 feet on either side of the line the strength drops significantly to 2 kV or less.

An experiment assessing exposure hazards of high voltage currents sponsored by the Electric Power Research Institute in 1978 and carried out by Battelle Pacific Northwest Laboratories (Ref. 7.1) suggests that there may be some reproductive and developmental effects in at least two species of animals. In this instance, pigs (Hanford swine) were subjected to fields of 30 kV/meter (a simulation of the electric field of 10 kV/meter a human would experience at the top of his head while standing beneath a 765 kV line) for approximately 20 hours a day. Three successive generations of swine were produced under these conditions and studied for effects. During the second breeding of the first generation of swine, the litters of exposed sows experienced a higher incidence of fetal malformations than those of the control sows. No such effects had been observed during their first breeding. When the second generation of sows were bred for the first time, this generation also exhibited the same higher incidence of fetal malformations. When these second generation sows were bred, the second time, however, there was no difference between the exposed and control group. With such inconsistencies, interpreting the results became difficult, although it was conceded that the differences between the control and exposed groups were statistically significant.

A team of independent teratologists invited to review the Battelle data concluded the study had not conclusively demonstrated the existence of a relationship between electric field exposure and fetal malformations. Questions were raised concerning the efficacy of using Hanford swine, which have an inherently high fetal malformation rate, and whether there were sufficient litters produced to assess the abnormalities between generations as well as those between the control and exposed groups. The experiment was subsequently repeated with modifications--this time using rats. In this experiment, the second generation of exposed female rats, unlike the second generation of swine, showed a higher incidence of fetal malformations than the control group in their second litters. Again, more confusing and unsettling data.

On July 8, 1987 one of the most comprehensive treatises yet produced on the biological effects of electrical and magnetic fields was released for public perusal. This report entitled, Biological Effects of Power Line Fields, (Ref. 7.3) was the final product of an administrative law opinion rendered by the New York State Public Service Commission on June 19, 1978. The interesting history of the development of this document warrants some further discussion.

In 1973, the New York State Public Service Commission received two applications for the construction of two 765 kV power transmission lines. The New York Power Authority proposed to construct a new line from Massena, New York near the Canadian border to Utica, New York. The Rochester Gas and Electric Corporation and the Niagara Mohawk Power Corporation proposed a second line from Rochester, New York to Oswego, New York. Substantial public interest and concern about possible health hazards were raised and due to the similarity of the proposals the public hearings on each line were consolidated into a joint hearing by the assigned Administrative Law Judges.

A final opinion (No. 78-13) was issued by the New York Public Service Commission on June 15, 1978.

It approved the construction of the two 765 kV lines with the following provisions:

1. A research program was to be established to determine possible human health risks arising from the

electric and magnetic fields of overhead power transmission lines and,

2.  A 350 foot right of way corridor surrounding each 765 kV powerline was established within which residences would not be allowed.

An agreement was subsequently signed between the New York Power Authority and the New York Public Service Commission of February 7, 1980.  This agreement provided for the establishment of a $5,000,000 research program to be funded by contributions from the New York Power Authority and seven other investor owned utilities.  This massive project was to be administered by the New York State Department of Health under the guidance of an impartial panel of nine scientific experts with recognized expertise in such areas as Anatomy, Physics, Biochemistry, Pharmacology, Genetics, Psychology, Neurology, Epidemiology, Electrical Engineering and Bioengineering.  The project became known as the New York State Powerlines Project.

The panel's first responsibility was to examine the state of knowledge on health effects of electric and magnetic fields typical of those in the vicinity of powerlines.  Their second responsibility was to identify possible areas of research related to potential health hazards.  The panel subsequently issued a request for research project proposals in the following seven areas:

1.  Genetic, Cytogenetic, Teratogenetic and Reproductive studies.

2.  Cell and Organ Culture studies.

3.  In Vivo Animal Physiology and Pathophysiology.

4.  Animal and Human Neurobiology.

5.  Animal and Human Behavior.

6.  Multidisciplinary Human Studies With Controlled Exposure Conditions.

7.  Epidemiology of Human Populations.

Of the total 164 proposals eventually submitted the following 16 projects were selected for funding:

1.  "Biological Effects of Extremely Low Frequency Electric and Magnetic Fields on Ocular Tissue: An In Vitro Study".

2.  "Mutagenicity and Toxicity of Electric and Magnetic Fields".

3.  a.  "In Vitro Genetic Effects of Electromagnetic Fields".

b.  "The Effects of Low Level Electromagnetic Fields on Cloning of Two Human Cancer Cell Lines: (Colo 205 and Colo 320)".

4.  "Effects of 60 Hz Electric and Magnetic Fields on the Developing Rat Brain".

5.  "Influence of 60 Hz Fields on Human Behavior, Physiology and Biochemistry".

6.  "Effects of 60 Hz Electromagnetic Fields on Calcium Efflux and Neurotransmitter Release".

7.  "Reproductive Integrity of Mammalian Cells Exposed to 60 Hz Electromagnetic Fields".

8.  "ELF Low Intensity Magnetic Fields and Epilepsy".

9.  "Effect of 60 Hz Electric and Magnetic Field on Neural and Skeletal Cells in Culture".

10.  "Behavioral Effects of ELF".

11.  "Childhood Cancer and Electromagnetic Field Exposure".

12.  "Epidemiological Studies of Cancer and Residential Exposure to Electromagnetic Fields".

13.  "Effects of Electromagnetic Fields in Primate Circadian Rhythms".

14.  "Investigation of Potential Behavioral Effects of Exposure to 60 Hz Electromagnetic Fields".

15.  "Biological Functions of Immunologically Reactive Human and Canine Cells Influenced by In Vitro Exposures to Electric and Magnetic Fields".

16.  "Chronic Effects of 60 Hz Electric and Magnetic Fields on Primate Central Nervous System Function".

I have taken the time to summarize the history of this project to demonstrate its comprehensiveness and the competent manner in which it was carried out. I do this because some of the final conclusions are especially significant and, therefore, the reader needs some assurance that those conclusions have substantial validity. I urge you to read the final report in its entirety, however, I am listing below certain excerpts from the Summary section of the report which focus on 3 effects with possible implications for human health.

"A.  Magnetic Fields

It is clear from the results of the studies sponsored by the Project, as well as from many other recent studies, that both 60 Hz electric and magnetic fields can effect certain biological systems. Magnetic fields effects were found in a number of projects in this program.

B.   Neurobiology and Behavior

At the onset of this project, there was serious question as to whether there was demonstrable neurobiological or behavioral effects of exposures to electrical and/or magnetic fields.  Data accumulated by our contractors as well as a rapidly building literature on retinal magnetoreceptors in birds and mammals leaves little doubt that such effects can be observed in well designed experiments.

C.   Cancer

Previous epidemiologic studies on adult and childhood cancer have been questioned because of serious methodological shortcomings  The results of the Savitz study on childhood cancer changes the situation considerably because it was designed to minimize flaws in previous studies, and because it was conducted under the supervision of a panel of independent scientists.  Even though the Savitz study also has certain limitations, it indicates an excess risk for childhood cancer, in particular leukemias, associated with high current wiring configuration near the homes.

Although this study basically confirms the results of previous studies, the causal relationship is still no more than a hypothesis.  However, the basis for the hypothesis is now stronger".

Undoubtedly this experimentation and debate will continue for some time.  Planners, however, may not have such luxury of time.  Consequently, planning agencies may wish to establish conservative, minimum buffer zones (for the purposes of placing dwellings at least) based upon their own interpretation of the literature or simply based upon the easily measurable or calculated diminution of the electric field away from the transmission line.  In the case of the 765 kV line, this would amount to about 175 feet on either side of the high voltage line (Figures 7.1 through 7.5).

Table 7.1:  Measured Electric Field Strength  as a  Function of Height
Above Ground

| Height Above Ground (ft) | Field Strength (kV/m)* |
|:---:|:---:|
| 1 | 10.1 |
| 2 | 10.4 |
| 3 | 10.0 |
| 4 | 10.1 |
| 5 | 11.0 |
| 6 | 12.5 |
| 7 | 13.0 |
| 8 | 13.0 |

Data taken under 765 kV Marysville Line.

*Value determined for vertical component of gradient.

Figure 7.1:  Measured and calculated Ey for Marysville 765 kV line (note upper curve gives values for field strength beginning at 120 feet, the lower from the center line to 120 feet).

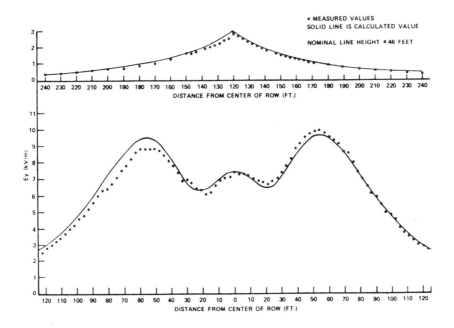

Source for  Table 7.1 and Figure 7.1:  Ref. 7.4

Figure 7.2: Electric field strength profile for 765 kV single circuit line (note that two ordinate scales are used, the left for distances to 125 feet, the right for distances greater than 125 feet).

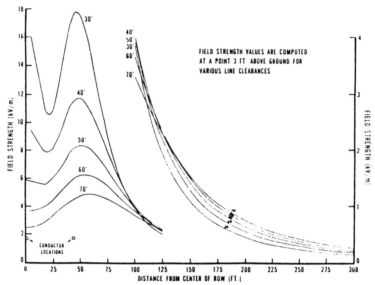

Figure 7.3: Electric field strength profile for 500 kV single circuit line (note that two ordinate scales are used, the left for distances less than 100 feet, the right for distances greater than 100 feet).

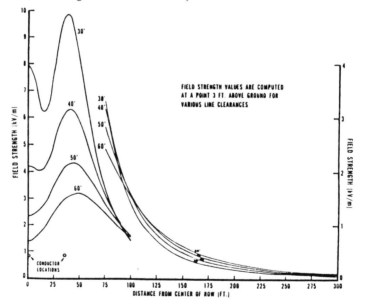

Source for Figures 7.2 and 7.3: Ref. 7.4

Figure 7.4: Electric field strength profile for 345 kV single circuit line (note that two ordinate scales are used, the left for distances less than 100 feet, the right for distances greater than 100 feet).

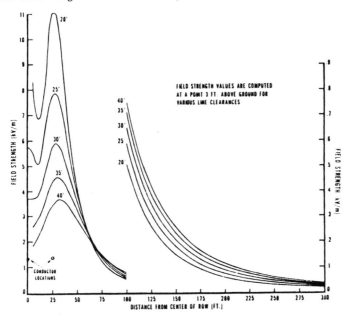

Figure 7.5: Electric field strength profile for 345 kV double circuit line (note that two ordinate scales are used, the left for distances less than 100 feet, the right for distances greater than 100 feet).

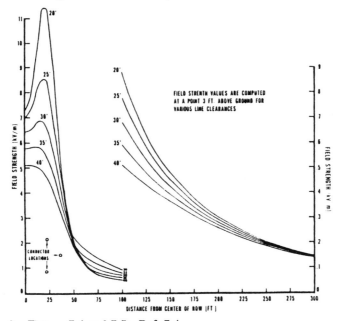

Source for Figures 7.4 and 7.5: Ref. 7.4

## REFERENCES

7.1   Electric Power Research Institute, EPRI Journal, July/August 1984, page 15.

7.2   Haupt, Roy C. and James R. Nolfi, "The Effects of High Voltage Transmission Lines on the Health of Adjacent Resident Populations", American Journal of Public Health, January, 1984, Vol. 74, No. 1.

7.3   New York State Department of Health, "Biological Effects of Powerline Fields, New York State Powerlines Project, Scientific Advisory Panel Final Report", July 1, 1987.

7.4   United States Environmental Protection Agency, "An Examination of Electric Fields Under EHV Overhead Power Transmission Lines", EPA-520/2-76-008, 1977.

Figure 7.6: Establishing buffer zones to limit the construction of residential dwellings within close proximity to these high field strength electrical transmission lines appears to be a prudent action for today's land planners.

# 8

# Agricultural Land

## AGRICULTURE UNDER FIRE

America is a land of contradictions and a premier example of this is the nation's policies toward its farmland. On the one hand, the federal government until very recently, encouraged and even financed the drainage and conversion of marginally acceptable land, including wetlands, to agricultural uses. At the same time, elsewhere in the nation, prime agricultural land was being buried, perhaps forever, under miles of concrete and asphalt roadway and beneath the foundations of thousands of dwellings and office buildings.

More than one-half of all Americans lived on farms in 1862. By 1920 this percentage had dropped to 30% and by 1940 to about 20%. Today, less than 4% of the American population lives on the nation's 22 million farms and less than 2 percent of the nation's work force is engaged in farming. As a consequence, few Americans have any first hand knowledge of modern agriculture or its problems. Nor, in my opinion, do they seem to care. Because American agriculture has been so successful in supplying the nation with an abundance of products, such fecundity is often taken for granted. Americans, however, are biting the very hand that feeds them in many respects.

Besides providing food for American households, agricultural land offers other amenities. First, it requires few municipal improvements (infrastructure) thereby reducing local operating budgets. Secondly, it preserves open space and provides aesthetics and habitat for innumerable species of wildlife. It is this second combination of attributes that makes agricultural land so attractive to Americans in general

193

and urbanites and land developers in particular.   In many areas the value of this land for development purposes has surpassed any possible value the land may have had for agricultural purposes.  As a result, some of the nation's best farmland is rapidly being sold and converted to residential subdivisions and suburbs. Accompanying this land transition is the relocation of a population of Americans unaccustomed to either rural living or normal agricultural practices.  Previously routine farming practices such as the spraying of pesticides and herbicides, the application of animal manures and the tilling of soil must now be carried out in close proximity to numerous human receptors. As a result, farms and residential subdivisions have become incompatible neighbors.

These new rural landholders often complain of such nuisances as excess dust from tilling of the soil, animal noises and odors from manure spreading. Less frequent, but more serious complaints include actual or perceived personal injury from pesticide or herbicide sprays, injury to property from loose livestock and contamination of the ground water.  Farmers are also suffering from these encounters, and it is they who often suffer the most serious injury.  Fences used to contain livestock are frequently cut or broken.  Hikers, horseback riders, motorbikes and all-terrain vehicles tear up newly planted crops, compact soils, scare livestock and cause soil erosion.  Free-roaming pets, especially dogs, intrude into pastures to harass and kill livestock.

A 1986 court case amply demonstrates why farmers have strong incentive to sell en masse in certain locales.  This particular case involved a 200 acre dairy farm abutted along its easterly boundary by a large 250 unit townhouse development.  The farm had been owned by the same family for over 150 years and the current owners vowed to preserve the farm intact even while adjoining farms succumbed to lucrative cash deals being offered by a plethora of anxious land developers.

Approximately one-half of the units had been built when several complaints were lodged against the farm family by a newly formed home owners association of the adjoining townhouse development.  The complainants pleaded impassionately with the county judge that the farmer's electric fences located on the farm's perimeter created a danger to their children and pets. Apparently several unleashed dogs on forays into the farm property had already encountered the mild shock of the electric fencing, much to their owner's dismay. In an opinion I still find rather incredulous, the county judge ordered the farmer to remove the fence and to relocate it at least 150 feet back from his boundary lines. He also barred the farmer from replacing the relocated fence with any similar deterrent fencing such as barbed wire. Finally, he was permanently enjoined from using this newly created 150 foot buffer zone for anything other than the growing of crops. I can tell you today that this farm no longer exists.

These new migrants to rural America come with unrealistic expectations or total insensitivity to farmers and their land. On the one hand, they welcome the open space that farms provide and preserve, but then just as quickly condemn the practices the farmers must carry out to keep their operations profitable. I am not absolving farmers completely. There are cases where farmers have been legitimately criticized and cited for polluting surface streams and ground water and for improperly storing and applying liquid manures, and for misapplying pesticides. In spite of these shortcomings, agriculture is a vital national industry, at least until such time as some national policy dictates that every household must henceforth be entirely self-sufficient. Wouldn't that drastically alter the complexion of land use planning in America. In the meantime, land planners will continue to be involved in resolving conflicts between agricultural land uses and residential and other uses. The preservation of the agricultural industry must remain a top priority on the agenda of the nation's land planners.

## PROTECTING AGRICULTURAL USES

Land planners have some efficacious options available to assist them in the protection of farming and agricultural land. The first of these is the Farmland Assessment option. Under this program farmland is accorded a special status for real estate tax assessment purposes. For instance, in New Jersey, which has had such a program since 1964, all land, 5 acres or more in size and actively devoted to agricultural or horticultural uses is eligible for a special Farmland Tax Assessment. This special program uses productivity values rather than market value as the basis for assessment. As an example, one acre of woodland assessed under residential market value would be valued at $30,000. Under the Farmland Assessment Program, using productivity value, the same acre would be valued at $21. Using a tax rate of $2.50 per 100 dollars of assessed value, this amounts to a tax saving of about $749.47. This program has been a major factor in prolonging farming in the state. Without it, New Jersey would probably have ceased being the Garden State more than a decade ago.

Some claim that the program has been abused by land speculators who buy large tracts of farmland and lease it back to farmers to preserve this substantial reduction in real estate taxes. While this may be true, the land does remain as part of the state's ever dwindling, open space while at the same time providing habitat for a variety of wildlife. To partially balance this preferential treatment, all land enrolled in this program is subject to rollback taxes if sold for other than agricultural uses. That is to say that at the time of sale of the property, the seller or buyer must pay the difference in assessment between productivity value and market value for the prior two to three years.

A second option available to planners to protect valuable agricultural land is Farmland Preservation. Under this program, usually the result of a state referendum or state legislation, monies are appropriated for the direct purchase of permanent restrictions or easements on existing agricultural land. Such deed restrictions prohibit the use of the land for other than agricultural uses in perpetuity.

Such purchases are generally restricted to areas previously established in local land use master plans as farmland preservation districts. Participation in the program is voluntary and this can interfere with implementation since some of the landowners in the preservation zone may not be anxious to sell such easements. Consequently, all of the acreage designated to be acquired may not be available. The prices of these easements are established by appraisal and commonly fall between the current market value of the land under present zoning and the agricultural productivity value.

A third and similar option is the transfer of development rights program or TDR. As in the farmland preservation program outlined previously, the municipality would place certain agricultural land in a preservation zone and prohibit development. This time, however, there would be no simultaneous fee simple purchase of easements or restrictions. Rather reliance would be upon the private purchase and transfer of development rights attached to the land. Development rights are just one of many rights that may be sold separately from the land. Other similar rights include mineral, timber and air rights--all of which may be transferred separately while the landowner retains ownership of the land.

Concurrent with the establishment of the preservation zone, the community must also establish a transfer zone to which these development rights can be transferred, otherwise the TDR system will not work. (It is assumed, of course, that this transfer zone will have adequate infrastructure and assimilative capacity to support the increased density.) Consequently, a developer who wishes to build at a higher density in the transfer zone must first purchase development rights from the property owners in the agricultural preservation zone.

A simple example would perhaps best illustrate the concept. A farm consisting of 150 acres with two residential structures on it is placed within a farmland preservation zone. Under present zoning, a density of one dwelling unit (DU) is permitted per acre. Consequently, the farm's gross development potential would be 150 residential DU's.[1] From this we subtract the two existing DU's leaving a net total of 148 units or rights. The companion transfer zone allows a standard building density of two units per acre with an option to go as high as six units per acre provided development rights are purchased from the farm(s) in the preservation zone. Consequently, a builder looking to build dwellings at the higher density in the transfer zone must purchase a development right from the farmland preservation zone for each additional DU proposed above the standard two per acre. The purchase price of the development rights is arrived at through the bargaining process of the open market.  The major advantage of this option is that there are no public monies required to support the program.

A fourth option to aid the besieged farmer is the passage of legislation generally known as "Right-To-Farm" laws.  These statutes bar the filing of nuisance lawsuits against farmers by adjoining property owners if the farm preceded the development around it.  These laws, however, do not indemnify the farmer completely if he operates in a careless or reckless manner.  He must utilize currently acceptable farming practices.  Examples of these available options can be found in Appendix 1, Part E.

Finally, some states have also adopted more stringent trespassing laws that are specifically designed to control trespass on agricultural or horticultural land.

## REVERSING THE TREND

As discussed earlier, American agriculture must also put its house in order.  The contamination of ground water with pesticides and nitrates from fertilizers, soil erosion (Figure 8.1) and runoff of manures and silt into surface waters, remain significant problems on America's agricultural land.  As a nation we must share some of the blame for the current environmental transgressions of American farmers.  It was we who encouraged and fiscally supported agriculture's conversion to highly specialized, often environmentally indifferent farming methods such as the extensive monoculture of grains, the use of large animal feedlot operations and the use of enormous quantities of synthetic fertilizers and pesticides.  We must now similarly encourage and support alternative technology, some of which may very well be the same methods employed prior to the 1950's and the chemical dominated era of American agriculture.  The latest provisions of the federal 1985 Food Security Act will certainly provide some incentive in this regard.  This latest Farm Bill reemphasizes the inclusion of conservation programs in current farm practice.  Four key provisions of that bill are:

1.   Conservation Reserve

This program offers farmers help in retiring highly erodible cropland.  The Agricultural Stabiliza-

Figure 8.1: This photo demonstrates one of the major problems with agricultural operations. Lacking adequate soil erosion control measures this farm's highly erodible soils are washed freely from fields to nearby roads and waterways during rainstorms.

tion and Conservation Service (ASCS) will share up to half of the cost of establishing permanent grasses, legumes, trees, windbreaks, or wildlife plantings on highly erodible cropland.(2)

2.   Conservation Compliance

This program applies to farmers who continue to plant annually tilled crops on highly erodible soils. To remain eligible for certain USDA benefits farmers must develop and be actively applying a locally approved conservation plan for those highly erodible soils by January 1, 1990. Farmers must then have such a plan fully implemented by January 1, 1995.

3.   Sodbuster Provision

The Sodbuster provision applies to those farmers who plant annually tilled crops on highly erodible soils that were not used for crop production during the period from 1981 through 1985.

If these soils are now plowed, they must do so under a conservation system approved by the local conservation district to remain eligible for USDA program benefits.

4.    Swampbuster Provision

The swampbuster provision applies to farmers who convert naturally occurring wetlands to cropland after December 23, 1985. With some exceptions, farmers must discontinue production of annually tilled crops on newly converted wetlands to remain eligible for certain USDA programs.

State and local planning agencies can and should complement these federal initiatives with their own programs. Possible options include:

1.    The mandatory participation in a state or local soil conservation program as a condition for receiving preferential farmland tax assessment.

2.    The adoption of ordinance regulating the maximum number of livestock allowable per acre with the provision that greater densities could be obtained by those farms which implement best management practices for control of non-point source pollution.

3.    Requiring the installation of monitoring wells at selected locations on each farm where synthetic pesticides or fertilizers are used.

4.    Requiring that livestock be fenced away from surface streams.

5.    Requiring the submission of an annual management plan detailing the location and rates of application of fertilizers and manures.

## FOOTNOTES

### CHAPTER 8

(1)    It should be noted that additional acreage and subsequently additional DU's may be subtracted out when critical areas such as wetlands, steep slopes and floodways are eliminated as buildable areas.

(2)    As of August, 1988, an estimated 12 million hectares (30 million acres) had already been placed into the Conservation Reserve program under 10 year contracts. This is about 75% of the U.S. government's goal to include 16 million hectares (40 million acres) by the end of 1990. Three-fourths of the way to its goal the program has already reduced the national annual soil erosion rate by more than 800 million tons (Ref. 8.5).

## REFERENCES

## CHAPTER 8

8.1    Chavooshian, B. Budd, Thomas Norman, Esq., Dr. George H. Nieswand, "Transfer of Development Rights: A New Concept in Land Use Management", Rutgers University, Leaflet 492-B, undated.

8.2    Haar, Charles M., Steven G. Horowitz, Daniel F. Katz, "Transfer of Development Rights: A Primer", Lincoln Institute of Land Policy, 1980.

8.3    Urban Land Institute, Urban Land Magazine, Vol. 34 No. 1, January, 1975.

8.4    United States Environmental Protection Agency, EPA Journal, Vol. 14, No. 3, April, 1988.

8.5    Brown, Lester R. and Alan Durning, Christopher Flavin, Lori Heise, Jodi Jacobson, Sandra Postel, Michael Renner, Cynthia Pollack Shea, Linda Starke, "State of the World--1989", Worldwatch Institute, W.W. Norton and Co., New York, 1989.

# 9

# Mines and Quarries

## A LASTING LEGACY

Nothing causes greater apprehension amongst local planning agencies than an application to either open or reactivate a mine or quarry. This is especially true in areas of dense population.

The public perception of the mining and quarrying industries as environmentally harmful businesses has some basis in fact. The coal mining industry, in particular, has left behind a nationally renowned legacy of acid contaminated drainage and barren landscapes. While many of these areas are now being restored, it will take much longer to do likewise with the public trust.

## PROBLEM AREAS

Various materials in addition to coal are routinely extracted from the American landscape. These include such items as sand and gravel, copper, uranium or iron ore, clay, marble, slate and road stone. While the mechanics of removal may vary somewhat the resulting environmental impacts are surprisingly similar with soil erosion and contaminated storm water runoff being the most common problems. A substantial amount of literature is available on methods to control soil erosion and storm water runoff and the reader is advised to consult the numerous references available (see also Chapter 2). Other problems frequently associated with mining and quarrying operations include:

1. The interception and diversion of the ground water table causing downstream wells and surface streams to dry up.

2. The disturbance of nearby foundations and other structures caused by the detonation of explosive charges.

3. The creation of steep, nearly vertical slopes.

4. The creation of deep, water filled, cavities and ponds.

5. The generation of acidic (low pH) and sometimes basic (high pH) and heavy metal contaminated runoff.

6. Wind blown dust and solids.

7. Oil and grease runoff from rock crushers to surface streams.

8. Noise from drilling equipment and heavy trucks hauling materials.

9. Increased heavy truck traffic on local roadways resulting in pavement deterioration and the deposition of products onto the pavement.

10. The creation of subsurface vertical and horizontal shafts and tunnels which can eventually collapse and cause subsidence of the ground surface.

11. The abandonment of unsealed ore extraction wells which can contribute to ground water pollution.

12. Increased storm water runoff and damage to downstream properties and drainage structures.

13. The release of suspended solids (turbidity) to surface streams from wash water flowing off of piles of product cleaned before shipment or from water suppression systems used to control fugitive dust at rock crushing operations.

## PLANNING TO MITIGATE THE IMPACTS

The apprehensiveness experienced by local planning agencies when confronted with an application to open or reactivate a mine or quarry is understandable. Strong citizen opposition is likely, particularly if the proposed operation is within close proximity to a densely populated area or if there have been prior problems with mining or quarrying activities.  In addition, the applicant may deluge the planning agency

Figure 9.1: Noise and dust from large trucks traveling to and from a quarry can severely impact local roads and nearby residences.

Figure 9.2: As evidenced by this photo water pollution control measures at quarries and mines are often crude and woefully inadequate to protect water quality in adjoining waterways.

with an array of technical reports and specialists to buttress the proposed project.

The agency need not feel intimidated, however, for while they will have overall responsibility, some state and even federal agencies will undoubtedly be assisting in the review process. In particular, all point source discharges of wastewater to surface streams, including mine dewatering discharges, will require at least an application, if not the actual issuance of a federal or state NPDES permit. The state and federal bureaus of mine safety may also have some jurisdiction and be required to review and approve the proposed operation.

In any event, regardless of what other agencies participate in the review process, the local planning agency will need its own experts to evaluate the merits of the project and to interpret any data submitted by the applicant. If such expertise cannot be found amongst the experts normally hired by the planning agency, then outside experts must be retained. In addition, the very nature of these projects require that an environmental assessment and an environmental impact statement be prepared and submitted by the applicant.

An astute applicant will have already prepared such documents without a formal directive from the planning agency. The following minimum conditions for approval may be conveyed to the applicant immediately upon application since these are now generally expected practices in approved mining and quarrying operations:

1.    An undisturbed buffer zone is to be established around the complete periphery of the mine or quarry property. The width of this zone will be determined by the planning agency based on predicted noise levels, topography, microclimate and the distance(s) to the nearest receptor. This buffer zone may be of a varying width if needed.

2.    Fencing of sufficient durability and height (preferably chain link) is to be installed around the periphery of the property to prevent inadvertent and potentially dangerous intrusions by humans and animals.

3.    A comprehensive soil erosion and sediment control plan is to be submitted detailing the short term and long term practices to be implemented during

the life of the mining or quarrying operations.   A
performance bond for these practices is also to be
submitted.

4.   A proposed final closure plan is to be included detailing the ultimate disposition and restoration of the site once mining and quarrying operations
have ceased.   As in item 3, a performance bond is to
be submitted.

5.   All surface streams traversing a proposed
quarrying or mining site that have the potential to be
impacted must be monitored.   This will require the
establishment of permanent monitoring stations at various locations upstream and downstream of the site to
measure the impact, if any, on surface water quality
or quantity.   These monitoring locations should be
established jointly, however, where disagreement
occurs, the approving agency will have the final say.
Sampling will usually be done on a daily basis.
Parameters to be measured will be determined on a case
by case basis.

Where subsurface chemical extraction of ores or
other materials is accomplished by the use of wells
then a series of upstream and downstream monitoring
wells should likewise be installed to monitor any
changes in ground water quality.

## SOME BENEFICIAL USES OF THE ABANDONED MINE OR QUARRY

There are obvious economic benefits to society
derived from the materials extracted from the nation's
mines and quarries.   However, the residual materials
and landscape left behind upon cessation of operations
are often regarded as complete liabilities.   While
this may be true in some instances, it is not a valid
assumption for all situations.   In this section, we
will discuss some of the less obvious but beneficial
uses to be made of abandoned mines and quarries.

A common benefit of open pit rock or mineral ore
quarries or sand and gravel mines, for example, is the
creation of ponds or water filled pits which may be
suitable for swimming or for fishing if stocked with
fish.   This of course, assumes that the quality of the
water has not been degraded by the previous mining
activities.   Open pit mines and quarries can also be
donated for various public uses, including land based
recreational activities or occasionally as a shooting

range for local, county or state police officers.  In the past, such open pits have also been used by municipal and county governments for landfilling solid waste.  This use is no longer recommended unless there has been a comprehensive pre-evaluation of the site's suitability followed by implementation of state of the art environmental protection practices (see Chapter 11).  Interestingly enough, the current national priority to phase out and secure hundreds of thousands of old landfills and hazardous waste disposal sites may in fact create the need to reopen abandoned clay mines or even create new ones in order to satisfy an increased demand for impervious covering materials or bottom liners.

Some rock quarries have also been purchased by private land developers and converted to very attractive residential housing sites.  Such proposals, however, need extraordinary scrutiny by the land planner. The construction of building foundations and streets and placement of underground storm water and sanitary sewer lines is no small undertaking, often requiring significant amounts  of blasting and rock removal. Adjoining property owners will have legitimate concerns about damage to their foundations and potable water wells from the concussions of the blasting.  The posting of a bond to cover potential offsite damages will often be required.  Additionally, some rock quarries, because of the type of rock present, are not suitable sites for residential buildings.  Fractured basalt, for example, weathers severely at its surface and persistent slumping and slides are common.  Similarly, quarries in cavernous limestone bedrock pose similar threats of subsidence.

## REFERENCES

9.1    United States Environmental Protection Agency, "Erosion and Sediment Control, Surface Mining in the Eastern U.S.", EPA-625/3-76-006, October, 1976.

9.2    New Jersey State Soil Conservation Committee, "Standards For Soil Erosion and Sediment Control in New Jersey", June 14, 1972.

# 10

# Underground Storage Tanks

**DEFINING THE PROBLEM AND REGULATION**

It is indeed rare to find a land planner in contemporary America who is unable to recount at least one horror story of ground water pollution caused by leaking underground storage tanks. Thousands of incidences of ground water contamination caused by leaking underground storage tanks have been documented in every part of the nation. No one, however, at this point in time knows the exact number of underground storage tanks that are leaking, nor does anyone know with exactness how many will leak in the decades forthcoming.

The total number of underground storage tanks in the United States that contain petroleum products and hazardous substances (the primary contributors to ground water pollution) is likewise not known with full certainty, however, current estimates range from 1.5 to 3.5 million. The national magnitude and seriousness of the problem prompted the passage of federal legislation to deal with it in 1984. The 1984 Amendments to the Resource Conservation and Recovery Act (RCRA) of 1976, particularly Subtitle C and Subtitle I, provide a regulatory program for underground tanks that store petroleum products (including crude oil and gasoline) and substances defined as hazardous under the Comprehensive Environmental Response, Compensation and Liability Act (CERCLA) of 1980 (also referred to as the "Superfund Act"). An underground storage tank is defined under the 1984 Amendments as any tank with at least 10% of its volume below ground <u>including any pipes attached to the tank</u> (emphasis added) [1].

Subtitle I specifically exempted the following types of tanks from these 1984 Amendments:

1. Farm and residential tanks holding less than 1100 gallons of motor fuel and used for noncommercial purposes.

2. Tanks storing heating oil for burning on the premises where stored.

3. Septic tanks.

4. Pipelines regulated under other laws.

5. Systems for collecting storm water and wastewater.

6. Flow-through process tanks.

7. Liquid traps or associated gathering lines related to operations in the oil and natural gas industry.

8. Tanks in an underground area, such as a basement, if the tank is upon or above the surface of the floor.

On May 7, 1985 a provision banning the installation of underground tanks that do not meet certain minimum requirements went into effect as a result of the RCRA Amendments. This provision stated that no person could henceforth install an underground storage tank unless:

1. It was designed to prevent releases of the stored substances due to corrosion or structural failure for the operational life of the tank, and

2. (2)It was cathodically protected against corrosion or constructed of either noncorrosive material or steel clad with a noncorrosive material or was designed specifically to prevent the release or threatened release of stored substances, and

3. The material used in the construction or lining of the tank was compatible with the substance to be stored.

Of special interest to land (3)planners was a series of notification provisions in the 1984 Amendments that placed the following requirements on owners of operational tanks, sellers of tanks, distributors of regulated substances and owners of tanks taken out of operation within the past ten years, but still in the ground:

1. Between December 8, 1985 and June 8, 1987 any person who deposited regulated substances into an underground storage tank was to inform the tank owner of the requirement to notify the state agency of such ownership.

2.   By May 8, 1986 the owners of existing underground storage tanks were to submit those notification forms to the appropriate state agency.

3.   By May 8, 1986 owners of underground storage tanks taken out of operation after January 1, 1974, but still in the ground, were to notify the state agency of each tank's age, the date taken out of service and the type and quantity of substances left in the tank.

4.   After May, 1986, owners of newly installed underground storage tanks must notify the state agency within 30 days after bringing the tank into use.

Several states have or are developing their own regulatory programs for underground storage tanks. You should contact the designated state environmental protection agency in your area to determine if they have adopted their own underground storage tank program. Also of added interest to land planners is the provision in the new federal law for inspection and enforcement authority for both federal and state personnel. Consequently, authorized federal and state inspectors may inspect and sample tanks, request pertinent information from tank owners and monitor and test tanks and the surrounding soils, air and surface and ground water.

In some of the states which have already adopted their own leaking underground storage tank program (New Jersey for example) the removal and replacement of underground tanks has become a commonly observed activity. Gasoline stations, in particular, are making substantial efforts to rid themselves of aging tanks. The enormity of the liabilities associated with ground water contamination has proven to be one of the program's most successful and unplanned incentives. For instance, pre-1984 tank replacement and/or repair procedures consisted of tank excavation, tank and piping replacement or repair and regrading. Costs generally ranged between $5,000 and $8,000 for a typical 10,000 gallon steel tank. Little attention was paid to ground water pollution. Today, those procedures have been enlarged and now generally consist of tank testing, tank excavation, a comprehensive soil analysis, a comprehensive ground water analysis, contaminated soil removal and disposal, ground water cleanup, tank replacement and regrading. The costs have similarly increased and now may range between

$100,000 and $250,000 for the same 10,000 gallon tank[4].

As land planners, you will be called upon to approve, direct or oversee installation or removal of these underground storage tanks with increasing regularity. These new legislative requirements coupled with the complexity of remediating contaminated sites makes it imperative that land planners acquire a more thorough knowledge of the topic. In the following sections of this chapter we hope to initiate the transfer of some of that information.

## TANK TYPES AND COSTS

During the 1950's and 1960's uncoated carbon and asphalt coated steel tanks were the most commonly installed underground tanks in the United States. These tanks, however, exhibited great susceptibility to corrosion with some beginning to leak in as little as 2 to 3 years after installation. Technological improvements have substantially increased the corrosion resistance of today's steel tank with low carbon steel, galvanized steel and stainless steel 304 and 316 now available as superior substitutes.

In addition, tanks are now available in a wide variety of materials considered as non-corrosive. These include fiberglass reinforced plastic, Teflon, polyvinylchloride, polypropylene, polyethylene, polymethylacrylate, silicon and neoprene. See Table 10.1 for some additional information on tank costs.

## CAUSES OF TANK LEAKS

As mentioned earlier corrosion is the major factor contributing to leaks in metallic tanks and piping systems. This corrosion can occur both internally and externally with external corrosion being the most common.

Corrosion of metals is an electrical process and may be either galvanic or electrolytic in nature. Galvanic corrosion is a self generated activity resulting from differences in electrical potential that develop when a metal is immersed in an electrolyte. For example, when two dissimilar metals are connected and placed in an electrolyte (soil in the case of underground tanks) current will be generated. This

## Table 10.1:  Some Typical Tank Prices

| TYPE | COST* (approximate) | WARRANTY [5] |
|------|---------------------|------------|
| 1. ASPHALT COATED STEEL | $3,000 | 1 Year Unconditional |
| This tank only provides a limited amount of corrosion protection. | | |
| 2. FIBERGLASS COATED STEEL | $6,000 | 20 Year Unconditional |
| This is a standard steel tank coated on the outside with a 100mm thickness of fiberglass reinforced plastic. It is susceptible to corrosion if the fiberglass coating is scratched or if there are pinholes existing in the coating itself. | | |
| 3. FIBERGLASS COATED DOUBLE-WALLED STEEL | $14,000 | 20 Year Unconditional |
| This tank consists of one steel tank inside another with a fiberglass coating over the outside tank. It too is susceptible to corrosion if the fiberglass coating is scratched or if pinholes exist in the coating itself. | | |
| 4. EPOXY COATED STEEL WITH SACRIFICIAL ANODE | $4,500 | 20 Year Limited |
| A coating of epoxy covers the outside of the steel tank. A sacrificial anode is connected to the tank to protect against corrosion caused by pinholes or scratches in the epoxy coating. | | |
| 5. FIBERGLASS (STANDARD) | $4,500 | 20 Year Unconditional |
| Eliminates corrosion, however, lacks structural strength of steel and consequently requires more stringent installation to avoid tank rupture. | | |
| 6. FIBERGLASS (ALCOHOL BLEND) | $5,000 | 1 Year Unconditional |
| Unlike the standard fiberglass tank, this tank can store ethanol and methanol blends. | | |

*Cost for 10,000 gallon tank F.O.B. manufacturer

Source: Adapted from Ref. 10.1. Copyright held by Pollution Engineering, Northbrook, IL 60062.

current subsequently flows into the electrolyte (soil) from the corroding metal (called the anode). The current will then flow over to the non-corroding metal (the cathode) and back through the connection between the two metals (Figure 10.1). Galvanic corrosion can also occur in a single metal if there are variations in surface conditions. In the case of underground tanks, this can be as simple as having a patch of clay stuck to one side of a tank while the other side is in contact with sand or silt (Figure 10.2).

Electrolytic corrosion, on the other hand, is the result of an induced electrical current from an external source entering and then leaving a metallic structure by way of the electrolyte (soil). Again, corrosion occurs at the anode where the current leaves the metallic surface. Corrosion can be very localized or general.

Since soil acts as the electrolyte in the corrosion of these underground metallic tanks, we should know some of the more important edaphic conditions that will affect the corrosion process. These include:

1. <u>Soil acidity</u> - acidic (low pH) soils are generally more corrosive than neutral (pH 7) or alkaline (high pH) soils. With amphoteric metals such as zinc and aluminum, however, highly alkaline soils may be just as corrosive as acidic soils.

2. <u>Temperature</u> - the rate of corrosion tends to increase as the soil temperature increases.

3. <u>Soil resistivity</u> - soil resistivity is the measure or resistance of soil to the flow of electric current. The higher the resistivity, the less corrosive is the soil[6].

4. <u>Soil moisture</u> - moisture in the soil acts to reduce soil resistivity and conversely enhances conductivity.

5. <u>Soil bacteria</u> - under anaerobic conditions bacteria can accelerate the corrosion process.

6. <u>High water table</u> - increases conductivity and causes empty or nearly empty tanks to float[7].

7. <u>Salinity of ground water</u> - enhances conductivity.

8. <u>Quantity of oxygen in the unsaturated soil zone</u>.

Other non-edaphic factors affecting corrosion include:

Figure 10.1

# END VIEW

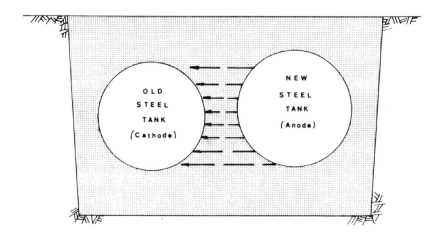

Figure 10.2

# END VIEW

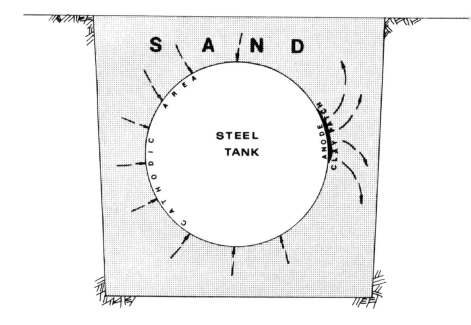

1.      The presence of adjacent buried metal
tanks or structures.
2.      The presence of stray underground currents
from nearby electrical facilities.

Leaks from tanks made of non-corroding materials
generally result from one or more of the following:
(a) improper handling during installation resulting in
fractures and cracks; (b) improper bedding of the tank
resulting in failure when the weight of the product is
introduced; (c) improper anchoring in areas of high
water table; and (d) softening, cracking or swelling
due to some incompatibility between the tank materials
and the stored product.

## LEAK DETECTION

Prior to our heightened awareness of the hazards
associated with leaks from underground storage tanks,
leak detection was largely unplanned.   As a result,
tank owners were often unaware of leaks until confronted with peripheral or secondary symptoms such as:
1.   Discrepancies in product inventories.
2.   Customer complaints about water in the prod-
     uct.
3.   Product appearance in a nearby surface
     waterway.
4.   Neighborhood complaints about polluted wells
     or odors in sewer lines or basements.
5.   Water found in the storage tank.
6.   Settling or slumping of the soil around the
     tank(s).

The U.S. Environmental Protection Agency in
cooperation with the American Petroleum Institute
(API) and the Petroleum Equipment Institute (PEI) has
recently compiled a list of available and developing
leak detection methodologies.  Approximately 36 methods have been identified thus far (Ref. 10.4, Table
10.2).  Due to certain similarities in these methods,
they can be assigned to the following four generic
categories:
1.      Volumetric (quantitative) leak testing and
leak rate measurement

These methods of leak detection are based upon
the detection of changes in product volume by the measurement of such parameters as liquid level, temperature, pressure and density.   Such volumetric leak

## Table 10.2: Summary of Leak Detection Methods

PART 1  VOLUMETRIC TYPE

| Method | Principle | Claimed Accuracy (gal/h) | Cost of Testing† | Total Down Time for Testing | Requires Empty-Full Tank for Test |
|---|---|---|---|---|---|
| *Inventory monitoring* | | | | | |
| Gage stick | Measuring project level with dip stick when station is closed | Gross | Minimal | Variable† | No |
| MFP-414 TLG leak detector | Monitoring product weight by measuring pressure and density at the top, middle, and bottom of tank | Sensitive to 0.1% of product height change | $5000 to $6000 (equipment) | None | No |
| TLS-150 | Using electronic level measurement device or programmed microprocessor inventory system | Sensitive to 0.1-in. level change | $5000 (equipment) | None | No |
| *Monitoring of leak effects* | | | | | |
| Collection sumps | Using collection mechanism of product in collection sump through sloped floor under the storage tank | Does not provide leak rate | Provided by contractor | None | No |
| Dye method | Hydrocarbon detection by use of soluble dye through perforated pipe | Does not provide leak rate | Provided by contractor | None | No |
| Ground-water and soil sampling | Water and soil sampling | Does not provide leak rate | Provided by contractor | None | No |
| Interstitial monitoring in double-walled tanks | Monitoring in interstitial space between the walls of double-walled tanks with vacuum or fluid sensors | Does not provide leak rate | Provided by tank manufacturer | None | No |
| L A S P | Diffusion of gas and vapor to a plastic material | Does not provide leak rate | Provided by contractor | None | No |

(continued)

Table 10.2: (continued)

PART 1 Continued

| Method | Principle | Claimed Accuracy (gal/h) | Cost of Testing† | Total Down-Time for Testing | Requires Empty/Full Tank for Test |
|---|---|---|---|---|---|
| Mooney tank test detector | Measuring level change with a dip stick | 0.02 | $250/tank 1- to 2-h test) | 14 to 16 h‡; 12 to 14 h waiting after fillup. | Full |
| PACE tank tester | Magnification of pressure change in a sealed tank by using a tube (based on manometer principle) | Less than 0.05 | Not commercial | 14 h | Full |
| PALD-2 leak detector | Pressurizing system with nitrogen at three different pressures; level measurement by an electro-optical device; estimate of leak rate based on the size of leak and pressure difference across the leak | Less than 0.05 | Not commercial | 14 h (preferably 1 day before, 1-h fill testing, includes sealing time) | Full |
| Pneumatic testing | Pressurizing system with air or other gas; leak rate measurement by change in pressure | Gross | Low | Several hours | No |
| Tank auditor | Principle of Buoyancy | 0.00001 in the fill pipe; 0.03 at the center of a 10.5-ft-diameter tank | $400/tank | 1.5 to 3 h | Typically full |
| Two-tube laser interferometer system | Measuring level change by laser beam and its reflection | Less than 0.05 | Not commercial | 4 to 5 h§ | No (at existing level) |

† Charges could be negotiated with manufacturer or different numbers of tank testing and different tank specifications
‡ Including the time for tank end stabilization wh. testing with standpipe.
§ Including 1 to 2 h for reference tube temperature equilibrium.

(continued)

## Table 10.2: (continued)

PART 2  NONVOLUMETRIC TYPE

| Method | Principle | Claimed Accuracy (gal/h) | Cost of Testing† | Total Down-Time for Testing | Requires Empty/Full Tank for Test |
|---|---|---|---|---|---|
| Acoustical Monitoring System (AMS) | Sound detection of vibration and elastic waves generated by a leak in a nitrogen-pressurized system; triangulation techniques to detect leak location. | Does not provide leak rate; detects leaks as low as 0.01 gal/h | Not commercial | 1 to 2 h | No |
| Leybold-Heraeus helium detector, Ultratest M2 | Rapid diffusivity of helium; mixing of a tracer gas with products at the bottom of the tank; helium detected by a sniffer mass spectrometer | Does not provide leak rate; helium could leak through 0.005-in. leak size | By contractor | None | No |
| Smith & Denison helium test | Rapid diffusivity of helium; differential pressure measurement; helium detection outside a tank | Provides the maximum possible leak detection based on the size of the leak (does not provide leak rates); helium could leak through 0.05-in. leak size | By contractor | Few - 24 h (excludes sealing time) | Empty |
| TRC rapid leak detector for underground tanks and pipes | Rapid diffusion of tracer gas; mixing of a tracer gas with product; tracer gas detected by a sniffer mass spectrometer with a vacuum pump | Does not provide leak rate; tracer gas could leak through 0.005-in. leak size | By contractor | None | No |
| Ultrasonic leak detector (Ultrasound) | Vacuuming the system (5 psi); scanning entire tank wall by ultrasound device; noting the sound of the leak by headphones and registering it on a meter | Does not provide leak rate; a leak as small as 0.001 gal/h of air could be detected; a leak through 0.005 in. could be detected | Not commercial | Few hours (includes tank preparation and 20-min test) | Empty |
| VacuTect (Tanknology) | Applying vacuum at higher than product static head; detecting bubbling noise by hydrophone; estimating approximate leak rate by experience | Provides approximate leak rate | $500/tank | 1 h | No |
| Varian leak detector (SPY2000 or 938-41) | Similar to Smith & Denison | Similar to Smith & Denison | Varies by contractor | Few - 24 h (excludes sealing time) | Empty |

† Charges could be negotiated with manufacturer for different numbers of tank testing and different tank specifications

(continued)

Table 10.2: (continued)

PART 3 OTHER TYPES

| Method | Principle | Claimed Accuracy (gal/h) | Cost of Testing† | Total Down-Time for Testing | Requires Empty/Full Tank for Test |
|---|---|---|---|---|---|
| Ainlay tank integrity testing | Pressure measurement by a coil-type manometer to determine product level change in a propane bubbling system | 0.02 | $225/day + exp. (3 tanks/day) | 10 to 12 h (filled a night before 1.5 h testing) | Full |
| ARCO HTC underground tank detector | Level change measurement by float and light-sensing system | 0.05 | | 4 to 6 h | No |
| Certi-Tec testing | Monitoring of pressure changes resulting from product level changes | 0.05 | | 4 to 6 h | Full |
| "Ethyl" tank sentry | Level change magnification by a "J" tube manometer | Sensitive to 0.02-in. level change | $300/tank | Typically 10 h | No |
| EZY-CHEK leak detector | Pressure measurement to determine product level change in an air bubbling system | Less than 0.01 | $300/tank | 4 to 6 h (2 h waiting after fillup, 1-h test) | Full |
| Fluid-static (standpipe) testing | Pressurizing of system by a standpipe, keeping the level constant by product addition or removal, measuring rate of volume change | Gross | Low | Several days | Full |
| Heath Petro Tite tank and line testing (Kent-Moore) | Pressurizing of system by a standpipe, keeping the level constant by product addition or removal; measuring volume change; product circulation by pump | Less than 0.05 | $75/1000 gal | 6 to 8 h | Full |
| Helium differential pressure testing | Leak detection by differential pressure change in an empty tank; leak rate estimation by Bernoulli's equation | Less than 0.05 | | Minimum 48 h | Empty |

(continued)

Table 10.2: (continued)

PART 3 Continued

| Method | Principle | Claimed Accuracy (gal/h) | Cost of Testing[†] | Total Down-Time for Testing | Requires Empty/Full Tank for Test |
|--------|-----------|--------------------------|-------------------|------------------------------|------------------------------------|
| Leak Lokator test hunter—formerly Sunmark Leak Detection | "Principle of Buoyancy." The apparent loss in weight of any object submerged in a liquid is equal to the weight of the displaced volume of liquid | 0.05 even at product level at the center of a tank | $500/tank | 3 to 4 h | Typically full |
| Observation wells | Product sensing in liquid through monitoring wells at areas with high ground water | Does not provide leak rate | Varies by contractor | None | No |
| Pollulert and Leak-X | Difference in thermal conductivity of water and hydrocarbon through monitoring wells | Does not provide leak rate | Varies by contractor | None | No. |
| Remote infrared sensing | Determining soil temperature characteristic change due to the presence of hydrocarbons | Does not provide leak rate | Varies by contractor | None | No |
| Surface geophysical methods | Hydrocarbon detection by ground-penetrating radar, electromagnetic induction, or resistivity techniques | Does not provide leak rate | Varies by contractor | None | No |
| U-Tubes | Product sensing in liquid; collection sump for product directed through a horizontal pipe installed under a tank | Does not provide leak rate | Varies by contractor | None | No |
| Vapor wells | Monitoring of vapor through monitoring well | Does not provide leak rate | Varies by contractor | None | No |

[†] Charges could be negotiated with manufacturer for different numbers of tanks and different tank specifications.

Source:  Ref. 10.5

detection methods can reportedly detect leakage rates as low as 0.02 gal/hr.

2.    Nonvolumetric (qualitative) leak testing

Testing for leaks by this method usually involves the use of a tracer material--frequently helium. After pressurizing the suspect tank with helium, either the loss in pressure is measured or an external mass spectrometer is used to detect the tracer gas outside the tank. Other nonvolumetric methods include pressurizing the tank with gas or placing it under a vacuum and then using sound detection devices to listen for the hiss of escaping gas or the sound of incoming gas bubbling through the liquid product.

Some concerns with the use of these types of testing methods include the potential to accelerate existing leakage and the possibility of an explosion. Dye tracers may also be used.

3.    Inventory control or inventory monitoring

This is the simplest method of leak detection. Inventory monitoring can be performed utilizing a gauge stick, electronic level measurement, or by using pressure and density measurements to determine weight changes in the tank's contents.

4.    Monitoring of leak effects

These methods monitor the environmental effects outside the tank once it has begun to leak. This type of monitoring includes the installation of monitoring wells and the subsequent collection and analysis of samples. These methods do not provide information on leakage rates or the size of the leak.

It should be noted that the accuracy of each detection method is affected by a variety of factors some of which are shown on Table 10.3.

## MINIMUM SAFEGUARDS

Certain minimum safeguards are now mandated for the installation or replacement of underground storage tanks. Metal tanks in particular will need some form of external corrosion control--a common method of which is the application of protective surface coatings. These coating materials may range from the questionably efficacious asphalt to the more contemporary fiberglass reinforced plastic or epoxy.

Table 10.3:  Variables Affecting Accuracy of Leak Detection

| Variable | Impact |
|---|---|
| Temperature change | Expansion or contraction of a tank and its contents can mask leak and/or leak rate |
| Water table | Hydrostatic head and surface tension forces caused by ground water may mask tank leaks partially or completely |
| Tank deformation | Changes or distortions of the tank due to changes in pressure or temperature can cause an apparent volume change when none exists |
| Vapor pockets | Vapor pockets formed when the tank must be overfilled for testing can be released during a test or expand or contract from temperature and pressure changes and cause an apparent change in volume. |
| Product evaporation | Product evaporation can cause a decrease in volume that must be accounted for during a test |
| Piping leaks | Leaks in piping can cause misleading results during a tank test because many test methods cannot differentiate between piping leaks and tank leaks. |
| Tank geometry | Differences between the actual tank specifications and nominal manufacturer's specification can affect the accuracy of change in liquid volume calculations. |
| Wind | When fill pipes or vents are left open, wind can cause an irregular fluctuation of pressure on the surface of the liquid and/or a wave on the liquid-free surface that may affect test results |
| Vibration | Vibration can cause waves on the free surface of the liquid that can cause inaccurate test results |
| Noise | Some nonvolumetric test methods are sound-sensitive, and sound vibrations can cause waves to affect volumetric test results. |
| Equipment accuracy | Equipment accuracy can change with the environment (e.g., temperature and pressure) |
| Operator error | The more complicated a test method, the greater the chance for operator error, such as not adequately sealing the tanks if required by the test method in use |
| Type of liquid stored | The physical properties of the liquid (including effects of possible contaminants) can affect the applicability or repeatability of a detection method (e.g., viscosity can affect the sound characteristics of leaks in acoustical leak detection methods). |
| Power vibration | Power vibration can affect instrument readings. |
| Instrumentation limitation | Instruments must be operated within their design range or accuracy will decrease |
| Atmospheric pressure | A change in this parameter has the greatest effect when vapor pockets are in the tank, particularly for leak-rate determination. |
| Tank inclination | The volume change per unit of level change is different in an inclined tank than in a level one |

Source:  Ref. 10.5

These coatings are designed to seal off any potential anodic and cathodic areas of the tank's external surface from the corrosive electrolyte (in this case soil). To be effective these coatings have to have a perfect seal. Utilization of this method of corrosion control requires the exercise of extreme care during installation to avoid scratches or marks in the surface coating.

Another method of corrosion control is the application of cathodic protection. Cathodic protection is a process in which an electrical current is applied uniformly to a metal tank to overcome the current flowing off of it. The amount of current required to protect the entire tank is proportional to the amount of bare metal exposed to the electrolyte (soil). Coating the tank as in the first method, reduces its bare surface area which consequently decreases the amount of current needed to protect any bare metal surfaces. There is at present some confusion surrounding the reliability and efficacy of this method caused by a lack of technical understanding and in part by the exaggerated claims of some vendors. For instance, merely specifying a tank equipped with sacrificial anodes does not guarantee that you will have complete cathodic protection.

There are two basic types of cathodic protection systems, the sacrificial anode system and the impressed current system.

The sacrificial anode system generates a protective current as the anodes corrode (Figure 10.3). Aluminum, magnesium and zinc are common types of metals used for sacrificial anodes. There are some major limitations with the use of sacrificial anode systems. First, their driving potential or voltage is limited. A zinc anode for instance, can only produce a potential of 0.25 volts which may be completely ineffective in a soil with a resistivity greater than 2000 ohm-cm. Likewise, magnesium anodes produce a potential of between 0.10 and 0.40 volts. In soils with a resistivity above 10,000 ohm-cm they may, much like zinc, be similarly ineffective. These low potential voltages are also generally ineffective if they are in electrical contact with other buried metallic objects. Further, these sacrificial anodes often waste much of their mass without generating usable protective cur-

Figure 10.3

# END VIEW

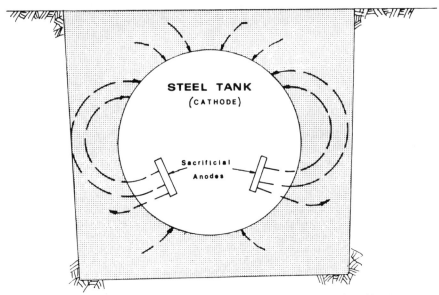

Figure 10.4: This figure illustrates one method of measuring the effectiveness of corrosion protection systems on metal tanks using copper-copper sulfate electrodes. The lower readings at the middle of the tank indicate that the soil is impeding the flow of protective current from the sacrificial anodes. Consequently corrosion in this area is likely.

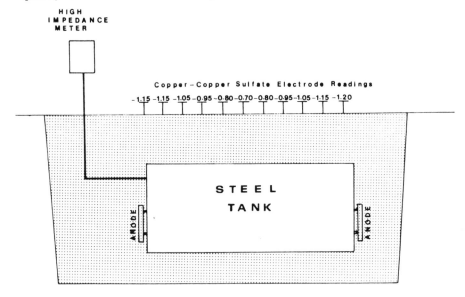

Source for Figures 10.3 and 10.4:  Ref 10.10

rent. Magnesium anodes for instance, have a 50% efficiency while zinc is about 90% efficient in generating
usable protective current.

The impressed current system, the second method
of cathodic protection, distributes current directly
by a transformer-rectifier to buried anodes which are
not designed to be sacrificial. Rather, the rectifier
supplied direct current produces a continuous potential and current to the anodes through the soil
(electrolyte) and onto the protected tank. Since the
anodes no longer have to be sacrificed, they may be
constructed of materials other than zinc, aluminum and
magnesium. Cast iron, graphite, high silicon
chromium, cast iron and platinum have all been used
with the impressed current system. An advantage of
this system is the ability to raise or lower the voltage of the DC power source as site conditions vary
(e.g. soil moisture increases or decreases, road salts
wash into the soil, etc.). Some disadvantages of the
impressed current system are the possible loss of protective current due to a power failure; damage to the
buried tank by excessive voltages, and damage to other
nearby buried metal structures by stray currents.

No single cathodic protection system will fit
all situations. Each site must be carefully evaluated
on a case by case basis. Major environmental factors
affecting the design of cathodic protection systems
include: soil resistivity, soil pH, soil chloride and
sulfate concentrations, the presence of microbiological activity and stray electrical currents. All of
this information should be made available to the reviewing and permitting agency whenever a design for a
cathodic protection system is submitted for review.
If there is still some uncertainty about the final design, you may wish to request an as-built verification
study after the tank has been installed. This can be
accomplished by the use of copper-copper sulfate electrodes placed in contact with the ground (at approximately 5 foot intervals) along the entire length of
the tank and a high impedance volt meter (Figure
10.4). The electrodes in this instance act as one
half cell of a battery while the steel tank and soil
act as the other half. The adequacy of the cathodic
protection system can then be measured against any of
the following three criteria:

### Table 10.4:  Comparison of Containment Systems for Underground Storage Tanks*

| Type of System | Advantages | Disadvantages | Relative Cost |
| --- | --- | --- | --- |
| Clay Liners | If available close to the site, clay provides the least expensive liner; use of clay is a well-established practice and standard testing procedures are available | Subject to drying and cracking; therefore, must be protected with soil cover; subject to leaching of components when exposed to ground water or other solutions; subject to ion exchange when exosed to water containing acids, alkalis, or dissolved salts; subject to destabilization when exposed to some organic solvents | Low |
| Polymeric Liners | Provide well-established solution to problem of containing petroleum products; particularly good for temporary storage; high resistance to bacterial deterioration | Require subgrade preparation and sterilization to reduce risk of puncture; must be protected for damage, particularly that due to vehicular traffic; must be protected from sunlight and ozone; may be attached by hydrocarbon solvents, particularly those with high aromatic content; good oil resistance and good low-temperature properties do not normally go hand in hand | Moderate to high |
| Soil Cement | Is durable; resistant to aging and weathering | Subject to degradation due to frost heaving of subgrade; in-place soil normally used; permeability varies with the type of soil | Moderate |
| Bentonite | Has low permeability; does not deteriorate with age; is self-sealing | Untreated bentonite may deteriorate when exposed to contaminant; requires protective soil cover, typically 18 inches; subject to destabilization when exposed to some organic solvents | Moderate |
| Concrete Vaults | Has good strength and is durable | Requires surface coating to ensure impermeability; subject to cracking when exposed to freeze/thaw cycles | High |
| Double-Walled Tanks | Constructed of material (FRP or coated steel) that is resistant to the stored product and to external corrosion; tank design includes leak-detection system | Some models only available in tank sizes up to 4,000 gal. | High |

* Data from State of New York, 1983.

Source:  Ref. 10.5

1.    A  potential  of  at  least  -0.85  volts
measured between the surface of the tank and a copper-
copper sulfate electrode in contact with the soil.

This 0.85 volt criteria is based upon the observation that the anodic areas on a steel structure are
usually never more negative then -0.80 volts with respect to copper-copper sulfate.  By applying a protective  current  with  a  potential  of  -0.85  or  greater
along the entire tank surface, the potential anodic
sites are converted to cathodic sites.  Consequently,
corrosion in theory at least, is not possible.  Please
note that certain areas along the tank surface shown
in Figure 10.4 show a reading of less than -0.85.
This indicates that the cathodic protection system is
not providing sufficient protection to these areas.

2.    A negative potential shift of at least 300
mV measured between the surface of the tank and a
copper-copper sulfate electrode in contact with the
soil.

3.    A negative polarization potential shift of
at least 100 mV between the surface of the tank and a
copper-copper sulfate electrode in contact with the
soil[8].

It is recommended that all three criteria be
measured during any as-built verification study.

An additional safeguard that should be included
in the design of every buried tank is a spill containment system.  While the materials used for the construction of these systems varies considerably (Table
10.4) they all function in a similar manner--by encap-
sulating the tank and isolating it and its contents
from the surrounding environment.

## BEHAVIORAL CHARACTERISTICS OF CONTAMINANTS
## INTO SOIL AND WATER

It will be some time before all of the nation's
leaking tanks are removed or replaced.  In the interim, leaks from underground tanks will continue to
pollute the nation's ground water.  We must, therefore, acquaint ourselves with some of the behavioral
characteristics of leaks into soil and water.  By doing so, we will be better prepared to assess the adequacy of any proposed monitoring or remediation program submitted for our review and approval.

The extent to which a leak from an underground storage tank migrates depends largely on 3 factors: the quantity of liquid released; the physical properties of the substance; and the structure of the soil(s) or rock through which the material or pollutant passes. Figure 10.5 shows a typical subsurface environment beneath an underground storage tank. Note in particular, the presence of an unsaturated zone (called the vadose zone) and a saturated zone (the top of which is the ground water table). In between the two zones lies the capillary fringe zone where water is drawn up into the unsaturated zone from the water table by capillarity.

When a liquid contaminant is released from an underground tank lacking any secondary contaminant it begins to percolate downward by gravity through the vadose zone toward the water table[9]. If the quantity of material leaked is small, it may remain within the interstices (pores) of the soil in the vadose zone until such time as percolating rainwaters flush or carry it deeper. A large release, on the other hand, may flow immediately through the vadose zone to the top of the water table.

It is at the capillary zone that we see the first significant change in direction of movement of the migrating contaminant. Here, the primary movement is in a lateral direction[10] following the slope (hydraulic gradient) of the ground water table (Figure 10.6). This lateral flow is frequently referred to as a plume. If the slope of the water table is steep, the plume will usually be narrow in width. A flatter slope will conversely result in a wider plume. An idealized version of a hydrocarbon leak into soil is shown in Figure 10.7. We know, however, that this is usually not representative of the real world situation.

Along with these complex variations in the upper surface of the lithosphere, we must also consider the characteristics of the contaminant. Does it dissolve readily in water? What about its specific gravity? Does it volatilize? All of these factors will affect the ultimate migration path of the contaminant through the soil and ground water. Immiscible substances that have a specific gravity less than 1.0 (lighter than water) will usually accumulate near the top of the saturated zone. Immiscible substances with a specific

Figure 10.5: Typical subsurface environment beneath an underground storage tank.

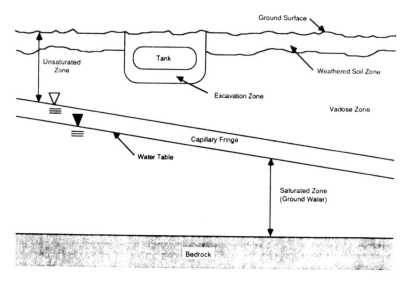

Figure 10.6: Typical flow pattern for hydrocarbons in the capillary zone.

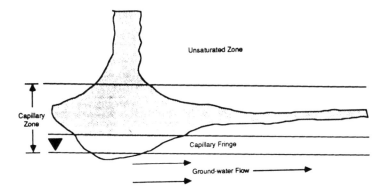

Source for Figures 10.5 and 10.6:  Ref. 10.5

Figure 10.7: Idealized version of a hydrocarbon leak into soil.

BEHAVIOR OF HYDROCARBONS IN SOIL FOLLOWING AN INITIAL
HIGH VOLUME RELEASE.

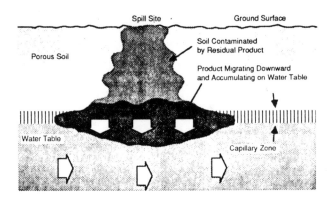

BEHAVIOR OF HYDROCARBONS AFTER RELEASE HAS STABILIZED

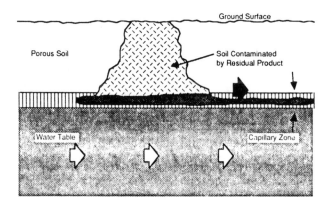

Source:  Ref. 10.5

gravity greater than 1.0 (heavier than water) will usually be found deeper in the saturated zone.

We must also consider whether the contaminant has a vapor or gaseous phase, since vapors can migrate both vertically and horizontally (depending on the specific vapor pressure of the material) and infiltrate storm sewers, sanitary sewers, basements and even wells. Such vapor phases usually begin when the contaminant enters the unsaturated zone. The production and migration of vapors from underground storage tank leaks in the United States is well documented. Vapors less dense than air will migrate vertically while vapors more dense than air usually move horizontally paralleling the path of the liquid phase. Consequently, you can see that cleanup or remediation of a site may include not only ground water cleanup but purging or cleansing of the soil in the unsaturated zone of absorbed residual liquids and vapors as well.

## POST LEAK ACTIONS

When confronted with the confirmation of an underground leak, or at least the strong suspicion of one, how do we respond? Ideally, there will be a two-tiered response with the possibility that both tiers may be carried out simultaneously. The initial phase termed the short term response phase will be concerned largely with:

1.    The accurate identification of the leaking material.

2.    A determination of possible migration pathways and potential receptors, and,

3.    Attempts to quickly terminate the leak at its source.

Item 3 in this group will usually receive priority attention. Typical actions undertaken in this phase include:

1.    Removal of the remaining product from the leaking tank, assuming the responsible tank can be readily identified.

2.    Inspection, testing, repair or complete removal of the suspect tank.

3.    Physical and visual examination of the soil surrounding the tank by backhoe either before or after the tank is removed.

4.    Removal of any pooled product from beneath the tank.

5.    Removal of any soil surrounding the tank that appears visually to be contaminated.

6.    Collection of soil or ground water samples from the tank pit once the tank is removed.

In addition, some investigation will be made as to what general geological conditions exist beneath the leaking tank and the surrounding area. This information will be gathered from existing sources such as: well logs for wells drilled previously in the area, from general or special geological reports and publications; and from inferences from the surface topography. Existing wells in the area will also be utilized, if access can be gained, for the collection of ground water samples and the measurement of depths to the ground water table to determine the approximate direction of flow of the ground water.

The second phase called the long term phase is a more detailed examination of the effects of the leakage with the emphasis placed on assessing the actual aereal extent of the pollution, the potential hazards to human health and safety, and the evaluation of the appropriateness of various methods of remediation. Generally, information collected in this phase will fall within the following three categories:  hydrogeological data; geographic and topographic data; and, land, water and demographic data. Information typically collected for each category is listed below:

Hydrogeological Data:

1.   Soil profiles
2.   Soil physical properties
3.   Soil chemistries
4.   Depth to bedrock
5.   Depth to ground water
6.   Physical properties of the aquifer
7.   Ground water flow rate
8.   Ground water flow direction and ground water contour map
9.   Recharge areas
10.  Recharge rates
11.  Aquifer characteristics
12.  Ambient ground water quality

Geographic and Topographic Data:

1.   Precipitation
2.   Evapotranspiration rate

3.   Topography
4.   Accessibility
5.   Site size
6.   Proximity to surface water courses
7.   Proximity to human interfaces

<u>Land, Water and Demographic Data</u>:

1.   Current water use patterns
2.   Future water use patterns
3.   Current land use patterns
4.   Growth projections
5.   Population currently affected
6.   Alternative water supplies
7.   Ground water quality standards to be met
8.   Existing private or public wells affected or threatened
9.   Land available for implementing corrective action
10.  Location of public utilities such as sanitary sewers, storm sewers and electric lines
11.  Concentrations of product in the ground water and soil including vapors

Because data collection is much more extensive in this phase the existing sources utilized in phase 1 may be inadequate for phase 2.  Therefore, it is likely that a series of new monitoring wells will have to be installed and used for supplemental data collection.  Their placement must be determined on a case by case basis.  In almost every instance, however, at least one of those wells will be located at a point upstream of the leaking tank to measure ambient or background water quality.  Furthermore, in areas of complex geology, these wells may be placed in clusters to assess various levels of the saturated zone (Figures 10.8 and 10.9).

Other geophysical equipment may also be used in this phase to measure such things as electrical resistivity and electromagnetic conductivity--both of which are useful in defining plumes of leakage, particularly of inorganic liquids.  In addition, soil vapor concentrations may be measured by one of the following four methods described below (Ref. 10.5):

Figure 10.8: Typical well cluster configuration.

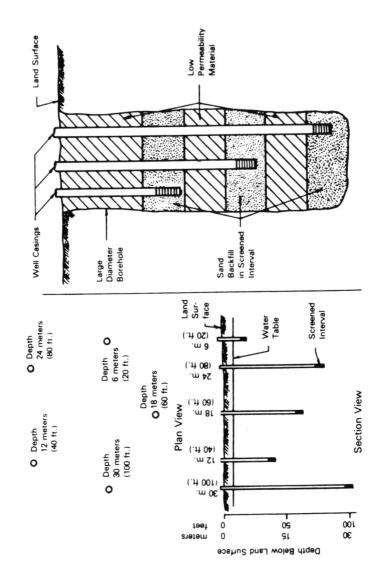

Source:   Ref. 10.9

Figure 10.9:  Typical monitoring well screened over a single vertical interval.

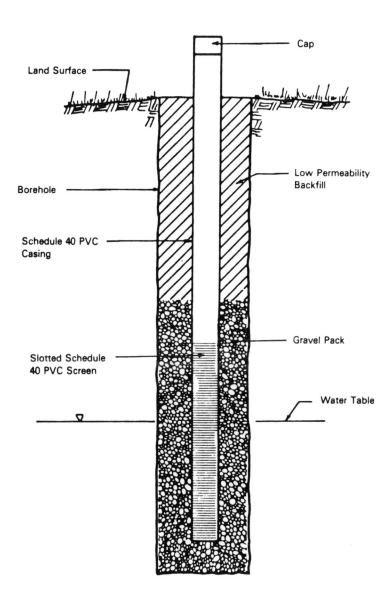

Source:  Ref. 10.9

## Surface Flux Chambers

The surface flux chamber is an enclosed device used to sample the gaseous emissions from a defined area on the ground surface. Clean, dry, sweep air is added to the chamber at a controlled and measured rate, and the concentration of the species of interest is measured at the exit of the chamber. Analysis is by portable gas analyzer or GC.

## Downhole Flux Chambers

The downhole flux chamber operates on the same principle as the surface flux chamber. It is either driven into the ground or placed down an augured hole. Sweep air is added, and the exit gas concentration is monitored. Analysis is by portable gas analyzer or GC. The downhole flux chamber is more sensitive than the surface flux chamber because sampling occurs at depths below any surface interferons and where soil gas concentrations are normally at higher levels.

## Accumulator Devices

Use of the accumulator devices involves collecting and concentrating soil gases to provide a time-integrated sample. One variation is to collect gases on activated charcoal bonded to a curie-point wire. Analysis is by mass spectrometry. A second variation is to pump gases from an enclosure or ground probe and then collect the gas in an absorbent. The primary advantage of this technique is that the sampling time can be adjusted to reach the desired sensitivity.

## Ground Probes

The use of ground probes involves placing a tube into the ground to the desired sampling depth. Openings in the tube near the leading edge allow the entry of soil gases. A port in the upper end of the tube allows sample gas to be extracted. Analysis is typically by GC. The technique is relatively sensitive, and sampling depth can be varied to reach below interferents or low-concentration areas.

## REMEDIATION TECHNOLOGIES

The number of remedial technologies available to effectuate a cleanup of a leak are far too numerous to cover in detail in this one chapter.  I would, therefore, direct you to the references cited at the end of this chapter for further information on available corrective technologies.  In the interim, Tables 10.5 through 10.7 may provide you with some initial general information on available technologies and their possible applications.

## SPECIAL CONSIDERATIONS

### 1.    Contaminated Soil Disposal

As more leaking tanks are removed, the volume of excavated and contaminated soil grows larger (Figure 10.10) and there is a question as to whether all of this soil can be accommodated in the nation's dwindling number of hazardous waste and sanitary landfills--the most common method of disposal at present.  Consequently, alternatives to landfilling should be considered wherever possible.  Alternatives to landfilling presently include:

1. Water washing or flushing
2. Solvent extraction
3. Vapor extraction
4. Drying
5. Incineration
6. Solidification
7. Encapsulation

The efficacy of each of these alternatives continues to be evaluated, particularly by the U.S. EPA.  Some recent data already available (Ref. 10.11) provides some insight on at least three of these methodologies that should be of some interest to land planners.

### Water washing or flushing

Water washing with extractive agents is applicable for cleaning nonvolatile hydrophilic and hydrophobic organics and heavy metals from soil.  In cleaning soil by this method, large objects are removed from the excavated soil by screening.  The screened soil is then mixed thoroughly with water solutions of any of a number of extraction agents (e.g. caustic, lime,

Table 10.5:  Initial Response Corrective Action Options

| Situation | Tank repair/removal | Free product recovery | Ground-water recovery and treatment | Subsurface barriers | Soil excavation | Vapor migration control and collection | Sediment removal | Surface water diversion drainage | Alternative/treatment central water supply | Alternative/treatment point-of-use water supply | Restoration of utility, water, and sewer lines | Evacuation of nearby residents | Restricted egress/ingress |
|---|---|---|---|---|---|---|---|---|---|---|---|---|---|
| **Ground-water contamination** | | | | | | | | | | | | | |
| Existing public or private wells | • | • | • | • | | | | | | | | | |
| Potential future source of water supply | • | • | • | • | | | | | | | | | |
| Hydrologic connection to surface water | • | • | • | • | • | | | | | | | | |
| **Soil contamination** | | | | | | | | | | | | | |
| Potential for direct human contact: nuisance or health hazard | • | | | | • | | • | • | | | | | |
| Agricultural use | | | | | | | | | | | | | |
| Potential source of future releases to ground water | • | • | • | • | • | • | • | • | | | | | |
| **Surface-water contamination** | | | | | | | | | | | | | |
| Drinking water supply | • | | | | | | | • | • | • | • | | |
| Source or irrigation water | • | | | | | | | • | • | • | • | | |
| Water-contact recreation | • | | | | | | | • | | | | | • |
| Commercial or sport fishing | • | | | | | | | | | | | | • |
| Ecological habitat | • | | | | | | | | | | | | |
| **Other hazards** | | | | | | | | | | | | | |
| Danger of fire or explosion | • | | | | | • | | | | | | • | • |
| Property damage to nearby dwellings | | | | | | • | | | | | | • | • |
| Vapors in dwellings | • | | | | | • | | | | | | • | • |

Source:  Ref. 10.5

## Table 10.6:  Permanent Corrective Action Options

| | Removal/excavation of soil and sediments | Onsite and offsite treatment and disposal of contaminants | Free product recovery | Ground-water recovery systems | Subsurface barriers | In situ treatment | Ground-water treatment | Vapor migration control, collection, and treatment | Surface water/drainage controls | Restoration of contaminated water supplies and sewerlines |
|---|---|---|---|---|---|---|---|---|---|---|
| | | | | Corrective action categories | | | | | | |
| Volatilization of chemicals into air | | | | | | | | ● | | |
| Hazardous particulates released to atmosphere | ● | | | | | | | ● | | |
| Dust generation by heavy construction or other site activities | | | | | | | | | | |
| Contaminated site run-off | ● | | | | | | ● | | ● | |
| Erosion of surface by water | | | | | | | | | ● | |
| Surface seepage of released substance | ● | | ● | | | | | | ● | |
| Flood hazard or contact of surface water body with released substance | ● | | | | | | | | ● | |
| Released substance migrating vertically or horizontally | | | ● | ● | ● | | | | | |
| High water table which may result in ground-water contamination or interfere with other corrective action | | | | ● | ● | | | | | |
| Precipitation infiltrating site and accelerating released substance migration | | | | | | ● | | | ● | |
| Explosive or toxic vapors migrating laterally underground | | | | | ● | | | ● | | |
| Contaminated surface water, ground water, or other aqueous or liquid waste | ● | ● | | ● | ● | ● | ● | | | |
| Contaminated soils | ● | ● | | | | ● | | | | |
| Toxic and/or explosive vapors that have been collected | | ● | | | | | | ● | | |
| Contaminated stream banks and sediments | ● | ● | | | | | | | | |
| Contaminated drinking water distribution system | | ● | | ● | ● | | ● | | | ● |
| Contaminated utilities | | | | | | | | ● | | ● |
| Free product in ground water and soils | ● | ● | ● | ● | ● | ● | ● | | | |

Source:  Ref. 10.5

Table 10.7:  Potential Corrective Action Technologies Relative to Release
Volume and Chemical Characteristics

| Technology | Small- to moderate-volume recent gasoline/petroleum release (gas station/tank farms) | Large-volume or long-term chronic gasoline/petroleum release (gas station/tank farms) | Release from tanks containing hazardous substances (organic) | Release from tanks containing hazardous substances (inorganic) |
|---|:---:|:---:|:---:|:---:|
| **Removal/excavation of tank, soil, and sediment** | | | | |
| Tank removal | • | • | • | • |
| Soil excavation | • | • | • | • |
| Sediment removal | | • | • | • |
| **Onsite and offsite treatment and disposal of contaminants** | | | | |
| Solidification/stabilization | | • | • | • |
| Landfilling | | | | |
| Landfarming | • | • | • | • |
| Soil washing | | | | • |
| Thermal destruction | | • | • | |
| Aqueous waste treatment | • | • | • | • |
| Deep well injection | | | | |
| **Free product recovery** | | | | |
| Dual pump systems | • | • | • | • |
| Floating filter pumps | • | • | • | • |
| Surface oil/water separators | • | • | • | |
| **Ground-water recovery systems** | | | | |
| Ground-water pumping | • | • | • | • |
| Subsurface drains | • | • | • | • |
| **Subsurface barriers** | | | | |
| Slurry walls | | • | • | • |
| Grouting | | • | • | • |
| Sheet piles | | | | |
| Hydraulic barriers | • | • | • | • |
| **In situ treatment** | | | | |
| Chemical treatment | | | | |
| Physical treatment | | | | |
| Soil flushing | • | • | • | • |
| Biostimulation | • | • | • | • |
| **Ground-water treatment** | | | | |
| Air stripping | • | • | • | |
| Carbon adsorption | • | • | • | |
| Biological treatment | | | | |
| Precipitation/flocculation/sedimentation | • | • | • | |
| Dissolved air flotation | | | | • |
| Granular media filtration | | | | • |
| Ion exchange/resin adsorption | | | | • |
| Oxidation/reduction | | | | • |
| Neutralization | | | | • |
| Steam stripping | | | • | |
| Reverse osmosis | | | | • |
| Sludge dewatering | | | | • |
| **Vapor migration control, collection, and treatment** | | | | |
| Passive collection systems | | • | • | |
| Active collection systems | | • | • | |
| Ventilation of structures | • | • | • | |
| Adsorption | | | • | |
| Flaring | | | | |
| **Surface water/drainage controls** | | | | |
| Diversion/collection systems | • | • | • | • |
| Grading | • | • | • | • |
| Capping | | • | • | • |
| Revegetation | • | • | • | • |
| **Restoration of contaminated water supplies and sewer lines** | | | | |
| Alternative central water supplies | | • | • | • |
| Alternative point-of-use water supplies | • | • | • | • |
| Treatment of central water supplies | | • | • | • |
| Treatment of point-of-use water supplies | | • | • | • |
| Replacement of water and sewer lines | | • | • | • |
| Cleaning/restoration of water and sewer lines | • | • | • | • |

Source:  Ref. 10.5

Figure 10.10: A common sight at American gas stations. Millions of tons of contaminated soil excavated during the removal of leaking underground storage tanks pose additional solid waste disposal problems for America's land planners. With landfill space dwindling rapidly, alternate technologies that could cleanse the soil for later reuse need to be explored.

slaked lime, alkali-based washing compounds, sulfuric, nitric, phosphoric or carbonic acids, surfactants or chelating agents) to remove the contaminants from the soil.  Oxidizing agents such as hydrogen peroxide or sodium hypochlorite are also sometimes added to facilitate the washing process.  This mixing is followed by an initial solid/liquid separation where the coarser fraction of the soil is separated.  The extracted solution with the major quantity of contaminants and the smaller soil particles (clay and fine silt) undergoes further solid/liquid separation where the remaining fine soil fractions are further separated.  The extraction liquid is then cleaned and recycled.

Solvent extraction
        This method can be used to clean soil contaminated with concentrations of nonvolatile hydrophobic organics.  The type of solvent selected to flush the soil will depend primarily on the chemical structure of the contaminant and soil type.  Leaching and immersion are the two general extraction techniques.  Similar to the water washing method leaching is accomplished as a batch process.  The screened soil is placed in a tank and solvent is sprayed over it.  The solvent then leaches the contaminant from the soil. This leaching process is generally not viable for low solubility contaminants and fine clay and silt soils because of the slow mass transfer rates.  In these cases the soil must be dispersed in the solvent.  This is usually accomplished by depositing the soil in an agitated tank filled with solvent.  When the extraction equilibrium is reached the agitation is stopped and the solids allowed to settle.  For coarse, sandy soils the solvent is separated by a gravity drain. For finer soils the additional step of centrifugation or filtration may be required.  Any residual solvent in the separated soils is removed by gas, steam or vapor stripping.  Contaminants in the solvent are generally removed by distillation based on differences in boiling points.

Vapor extraction
        Vapor extraction is generally used to remove volatile organic compounds (VOCs).  To strip VOCs from the soil they must be vaporized.  Air and steam are the most commonly used stripping gases.  This strip-

ping can be done at ambient temperatures or after the addition of heat. When steam is used as a stripping medium, the steam can be removed by condensation thereby producing a relatively concentrated vapor of VOCs.

Vapor extraction can be employed with either in place soils or excavated soils. It should also be noted that any system designed to dry soils can also strip VOCs.

## FOOTNOTES

## CHAPTER 10

(1)    As a result of this latter language even above ground tanks with extensive underground piping could be regulated.

(2)    Where the soil resistivity at the installation site is greater than 12,000 ohm-cm, a storage tank without corrosion protection can be installled.

(3)    These notification provisions are of particular importance to local land planners because this notification is to be made to designated state agencies, thereby encouraging a more localized information gathering process.

(4)    Extensive ground water and soil contamination at some sites have boosted the price to between $10,000,000 and $25,000,000.

(5)    May vary with manufacturer.

(6)    The National Fire Protection Association's (NFPA) Uniform Fire Code defines corrosive soils as those having a resistivity of 14,000 ohm-cm or less. See Footnote (2).

(7)    This author does not recommend the installation of underground tanks where a high water table exists unless provisions have been made to preclude the ground water from the site completely.

(8)    The 100 mV polarization shift is determined by interrupting the flow of current between the steel tank and its anodes. The potential of the steel tank is measured immediately after

the current is interrupted. The measured potential will decay as the polarizing film created by chemical reactions at the metal surface slowly dissipates. The potential is again noted after the readings have stopped decreasing. If the difference between the two measured potentials exceeds 100 mV, the tank is considered cathodically protected (Ref. 10.1)

(9)     You will recall my earlier recommendation to refrain from placing underground tanks in areas with a high ground water table since a leak can introduce contaminants directly into the ground water.

(10)    Some multi-directional lateral spreading can also occur if the downward, vertical flow of contaminants exceeds the lateral, downstream contaminant flow.

### REFERENCES

### CHAPTER 10

10.1    Cheremisinoff, P.N., J.G. Casana, and R.P. Ouellette, "Special Report: Underground Storage Tank Control", Pollution Engineering, February, 1986.

10.2    Cheremisinoff, P.N., J.G. Casana, and H.W. Pritchard, "Special Report: Update on Underground Tanks", Pollution Engineering, August, 1986.

10.3    Woods, Paul H., Jr., and Dale E. Webster, "Underground Storage Tanks: Problems, Technology and Trends", Pollution Engineering, July, 1984.

10.4    United States Environmental Protection Agency, "Project Summary, Underground Tank Leak Detection Methods: A State-of-the-Art Review", EPA/600/S2-86/001, July, 1986.

10.5    United States Environmental Protection Agency, "Underground Storage Tank Corrective Action Technologies", EPA/625/6-87-015, January, 1987.

10.6    Petroleum Equipment Institute, "Recommended Practices For Installation of Underground Liquid Storage Systems", 1990.

10.7    Cheremisinoff, Paul N., John G. Casana and Robert P. Ouellette, "Underground Storage Tanks Guidebook", Pudvan Publishing Co., Northbrook, Illinois, 1987.

10.8    Niaki, Shahzad and John A. Broscious, "Underground Tank Leak Detection Methods", Noyes Publications, Park Ridge, N.J., 1987.

10.9    United States Environmental Protection Agency, "Handbook, Remedial Action at Waste Disposal Sites", EPA-625/6-82-006, June, 1982.

10.10   Cignatta, John V., "UST's and Corrosion Science", Pollution Engineering, February, 1987.

10.11   United States Environmental Protection Agency, Raghavan, R., E. Coles and D. Dietz, Project Summary, "Cleaning Excavated Soils Using Extraction Agents: A State-of-the-Art Review", EPA/600/S2-89/034, Jan., 1990.

# 11

---

# Solid Waste Disposal

---

## HISTORICAL PERSPECTIVE

By the time this text is in your hands another significant number of American politicians will have won or lost election to public office based solely on their platform regarding the topic covered in this chapter--namely solid waste disposal. Public emotion on the issue has been so fervent, in fact, that in some instances public officials have been threatened with bodily harm and even death by their constituents. What has provoked the American people to such passion is the nation's deteriorating solid waste disposal infrastructure and the lack of confidence in the remedies being offered by the agencies empowered to correct it.

Americans generate a staggering eleven billion plus tons of solid waste a year[1] which includes 7.6 billion tons of industrial non-hazardous waste; 136 million tons of hazardous waste; 160 million tons of municipal waste and 7.5 million tons of sewage treatment plant sludges.

Landfilling (originally land dumping) remains the dominant form of solid waste disposal in the United States. Ninety percent of America's municipal solid waste is landfilled and about 5% is incinerated. Of the estimated 10,000 landfills still operating in the United States (6,000 of which are municipal) only 15 percent have some kind of containment system or liner to help prevent contaminants (called leachate) from migrating away from the landfill and polluting ground and surface waters. Existing landfills are being closed on the average of one per day and the diminution of landfill space is most acute in the nation's metropolitan areas and largest cities.

Philadelphia, Pennsylvania, for example, which had it's last major landfill closed in 1984 now trucks some of its garbage to Dundalk, Maryland, over 100 miles away. The city of San Francisco is paying $2.5 million just to reserve future dumping space in Solano County, about 40 miles away. New Jersey, which ten years ago had about 400 landfills now has only 90 and must export 50 to 60 percent of its garbage as far away as West Virginia, Michigan, Ohio and Kentucky. It's neighbor, New York, predicts that all of its landfill space will be exhausted by the year 2000.

Out of state shipments of solid waste, however, may be a short-lived solution. Recipient states like Alabama, Pennsylvania and Ohio have began to seriously challenge the importation of solid and hazardous waste into landfills within their borders (Ref. 11.33). Alabama for instance passed legislation in 1989 banning hazardous waste shipments into that state from states without treatment or disposal facilities of their own. This was the result of a significant increase in the amount of hazardous waste shipped to its Emelle landfill between 1978 and 1988 (from 200 million pounds in 1978 to 1.1 billion pounds in 1988). The governor of Pennsylvania similarly alarmed by the excessive importation of out of state solid waste issued an executive order on October 17, 1989 banning all new landfill construction. Many more states are expected to adopt this parochial attitude toward out of state trash.

Obviously, the present solid waste disposal crisis did not originate overnight. Those of you who grew up in one of America's rural communities in the mid-1940's and early 1950's can probably recall the accepted solid waste disposal practices of that era. Household waste, old furniture, tires and countless other items were simply heaved over the nearest stonewall or embankment. Pits, sinkholes and other surface depressions likewise became made to order receptacles for trash. Whatever wasn't tossed, was usually burned, sometimes in a barrel but more frequently on the bare ground.

Our urban neighbors often lacking the luxury of plentiful amounts of open space were a bit more frugal in their methods of disposal and often required waste to be dumped only in specific areas usually referred to as town dumps (not to be confused with the more

sophisticated sanitary landfill). Old mining sites,
rock quarries and "useless" wetlands and swamps along
surface waterbodies became favored sites for such
dumps. Left uncovered and exposed to the elements,
these mounds (sometimes mountains) of concentrated
refuse filled the air with noxious odors and became
havens for an assortment of pests including rats,
feral dogs, cockroaches and flies. More important,
however, was the production of a highly polluted
liquid called "leachate" that was formed when rainwater percolated through the mounded and rotting
garbage. Seeping into the ground beneath the discarded refuse and eventually into the ground water, it
was initially an unrecognized threat. Flowing into
the surface waterbodies, however, its presence was
more evident, exhibiting its effects in fish kills and
degradation of the stream bottom. American industry
parroting these government sanctioned practices
created dumps of their own or convinced municipal
officials to accept their wastes at the local dump.
Regrettably, in either case, this indiscriminate disposal of industrial waste added an assortment of toxic
and hazardous materials that significantly enhanced
the toxicity of the resulting leachate and complicated
the problems that would be associated with leachate
migration and today's cleanup operations.

Some cities supplemented dumping with incineration utilizing crude incinerators that lacked little
if any air pollution control devices and emitted noxious plumes of dark smoke and particulates through
their tall smokestacks.

As little as 30 years ago, those responsible for
solid waste disposal would have been primarily representatives of local government. Not so today. Now,
citizen action groups, environmental organizations,
county, state and federal governments, legislators and
a variety of industries and their lobbying organizations all participate. Some, like the federal government, have arrived by legislative mandate, but most
others have joined out of fear or concern or both.
Opinions on solutions vary considerably amongst these
newest participants, with some describing their co-
participant's position as irrational, inflexible,
polarized or even fanatical. Physical confrontations
have occurred on more than one occasion. What has
evoked such strong reactions and differences of opin-

ions to a problem that as a nation we hardly cared
about 30 years ago? The answer, as with most of our
environmental problems, is complex. First, we have
nearly twice as many Americans as we did in 1940. As
a consequence, the amount of solid waste we produce
has increased significantly and little planning was
done to prepare for it (Table 11.1). Simultaneously,
the amount of solid waste generated by each American
has also risen, from 2.65 pounds per day in 1960 to
3.58 pounds per day in 1986 (Ref. 11.1).

These increases in population have not been uniformly dispersed, but have been clustered around existing metropolitan centers (many near coastal areas
as we shall see in Chapter 13), thereby accelerating
the depletion of existing solid waste disposal infrastructure (primarily landfills, dumps and incinerators). Thirdly, and I think most importantly, time
and technological advancements have revealed the
hazards of our previous indiscriminate disposal
practices.

As a result, we must now divert present
resources away from current solid waste disposal
problems to try to restore thousands of old dump sites
which, if left unremedied, will continue to threaten
our health and the health of future generations of
Americans. Unfortunately, the initial governmental
responses to these discoveries have done little to
persuade public opinion that the nation's regulatory
agencies are as adroit as they profess to be, either
in environmental protection or environmental
remediation. Just how expensive will these remedies
be? Certainly in the hundreds of billions of dollars.
For example, the costs to remedy the six acre Lipari
landfill in New Jersey (a Superfund site) which
accepted liquid and semi-solid chemical waste from
1958 through 1971 will be an estimated $40 million.
The 220 acre Kin-Buc landfill, another federal Superfund site, also in New Jersey and the recipient of 70
million gallons of liquid hazardous waste has cost
estimates for remediation that are even more
staggering. To excavate all of the waste on this site
and incinerate it, would take an estimated 35 years
and cost in excess of four billion dollars. These
billions of dollars may not, however, produce the type
of permanent remedies most Americans envision or
desire. This has led to a major schism between the

## Table 11.1: Materials Discarded Into the Municipal Waste Stream, 1960 to 2000 (in millions of tons)

| Materials | 1960 | 1965 | 1970 | 1975 | 1980 | 1981 | 1982 | 1983 | 1984 | 1985 | 1986 | 1990 | 1995 | 2000 |
|---|---|---|---|---|---|---|---|---|---|---|---|---|---|---|
| Paper and Paperboard | 24.5 | 32.2 | 36.5 | 34.4 | 42.0 | 43.6 | 41.4 | 45.8 | 49.4 | 48.7 | 50.1 | 54.9 | 60.2 | 66.0 |
| Glass | 6.4 | 8.5 | 12.5 | 13.2 | 14.2 | 14.3 | 13.8 | 13.3 | 12.8 | 12.2 | 11.8 | 12.3 | 12.2 | 12.0 |
| Metals | | | | | | | | | | | | | | |
| Ferrous | 9.9 | 10.0 | 12.4 | 12.0 | 11.2 | 11.1 | 11.0 | 11.1 | 11.0 | 10.4 | 10.6 | 11.1 | 11.3 | 11.3 |
| Aluminum | 0.4 | 0.5 | 0.8 | 1.0 | 1.4 | 1.4 | 1.3 | 1.5 | 1.5 | 1.6 | 1.7 | 2.0 | 2.4 | 2.7 |
| Other Nonferrous | 0.2 | 0.2 | 0.3 | 0.3 | 0.4 | 0.4 | 0.3 | 0.3 | 0.3 | 0.3 | 0.3 | 0.3 | 0.3 | 0.4 |
| Plastics | 0.4 | 1.4 | 3.0 | 4.4 | 7.6 | 7.8 | 8.4 | 9.1 | 9.6 | 9.7 | 10.3 | 11.8 | 13.7 | 15.6 |
| Rubber and leather | 1.7 | 2.2 | 3.0 | 3.7 | 4.1 | 4.1 | 3.8 | 3.4 | 3.3 | 3.4 | 3.9 | 3.5 | 3.6 | 3.8 |
| Textiles | 1.7 | 1.9 | 2.0 | 2.2 | 2.6 | 3.4 | 2.8 | 2.8 | 2.8 | 2.8 | 2.8 | 3.0 | 3.1 | 3.3 |
| Wood | 3.0 | 3.5 | 4.0 | 4.4 | 4.9 | 4.4 | 5.0 | 5.2 | 5.1 | 5.4 | 5.8 | 5.3 | 5.7 | 6.1 |
| Other | 0.0 | 0.0 | 0.1 | 0.1 | 0.1 | 0.1 | 0.1 | 0.1 | 0.1 | 0.1 | 0.1 | 0.1 | 0.1 | 0.1 |
| TOTAL NONFOOD PRODUCT WASTES | 48.2 | 60.5 | 74.7 | 75.6 | 88.6 | 90.5 | 87.8 | 92.6 | 95.9 | 94.5 | 97.4 | 104.2 | 112.5 | 121.3 |
| Food Wastes | 12.2 | 12.4 | 12.8 | 13.4 | 11.9 | 12.1 | 12.0 | 12.0 | 12.2 | 12.3 | 12.5 | 12.5 | 12.4 | 12.3 |
| Yard Wastes | 20.0 | 21.6 | 23.2 | 25.2 | 26.5 | 26.7 | 27.0 | 27.5 | 27.8 | 28.0 | 28.3 | 29.5 | 31.0 | 32.0 |
| Miscellaneous Inorganic Wastes | 1.3 | 1.6 | 1.8 | 2.0 | 2.2 | 2.3 | 2.4 | 2.4 | 2.4 | 2.5 | 2.6 | 2.8 | 3.0 | 3.2 |
| TOTAL WASTES DISCARDED* | 81.7 | 96.1 | 112.5 | 116.2 | 129.2 | 131.6 | 129.1 | 134.5 | 138.3 | 137.3 | 140.8 | 149.0 | 158.9 | 168.8 |
| ENERGY RECOVERY** | 0.0 | 0.2 | 0.4 | 0.7 | 2.7 | 2.3 | 3.5 | 5.0 | 6.5 | 7.6 | 9.6 | 13.3 | 22.5 | 32.0 |
| NET WASTES DISCARDED | 81.7 | 95.9 | 112.1 | 115.5 | 126.5 | 129.3 | 125.6 | 129.5 | 131.8 | 129.7 | 131.2 | 135.7 | 136.4 | 136.8 |

*Wastes discarded after materials recovery has taken place.
**Municipal solid waste consumed for energy recovery. Does not include residues.

Details may not add to totals due to rounding.

Source: Adapted from Ref. 11.2

agencies charged with carrying out the cleanups and the other numerous participants. The American public wants remediation schemes that restore sites to pre-dump conditions--a formidable task when one considers the difficulties that must be faced to accomplish this. Even such simple technology as removing the obviously dumped materials and a portion of the visually discolored soil beneath them is difficult to carry out thanks to a paucity of secured (lined) landfills into which these materials may be re-interred. Furthermore, if we choose to burn it, there is a similar dearth of approved hazardous waste incinerators plus the fact that we would still have an ash residue (albeit much less in volume) that would face the same lack of secured landfill space[2].

The prior century's wastes are not the only competitor for the nation's remaining landfill space. Millions of tons of contaminated soil being removed under the auspices of state and federal leaking underground storage tank programs are also being directed to the nation's remaining landfills, as are increasing amounts of municipal sewage treatment plant sludges generated by the federal Clean Water Act's demands for improved removal of contaminants from the nation's wastewaters.

We now have the capability to detect the presence of some chemical substances in the air, soil and water at these old dump sites down to levels as low as a part per trillion, and most recently, parts per quadrillion. The public, therefore, expects that we can predict with similar preciseness the levels at which absolutely no harm to human health or the environment will occur. These levels would subsequently become the standards which any cleanup would have to meet. When such levels cannot be provided the alternative often becomes no detectable levels of anything, thereby fomenting further dissension between the regulatory agencies and the public.

Additionally, our continuing proclivity to correct environmental problems by the mass production of environmental legislation inundates the American public with more unfamiliar terminologies and technologies plus an ever-expanding agenda of objectives. Nowhere is this continuous expansion of objectives better exemplified than in the 1965 federal Solid Waste Act and its successors, all of which constitute

the nation's primary federal legislation regarding
solid waste disposal.

## Purposes--1965 Solid Waste Disposal Act:

"(1)      to initiate and accelerate a national
research and development program for new and improved
methods of proper and economic solid-waste disposal
including studies directed toward the conservation of
natural resources by reducing the amount of waste and
unsalvageable materials and by recovery and utilization of potential resources in solid wastes; and

(2)      to provide technical and financial
assistance to State and local governments and interstate agencies in the planning, development and
conduct of solid-waste disposal programs."

## Purposes--1970 Resource Recovery Act:

"(1)      to promote the demonstration, construction and application of solid waste management and resource recovery systems which preserve and enhance the
quality of air, water and land resources;

(2)      to provide technical and financial assistance to States and local governments and interstate agencies in the planning and development of
resource recovery and solid waste disposal programs;

(3)      to promote a national research and
development program for improved management techniques, more effective organizational arrangements,
and new and improved methods of collection, separation, recovery and recycling of solid wastes, and the
environmentally safe disposal of nonrecoverable
residues;

(4)      to provide for the promulgation of
guidelines for solid waste collection, transport, separation, recovery, and disposal systems; and

(5)      to provide for training grants in occupations involving the design, operation and maintenance of solid waste disposal systems."

## Purposes (now termed objectives)--1976 Resource Conservation and Recovery Act:

"The objectives of this Act are to promote the
protection of health and the environment and to
conserve valuable material and energy resources by--

(1)      providing technical and financial
assistance to State and local governments and inter-

state agencies for the development of solid waste management plans (including resource recovery and resource conservation systems) which will promote improved solid waste management techniques (including more effective organizational arrangements) new and improved methods of collection, separation, and recovery of solid waste and the environmentally safe disposal of nonrecoverable residues;

(2)    providing training grants in occupations involving the design, operation and maintenance of solid waste disposal systems;

(3)    prohibiting future open dumping on the land and requiring the conversion of existing open dumps to facilities which do not pose a danger to the environment or to health;

(4)    regulating the treatment, storage transportation and disposal of hazardous wastes which have adverse effects on health and the environment;

(5)    providing for the promulgation of guidelines for solid waste collection, transport, separation, recovery and disposal practices and systems;

(6)    promoting a national research and development program for improved solid waste management and resource conservation techniques, more effective organizational arrangements and new and improved methods of collection, separation and recovery and recycling of solid wastes and environmentally safe disposal of nonrecoverable residues;

(7)    promoting the demonstration, construction and application of solid waste management, resource recovery and resource conservation systems which preserve and enhance the quality of air, water and land resources, and;

(8)    establishing a cooperative effort among the Federal, State and local governments and private enterprise in order to recover valuable materials and energy from solid waste."

Purposes/Objectives--1976 Resource and Conservation and Recovery Act as Amended in 1984 and 1986:

"Sec. 1003.    (a)    Objectives--The objectives of this Act are to promote the protection of health and the environment and to conserve valuable material and energy resources by--

(1)        providing technical and financial assistance to State and local governments and inter-state agencies for the development of solid waste man-agement plans (including resource recovery and resource conservation systems) which will promote improved solid waste management techniques (including more effective organizational arrangements), new and improved methods of collection, separation and recov-ery of solid waste and the environmentally safe dis-posal of nonrecoverable residues;

(2)        providing training grants in occupa-tions involving the design, operation and maintenance of solid waste disposal systems;

(3)        prohibiting future open dumping on the land and requiring the conversion of existing open dumps to facilities which do not pose a danger to the environment or to health;

(4)        assuring the hazardous waste management practices are conducted in a manner which protects human health and the environment;

(5)        requiring that hazardous waste be prop-erly managed in the first instance thereby reducing the need for corrective action at a future date;

(6)        minimizing the generation of hazardous waste and the land disposal of hazardous waste by en-couraging process substitution, materials recovery, properly conducted recycling and re-use, and treat-ment;

(7)        establishing a viable Federal-State partnership to carry out the purposes of this act and insuring that the Administrator will, in carrying out the provisions of Subtitle C of the Act, give a high priority to assisting and cooperating with States in obtaining full authorization of State programs under Subtitle C;

(8)        providing for the promulgation of guidelines for solid waste collection, transport, sep-aration, recovery and disposal practices and systems;

(9)        promoting a national research and development program for improved solid waste manage-ment and resource conservation techniques, more effec-tive organizational arrangements and new and improved methods of collection, separation and recovery and recycling of solid wastes and environmentally safe disposal of nonrecoverable residues;

(10)      promoting the demonstration, construction and application of solid waste management, resource recovery and resource conservation systems which preserve and enhance the quality of air, water and land resources; and

(11)      establishing a cooperative effort among the Federal, State and local governments and private enterprise in order to recover valuable materials and energy from solid waste.

(b)      National    Policy--The    Congress    hereby declares it to be the national policy of the United States that, wherever feasible, the generation of hazardous waste is to be reduced or eliminated as expeditiously as possible.  Waste that is nevertheless generated should be treated, stored or disposed of so as to minimize the present and future threat to human health and the environment."

You will note in the successively amended goals to the original 1965 Solid Waste Act an increasing emphasis on alternatives such as resource recovery, recycling and waste minimization.  Without the implementation of such alternatives there is little hope of overcoming even our present solid waste disposal problems let alone remediating the legacies left to us by a previous generation.

There will still be a need, of course, for landfilling and incineration, but not to the extent that we have relied upon them in the past.  However, if current public opinion is some portions of the United States were to prevail, there would be no new landfills or incinerators whatever, an unrealistic point of view, in my opinion.

In any event, our immediate emphasis must be on rearranging the nation's previous priority in methods for solid waste disposal.  In the remainder of this chapter we will discuss those shifts in priorities generally in the order I believe they need to be implemented.

## WASTE MINIMIZATION AND SOURCE REDUCTION

It has always been this writer's opinion that potential environmental problems can best be handled at their source.  The further the problem moves from its point of origin the greater is the potential extent of its impact.  This is why at source waste mini-

mization and/or resource recovery need to be a priority activity in the nation's solid waste disposal
programs.   These concepts are not new, having been
utilized by certain industries for quite some time,
though not necessarily for reasons of environmental
protection.   Representative examples of such activities include:

     (a)   Improving housekeeping to avoid excess
spillage of raw materials and/or finished products.

     (b)   Process modification to eliminate certain
wastes completely.

     (c)   Sale or exchange of spent wastes (perhaps
acids or solvents) to be used by others.

     (d)   Separation of hazardous and non-hazardous
wastes.

     (e)   Changing the raw materials used, perhaps
substituting recycled materials for raw materials.

     (f)   Redistillation of spent solvents back to
an acceptable purity so that they may be re-used.

## RECYCLING

    If anything positive can be said about the
nation's solid waste disposal crisis, it is the fact
that it has forced that nation to finally accept recycling as an integral part of any solid waste disposal
program.   The removal, collection and recycling of
such items as paper, aluminum cans, plastics and glass
can reduce the nation's total volume of solid waste by
as much as 25 to 90 percent.   Japan, for instance,
recycles 50% of its municipal refuse.   This is not to
say that recycling has been non-existent in America.
To the contrary, a substantial number of communities,
some industries and a few states have been actively
recycling such things as paper, glass, aluminum cans
and small amounts of plastics for a decade or more.
For example, 20 million tons of waste paper and 75,000
tons of plastic bottles were recycled in the United
States in 1987.   The lack of a consistent market has
been cited as the major obstacle to the adoption of a
national recycling program.   Adjectives, historically
used to describe the available markets for recyclables
included volatile, weak or non-existent.   Today the
demand for recycled materials is increasing both in
the United States and overseas.   As reported recently
in the Wall Street Journal the leading exports out of

the ports of New York City and Boston were scrap metals and waste paper.

Over 50% of the nation's waste aluminum (largely used beverage containers) was recycled in 1987. Reynolds Aluminum, Alcan Aluminum and the Aluminum Company of America have invested heavily in recycling plants in the last decade. They had a strong incentive. It is cheaper and easier to recycle aluminum cans than it is to mine and process bauxite, the raw ore used in making aluminum. At present, the aluminum industry has 10,000 recycling centers in the United States. They are optimistic that 80% or more of the used beverage containers in America can be recycled on a yearly basis by the year 2000.

Aluminum recycling may in fact be too successful. The Highway Division of the Oregon Department of Transportation reported that in 1988, $60,000 worth of aluminum bridge railings and signs from state highways were stolen by thieves who subsequently sold it as scrap at recycling centers.

Glass remains a popular recyclable. It is easy to sort, clean and remelt and offers glass manufacturers the opportunity to save substantially on energy costs thanks to reduced furnace temperatures during processing. It is the domestic demand for glass that presently sustains the market. Overseas shipments and subsequent foreign demands are stifled largely by the heavy weight of the material.

The Glass Packaging Institute estimates that 25% of the glass bottles manufactured in the United States in 1988 were made of recycled glass. The glass industry favors mandatory curbside collection programs as the best method for recycling glass.

The production of plastics has more than doubled since 1975, and this increase is clearly reflected in the amount of plastic materials that are now found in our refuse and in our environment in general. In fact, this proliferation of plastics has created a sub-crisis of global proportions within the overall solid waste disposal crisis. Nowhere has this been better dramatized than in the world's oceans where recent studies indicate that 80% of the floating debris now sighted at sea is plastic material. It is estimated that over one million pounds of plastics are dumped into the sea annually. Sea turtles, mistakenly ingesting plastic bags for jellyfish are suffocating and starving to death. Fish, whales, seals and fish are likewise dying as a result of ingestion or entrapment in a variety of plastics such as six pack carriers, discarded fish nets and fishing line. Shorelines even in the most remote parts of the world are strewn with layers of indestructible plastics. A single, three hours long cleanup of 157 miles of Texas shoreline in September of 1987 yielded 31,773 plastic bags, 30,295 plastic bottles, 15,631 plastic six pack rings, 28,540 plastic lids, 1,914 disposable diapers, 1,040 plastic tampon applicators and 7,460 plastic milk jugs.

There is at present a great deal of controversy over the recyclability of plastics. The major obstacles to recycling are the costs of identifying and separating the wide variety of plastics now manufactured--many of which have chemically incompatible parent resins. While they may be collected together initially, they will ultimately have to be separated according to resin compatibility.

Communities frustrated by the lack of an immediate solution have introduced legislation banning plastic packaging altogether--a trend that seems to be gaining national and international momentum much to the chagrin of the plastics industry. For instance, in January 1987, Florence, Italy banned non-degradable plastic containers and the European Economic Community has called for a ban on the sale and use of all plastic bags by 1991.

While the plastic industry continues to argue that this problem is really the lax enforcement of anti-litter and pollution laws, it has begun to take some positive steps to encourage recycling of its products.

First, it has adopted a uniform coding system to identify the different resins in plastic bottles and set a recycling goal of 50% for plastic soft drink containers by 1992. The coding system is a triangular arrow with a number in the center and letters underneath. There are seven codes:

    a.   1, PET (polyethylene terephthalate)
    b.   2, HDPE (high-density polyethylene)
    c.   3, V (Vinyl)
    d.   4, LDPE (low-density polyethylene)

e.   5, PP (polypropylene)

f.   6, PS (polystyrene)

g.   7, Other

Secondly, it has established research groups to explore new uses for recycled plastics.   The Center for Plastics Recycling Research at Rutgers University in New Jersey is one of the first.

These initiatives are encouraging, but a good deal more effort is needed before the recycling of plastics becomes as common as the recycling of aluminum cans and glass.   One solution showing great promise is the production of biodegradable forms of plastics--a topic we will discuss next.   In the meantime, the 100th United States Congress has approved an international agreement (MARPOL Annex V) banning plastic disposal in the marine environment and enacted PL 100-200, (Title II) banning plastics disposal in the United States 200 mile Exclusive Economic Zone.

The key to establishing a permanent recycling program in the United States is the establishment of a permanent market for these recycled materials.   Mandatory government procurement programs for recycled materials would be of immense assistance in this regard. In April, 1988 such a program received a big boost when a consent order was signed by a federal judge forcing the U.S. Environmental Protection Agency to issue guidelines for government procurement of recycled paper, re-refined lubricating oil, re-manufactured tires and building insulation made from reclaimed material.   This consent order settled a suit brought against the U.S. EPA by the National Recycling Coalition, the Environmental Defense Fund, the Environmental Task Force and the Coalition for Recyclable Waste--all of which claimed that the EPA had repeatedly failed to issue procurement guidelines by the deadlines set forth in the 1976 Resource Conservation and Recovery Act.   This mandatory procurement program applies not only to federal offices, but to states and municipalities who receive federal monies, as well.

For further information on recycling markets contact:

Metal Scrap, Paper   Institute of Scrap Recycling Industries, Inc., 1627 K Street NW, Washington, D.C., 20006.

Paper   Paper Recycling Committee, American Paper Institute, 260 Madison Avenue, New York, NY   10016, or

Garden State Paper Co., River Drive, Center 2, Elmwood Park, NJ 07407.

Plastics   National Association for Plastic Container Recovery, P.O. Box 7784, Charlotte, NC 28241, or

Society of the Plastics Industry, 1275 K Street NW, Suite 400, Washington, D.C. 20005.

Bottles   Empire Returns Corp., 100 Mushroom Blvd., Rochester, NY 14623.

Glass   Glass Packaging Institute, 1133 20th Street NW, Washington, D.C. 20036.


## BIODEGRADABLE PLASTICS

The technology to produce degradable plastics has been available for nearly two decades. Unfortunately, there has been little incentive until now to mass produce products with this capability. Plastics can be manufactured to degrade biologically when in contact with soil bacteria and other microorganisms or to photodegrade when exposed to sunlight. There are three photodegradable technologies currently available, all of which rely upon some introduced additive. The first utilizes an ethylene carbon monoxide (ECO) additive which is sensitive to ultraviolet light. With continuous exposure to ultraviolet light, the ECO in the plastic product is embrittled and with subsequent exposure to wind, rain, snow, etc. the plastic breaks down into smaller and smaller pieces. This additive was first test-marketed in 1962, and is used today to produce degradable six-pack ring carriers. Degradation time is usually within several months. A second technology utilizes ketone carbonyl which is also sensitive to ultraviolet light. This technology is offered commercially for license by Eco Plastics, Ltd. of Ontario, Canada.

The third technology uses a phenolic compound that reacts similarly to the other two additives.

Biodegradation of plastics is generally accomplished through the use of cornstarch or corn oil. These additives, sometimes called fillers, become food for microorganisms when the plastic is covered by or in contact with the soil. As with the photodegradation technologies, the plastic molecules are subsequently broken down into smaller and smaller pieces.

This technology was commercialized in the United Kingdom by Colorall, Ltd. The worldwide rights have subsequently been acquired by the St. Lawrence Starch Company, Ltd. in Ontario, Canada which manufactures a plastic called ECOSTAR.

A polyethylene plastic bag with a 15% starch additive by weight can be expected to decompose within 6 months. These biodegradation technologies do not come cheap. Polyethylene containing 60% starch and corn oil sells for about double the price of polyethylene. There is a common problem with all of these technologies at present. While these methods have the capability to reduce the sheer bulk of our plastic wastes, we do not know how all of this degraded plastic residue will ultimately affect our environment. Ideally we would prefer that the degradation of the plastic material result in end products that can be safely assimilated by the environment. A United Kingdom firm, Imperial Chemical Industries, claims to have developed such a plastic polymer called PHBV, whose end products are carbon dioxide and water.

## SPECIAL CONSIDERATIONS--GENERAL

### 1.    Household Hazardous Wastes

An increasing number of American communities have instituted or are presently considering programs to separate household hazardous waste (HHW) from municipal solid waste. The most common methods used so far to accomplish this have been the use of special collection days or the establishment of collection centers where individual citizens can bring their hazardous household waste to be disposed of without charge. Items included in this category include paints, old medicines, herbicides, pesticides, solvents, acids, poisons and even old batteries. While the U.S. EPA has determined that these items make up only 0.34 to 0.40 percent of the total amount of household refuse discarded, their presence can affect the quality of air emissions if incinerated or the quality of leachate if landfilled. Some engineers argue, however, that separation is not really necessary. They contend that if the HHW is incinerated along with other municipal solid waste (a ratio of 200 to 1) that emissions will be no worse than those emitted from state-of-the-art high temperature hazardous waste

incinerators. They further argue that the recent
changes in Subtitle D of RCRA that require all landfills to be lined and the resulting leachate to be
collected and treated will accommodate whatever contributions HHW makes to leachate quality. This author
does not agree with either of these opinions. First,
the continued dumping of HHW materials along with
other municipal refuse will continue to be a threat to
those who must collect it. Explosions, fires, pesticide induced illnesses and skin rashes and burns have
all been directly caused by the presence of HHW.
Secondly, we need to continue to encourage the concept
of personal responsibility in the ultimate disposal of
our waste products. Allowing such promiscuous disposal of HHW to continue runs counter to this philosophy. Therefore, planners should continue to encourage
the separate disposal of household hazardous waste.

## 2.   Medical Waste

The summers of 1987 and 1988 will be long remembered by those who visited the coastal areas of the
northeastern United States seeking sun, sand and fun.
Unfortunately, there will not be many fond memories.
Blood filled vials, syringes, plastic catheters and
I.V. bags and even human body parts became an unwelcome and all too frequent addition to many of the
northeast's coastal beaches. The appearance of these
items was yet another clue that the nation's solid
waste disposal problems had reached crisis proportions. It also dramatized the fact that the nation
lacked a national regulatory program, or for that fact
any overall strategy whatever for the safe disposal of
this offensive and potentially hazardous material.
Prompted by the confirmation that some of the blood
bearing vials were AIDS and hepatitis contaminated, at
least two states, New York and New Jersey, instituted
emergency regulations to begin the tracking of medical
waste. New York's emergency rules which went into
effect in early August, 1988 required all clinical
laboratories and health care facilities (generating
100 kilograms or more of medical waste a month) to
separately bag all their infectious waste and to label
those bags with their address.

All of this contaminated waste is to be manifested to document who was the hauler and where it was
ultimately disposed of. New Jersey's rules effective
since late August, 1988 require a manifest system for

all large generators (of the same 100 kilogram or more a month limitation as New York) of infectious waste. All generators must also keep a daily log. The governor of Rhode Island, concerned as well, established a task force which contracted with a consultant to determine the origin of medical waste on their beaches by the use of numerical model simulation technique. Interestingly, the final report[3] indicated that the medical waste which washed up on Rhode Island's beaches most likely originated in the New York Bight apex.

It was these recent medical waste wash-ups that finally prompted federal attention and as a result Congress passed and then President Reagan signed the Medical Waste Tracking Act of 1988 into law on November 2. This bill establishes a demonstration program for tracking medical waste in New York, New Jersey, Connecticut, the Great Lakes area and any other states that choose to be a part of this program.

The U.S. Environmental Protection Agency will develop a tracking system for those medical wastes determined to pose a hazard to human health and the environment. Expected benefits are the development of a uniform tracking form and the development of regulations governing waste segregation, labeling and containerization. The small quantities of medical waste produced by doctors offices, clinics and household users would not be governed by this act, at least at present.

Of the 160 million tons of municipal solid waste generated in the United States each year, 3.2 million tons is medical waste generated by hospitals. The U.S. EPA estimates that 10 to 15% of this waste is infectious. EPA presently identifies six types of waste as potentially infectious:

(1)   Cultures and stocks of infectious agents and associated biologicals

(2)   Pathological wastes

(3)   Human blood and blood products

(4)   Contaminated sharps (e.g. needles and scalpels)

(5)   Contaminated animal carcasses, body parts, bedding, and

(6)   Isolation waste

In addition to this infectious waste, there are other types of medical waste that needs to be disposed of. These include general waste such as paper, food waste, floor sweepings and special waste such as radioactive materials and RCRA designated hazardous waste.

This latter category is generally produced in small quantities which require special treatment and permits. In most instances, they are disposed of through licensed contractors. All of the other forms of medical waste, however, are either incinerated (80%) onsite or offsite, sterilized (15%) in an autoclave or landfilled (5%).

Currenntly, many hospitals are choosing to dispose of all or most of their solid waste by onsite incineration for the following reasons:

(1)    Increasing reluctance by landfills to accept medical waste.

(2)    Liability considerations due to possible transmission of AIDS or other viral diseases.

(3)    Rising costs for the transportation and handling of these wastes for offsite disposal.

Historically, hospital incinerators for pathological and other medical wastes were small units with desiccating hearths operating under excess air conditions which resulted in high particulate emissions. Thanks to more stringent air pollution regulations promulgated under the Federal Clean Air Act, such excessive particulates became unacceptable. This resulted in a changeover to the controlled air incinerator which has a primary and secondary combustion chamber resulting in a more efficient incineration. While these controlled air incinerations proved capable of limiting particulates, odors and opacity they are having difficulty meeting the most recently adopted performance standards which are directed toward the further control of all of the above items plus organic compounds and the products of incomplete combustion. The U.S. EPA has shown that hospital waste may contain as much as 20% to 30% plastics which, when incinerated, produce HCl (hydrogen chloride) gas and toxic organic compounds.

In light of these most recent developments, planners can expect to receive a substantial number of proposals to retrofit old incinerators and build new ones. The factors to be considered when evaluating these proposals are as follows:

(1)    The overriding issue for urban or inner city hospitals is finding adequate space for the incinerator, its stack and the air pollution control equipment.  Some city hospitals have considered the roof as a possible location, however, even if structurally possible, the logistics of delivering the waste to the roof and removing the ash would be questionable.  Better areas would be an existing boiler room or part of an outside parking area.

(2)    Existing chimneys are seldom suitable for the high temperature gases discharged from an incinerator and in most instances should not be used.

(3)    Any new chimney may be aesthetically objectionable.  It may also be dangerous if not elevated above existing roof air intakes.

(4)    To meet the most recent performance standards, all new medical waste incinerators should have:

(a)    Higher operating temperatures in the primary combustion chamber ($1500^{\circ}$ to $1800^{\circ}$F) to maximize destruction of pathogenic organisms.

(b)    More oxygen available in the primary combustion chamber to minimize the formation of products of incomplete combustion.

(c)    Higher operating temperatures in the secondary combustion chamber ($1800^{\circ}$ to $2000^{\circ}$F) with increased gas retention times (1-2 seconds) to insure destruction of any remaining pathogens and organic compounds.

(d)    Additional mixing devices in the secondary combustion chamber to increase destruction efficiency of organic compounds, and

(e)    BAT air pollution control devices to control particulates and acid gases.

In concluding this section, it is interesting to note that an alternative to the incineration of infectious medical waste has recently debuted.  The system pulverizes the medical waste to a granular material after it has been treated with a chlorine solution, thereby reducing its bulk and pathogenicity.  Once treated, the manufacturer, Medical Safe Tec, Inc. of Indianapolis, Indiana claims the material is suitable for landfill disposal.  Whether it becomes a competitor with incineration remains to be seen.

**INCINERATION**

Incineration is the second most common method of waste disposal in the United States, having the desirable attribute of being able to quickly reduce the volume of solid waste by 70-90%. Incineration is defined generally as a controlled combustion process that reduces solid, liquid or gaseous combustible wastes to carbon dioxide, other gases and a relatively noncombustible residue. The earliest forms of incineration, however, were far from controlled and in fact were little more than open burning either on level ground in pits or in the now banned tepee burners. The products of combustion were vented directly into the atmosphere with little concern given to the air pollution it caused.

Alfred Fryer is credited with constructing the first true incinerator (called a destructor) for municipal waste in Nottingham, England in 1874. The first American municipal waste incinerator was constructed on Governor's Island in New York Harbor in 1885. These early incinerators left much to be desired with regard to pollution of the air. Their tall stacks could often be found emitting highly visible plumes of dark smoke and particulates into the atmosphere. It is this dark image of early municipal incinerators that many American still conjure up whenever modern methods of incineration are suggested as alternatives to solid waste disposal.

In the hundred years since the first incinerator was constructed on American soil, significant improvements have been made in their efficiency. Present public attitudes towards this form of disposal generally range from extreme wariness to outright rejection--justifiable responses perhaps to the less than admirable past performance record of some of our existing facilities and the general paucity of knowledge about incinerator types, their capabilities and air emission characteristics, hazards and controls. In spite of the public hesitancy to embrace incineration, it is clear that the number of incinerators in America will likely triple in number from the present 100 facilities by the year 2000--thanks largely to intensifying governmental restrictions on landfilling. Incineration seems to offer a quick fix. Land planners can therefore expect to be called upon to review

an increased number of proposals for incinerator construction and in some cases they may even be the
proponents. The evaluation of these proposed incinerator projects will be a formidable undertaking even
without the expected public opposition. Table 11.2,
for example, lists just a few of the wide variety of
incinerator designs currently available. Because of
the availability of other texts dealing exclusively
with incinerator design (see references at the end of
this chapter) we will not examine each incinerator
design at this time.

We can, however, address in some detail what is
probably the major concern of the opponents of
incineration--namely the quality of the emissions from
incinerator smokestacks (see also Special Consideration Section).

Published reports of studies on emissions from
existing incinerator smokestacks (Ref. 11.34, 11.35,
11.36) both in the United States and Europe have indicated that potentially hazardous heavy metals such as
lead, nickel, mercury, cadmium, various polychlorinated dibenzo-p-dioxins (PCDD's), polychlorinated
dibenzofurans (PCDF's) and acidic gases may all be
present as air pollutants. The presently preferred
inference from such studies, particularly from a
nation presently consumed by cancer phobia, is that
such hazardous air pollutants can be expected from
smokestacks of all incinerators. We know, of course,
that such a broad extrapolation is simply not true.
The final emissions from incinerator smokestacks are
the result of a variety of factors including:
1) Waste composition
2) Incinerator design
3) Rate of introduction of waste
4) Method of introduction of waste
5) Burning conditions (i.e. temperature,
   time, turbulence)
6) Maintenance schedule for all equipment
7) Type of air pollution control equipment

Of these seven factors, this author believes
that altering the composition of the waste offers the
easiest and least expensive means of minimizing the
amount of hazardous air pollutants eventually generated by incineration. By alteration, we mean removing
items which are documented generators or precursors of
hazardous or toxic air pollutants through recycling or

## Table 11.2:  Some Selected Incinerator Types

A. **For General Solid Waste Incineration**

1. Single Chamber Incinerators--Simplest of all designs.  Has a singular chamber. Generally will not meet current air pollution standards.  Frequent problems with materials handling, firing and ash removal.

2. Multiple Chamber Incinerators--Has a primary and secondary combustion chamber.  The secondary chamber provides additional residence time for the further combustion of any incompletely burned gases or combustible solids discharged from the primary chamber.  There are two basic types--the retort system which uses baffles to route combustion gases through both vertical and horizontal 90 degree turns and the in-line system where combustion gases flow straight through the incinerator.

3. Pyrolitic Incinerators--Utilizes the thermal process of pyrolysis which is the destructive distillation of solid, normally carbonaceous material in the presence of heat and the absence of stoichiometric oxygen.  The products of combustion can be extremely varied depending upon the temperature, pressure and amount of oxygen in the combustion chamber.  Two well known pyrolitic incineration systems are the PUROX system developed by Union Carbide Corporation and the TORRAX system developed by the Carborundum Corporation.

4. Controlled Air Incinerators--Also known as starved air incinerators.  Has a primary and secondary combustion chamber.  Waste is introduced in the primary chamber and only enough air (usually 70% to 80% of the stoichiometric air required) is provided to allow pyrolytic burning.  In the secondary chamber excess air (typically 140% to 200%) is added to effect complete combustion of any remaining gases and combustible solids.

B. **For Sludge Incineration**

1. Multiple Hearth Incinerators--Commonly referred to as the Hereschoff furnace. Generally has 5 to 9 vertically stacked hearths (Figure 11.1) along which sludge is passed from top to bottom until it is reduced to an ash.  Some of the ash may become airborne and there may be occasional problems with odor.

2. Fluid Bed Incinerators--Uses a bed of sand fluidized by the passage of air (Figure 11.2).  This fluidization expands the sand volume by 30-60% and provides maximum contact of the sludge with air.

3. Electric Furnace Incinerator--Also known as the radiant heat or infrared furnace. Uses a conveyor belt system to transport a thin layer of sludge (about 1" thick) through a rectangular chamber.  Heat is supplied by electric infrared heating elements.  May not be feasible where electrical costs are high especially if used with wet sludges.

4. Cyclonic Furnace Incinerator--This is a cylindrically shaped furnace with a single movable hearth.  Sludge is fed into the furnace with a screw feeder at the periphery of the hearth (Figure 11.3).

5. Rotary Kiln Incinerator--This is the most universal design.  It is used for disposing of a wide variety of solids and sludges as well as liquid and gaseous wastes.  Basic components include a horizontal cylinder (kiln) which turns about its horizontal axis.  As the kiln rotates the waste is exposed to oxygen and heat and burns to ash.  One problem with the rotary kiln incinerator is leakage at its ends.

Figure 11.1:  Cross section of a typical multiple hearth incinerator.

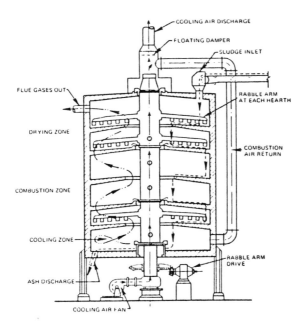

Figure 11.2:  Cross section of a fluid bed reactor.

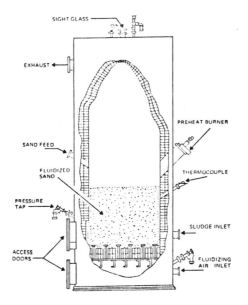

Source for Figures 11.1 and 11.2: Ref. 11.32

Figure 11.3:  Cyclonic furnace.

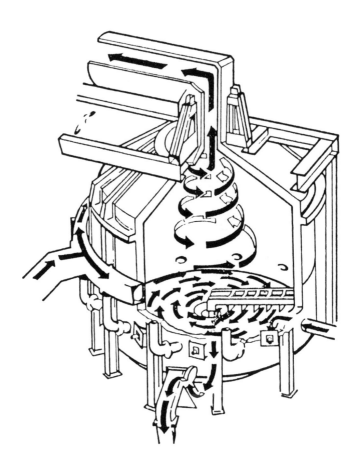

Source:  Ref. 11.32

resource recovery prior to incineration.  For each of the three major categories of solid waste presently incinerated in the United States, namely sewage treatment plant sludges, municipal refuse and hazardous wastes, this may not always be possible.

## Municipal Refuse Incineration

Municipal refuse offers the greatest potential for altering waste composition.  Items routinely found in municipal refuse such as ferrous metals, plastics, household chemicals, tires and batteries can be major contributors of hazardous materials upon incineration. Plastics, for instance, can generate hydrogen chloride gas and chromed or galvanized metal products can produce metal oxides of lead, mercury, chromium and cadmium.  These metal oxides, which range from submicron to a few microns in size, can be subsequently inhaled and readily transported to the lower part of the lung where they are quickly absorbed by the blood. These pollutant contributing materials can be removed with little difficulty following procedures already in place for recycling and/or recovering glass, aluminum and paper.  No planner should therefore, approve the construction and operation of any municipal refuse incinerator unless a comprehensive recycling or resource recovery program for all of the aforementioned materials is included[4].

Do not be misled, however, into thinking that the elimination of all of the previously described constituents will result in totally clean smokestack emissions.  Incinerated municipal waste even without such items still produces emissions of carbon dioxide, nitrogen, oxygen, water vapor, sulfur oxides, nitrogen oxides and carbon monoxide.  Some of these nitrogen oxides are formed from the air used in combustion which is 79% by volume nitrogen gas.  Sulfur oxides and nitrogen oxides have been linked to the formation of acid rain and ozone respectively.  The sulfur oxide emissions from municipal waste incinerators are often dramatized by opponents as comparable to fossil fuel emissions which are under withering criticism as major contributors to acid precipitation.  Municipal solid wastes, however, have sulfur contents on the order of about 0.1 to 0.2 percent by weight.  Fossil fuels, such as coal and oil, on the other hand, have sulfur contents ranging from about 1 to 3 percent by weight which is 5 to 30 times the amount found in municipal solid waste.

## Sewage Treatment Plant Sludge Incineration

Municipal sewage treatment plant sludges and hazardous wastes are not as readily amenable to alterations of their compositions. Sewage sludge, for example, an end product of wastewater treatment, has become more complex as stricter regulations under the federal Clean Water Act mandate more complete wastewater treatment. The potential precursors of air pollutants in sludge, unlike those in municipal refuse, are generally molecular in size. Hence, their removal goes beyond simple physical removal. As with municipal refuse the formation of heavy metal oxides during incineration is a concern. Metallic cations such as Zn, Cu, Ni, Pb, Cd, Cr and Hg will be concentrated in the sludge during the sewage treatment or sludge stabilization processes (Table 11.3). Normally this concentration of metals will take place through the formation of insoluble metal precipitates such as hydroxides, oxides, carbonates, phosphates and sulfides or through absorption on organic particulates. Concentrations of these metals can be substantial for municipal sewage treatment plants that receive industrial wastes. Further treatment of sludge to remove excessive metal concentrations prior to incineration would not be practical or cost effective. This also applies to many other materials found in sewage sludge such as various organic chemicals (Tables 11.4 and 11.5), pesticides and polychlorinated biphenyls. For all of these materials the only practical method for decreasing their concentrations would be to require all industrial contributors to pre-treat their wastewater prior to discharge to the sewage treatment plant. Prevention, as we have stated throughout this text is still the key.

## Hazardous Waste Incineration

As mentioned earlier the U.S. EPA estimates (1981 figures) that American industries generate about 136 million tons or 40 billion gallons of hazardous waste annually. Most of that waste was generated by

Table 11.3:  Concentrations of Selected Heavy Metals in Sewage Sludge*

| Component | Type[†] | Number | Range | Median | Mean |
|-----------|---------|--------|-------|--------|------|
| | | | - - - - - (mg/kg)- - - - - | | |
| Pb, mg/kg | Anaerobic | 98  | 58-19,730  | 540   | 1,640 |
|           | Aerobic   | 57  | 13-15,000  | 300   | 720   |
|           | Other     | 34  | 72-12,400  | 620   | 1,630 |
|           | All       | 189 | 13-19,700  | 500   | 1,360 |
| Zn, mg/kg | Anaerobic | 108 | 108-27,800 | 1,890 | 3,380 |
|           | Aerobic   | 58  | 108-14,900 | 1,800 | 2,170 |
|           | Other     | 42  | 101-15,100 | 1,100 | 2,140 |
|           | All       | 208 | 101-27,800 | 1,740 | 2,790 |
| Cu, mg/kg | Anaerobic | 108 | 85-10,100  | 1,000 | 1,420 |
|           | Aerobic   | 58  | 85-2,900   | 970   | 940   |
|           | Other     | 39  | 84-10,400  | 390   | 1,020 |
|           | All       | 205 | 84-10,400  | 850   | 1,210 |
| Ni, mg/kg | Anaerobic | 85  | 2-3,520    | 85    | 400   |
|           | Aerobic   | 46  | 2-1,700    | 31    | 150   |
|           | Other     | 34  | 15-2,800   | 118   | 360   |
|           | All       | 165 | 2-3,520    | 82    | 320   |
| Cd, mg/kg | Anaerobic | 98  | 3-3,410    | 16    | 106   |
|           | Aerobic   | 57  | 5-2,170    | 16    | 135   |
|           | Other     | 34  | 4-520      | 14    | 70    |
|           | All       | 189 | 3-3,410    | 16    | 110   |
| Cr, mg/kg | Anaerobic | 94  | 24-28,850  | 1,350 | 2,070 |
|           | Aerobic   | 53  | 10-13,600  | 260   | 1,270 |
|           | Other     | 33  | 22-99,000  | 640   | 6,390 |
|           | All       | 180 | 10-99,000  | 890   | 2,620 |

* Concentrations are on a dried solid basis.

† "Other" includes lagooned, primary, tertiary, and unspecified sludges.  "All" signifies data for all types of sludges.

Source:  Ref. 11.7

Table 11.4: Characterization of Organic Compounds in 238 Sludges Collected from Treatment Plants in Michigan

| | - - - - - - - - - -mg/kg (dry weight)[#] - - - - - - - - - | | | |
| Compound | Detection Limit[*] | Range | Mean[†] | Median |
| --- | --- | --- | --- | --- |
| Acrylonitrile | 4 (25/155) | 4-82 | 16±19 | 7 |
| Chlorobenzene | 60 (3/158) | 60-846 | 337±441 | 106 |
| p-chlorotoluene | 59 (6/158) | 93-324 | 153±87 | .121 |
| o-dichlorobenzene | 6 (15/215) | 6-809 | 89±209 | 16 |
| m-dichlorobenzene | 5 (44/216) | 6-1,651 | 119±327 | 22 |
| p-dichlorobenzene | 10 (18/216) | 10-633 | 77±151 | 23 |
| 1,2-dichloropropane | 0.08 (91/157) | 0.09-66 | 1.91±7.36 | 0.66 |
| 1,3-dichloropropane | 0.5 (40/158) | 0.6-309 | 18±51 | 3.2 |
| 1,3-dichloropropane | 0.1 (119/157) | 0.1-1,232 | 24±116 | 3.9 |
| Ethylbenzene | 0.08 (14/220) | 1.2-66 | 25±22 | 20 |
| Hexachloro-1,3-butadiene | 3 (1/217) | - | 4 | - |
| Hexachloroethane | 0.05 (40/217) | 0.05-16.5 | 0.7±2.6 | 0.2 |
| Pentachloroethane | 0.4 (5/199) | 0.4-9.2 | 2.7±3.7 | 1.3 |
| Styrene | 90 (6/219) | 99-5,848 | 1,338±2,249 | 405 |
| Tetrachloroethylene | 10 (108/128) | 1-1,218 | 68±132 | 29 |
| 1,2,3-trichlorobenzene | 1 (7/216) | 1-152 | 25±56 | 1 |
| 1,2,4-trichlorobenzene | 3 (17/217) | 3-51 | 14±12 | 13 |
| 1,3,5-trichlorobenzene | 50 (0/217) | - | - | - |
| 1,2,3-trichloropropane | 4 (2/141) | 9-19 | 14±7 | 14 |
| 1,2,3-trichloropropane | 3 (21/137) | 3-167 | 23±47 | 6 |
| o-chlorophenol | 0.03 (20/231) | | 13±23 | |
| m-chlorophenol | 0.03 (16/231) | 0.1-93 | 9±24 | 0.9 |
| p-chlorophenol | 0.03 (19/231) | 0.1-90 | 18±30 | 3.6 |
| o-cresol | 0.03 (16/231) | 0.2-183 | 25±52 | 2.0 |
| 2,4-dichlorophenol | 0.03 (17/230) | 0.2-203 | 25±54 | 4.8 |
| 2,4-dimethylphenol | 0.03 (41/231) | 0.09-87 | 6.5±14.9 | 2.2 |
| 4,6-dinitro-o-cresol | 0.06 (20/229) | 0.2-187 | 12.7±41 | 2.3 |
| 2,4-dinitrophenol | 0.18 (66/228) | 0.3-500 | 24±81 | 5.0 |
| Hydroquinone | 0.07 (61/229) | 0.1-223 | 8±29 | 2.6 |
| Pentachlorophenol | 0.03 (155/223) | 0.2-8,495 | 81±685 | 5.0 |
| Phenol | 0.03 (178/229) | 0.05-238 | 9±29 | 2.0 |
| 2,4,6-trichlorophenol | 0.06 (66/223) | 0.2-1,333 | 42±178 | 4.8 |

* Number in parenthesis is the number of sites having concentrations less than detection limit/total number of sites analyzed.

† Mean ± standard deviations.

# Concentrations on a dry solids basis.

Source: Ref. 11.7

Table 11.5:  Organic Compounds Detected in  Sludges*

| Compound | Percent Occurrence at Indicated Concentrations[†] | | |
|---|---|---|---|
|  | >1 mg/kg | >10 mg/kg | >50 mg/kg |
| Methane, dichloro- | 41 | 12 | 3 |
| Methane, trichloro- | 3 | 0 | 0 |
| Ethane, 1,1,1-trichloro- | 5 | 3 | 0 |
| Ethane, trichloro- | 26 | 9 | 3 |
| Ethane, tetrachloro- | 27 | 8 | 3 |
| Benzene, 1,4-dichloro | 36 | 18 | 5 |
| Ethylbenzene | 33 | 3 | 0 |
| Toluene | 59 | 35 | 11 |
| Phenol | 63 | 25 | 13 |
| Naphthalene | 65 | 33 | 15 |
| Phenanthrene | 60 | 20 | 8 |
| Phthalate diethyl | 43 | 23 | 13 |
| Phthalate, di-n-butyl | 63 | 25 | 13 |
| Phthalate, bis (2-ethyl hexyl) | 75 | 63 | 25 |
| Phthalate, butylbenzyl | 50 | 35 | 18 |
| All others | <50 | <25 | <15 |

* Survey of 25 cities located throughout the United States. Plant treated from 13,200 $m^3$/day to 1,170,000 $m^3$/day and percentage of industrial flow varied from 0 to 60 percent.

† Dry solids basis.

Source:  Ref. 11.7

the nation's chemical industries (71 percent). Surface impoundments (including landfills) and deep well injection were the most commonly used method of disposal. Unfortunately, many of these surface impoundments and wells were improperly located and poorly designed and pose a substantial threat to the nation's ground water. As a consequence, there has been a strong incentive to find more suitable methods of disposal. High temperature incineration already utilized by industry for some time has become a popular alternative.

From the public point of view, there is little, if anything, that can be done to improve the "quality" of hazardous wastes prior to incineration--an observation that has some validity. While Americans may accept or at least remain neutral toward the incineration of municipal solid waste and sewage sludge, they do not generally sustain either of those attitudes regarding the incineration of hazardous wastes. Proposals to incinerate hazardous waste continue to provoke bitter public debate and anti-incineration demonstrations--this despite the fact that destruction efficiencies of 99.99% have been demonstrated. The state of New Jersey's eight year saga[5] to locate just two such regional, hazardous waste incinerators within the states 7,800 square miles exemplifies the difficulties that land planners elsewhere in the nation may face.

While incineration is presently the best "destructive" method[6] for disposal of hazardous wastes it would be wise for the nation's planners to devote some of their efforts to reducing the total amount of hazardous waste produced.

Therefore, American industries must be encouraged to adopt waste minimization as a routine part of their operations. In the interim, emissions from hazardous waste incinerators need to be controlled by the application of best available technology (BAT) air pollution control equipment--a topic we will discuss next.

## Air Pollution Control

Incinerators produce air pollutants which must be removed or reduced in concentration prior to atmospheric discharge. The equipment used in this process provide the last defense, so to speak, between any hazardous and toxic emissions and the environment. Therefore, the selection of air pollution control equipment must be carefully supervised and scrutinized. While the levels of pollutant reduction will be dictated largely by the appropriate regulatory agency, we should not accept in any instance, equipment less than best available technology (BAT).

Let us now examine the wide variety of air pollution control equipment currently in use. Please note that some of the equipment to be described, although still in use at some locations, does not qualify as BAT. We are presenting information on them to enhance the reader's overall knowledge of the extreme variations of air pollution control technologies.

1.    Settling Chambers

These are basically long, boxlike structures that are ineffective for all but the heaviest or largest particles (usually larger than 40 microns). They are not a practical pollution control device other than one part of a larger system to help remove larger particles from the gas stream.

2.    Dry Impingement Separators

Impingement separators are essentially a series of baffles around which gaseous emissions are forced to flow. This baffling intercepts particulates and encourages them to drop out. This method is effective for particles above 15 microns.

3.    Dry Cyclonic Separators

This is an inertial separator. Gas entering the separator forms a vortex which eventually reverses and forms a second vortex that exits the cyclonic chamber. During this process particulates move toward the outside wall by inertia where they drop to be collected by an external receiver.

4.    Cyclone Collectors

These are basically cyclonic separators provided with a water spray to facilitate collection and removal of collected particulates.

5.    Cyclonic Scrubber

These units utilize a spin damper to create a contraction within the entering gas stream. Water is usually injected immediately upstream of the spin damper. Upon entering the cyclonic chamber the rapid expansion of the gas combined with the effect of the cyclonic action and water scrub separate particulates from the gas steam.

6.   Venturi Scrubber

This system relies on a venturi throat through which the gases are forced to pass. This causes them to contract thereby boosting their velocities to 200 to 600 feet per second. Once through the venturi, the gas enters an expansion chamber where particle separation occurs. Water supplements this contraction-expansion process and is injected just upstream of the venturi.

7.   Tray Scrubbers

Tray scrubbers contain perforated plates within a tower usually located immediately downstream of a venturi. A water level is maintained above the trays. Gas flows up through the openings of the plates counter to the flow of water trying to flow downward by gravity. The resulting turbulence of the two media effectively scrubs particulates from the gas.

8.   Electrostatic Precipitator

Sometimes called electro filters, these units force the gas stream past a series of discharge electrodes. The electrodes are negatively charged (in the range of 1000-6000 volts) resulting in a corona around each. As particulate matter passes through these fields, they are given a negative charge. A grounded surface or collector electrodes surrounding the discharge electrodes then subsequently collects these charged particles.

9.   Fabric Filters

These are essentially a series of permeable bags which allow the passage of gas, but not particulates. Filter fabrics are usually woven with relatively large spaces in excess of 50 microns, however, particles as small as or smaller than one micron are captured. This is the result of other forces such as impaction, diffusion, gravitational attraction and electrostatic forces generated by interparticle friction. Even the accumulating dust layer itself, acts as part of the filter. Materials frequently used to construct such filters include: cotton, wool, nylon, orlon, dacron, polypropylene, fiberglass and Teflon.

10.   HEPA Filters

Known as high efficiency particulate air filters, these extremely efficient filters were originally developed for control of particulates from nuclear energy facilities. They can remove over 99.7% of the particles in a gas stream that are 0.3 microns or greater in size.

Air pollution control efficiencies for some of the above described equipment are shown in Table 11.6

## SPECIAL CONSIDERATIONS--INCINERATION

### 1.    Excessive Water Consumption

A problem frequently overlooked by proponents of incineration is the amount of water required to operate either air emission scrubbing equipment or to produce steam in those incinerators designed to produce electricity. One 3000 ton per day garbage to energy incinerator presently under construction in Bergen County, New Jersey for instance, is expected to consume 65,000 gallons of fresh water a day for these two purposes. All of this water will come from a local potable water purveyor.

Those areas of the nation, therefore, who have limited or constrained water resources may have to consider alternatives to wet scrubbing equipment for air pollution control and perhaps, seek alternatives to incineration or at least forgo garbage to energy facilities.

In addition, consideration must be given to the proper disposal of wastewater resulting from the operation of these air scrubbing devices and washdown activities, in general, at the incinerator. Because of the likelihood that hazardous materials will be included in such wastewater, the most appropriate method for disposal will be to a sanitary sewer system leading to a sewage treatment plant providing at least secondary treatment. Those wastewaters containing exceptionally high levels of hazardous pollutants will require some form of pretreatment. Communities under a sewage connection ban or with a poorly operating treatment plant may not be appropriate areas to locate incinerators.

### 2.    Conflict Between Incineration and Recycling

State and national initiatives to encourage recycling and resource recovery may be unwittingly jeopardized in some instances by incineration, especially where those facilities utilize refuse derived fuel (RDF). One New Jersey community recently found itself facing this very dilemma when its new 400 ton

Table 11.6:  Average Control Efficiencies of Selected Air Pollution Control
Systems

| | System Removal Efficiency (wt %) | | | | | | | |
|---|---|---|---|---|---|---|---|---|
| | Mineral Parti-culate | Carbon Mon-oxide | Nitro-gen Oxides | Hydro-carbons | Sulfur Oxides | Hydrogen Chloride | Poly-nuclear Hydro-carbons | Vola-tile Metals |
| None (flue setting only) | 20 | 0 | 0 | 0 | 0 | 0 | 10 | 2 |
| Dry expansion chamber | 20 | 0 | 0 | 0 | 0 | 0 | 10 | 0 |
| Wet bottom expansion chamber | 33 | 0 | 7 | 0 | 0 | 10 | 22 | 4 |
| Spray chamber | 40 | 0 | 25 | 0 | 0.1 | 40 | 40 | 5 |
| Mechanical cyclone (dry) | 70 | 0 | 0 | 0 | 0 | 0 | 35 | 0 |
| Medium-energy wet scrubber | 90 | 0 | 65 | 0 | 1.5 | 95 | 95 | 80 |
| Electrostatic precipitator | 99 | 0 | 0 | 0 | 0 | 0 | 60 | 90 |
| Fabric filter | 99.9 | 0 | 0 | 0 | 0 | 0 | 67 | 99 |

Source:  Adapted from Ref. 11.8, © Van Nostrand Reinhold Co., New York, 1984.

per day garbage to energy incinerator went on line.
Thanks to the efficiency of its local recycling
program, it was unable to generate sufficient refuse
to keep the new incinerator operating at optimum
temperatures.  As a consequence, it had to exclude
paper products from mandatory recycling to increase
the volume of refuse for fuel.  While this is disturbing in itself, the future recycling of plastics,
although not now required, may be jeopardized as well
since these materials are currently part of the fuel
utilized to sustain this incinerator.

## LANDFILLING

### Introduction

As discussed in the first part of this chapter,
landfilling has been the nation's most widely used
method of solid waste disposal.  Its historical popularity is based upon its relatively low costs and ease
of implementation; two attributes destined for obliv-
ion, thanks to stiffening federal and state legislation.  The types of materials that have been landfilled are too numerous to list here.  Suffice it to
say that they include nearly every sort of industrial,
commercial and municipal waste and not necessarily all
of it solid.

At one time, landfilling was considered the only
final method of solid waste disposal in that it generated no residues or end product which required further
disposal, unlike incineration or composting.  The in-
sidious spread of leachate contaminated ground and
surface water and methane gas explosions have caused
us to modify that philosophy substantially.

Our terminology for landfilling is likewise in
need of clarification.  The terms dump, landfill and
sanitary landfill have been used so interchangeably
that professional and lay persons alike remain confused over their meanings.  For the record, sanitary
landfilling is the only acceptable method for inter-
ring solid wastes.  All other terminologies and practices, not in conformance with the following criteria[7] defining sanitary landfilling can be discarded:
   1.   Vector breeding or their sustenance is pre-
        vented.

2.  Air pollution by dust, smoke, odors and gas is controlled.
3.  Fire hazards are controlled during the operational phase and are negligible in the completed landfill.
4.  Pollution of surface and ground water is precluded, and,
5.  All nuisance factors (i.e. noise and aesthetics) are effectively controlled.

Unfortunately, many of the nation's alleged sanitary landfills fail to meet all of these criteria with the largest percentage failing to comply with item 4. It is these serious omissions that have, in part, prompted the federal government to become more directly involved in the regulation of landfills. Retrofitting these short shrifted landfills with the appropriate safeguards presents a monetarily and physically formidable task--and one that is certain to drain our present resources. Lenient environmental safeguards have again proven to be no bargain. Having learned a lesson (hopefully) from these past transgressions we, as land planners, ought not to feel any remorse whatever when we insist on the maximum environmental safeguards at all new sanitary landfills or expansion of old ones. And there are landfills in our future, despite predictions to the contrary, although they may be substantially smaller than some envision. We need them for the immediate future as temporary storage areas of solid wastes until such time as alternate disposal methods such as composting and recycling are up and working.

We will need them in the longer term for interment of such items as incinerator ash and radioactive wastes, and, if need be, for the storage of cured compost. Let us now examine the actual process of sanitary landfilling from site selection to site closure.

## Legislation and Regulation

Prior to 1976, landfill regulation was largely a municipal and state responsibility[8]. With the passage of the federal Resource Conservation and Recovery Act (RCRA) of 1976, the U.S. EPA became an important part of the regulatory process.

Land planners will find the Subtitle D regulations of RCRA promulgated by the U.S. EPA most applicable in their day to day review of landfilling proposals. However, they should also stay abreast of Subtitle C regulations which set up a comprehensive regulatory scheme for the treatment, storage and disposal (including landfilling) of the nation's hazardous wastes.

Until 1988, the most significant sections of Subtitle D were found in 40 CFR Part 257 Criteria promulgated September 13, 1979 and slightly modified on September 23, 1981. These Criteria established minimum national performance standards to ensure that no reasonable probability of adverse effects on health or the environment would result from solid waste disposal facilities or practices. A facility or practice that meets these Criteria was to be classified a "sanitary landfill"[9]. Any facility failing to satisfy any of the Criteria is considered an "open dump" for the purposes of State solid waste management planning. Open dumps are subsequently regulated under 40CFR Part 256 which requires that all State plans must provide for the closing or upgrading of all such existing dumps within their boundaries.

The Part 257 Criteria[10] are briefly summarized below:

Section 257.3-1 specifies that facilities or practices in floodplains shall not interfere with the floodplain or result in washout of solid waste so as to pose a hazard to human life, wildlife or land or water resources.

Section 257.3-2 prohibits solid waste disposal facilities and practices that cause or contribute to the taking of any endangered or threatened species or result in the destruction or adverse modification of the critical habitats of such species.

Section 257.3-3 specifies that disposal facilities shall not cause a discharge of pollutants or dredged or fill material to waters of the United States that is in violation of Section 402 or 404 of the federal Clean Water Act (CWA).

Section 257.3-4 establishes ground water protection standards which require that facilities and practices not exceed the federal Safe Drinking Water Act (SDWA) maximum contaminant levels (MCL)[11] in any underground potable water source beyond the landfill boundary or beyond an alternative boundary specified by the state.

Section 257.3-5 requires that a facility or practice meet certain restrictions on the concentrations of cadmium and polychlorinated biphenyls (PCB's) contained in waste to be applied to land used for producing food chain crops.

Section 257.3-6 specifies that waste disposal facilities and practices must institute appropriate vector controls such as periodic application of cover material. This section also requires pathogen reduction processes for sewage sludges and septic tank pumpings applied to land.

Section 257.3-7 prohibits open burning of solid waste (with certain exceptions).

Section 257.3-8 requires control of explosive gases, fires, bird hazards to aircraft and public access to the facility.

The Hazardous and Solid Waste Amendments (HSWA) of 1984 made significant modifications to Subtitle D including the following:

1.    A new Section 4010 was added to RCRA which required the EPA to conduct a study of the extent to which the guidelines and Criteria (except Subtitle C guidelines and Criteria) under RCRA are adequate to protect human health and the environment from ground water contamination. It also required EPA to revise the Subtitle D Criteria for those facilities that may receive household hazardous wastes (HHW) or hazardous wastes from small quantity generators (SQG) exempted from Subtitle C Criteria. These revisions were to include as a minimum, ground water monitoring to detect contamination, the establishment of location standards for new and existing facilities and provisions for corrective actions.

2.    Section 4005 of RCRA was amended to require the States to establish a permit program or other system of prior approval to ensure that facilities that receive HHW or SQG hazardous wastes are in compliance with the existing Part 257 Criteria.

As a result of these 1984 Amendments, a substantial number of studies (over 20) were undertaken by the U.S. EPA to assist them in the formulation of these new mandated regulations. Some interesting statistics were derived from these studies and I have listed those of special significance below[12]:

1.    There are about 227,000 disposal facilities (excluding waste piles) covered by the Subtitle D
criteria in the U.S.    This total includes approximately 16,500 landfills, 191,500 surface impoundments
and 19,000 land application units.

2.    There are approximately 6,034 municipal
solid waste landfills (MSWLF's) nationwide (as of
1986).  These MSWLF's are owned predominantly by local
governments (80%) with the remainder owned by private
entities (15%), the Federal government (4%) and State
government (1%).

3.    Approximately 42% of these MSWLF's are
less than 10 acres in size and 52% dispose of less
than 17.5 tons per day.

4.    Only 25 to 30% of these MSWLF's have some
type of ground water monitoring system.  About 25% of
those landfills were reported to be violating a State
ground water protection standard.  Most of these facilities do not monitor for organic, hazardous contaminants.

5.    Approximately 845 MSWLF's were cited for
air-related violations (most often odors) and about
660 were cited for surface water contamination.

6.    Of the 163 existing MSWLF's studied by
EPA, 146 were found to be contaminating the ground
water and 73 facilities were found to be causing surface water contamination.

7.    Typically, those facilities causing ground
water contamination were more than ten years older
than facilities reporting no impacts.  Ground water
impacts appeared to be more severe in locations characterized by high net infiltration rates and high
ground water flow rates.  In addition, most facilities
with contaminated ground water were located close to
the ground water table or were underlain by highly
permeable soils or lacked or had limited engineering
controls.

8.    Methane gas from MSWLF's must be controlled to protect human health.  Where it is not controlled, fires and explosions have occurred resulting
in human deaths in at least five instances.

9.    Of the 850 sites listed or proposed for
listing on the Superfund National Priorities List
(NPL) 184 sites or 22 percent are MSWLF's.  In addition, of the 27,000 sites in the Superfund data base,

almost one fourth are MSWLF's.  Most of the MSWLF NPL sites were in operation prior to 1980.

10.   Approximately 6,800 publicly owned treatment works (POTW's) dispose of their sludge in MSWLF's.  This represents the sludge disposal practice used by 44 percent of all POTW's.

11.   Specific design and operating standards for MSWLF's vary greatly amongst the States and Territories.  For example, only 24 States and Territories (as of 1987) required liners, while 27 require leachate collection systems.  Thirty-eight States and Territories required ground water monitoring systems.

12.   The minimum distances presented by States and Territories between MSWLF's and habitable residences vary from 200 feet to three-quarters of a mile and the required distances from community water supplies range from 400 feet to one mile.

With these fact gathering studies completed, the U.S. EPA presented the first phase of a two phase promulgation of new regulations pertaining to landfills pursuant to the 1984 HSWA Amendments.  (See Federal Register Vol. 53, No. 168, 33314, August 30, 1988.) This first phase is applicable only to MSWLF's.  By its own admission, EPA has, in the August 3, 1988 promulgation "proposed revisions that go beyond these minimally required by HSWA (i.e. location restrictions, ground water monitoring and corrective action)."  Items it considers to have exceeded the statutory minimum include:

1)   An update of the design and operating criteria in the existing Part 257, and,

2)   New requirements for closure and post closure care and financial responsibility.

I urge you to obtain and review the complete set of regulations upon final promulgation.  You will note in your review the EPA's characteristic proclivity to balance the need to protect the environment with economic concerns--in this case, the practical capability of owners and operators of MSWLF's.  The reader is, therefore, cautioned to regard these regulations more as minimum criteria rather than those that afford maximum protection to the environment.

## Siting and Design of the Sanitary Landfill

While we have shown that we can construct land-fills virtually anywhere, we now know that there are areas of high risk that should never be utilized for this purpose. These include[13]:

1. Floodplains (using a 100 year design storm delineation as a minimum).

2. All wetlands defined by the U.S. Army Corps of Engineers and the U.S. EPA as follows: "Areas that are inundated or saturated by surface or ground water at a frequency and duration sufficient to support, and that under normal circumstances do support, a prevalence of vegetation typically adopted for life in saturated soil conditions. Wetlands include, but are not limited to swamps, marshes, bogs and similar areas."

3. All geological fault zones up to and including 200 feet on either side of the center.

4. Seismic impact areas which are defined by EPA "as areas having a 10 percent or greater probability that the maximum expected horizontal acceleration in hard rock, expressed as a percentage of the earth's gravitational pull (g) will exceed 0.10 g in 250 years."

5. Subsidence prone areas which are subject to a lowering or collapse of the land surface.

6. Areas susceptible to mass movement of soil and rock under gravitational influence.

7. Areas of unstable soils that lose their ability to support foundations as a result of swelling or shrinkage.

8. All karst terrains, which are areas where solution cavities and caverns develop in limestone or dolomitic materials.

9. All areas within 10,000 feet of an airport used by turbojets and all areas within 5,000 feet of an airport used by piston-type aircraft if a bird hazard to aircraft is expected[14].

Any proposed expansion of a pre-existing land-fill in any of the above areas should likewise be discouraged. To do otherwise would only exasperate a bad situation.

If you are successful in locating a potential site after all of the unsuitable areas have been excluded, you will be required to prepare a detailed characterization of the site's assets and liabilities examining such things as geology, geo-hydrology, hydrology, topography, soils, ground water quality, demographics (particularly the proximity of human habitation), accessibility, vegetative cover and microclimatology. An accompanying environmental impact statement will also be required.

Today's landfill designers may assume, unless public opinion and present government initiatives change drastically, that every new landfill site will require, as a minimum, a complete impervious bottom liner (possibly two if hazardous materials are expected), a leachate collection system (LCS) including treatment, a methane gas collection and/or venting system and a ground water monitoring system. Unfortunately, requiring these extensive environmental safeguards may preclude the use of some of the most commonly used methods of sanitary landfilling (i.e. progressive excavation, trench cut and cover, and slope cut and cover). The trench cut and cover, for instance, in which progressive parallel trenches are dug in existing soil and backfilled with waste would be incompatible with a whole-area, impervious bottom liner (unless of course there is an existing underlying clay strata). Rather, each trench would now be required to have its own individual liner--a costly, perhaps prohibitive proposition.

Other items of interest to the designer and planner in the site evaluation process include:

1)   <u>Availability and Suitability of Soil Cover Material</u>

Because soil cover is an integral part of the day to day operation of a sanitary landfill, a continuous supply will be needed. Utilizing soil sources on site reduces the overall costs dramatically. The (15) characteristics of a desirable soil cover material include easy workability, moderate cohesion and significant strength. A sandy loam or silty loam are very acceptable soils. Clean sands are less desirable because of their permeability which allows easy penetration by water. Similarly, pure clays are to be

avoided because of their poor workability when wet and
their tendency to shrink and form deep cracks upon
drying.

2.    Site Microclimatology

Microclimatological parameters of special interest to landfill design are annual precipitation, prevailing winds and temperature.  The amount, frequency
and intensity of precipitation at a landfill site
affect the amount of leachate produced.  Humid areas
generally produce more leachate.  Conversely,
extremely arid regions with their paucity of rainfall
make leachate collection systems seem superfluous[16].
The velocity and direction of the local prevailing
winds system are important for the proper placement of
fences to collect blowing paper and other litter.
Seasonal temperatures, particularly below freezing
temperatures, may make the excavation and handling of
soil impossible at certain times of the year.  As a
consequence, provisions may have to be made for temporary storage areas.

3.    Traffic Impacts on Existing Roadways

An analysis of the capacities of the existing
roadway systems and their structural design will be
extremely important.  Several hundred large trucks a
day hauling waste to a landfill can quickly weaken
pavement not designed for such heavy loading.  Further, local traffic patterns may be drastically
disrupted during the peak hours of deliveries to the
landfill due to queuing of trucks at the entrance.  In
addition, the designer must further consider some way
to curtail mud tracking from landfill roads onto the
existing road system.

**Minimum Safeguards--A Choice of Alternatives**

A variety of alternative materials and designs
are available to meet minimum safeguards outlined previously.  We will examine some of them here.

1)    Impervious Bottom Liners

The primary purpose of lining a sanitary landfill is to prevent leachate from seeping from the
waste and into the underlying subsoil and eventually
into the ground or surface water.  Some liners also
serve a dual function by absorbing or attenuating pollutants directly thereby restricting further movement.
This absorptive capability is dependent, as we shall

shortly see, upon the chemical composition of the liner material and its mass (thickness). Liners may be grouped into two major types: natural[17] (clay, clay soil) and synthetic[18] (flexible membrane liners).

The earliest landfill liners were nearly all of natural derivation and included various mixtures of soil and clay, bentonite clay, cement stabilized sand and wherever possible, naturally occurring clay formations. Bentonite clay was particularly popular. Clays, however, have a great deal of ion exchange capacity (Table 11.7) and as a consequence, are subject to ion exchange reactions. For example, acids tend to adsorb on clays and dissolve aluminum, iron and silica particles in the clay lattice structure. Similarly, alkaline solutions can also remove silica from the clay lattice. As a result, permeabilities can increase significantly.

Clay liners are also prone to differential settlement which can lead to localized cracking and subsequent leakage. However, a recently developed product called Clay Max LC ® manufactured by Clem Environmental Corp. in Fairmount, Georgia shows promise in overcoming this latter flaw. Clay Max's manufacturer describes its product as "a specially constructed, flexible, impermeable liner system which utilizes the mineral sodium bentonite clay and the geotextile polypropylene." The clay is actually sandwiched between a flexible, heavy layer of woven polypropylene and a thin polyester scrim. When in contact with water, the 1/4" thick clay sheets swell to 1/2" to 1" resulting in the equivalent permeability of 30 feet of compacted clay, according to its manufacturer.

A comprehensive list of synthetic liner materials is shown in Table 11.8. Due to space restrictions, we can only discuss a few of them here. Please refer to Table 11.9 for additional information on other selected synthetic liners.

High Density Polyethylene (HDPE) is a popular material that has a high resistance to acidic and caustic chemicals as well as oils. Its equivalent, Darcy's[19] law permeability to water is a million times less than that of a well compacted clay (2.7 x $10^{-13}$ cm/sec vs. 1 x $10^{-7}$ cm/sec). HDPE liners are extremely weather resistant, maintaining their design properties in some tests for over 40 years. Because

Table 11.7:  Attenuation and Permeability Properties of Clays

| % | Material | Cation exchange capacity meq/100g | Bulk density g/cm³ | Initial hydraulic conductivity[b] cm/sec |
|---|---|---|---|---|
| 0 | Montmorillonite | 0.0 | 1.71 | 1.27E-03 |
| 2 | Montmorillonite | 1.7 | 1.71 | 9.45F-04 |
| 4 | Montmorillonite | 3.3 | 1.77 | 4.34E-04 |
| 8 | Montmorillonite | 6.8 | 1.79 | 4.70E-04 |
| 16 | Montmorillonite | 13.3 | 1.87 | 1.22E-05 |
| 32 | Montmorillonite | 27.3 | 1.55 | 1.27E-06 |
| 64 | Montmorillonite | 50.7 | 1.23 | 3.05E-07 |
| 100 | Montmorillonite | 79.5 | 0.84 | 7.26E-07 |
| 2 | Kaolinite | 0.2 | 1.68 | 7.44E-04 |
| 4 | Kaolinite | 0.5 | 1.76 | 4.78E-05 |
| 8 | Kaolinite | 1.0 | 1.80 | 9.90E-04 |
| 16 | Kaolinite | 2.2 | 1.87 | 2.86E-05 |
| 16 | Kaolinite | - | 1.94 | 1.09E-06 |
| 32 | Kaolinite | 4.3 | 1.66 | 2.40E-06 |
| 64 | Kaolinite | 8.2 | 1.22 | 5.45E-07 |
| 100 | Kaolinite | 15.1 | 0.90 | 2.98E-07 |
| 4 | Illite | 0.7 | 1.80 | 8.17E-04 |
| 16 | Illite | 2.7 | 1.83 | 2.68E-05 |
| 8 | Montmorillonite + 8 Kaolinite | 7.6 | 1.95 | 5.35E-07 |
| 8 | Kaolinite + 8 Illite | 2.8 | 1.95 | 1.48F-06 |
| 8 | Kaolinite + 8 Illite + 8 Montmorillonite | 9.2 | 1.64 | 8.08F-06 |

[a] Quartz sand added to make 100%
[b] Exponential notation:  E-03 means x 10⁻³

Source: Ref. 11.29

Table 11.8:  Liner Materials and Codes

| CODE* | DESCRIPTION |
|-------|-------------|
| HDPE/EPDM | High-density polyethylene/EPDM alloy |
| HDPE | High-density polyethylene |
| LDPE (Copol) | Low-density polyethylene copolymer |
| EVA (Copol) | Ethylene vinyl acetate copolymer |
| CSPE | Chlorosulfonated polyethylene |
| CR | Polychloroprene |
| EPDM | Ethylene-propylene-diene terpolymer |
| IIR | Isobutene-isoprene copolymer (butyl rubber) |
| CO | Polyepichlorohydrin |
| ECO | Epichlorohydrin-ethylene oxide copolymer |
| CPE | Chlorinated polyethylene |
| PVC | Polyvinyl chloride |
| Polyester | Polyester elastomer |
| AC | Asphalt concrete |
| HAC | Hydraulic asphalt concrete |
| SC | Soil cement--95 parts soil, 5 parts clay, 10 parts cement, 9 parts water |
| SA | Soil asphalt--7 parts asphalt, 100 parts soil |
| ASPH | Asphalt |
| ECB | Ethylene-bitumen copolymer |
| PVC-CPE | Polyvinyl chloride-chlorinated polyethylene blend |
| CPE/PE/CPE | Laminate of CPE and PE |
| XR-5 | XR-5 ® (Seaman Corporation) |

*As used in the chemical resistance matrix.

Source:  Ref. 11.13

Table 11.9:  Characteristics, Advantages, Disadvantages of Certain Liner Materials

| Liner Material | Characteristics | Advantages | Disadvantages |
|---|---|---|---|
| Butyl rubber | Copolymer of iso-butylene with small amounts of isoprene | Low gas and water vapor permeability; thermal stability; only slightly affected by oxygen-ated solvents and other polar liquids | Highly swollen by hydrocarbon sol-vents and petro-leum oils; diffi-cult to seam and repair |
| Chlorinated polyethylene | Produced by chemical reaction between chlorine and high density polyethylene | Good tensile strength and elongation strength; resis-tant to many inorganics | Will swell in pre-sence of aromatic hydrocarbons and oils; high elonga-tion, poor memory |
| Chlorosulfonated polyethylene | Family of polymers prepared by reacting polyethylene with chlorine and sulfur dioxide | Good resistance to ozone, heat, acids, and alkalis; easy to seam | Tensile strength increases on aging; good tensile strength when sup-ported; poor resis-tance to oil |
| Epichlorohydrin rubbers | Saturated high molecular weight, aliphatic poly-ethers with chloro-methyl side chains | Good tensile and tear strength; thermal stability; low rate of gas and vapor permea-bility; weathering; resistant to hydro-carbons, solvents, fuels, and oils | Difficult to field seam or repair |
| Ethylene propylene rubber | Family of terpoly-mers of ethylene, propylene, and nonconjugated hydrocarbon | Resistant to dilute concentrations of acids, alkalis, silicates, phos-phates and brine; tolerates extreme temperatures; flexible at low temperatures; excellent resis-tance to weather and ultraviolet exposure | Not recommended for petroleum solvents or halogenated sol-vents; difficult to seam or repair; low seam strength |
| Neoprene | Synthetic rubber based on chloroprene | Resistant to oils, weathering, ozone, and ultraviolet radiation; re-sistant to punc-ture, abrasion, and mechanical damage | Difficult to seam or repair |
| Polyvinyl chloride | Produced in roll form in various widths and thick-nesses; polymeri-zation of vinyl chloride monomer | Good resistance to inorganics; good tensile, elongation, punc-ture, and abrasion resistant proper-ties; wide ranges of physical prop-erties; easy to seam | Attacked by many organics, includ-ing hydrocarbons, solvents and oils; not recommended for exposure to weathering and ultraviolet light conditions |
| Thermoplastic elastomers | Relatively new class of poly-meric materials ranging from highly polar to nonpolar | Excellent oil, fuel, and water resistance with high tensile strength and ex-cellent resis-tance to weath-ering and ozone | None reported |
| High Density Polyethylene | Blow or sheet extended P.E. | Good resistance to oils and chemicals; resistant to weathering; avail-able in 20 to 150 mils thicknesses; resistance to high temperature | Thicker sheets require more field seams; subject to stress cracking; subject to puncture at lower thick-nesses.  Poor tear propagation. |

Source:  Adapted from Ref. 11.31

of its excellent chemical resistance, heat seaming
rather than adhesives are used to join sheets during
installation.

Polyvinyl chloride (PVC) is another widely used
liner. It has good initial flexibility thanks to the
addition of plasticizers, whereas PVC by itself is
rigid. However, if the plasticizers leach out, the
PVC will become brittle. The main cause of plasticizer loss is volatilization caused by the heat of
the sun. The addition of carbon black (2%) during
formulation reduces aging caused by ultraviolet light.
Surprisingly, rodents and soil microorganisms find the
plasticizers attractive as a food source and are
another cause of PVC liner deterioration. PVC has
adequate resistance against inorganic compounds but
poor resistance against organic compounds again due to
the extraction of the plasticizers.

Hypalon®, a chlorosulfonated polyethylene (CSPE)
has also been used as a landfill liner. It is usually
reinforced with a scrim fiber laminated between two
layers of polymer. Hypalon thickness from the liner
surface to the scrim surface can be as thin as 10
mils, consequently any abrasion of the Hypalon could
permit leachate access to the scrim where it could be
transmitted throughout the membrane or beyond. This
is referred to as wicking.

CSPE can be made two different ways: as a
thermoplastic elastomer or as a cross-linked polymer.
Cross-linking CSPE improves its chemical resistance
and lowers the amount of swelling that occurs with
some leachates.

CPE (chlorinated polyethylene), liners never
gained widespread popularity due to their lack of
strength and durability. The chlorination of the
polyethylene itself destroys the semi-crystalline
structure of the polyethylene molecules. CPE is also
plagued by the loss of plasticizers similar to PVC
that are added to impart flexibility. As with PVC,
this plasticizer loss causes the material to stiffen
and become brittle. CPE is also reinforced with a
fabric scrim much like Hypalon.

Although LDPE (low density polyethylene) another
synthetic is widely used in the polymer industry, it
is not recommended for use in lining landfills. LDPE
has a lower level of density which indicates a lower
level of crystallinity. This decrease in crys-

tallinity affects its chemical resistance, making it susceptible to acids, caustics and oils. In addition, this lower crystallinity makes LDPE's methane permeation and water vapor transmission rates higher than HDPE.

I am frequently asked if any of these liners, particularly synthetics, will remain functional forever and my answer, of course is an unequivocal no. Such a liner has not yet been developed, nor do I expect one to be developed, in the near future. Our hope is that they will last at least until decomposition has been completed and the production of leachate becomes minimal or ceases altogether.

The absence of older, lined landfills makes it difficult to predict with absolute certainty, liner longevity. It is this uncertainty that contributes to public apprehension about continued use of sanitary landfills for solid waste disposal. A number of presently available structural and operational practices, however, may make sanitary landfilling more environmentally acceptable hence more publicly palatable. First, some engineers have proposed to shorten the time required for complete decomposition to occur by introducing water at the top of the completed landfill, letting it percolate through the waste to the leachate collection system and then reapplying it back to the top of the landfill. This would create, in effect, a closed loop treatment system. In theory, the introduction of this constantly percolating water will accelerate decomposition and the leaching of harmful materials from the waste. Secondly, some states now require the installation of a leak detection system (also known as a witness drain) below the liner and leachate collection system (Figure 11.4). This would facilitate the early detection of liner failure and encourage the rapid implementation of corrective measures. Finally, liner integrity and strength can be enhanced significantly by the placement of a geotextile fabric above and below it (Figure 11.5). These geotextiles protect the liner from abrasion and punctures by rocks and other materials (including installation equipment) while at the same time spreading static loads and bridging liner-weakening subgrade cracks and small voids. They also serve a secondary function by facilitating the transfer of leachate from the waste[20].

Figure 11.4: Witness drains may be located directly below the regular leachate collection pipes or between each set of leachate collection laterals as shown here.

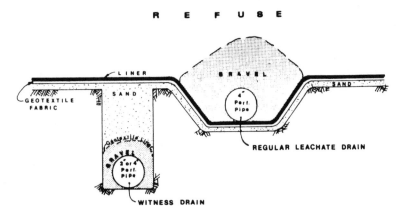

Figure 11.5: Enlarged detail showing geotextile fabric sandwiching of a landfill liner.

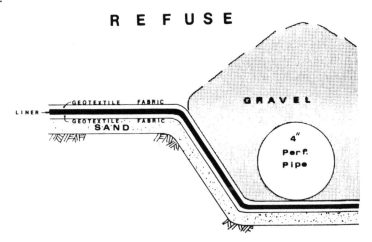

Figure 11.6: Typical cross-section of a subsurface leachate collection drain.

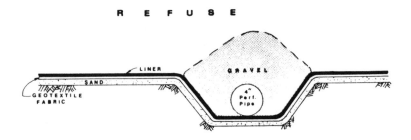

2)    Leachate Collection Systems

Installing a liner on the bottom of a sanitary landfill makes the installation of a leachate collection and removal system a necessity[21]. Without such collection and removal, leachate will accumulate on the liner, backing up into the waste (much like a bathtub) and result in seepage at the landfill surface or leakage through the liner due to excessive hydraulic pressure. A leachate collection system generally consists of a series of perforated drain pipes installed in gravel filled trenches set above the bottom liner (Figure 11.6). These collector drains (generally spaced 50 to 200 feet on center) subsequently drain the leachate, usually by gravity, to a sump or series of sumps where it is withdrawn for treatment and disposal.

The recommended slope of the bottom liner for this purpose is $\geq 2$ percent and for the collection pipes $\geq 0.005$ percent. Observation riser pipes should also be installed at strategic points along the collection system to facilitate the monitoring of leachate flows and the clean out of pipes in the event of clogging--a topic we will address in more detail shortly. Plastic (PVC and polyethylene) and fiberglass pipes are commonly used for these collector systems. However, if significant amounts of solvents are expected to be present concrete or cast iron may be the better choice. These latter materials may also be better able to withstand the pressure of the overlying waste, soil cover and heavy trucks.

Clogging of the subsurface leachate collection system is a major concern. It can prevent the flow of leachate both to and through the drains. The major mechanisms of clogging are physical, chemical, biochemical and biological. Physical mechanisms are the most common cause of drain failure and are usually due to pipe collapse, inadequate design capacity or sedimentation. Chemical mechanisms usually involve the formation of insoluble precipitates that are deposited on the inside surfaces of the drain pipes, or their perforations or in the filter envelopes (gravel or geotextiles) surrounding the pipe. Sulfides, silicates and calcium and manganese carbonates are the most common precipitates found. Biochemical mechanisms occur when inorganic precipitates, principally $Fe(OH)_3$ or $FeS$ are deposited on pipe surfaces and in

the envelope material.    There may also be filamentous
slimes.    Finally,  biological  clogging  occurs  when
organisms  grow  to  fill  the  interstices  in  the  drain
envelope,  similar to what occurs in the operation of a
trickling filter at a sewage treatment plant.

3)    Methane Gas Collection and Venting Systems

The processes of waste decomposition in a sanitary landfill are nearly identical to those that occur
in the composting process.    Unlike  composting,  however,  there is virtually no manipulation of the waste
environment  to  produce  uniform  conditions  that  maximize  rapid  aerobic  decomposition  and  minimize  production  of  nuisance  odors  and  potentially  dangerous  by-
products such as methane gas.    Wherever oxygen remains
abundant within the refuse, aerobic microbes dominate
and the major gas produced will be carbon dioxide.    In
those  sections  of  the  landfill  where  oxygen  is
depleted,  anaerobic  microbes  flourish  and  methane  gas
becomes the primary gas produced.    With waste decomposition  so  non-uniform,  it  is  common  to  find  both
carbon dioxide and methane gas being emitted simultaneously  (usually as a mixture) from the same landfill.
While the ratio of this mixture varies from landfill
to landfill and with time, field studies indicate that
a 50% methane and 50% carbon dioxide by volume mixture
is a reliable average.    One study at a sanitary landfill is Deptford Township, New Jersey[22] showed the
following  components  in  the  landfill  gas  from  that
site:

| Component | Volume % |
|---|---|
| Methane | 54 |
| Carbon Dioxide | 39 |
| Oxygen | 2 |
| Nitrogen | 3 |
| Water | 2 |
| Hydrogen Sulfide ($H_2S$) | 0.003 |

Methane  gas  and  carbon  dioxide  are  both  odorless, however, they are often "flavored" by a variety
of chemicals including aliphatic hydrocarbons, esters,
terpenes,  benzenes,  mercaptans,  sulfides  and  various
chlorinated hydrocarbons.    It is these compounds that
impart  the  characteristic  and  often  objectionable
odors to landfill gas.

While carbon dioxide is relatively innocuous, methane is not. It can be ignited in air at concentrations between 5 and 15 percent. The 5% is referred to as the lower explosive limit (LEL)[23] and as we shall see, forms the basis for regulating the concentrations of methane that may be emitted from a landfill before remedial action is required. Sadly, it has taken at least 5 deaths and an even greater number of injuries and extensive property damage to convince landfill owners and their regulators that the control of methane migration is an important consideration in landfill design.

Methane can migrate both vertically and laterally. Migration distances of up to 1000 feet laterally from the perimeter of a landfill have been documented. Methane's propensity to accumulate in confined areas such as basements, crawl spaces, pipes, tunnels and utility manholes makes the consideration of adjacent land uses critically important in sanitary landfill planning and design.

The U.S. EPA has set an explosive gas criterion in its RCRA Subtitle D regulations (257.3-8) that requires methane concentrations to be below the LEL of 5% at the property boundary[24].

A variety of collection and venting systems have been developed in response to the need to control the offsite migration of this methane bearing, landfill gas. This, in turn, has produced great interest in utilizing the recovered methane as a supplementary fuel. Now,? some designers are attempting to satisfy both needs with a singular system assuming that gas collection will simultaneously prevent landfill gas migration. This is an inaccurate assumption. One sanitary landfill in California, for example, with a gas recovery system in operation for several years was nonplussed to find that high concentrations of methane gas had still seeped several hundred feet beyond its property boundary[25]. Collection and venting systems can be significantly different for gas migration prevention than for gas recovery. The first priority is to be given to the former.

Gas collection and venting practices generally fall into two categories--passive and active. The simplest passive methods include construction of a very low permeability barrier around the landfill perimeter similar to, if not, the same as, that now required for leachate control. A second alternative is a gravel filled trench installed along the landfill periphery and vented to the atmosphere (Figure 11.7). A third method is the installation of individual gravel pack wells as shown in Figure 11.8.

Active systems are generally superior to passive systems and are more expensive to construct and operate. Two major active systems are presently in use; the extraction well or trench system and the air injection well or trench system (Figure 11.9). When using these systems specifically to control lateral offsite gas migration, they are to be installed along the periphery of the landfill.

The extraction well system consists of a series of vertical wells installed in refuse and connected by a common horizontal header pipe to an electrically driven suction blower. The blower creates a negative pressure in the extraction well which extends radially into the refuse around each well. This radial extension is called its zone of influence. Wells are spaced such that their zones of influence overlap. Migrating gas is drawn into the area of influence and ultimately into the extraction well. Care must be taken not to use excessive air flows to lessen the possibility of starting a fire in the refuse. Typically, a series of monitoring wells or probes (Figure 11.10) are installed along the property line or between the landfill and the area requiring protection to monitor the effectiveness of the system. All of the collection piping and wells must have sufficient flexibility to accommodate differential settlement of the refuse. Condensate in the header pipe caused by the cooling of the warmer, landfill gas must be accommodated. This is usually achieved by sloping the header pipes. The collected gases are either vented or flared (burned).

The extraction trench system is comprised of gravel filled interceptor trenches sealed at the surface and containing a perforated pipe connected to a collection header and suction blower. The blower creates a negative pressure in the intercepting trench which extends toward the refuse, thereby drawing the migrating gas to the perforated pipe. These systems are installed in natural soils and are best suited to shallow landfills (i.e., 20 feet deep or less). The gravel trench depth extends vertically from the ground

Figure 11.7:   Gravel trench gas venting system.

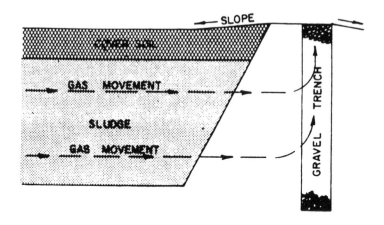

ALTERNATE   CONFIGURATIONS   OF   TRENCH   VENTS

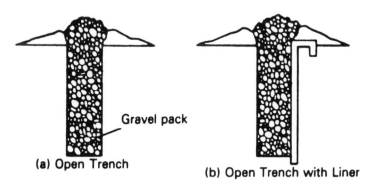

Source:  Ref. 11.14

Figure 11.8: Typical configuration of passive gas extraction systems.

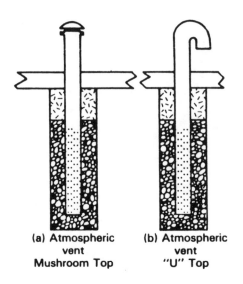

**(a) Atmospheric
vent
Mushroom Top**

**(b) Atmospheric
vent
"U" Top**

PASSIVE SYSTEM USING MULTIPLE VENTS

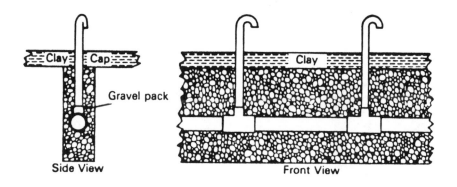

Source: Ref. 11.15

Figure 11.9:  Typical configuration of active gas extraction systems.

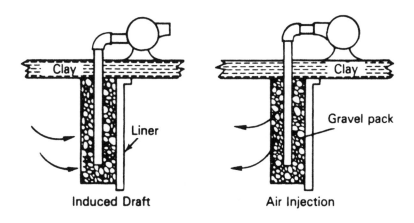

INDUCED DRAFT SYSTEM USING MULTIPLE VENTS

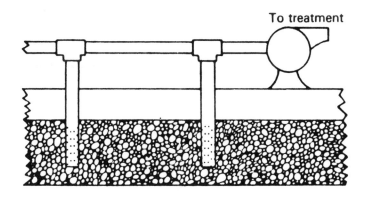

Source:  Ref. 11.15

Figure 11.10: Typical multi-level gas sampling probe installation.

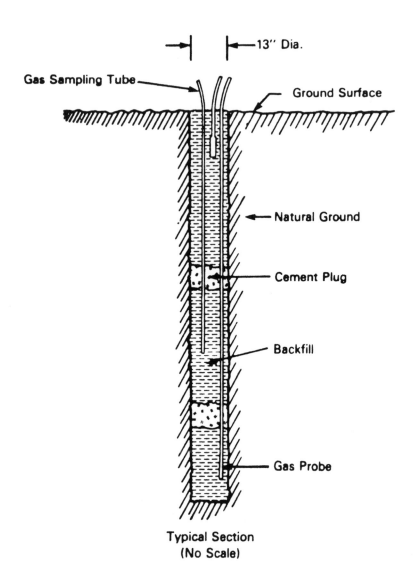

Typical Section
(No Scale)

Source: Ref. 11.15

surface to the full depth of the refuse or ground water whichever comes first.

Well injection systems consist of vertical wells
installed in natural soils between the landfill
perimeter or the property line and the area being protected. These systems are usually selected for landfill sites 20 feet deep or greater. The injection
wells are connected by a common header pipe to a
blower. Air is forced through the header pipes and
wells and into the surrounding soils, creating a zone
of positive pressure around each well. These zones of
positive pressure are designed to overlap thus creating an air curtain that dilutes any passing landfill
gas. Since this system is installed in natural ground
there are usually no problems with settlement. In
addition, forcing air into the system eliminates problems with condensate. More wells, however, are generally required than well extraction systems due to the
reduced permeabilities of the soil as compared to
refuse. Air injection systems do not require a vent
or flare.

The injection trench system is similar in design
to the extraction trench system. However, in this
instance, air is forced into a trench rather than
wells. Again a positive pressure barrier is created
and any migrating gas is subsequently diluted. These
systems are installed around shallow landfills (i.e.
20 feet deep or less).

In any active systems utilizing multiple wells,
the spacing of those wells is of critical importance.
As a general rule, 50 feet between wells would be
suitable (Ref. 11.15), however, this determination is
best made after an evaluation of onsite conditions
particularly that of refuse composition and soil
permeabilities.

Finally, some interesting case studies indicate
that at times even overlapping areas of influence may
fail to capture all of the laterally migrating landfill gas. Connelly's studies (Ref. 11.17) challenge
the basic design assumption that landfill methane
migrates at a gradual spatial increase. His sampling
results indicate that methane is released in pulses or
waves which surge out and then dissipate. These
pulses are particularly acute during periods of
rapidly falling barometric pressure; so acute, in
fact, that Connelly surmises that the quantity of

methane produced may exceed the capacity of the
extraction equipment to collect it.  If this is in
fact the case, then extraction wells may have to be
spaced considerably closer or the capacity of the
extraction system increased.

4)    Ground Water Monitoring Wells

Currently only 25 to 30% of the nation's municipal solid waste landfills have some type of ground
water monitoring system[27].  About 25% (roughly 500)
of these facilities are violating a state ground water
standard.  In addition, most facilities that do monitor do no testing for hazardous organic constituents.
Other glaring deficiencies in the nation's ground
water monitoring program beside the lack of any monitoring system at all are: an excessive reliance by the
State on facility self-monitoring, improper well location and construction, poor sampling techniques and
too few or inappropriate parameters being sampled.

Exactly how many wells are enough is a question
I am frequently asked.  Here, in New Jersey, the rule
of thumb has been a minimum of three--one upstream
(ground water wise) and two downstream.  In reality,
in some places, three to four times that amount might
be required.  The only correct way to estimate how
many will actually be needed is by an extensive geohydrological examination of the site and the development
of a ground water flow map.  The standard designs for
monitoring wells as recommended by the New Jersey
Department of Environmental Protection are shown in
Figures 11.11 and 11.12.  The installation of singular
wells may be a thing of the past, however.  Well clustering, wherein a group of wells (1 to 4 usually) are
set at various depths at the same location, has proven
to be the superior method for obtaining accurate
information on ground water contamination (Figure
11.13).

The overreliance on monitoring well data collected by landfill operators has been a major flaw in
the landfill monitoring program.  Sample collection,
even when carried out by hired "experts" is often
sloppy in spite of the extensive production of quality
assurance field procedure manuals developed by the
U.S. EPA and various states (see Appendix 1, Part D).
For instance, some samplers still do not evacuate the
recommended minimum volume of standing water in a casing (usually 3 to 5 times the volume) prior to sam-

Figure 11.11:  Monitoring well specifications for unconsolidated formations.

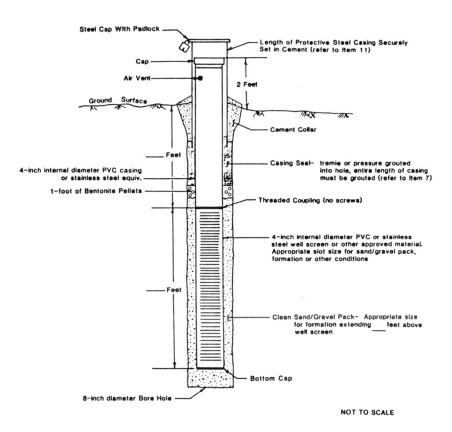

Steel Cap With Padlock

Length of Protective Steel Casing Securely
Set in Cement (refer to Item 11)

Cap

Air Vent

2 Feet

Ground   Surface

Cement Collar

____ Feet

Casing Seal– tremie or pressure grouted
into hole, entire length of casing
must be grouted (refer to Item 7)

4-inch internal diameter PVC casing
or stainless steel equiv.

1-foot of Bentonite Pellets

Threaded Coupling (no screws)

4-inch internal diameter PVC or stainless
steel well screen or other approved material.
Appropriate slot size for sand/gravel pack,
formation or other conditions

____ Feet

Clean Sand/Gravel Pack– Appropriate size
for formation extending ___ feet above
well screen

Bottom Cap

8-inch diameter Bore Hole

NOT TO SCALE

Source:  N.J. Department of Environmental Protection.

Figure 11.12:  Monitoring well specifications for bedrock formations.

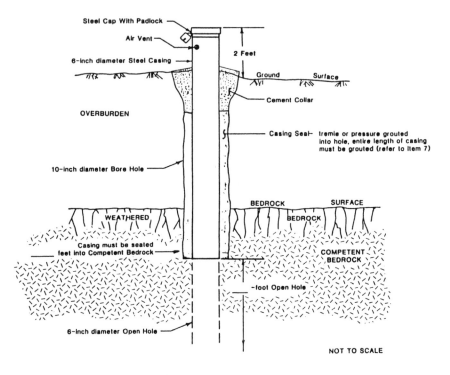

Source:  N.J. Department of Environmental Protection.

Figure 11.13:  Typical well cluster configuration.

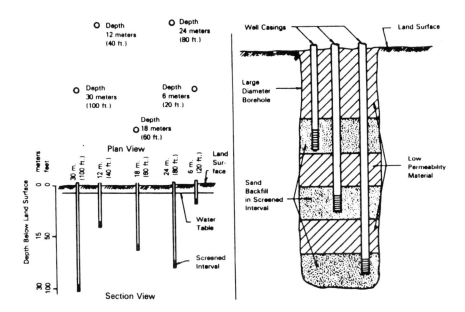

Source:  Ref. 11.15

pling to assure that a representative sample of the
ground water is withdrawn. Sampling equipment is
often improperly cleaned between samples and between
wells and inappropriate sampling equipment and sample
containers are often used.

Wisely, many states have begun long overdue,
sampling verification programs, collecting their own
samples from landfill wells to check the veracity of
self-monitoring data. Hopefully, they will also at
the same time evaluate well placement and well construction which are often uninspected. The monitoring
program for any sanitary landfill should include as a
minimum all of the parameters shown in Table 11.10
unless previously collected, reliable and site
specific sampling data indicates that all of these
parameters would be superfluous.

The frequency of sampling should be at least
quarterly so that any seasonal variations (i.e. summer
vs. fall vs. winter vs. spring) are measured. This is
especially important during the filling phase when the
waste material has not been fully encapsulated and is
therefore more susceptible to climatological
influences.

**Site Closure**

Closing the sanitary landfill requires the same
degree of care and attention that went into its planning and operation. Your primary responsibility will
be to provide a final cover for the entire top surface
of the landfill in order to exclude infiltration of
precipitation, control odors, prevent disturbance of
the emplaced refuse and to encourage stabilization
through re-vegetation. The stability of that cap
depends a great deal on how well you sorted
(discouraged heterogeneity) and compacted your waste,
especially in the final layers. Bulky items in particular, such as tires and wood should have been kept
out of the upper layers of each cell. Single layer
caps are not recommended due to the high probability
of settlement cracks (Figure 11.14).

Instead, a multi-layer cover is to be used.
Figure 11.15 demonstrates some representative multi-
layer covers. There are two basic components of such
multiple covers--a bottom layer of impermeable or very
low permeability clay or clay and soil mix (or syn-

Table 11.10:  List of Parameters to be Measured in Monitoring Wells as a
Minimum

```
 (1)   Ammonia (as N)
 (2)   Bicarbonate (HCO₂)
 (3)   Calcium
 (4)   Chloride
 (5)   Iron
 (6)   Magnesium
 (7)   Manganese, dissolved
 (8)   Nitrate (as N)
 (9)   Potassium
(10)   Sodium
(11)   Sulfate (SO₄)
(12)   Chemical Oxygen Demand (COD)
(13)   Total Dissolved Solids (TDS)
(14)   Total Organic Carbon (TOC)
(15)   pH
(16)   Arsenic
(17)   Barium
(18)   Cadmium
(19)   Chromium
(20)   Cyanide
(21)   Lead
(22)   Mercury
(23)   Selenium
(24)   Silver
(25)   The following volatile organic compounds:
```

| | |
|---|---|
| Acetone | Chloromethane |
| Acrolein | Dibromomethane |
| Acrylonitrile | 1,4-Dichloro-2-butane |
| Benzene | Dichlorodifluoromethane |
| Bromochloromethane | 1,1-Dichloroethane |
| Bromodichloromethane | 1,2-Dichloroethane |
| cis-1,3-Dichloropropene | 2-Hexanone |
| Trans-1,3-Dichloropropene | Iodomethane |
| 1,4-Difluorobenzene | Methylene chloride |
| Ethanol | 4-Methyl-2-pentanone |
| Ethylbenzene | 1,1-Dichloroethene |
| Ethyl methacrylate | trans-1,2-Dichloroethene |
| 4-Bromofluorobenzene | Styrene |
| Bromoform | 1,1,2,2-Tetrachloroethane |
| Bromomethane | Toluene |
| 2-Butanone | 1,1,1-Trichloroethane |
| (Methyl ethyl ketone) | 1,1,2-Trichloroethane |
| Carbon disulfide | Trichloroethene |
| Carbon tetrachloride | Trichlorofluoromethane |
| Chlorobenzene | 1,2,3-Trichloropropane |
| Chlorodibromomethane | Vinyl acetate |
| Chloroethane | Vinyl chloride |
| 2-Chloroethyl vinyl ether | Xylene |
| Chloroform | |

Source:  Ref. 11.30

Figure 11.15: Typical layered cover designs.

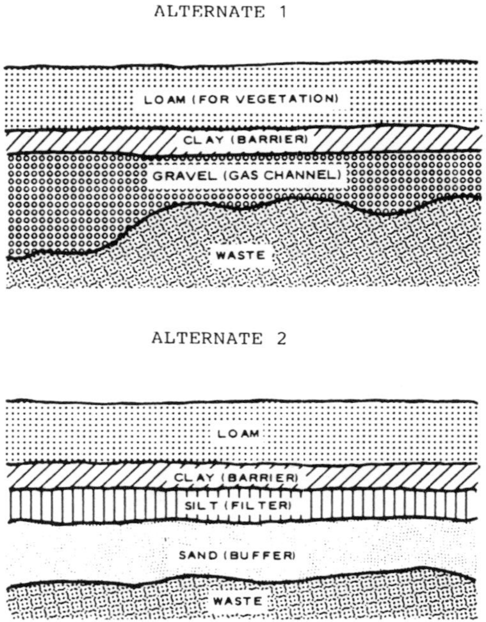

ALTERNATE 1

LOAM (FOR VEGETATION)
CLAY (BARRIER)
GRAVEL (GAS CHANNEL)
WASTE

ALTERNATE 2

LOAM
CLAY (BARRIER)
SILT (FILTER)
SAND (BUFFER)
WASTE

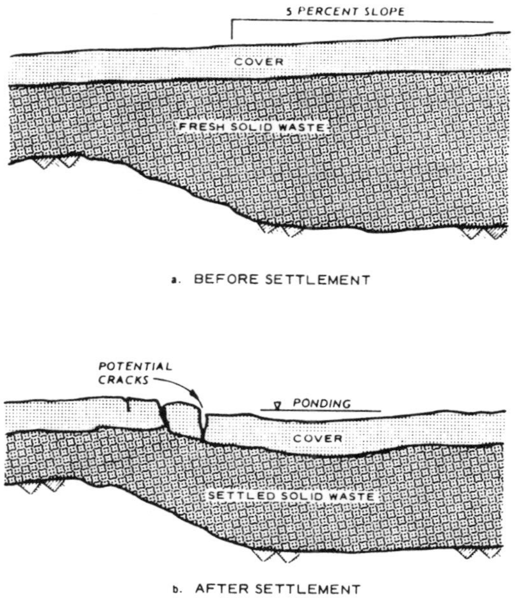

Source:  Ref. 11.14

Figure 11.14:  Disruption of landfill cover by excessive settlement.

5 PERCENT SLOPE
COVER
FRESH SOLID WASTE

a. BEFORE SETTLEMENT

PONDING
COVER
SETTLED SOLID WASTE
POTENTIAL CRACKS

b. AFTER SETTLEMENT

Source:  Ref. 11.14

thetic liner) and an upper layer of essentially topsoil. The bottom layer is expected to prevent infiltration by precipitation as well as prevent the release of landfill gas that could preclude re-vegetation[28] or cause odors. The upper layer, usually a minimum of 24 inches in depth, is for the re-establishment of vegetation[29] to stabilize the cover itself. While 24 inches of topsoil is recommended, an alternative mixture of 18" of silty sand overlain by 6" of topsoil is an acceptable substitute. Additional layers of various materials may be added between these by the engineer or planner for special purposes (Figure 11.16).

Mounding of the final cover is mandatory in order to assure a positive flow of surface water and precipitation away from the landfill. In addition, some provision will usually have to be made to carry leachate collection and gas venting and collection systems through the top cap where they will be readily accessible. Special attention must be paid to assuring that these structures are properly sealed at the point they protrude through the cap, otherwise they may prove to be points of leakage and infiltration.

Once capped, the landfill must be protected from unauthorized users such as all terrain vehicles, dirt bikes, automobiles and illegal dumpers. Even pedestrian traffic should be prohibited due to gas venting. Complete fencing of the landfill perimeter is the recommended choice, however, the posting of signs accompanied by the closing or gating of access roads may be an acceptable alternative.

Some landfills have opted for the use of security guards as well as fencing, especially where gas recovery or leachate collection and treatment systems are in operation.

## Final Use

It can never be too early to plan for the final use of a sanitary landfill once it has been closed. Landfills constructed under today's increasingly stringent standards will be easier to work into other uses, despite inherent problems such as settlement and landfill gas venting. It is the nation's older landfills that will prove the most challenging for land planners. Some of these existing landfills, however,

Figure 11.16:  Landfill cover modifications to accommodate tree planting.

ALTERNATE 1 IN CROSS SECTION-END VIEW

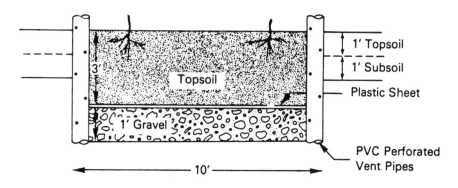

ALTERNATE 2 IN CROSS SECTION-END VIEW

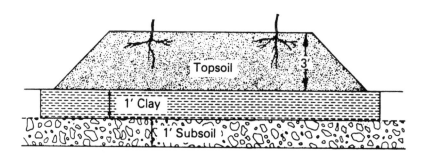

Source: Ref. 11.15

have already been converted, apparently quite successfully, to other uses including passive recreational areas, golf courses, ball fields and commercial and industrial developments. Their use for extensive residential developments is not recommended at this time, particularly if they are situated in geological fault areas, or areas prone to high earthquake activity. Landfills into which toxic wastes are known to have been dumped are to be particularly avoided for this use.

### SPECIAL CONSIDERATIONS--LANDFILLING

### 1.    Sludges and Bulk Liquids in Landfills

The intentional introduction of liquids and semi-liquids such as septage and sewage sludges into sanitary landfills is not recommended. In fact, this practice is already prohibited in twenty-one states and territories[27], with similar restrictions imminent for federal regulations under RCRA. Excess liquids on landfill liners can contribute to their failure due to increased hydraulic head and chemical reactions. In addition, some semi-liquids such as sludges and septage form layers that can retard the extraction of methane gas.

### 2.    Above Ground Landfills

Constructing above ground landfills is not a new concept. We have utilized this practice, although probably unwittingly, when we placed our solid wastes over areas such as swamps and wetlands whose moisture laden soils made excavation undesirable. With environmental safeguards now mandatory for all landfills, this concept offers some exciting advantages over the traditional below grade sanitary landfill. First, there would be no need to excavate below grade in order to install the liner and leachate collection system. Installation would, therefore, be faster and easier. Secondly, by placing all of the waste completely above ground we would increase the distance between the bottom of the landfill and the ground water table--an advantage in areas of a high seasonal water table. Thirdly, any gases generated by the waste could not move laterally through the subsurface to adjacent offsite areas.

Lastly, should we want to reclaim or recycle any of the waste in the future, removal would be much easier.  Two major disadvantages would be the aesthetics of such mounds if not properly landscaped, and the potential for soil erosion.  Despite these disadvantages, it may be a concept well worth considering for your community.

## COMPOSTING

### Historical Development

Composting has a variety of definitions.  Pavoni et al. (Ref. 11.9) for example, describes it "as the biochemical degradation of the organic fraction of solid waste material, having as its end product a humus-like substance that is used primarily for soil conditioning."  Goldstein (Ref. 11.10) defines it as "an aerobic, thermophillic (high temperature) degradation of putrescible material in refuse by microorganisms."  Whatever the definition, composting is simply man's enhancement of a process that occurs spontaneously in the natural environment--the best example of which is the decomposition of organic litter on the forest floor.  When this process is converted to commercial use by man, however, it is manipulated to provide optimum conditions to improve microorganism (those responsible for degradation) efficiency, thereby reducing the overall time normally required for the process to be completed (referred to as stabilization).

While composting has been practiced for centuries, it did not become a systemized process until the 1920's when Sir Albert Howard developed the Indore process.  The Indore process, unlike the methods of composting to be discussed later in this chapter, was an anaerobic process in which leaves, garbage and animal wastes were digested in soil pits for periods of up to six months.  The process, however, was later modified to include turning or mixing to encourage aerobic digestion.  Composting was introduced to the United States in 1949 with the development of the Frazier-Eweson process [30].  During the 1950's, a substantial amount of research on composting was carried out by Michigan State University, the University of California and the United States Public Health Service (Ref 11.9).  Between 1951 and 1969 approximately 18 composting facilities were constructed in the United

States with capacities ranging from 4 tons/day to 360 tons/day. While there were operational problems at these facilities (which included such things as difficulty in economically separating non-compostable refuse, deterioration of shredding equipment and high operational costs when compared to landfilling) it was the lack of a consistent and profitable market for the finished compost that ultimately dampened national enthusiasm for the process and sharply curtailed the further construction of new facilities. Goldstein's (Ref. 11.10) somewhat facetious definition of composting as, "a highly controversial method for treating garbage that should be 'looked into' when the dump is forced to close and incinerator bids come in $5 million higher than anticipated," best reflects the prevalent American attitude toward the process since then. When tipping fees at landfills were $2 to $3 a ton this flippancy toward composting was understandable. Today's rates of $40 to $130 per ton and higher have mellowed this flipness considerably.

Composting has been shown to be an effective treatment technology for a variety of solid wastes including household garbage, animal wastes, leaves, grass and other plant materials and most recently sewage treatment plant sludges and some industrial wastes. In fact, it is the successful composting of sewage treatment plant sludges, that has largely been responsible for keeping the technology alive and developing in the United States. While the lack of a consistent market for compost has not been completely eliminated, this no longer predominates as the major obstacle in implementing composting. In fact, the potential uses for the finished compost have increased substantially and now include among others, application to agricultural land as a soil fertilizer and soil conditioner, application as a final topsoil cover on landfills and application to mine spoil areas to aid the re-establishment of vegetation. In addition, it is this author's hope that with the assistance of the federal government we may someday be able to export our excess compost to other countries who wish to rejuvenate fertility exhausted soils.

Land planners can expect to see a dramatic increase in the number of facilities proposing to utilize this process in the next ten years. Consequently, it is important that they maintain some

familiarity with the process, including a knowledge of
state-of-the-art technologies, the types of materials
suited to the process and some of the potential problems and hazards that need to be considered in their
design. We will address all of these items in this
section.

We remind the reader that composting, as with
all other forms of solid waste disposal covered in
this chapter, should be carried out in conjunction
with a comprehensive recycling and/or resource recovery plan to eliminate plastics, paper (largely
newsprint), glass, ferrous and non-ferrous metals and
household hazardous wastes from the composting material. The purging of these materials will facilitate
the expansion of available markets.

## The Composting Process

While there are a variety of compost systems
(see Compost System Types) all of them generally
include five basic steps: 1) preparation, 2) digestion, 3) curing, 4) finishing, and 5) storage and/or
disposal.

1)   Preparation

The preparation step generally occurs after the
material to be composted is delivered to the composting plant. However, it has been argued that this step
actually begins much earlier, particularly with municipal garbage where recyclables and other non-composta-
bles are removed prior to collection. Once delivered
the particle size of the material must be reduced
(called comminution) to facilitate handling, mixing
and ultimately digestion. This is usually accomplished either with a hammermill or rasper. It should
be noted that a significant amount of the total energy
required for composting is consumed by comminution.
Hence, our earlier recommendation to remove recyclable
materials may help reduce overall energy consumption
as well as eliminate materials that could clog or
breakdown comminution equipment.

It is also at this step that a bulking agent
(wood chips, sawdust, straw, etc.) may be added to
improve porosity (particularly for wastewater treatment plant sludges) and facilitate aeration (oxygen
transfer) in the digestion phase. In addition, the
moisture content [31] (optimally between 50 and 60 per-

cent by net weight) and the carbon to nitrogen ratio
(C:N)(32) of the wastes are usually adjusted at this
stage.   It is interesting to note that a blend of
municipal wastewater sludge and municipal refuse can
reciprocally satisfy the pre-digestion adjustments of
these two wastes.

2)   Digestion

Digestion is the heart of the composting process
and entails the degredation of organic compounds by
naturally occurring microbes.  There is a variety of
microorganisms involved in the digestion process and
during this phase these microflora undergo both quantitative and qualitative changes (see Special Consid-
erations--Pathogen Destruction).

Parameters essential to proper digestion include
temperature, moisture and oxygen.  Of these, oxygen is
the parameter most often manipulated (usually by
induced aeration) by man.  Temperature, however, as we
shall shortly see, can also be a significant parame-
ter(33) by which to judge the efficacy and progression
of the digestion process.   In the initial stage of
digestion, mesophilic flora (organisms able to grow in
the $77°F$ to $113°F$ range) predominate and they are
responsible for most of the metabolic degradation.
This increased microbial activity elevates the temperature of the compost and as a result these mesophilic
flora are replaced by thermophilic (heat loving)
populations which metabolize optimally at temperatures
above $113°F$.  In windrow type, open composting systems
temperatures of $170°F$ have been recorded while in-
vessel (closed) systems have reached temperatures of
$180°F$.   These significant elevations in temperature
are controlled largely by the amount of oxygen available, hence man's emphasis on inducing aeration in the
digesting compost.  A drop in temperature can mean
that the composted material needs more aeration or
additional moisture.  It can likewise signal the cessation of decomposition.  In general, temperatures of
$140°F$ to $160°F$ should be maintained throughout the
entire mass of composting material to ensure adequate
digestion.

The digestion process can be considered complete
and the material satisfactorily stabilized when the
compost has the characteristics of humus, has no
unpleasant odors and high temperatures cannot be maintained in the pile even though optimum aeration and

moisture conditions exist. A number of chemical parameters are usually measured to verify that digestion has in fact been completed. Included are such parameters as chemical oxygen demand (COD) and total organic solids.

3)   Curing

Curing is the final step in stabilization of the compost and generally proceeds with little, further manipulation by man (except perhaps for stockpiling). Additional microbial activity continues during curing with some reheating of the pile occurring. As a consequence, additional volatile solids are removed, pathogenic organisms die off or are prevented from regrowing and the moisture content of the compost is reduced (if the piles are sheltered). Curing time may range from as little as 5 days to as long as 60 days.

4)   Finishing

It is at this step that the cured compost is passed through a screen to remove objectionable debris such as bits of glass, metal or plastic. The moisture content must also be adjusted at this step so that the moisture content does not exceed 30 percent by net weight (see Special Considerations--Pathogen Destruction). It is also at this step that the material may be bagged or pelletized to facilitate sale or storage.

5)   Storage or Disposal

Storage of the finished compost is usually necessary because the demand for compost is seasonal with peak demands occurring in the spring and fall. A composting plant should have a 6 month storage area unless available markets dictate otherwise. For a 300 ton per day composting facility this would require about 13 acres of storage area.

**Compost System Types**

A.   Open Systems
1.   Turned Pile System

This is the simplest type of composting system (Figure 11.17). Collected materials are assembled in piles (in small scale operations) or parallel windrows (in larger scale operations). A number of American cities have used this system for several decades to dispose of leaves, grass and other plant materials. When the volume of material was relatively small and

Figure 11.17: This turned pile system is being used quite successfully to compost nearly 10,000 cubic yards of leaves a year.

storage space was readily available there was
little incentive to speed up the decomposition
(digestion) of the emplaced materials. As the
volume of waste grew and space for deposition
diminished (due largely to population growth and
subsequent development) municipalities were compelled to accelerate the process. Consequently,
hand labor and eventually machines such as front
end loaders and backhoes were used to regularly
mix the composting material to accelerate the
time needed to fully decompose it.

Today this system is used with various
modifications. The addition of bulking agents
(wood chips, sawdust, straw, hay, peat moss,
among others) to provide additional pore space
for added aeration is perhaps the simplest and
most common modification. Some manufacturers
have also developed open top, steel containment
channels within which the waste is windrowed and
over which mechanical stirrers regularly pass to
mix the compost.

Problem Areas - Turned Pile Systems

1.   Turning alone does not ensure the attainment of
     satisfactory and consistent oxygenation. In as
     little as an hour after turning, oxygen levels
     within a pile can drop drastically thereby
     simultaneously reducing microbial degradation.
     As a consequence, the piles must be frequently
     turned making the system very labor intensive.

2.   Piles higher than about 3 meters (9-10 feet) are
     difficult to aerate and consequently may not be
     effectively sanitized.

3.   The material, unless covered, will be exposed to
     rain and snow thereby adversely affecting decom-
     position and producing leachate (runoff) due to
     the excess moisture. Conversely, in arid areas
     moisture loss may be excessive necessitating the
     addition of supplemental water. In addition the
     lack of a protective cover may make composting
     impractical in below freezing temperatures.

4.   Odors may be noticeable offsite when the piles
     are turned.

5.   Fugitive dust may also be produced during pile
     mixing which can migrate offsite. Such dust may
     carry spores of the fungus pathogen Aspergillus
     fumigatus.

2.    Static Pile System

In this system of composting, the waste material is not actively mixed. Some see this as a labor saving method. Biodegradation is facilitated by the introduction of air through the pile. There are at present 3 methods used to accomplish this. The first is called the "bottom suction" method (Figure 11.18). In this method air is drawn through the pile by the imposition of negative pressure from perforated pipes placed at the bottom of the compost pile. The second method is termed "bottom blowing" and as the name implies air is blown up through the pile (termed positive pressure) from perforated pipes underlying the compost. A third method called "alternate ventilation" employs a combination of bottom suction and bottom blowing.

Problem Areas - Static Pile Systems

1.    The height of the compost pile is a critical factor. When the height exceeds 2.5 to 3 meters (8 to 10 feet) uniform aeration becomes more difficult. In a pile that is excessively high the flow of induced air can lead to the development of short circuiting passageways that bypass sizable sections of the pile. The addition of a bulking agent helps to alleviate this problem if distributed uniformly throughout the pile.

2.    The external surface of the entire pile must be covered with an insulating layer (usually cured compost) to ensure that temperatures lethal to pathogens will be attained throughout the pile.

3.    In the bottom blowing system the air tends to cool and dry the bottom layers of the pile while the outer or upper layers remain warm and moist. Alternating between bottom blowing and bottom suction helps alleviate this problem.

4.    Induced air flow in the bottom suction system may enhance the production of odors, however this can be mitigated by covering the outlet pipe with screened compost.

5.    Uncovered piles will face the same problems with excess moisture and leachate production from precipitation described in the turned pile system(s).

**Figure 11.18:** Diagram—static pile system.

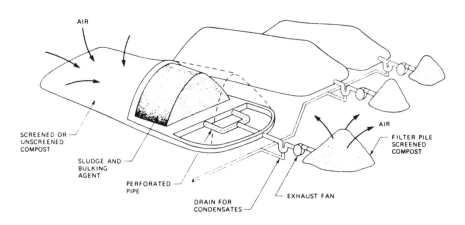

Source:  Ref. 11.27

**Figure 11.19:** In-vessel composting process system.

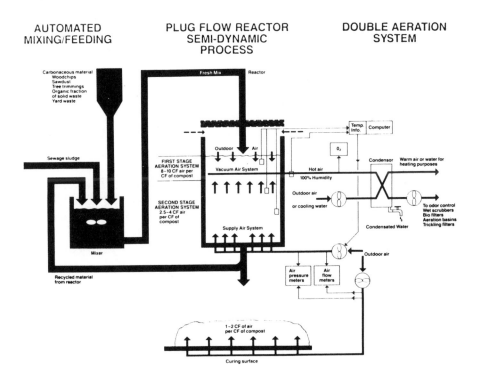

Source:  PURAC Engineering, Inc., Wilmington, DE 19808

B.    Closed (In-Vessel) Systems
      Some consider these systems (Figure 11.19) as a
significant improvement over the Open Systems.    In
some respects this is true.    For instance, they fully
enclose the material being composted thereby eliminating capricious weather conditions.    Further-more, some
types save space--an important consideration in heavily urbanized areas.
      Some of these systems also provide greater con-
trol over mixing and odors.    They are not, however,
entirely problem free, as we shall shortly see.
      1.    Continuous Vertical Reactors
            There are several types of vertical reac-
      tors available, however, they all generally
      operate in a similar manner.    Material is loaded
      into the reactor through the top and is
      eventually discharged from the bottom.    Oxygena-
      tion of the composting material is provided by
      forcing air up from the bottom of the reactor
      through perforations in the bottom of the floor.
      2.    Horizontal Reactors
            As the name implies the composting mate-
      rial is arranged along the length of a
      horizontal reactor.    This horizontal orientation
      mimics the turned pile and static pile methods,
      however, in this case the material is fully
      enclosed.    Air is supplied either by tumbling
      the material as it moves along inside the
      reactor or by forced aeration.
Problem Areas - Closed Systems
1.    In the continuous vertical reactor where air
      flows from bottom to top, hyperventilation can
      occur in the lower portion of the reactor where
      the need for oxygen is the least.    This can lead
      to excessive cooling and drying of the material
      which can in turn produce spontaneous combus-
      tion.    In the upper portion where ventilation is
      needed most oxygenation can be inadequate due
      both to the depth of the material (which should
      not exceed 2 to 3 meters) and the change in com-
      position of the induced air that takes place due
      to microbial activity as it rises.    You can
      therefore expect a decrease in oxygen ($O_2$)
      levels and an increase in carbon dioxide ($CO_2$)
      levels as the air rises.    As a consequence, by
      the time the air reaches the top of the pile the

$O_2$ concentration may be too low to support aerobic microbial metabolism. This results in anaerobiosis and temperatures too low to kill pathogens.

## SPECIAL CONSIDERATIONS--COMPOSTING

### 1.   Pathogen Destruction

A major challenge for the marketing and use of compost made from municipal solid waste, sewage treatment plant sludge and some industrial wastes is the assurance that the final product has been sanitized to kill pathogens. It is highly unlikely, however, that a compost product from these sources will ever be totally pathogen free (sterile).

Municipal solid waste and sewage treatment plant sludges can harbor large populations of pathogenic microorganisms such as bacteria, viruses, fungi and parasites (Table 11.11). Some of these pathogens (viruses and parasites, for example) cannot reproduce outside of a host and although they may survive for some time in a waste they cannot reproduce and increase in number. Other pathogens, however, such as bacteria and fungi, can reproduce and increase in number over time. The common assumption that composting and sterilization are synonymous is incorrect. The composting process can, however, be manipulated to provide conditions that substantially reduce the total number of pathogenic microorganisms as well as reduce the amount of assimilable organic matter (food) available for regrowth. Conditions of the composting process that may be manipulated to maximize sanitization are temperature, moisture and oxygen supply.

A)   Temperature

All pathogens have a temperature threshold with some sporogenous types able to survive at temperatures of $100^{\circ}C$ ($212^{\circ}F$) and above. Fortunately, those pathogens most dangerous to man are neither thermophilic nor sporogenous and can consequently be eliminated or at least reduced substantially by the milder heat levels encountered in the composting process. It is important, however, that the lethal temperature ($65^{\circ}C$ for 2 to 3 consecutive days recommended) be reached simultaneously

Table 11.11: Pathogens Commonly Found in Urban Municipal Waste and
Wastewater Sludges

| PATHOGEN | DISEASE |
|---|---|
| **Virus** | |
| Enterovirus | gastro-enteritis, heart disease, meningitis |
| Rotavirus | gastro-enteritis |
| Parovirus | gastro-enteritis |
| Adenovirus | respiratory tract infections, conjunctivitis |
| Hepatitis A virus | viral hepatitis |
| Polio virus | poliomyelitis |
| Ecovirus | meningitis |
| Coxsackie virus | meningitis |
| **Bacteria** | |
| Salmonella (1700 types) | typhus, salmonella |
| Shigellae | shigellosis |
| Mycobacterium tuberculosi | tuberculosis |
| Vibrio cholerae | cholera |
| Escherichia coli | gastroenteritis |
| Yersinia enterocolica | gastroenteritis |
| Clostridium perfingens | gangrene |
| Clostridium botulinum | botulism |
| Listeria manocytogenes | meningo-encephalitis |
| **Fungi** | |
| Candida sp. | systemic and skin mycoses |
| Tricosporon cutaneum | skin mycosis |
| Aspergillus fumigatus | lung mycosis |
| Tricophyton sp. | skin mycosis |
| Epidemophyton sp. Microspor | skin mycosis |
| Microsporum sp. | skin mycosis |
| **Protozoa** | |
| Eutamoeba | amebiasis |
| Giardia lamblia | giardiasis |
| Balantidium coli | balantidiasis |
| Naegleria fowleri | meningo-cephalitis |
| A. canthamoebe | meningo-cephalitis |
| **Helminths** | |
| Ascaris lumbricoides | ascariosis |
| Ancylostoma sp. | ancilastomiosis |
| Necator amaricanus | necatoriasis |
| Enterobius vermicularis | enterobiasis |
| Strongyloides stercoralis | strongyloidiasis |
| Toxocara sp. | larvae in the viscera |
| Trichuris trichiura | trichuriasis |
| Taenia saginata | tapeworm |
| Hymenolepsis nana | tapeworm |
| Echinococcus granulosus | enchinococcosis |
| Echinococcus multilocularis | |

Source: Adapted from Ref. 11.19

throughout the entire mass. These conditions, unfortunately, are often difficult to
achieve.

B) <u>Moisture</u>

Moisture increases the conductivity
of the mass and consequently a high moisture
content in the compost facilitates temperature transfer and enhances sanitization. In
the cured, stabilized compost, however, high
moisture conditions are not recommended nor
are they expected to be found. The optimum
moisture content in cured and stabilized
compost should be less than 30 percent. At
moisture levels of? less than 25 percent
microbial growth slows and eventually
ceases.

C) <u>Oxygen Supply</u>

The accumulation of heat and subsequent rise in temperature of the composting
material is the result of exothermic reactions from the bio-oxidation process.
Consequently, the attainment of high temperatures is dependent upon a sufficient supply
of oxygen for oxidation. An insufficient
supply of oxygen can lead to anaerobic
conditions which sharply curtail heat generation.

In addition to the above conditions there are
other significant factors that affect the sanitization
of compost. The first is microbial competition and
antagonism. The number of indigenous (native) saprophytes involved in the composting process is enormous.
In contrast, the number of pathogenic microbes is relatively minute. These extreme differences in population result in a high degree of antagonism and intense
competition for nutrients, thereby placing the
pathogens at a distinct disadvantage. This disadvantage is further enhanced by the fact that the composting material is not the native environment for the
pathogens. Secondly, when the finished compost is
applied to the soil any remaining pathogens are
further exposed to the homeostatic properties and
competing microorganisms indigenous to the soil.

Determining whether or not we have produced a
compost with an acceptable level of sanitization is
sometimes difficult. It would be impractical to test

for the presence of all possible pathogens. Consequently, we must resort to the use of indicator organisms. Presently recommended indicators, at least for bacterial pathogens, are salmonella, fecal coliform and fecal streptococcus. Research by de Bertoldi et al. (Ref. 11.19) has shown that if the temperature of the composting mass reaches $65^{\circ}C$ $(149^{\circ}F)$ and remains at that level for two days Salmonella sp. disappears completely. In addition, under those same conditions, the concentration of fecal coliform indicators is always less than $5 \times 10^2$ and fecal streptococcus is less than $3 \times 10^3$. Therefore these concentrations could be used as indicators of satisfactory sanitization.

## 2.   Composting of Grass Clippings

Encouraged by the general success of local leaf collection and composting programs, an increasing number of American communities are either contemplating or have already begun the mass collection and composting of grass clippings. The collection of these materials, however, poses problems that do not necessarily parallel those of leaf collection.

First, we now know that grass clippings should not, as a general rule, be composted by themselves. Their rapid decomposition leads to compaction, anaerobosis and ultimately severe odors. However, because of high nitrogen content they can be beneficially blended with partially composted[34] leaves. The exact ratio of this blend has yet to be determined. Dr. Peter Straw of Rutgers University has experimented with mixture ratios of 3:1 (partially composted leaves to grass) (Ref. 11.20).

Secondly, the collection of grass clippings may be seasonal or year round. If seasonal the period of collection will, in most instances be of a longer duration than that for leaf collection. Consequently, labor requirements will be increased as it will for year round collection.

Thirdly, the intense application of fertilizers, pesticides and herbicides to American lawns may result in significant residues of these materials on grass clippings, which may pass relatively unaltered through the composting process. One community in fact, Emmaus, Pennsylvania no longer accepts grass clippings in its waste composting program for this very reason.

Finally, the implementation of grass clipping collection may need some additional incentives to encourage participation, particularly if the program is voluntary. The experience of one city, Omaha, Nebraska, has shown that renting 90 gallon containers to participating residents and collecting clippings on Mondays encouraged participation significantly. Providing these containers to the residents negated the need for them to purchase and fill multiple plastic bags. Collecting clippings on Mondays facilitated quick removal of the material since most lawns are traditionally cut on weekends.

**FOOTNOTES**

**CHAPTER 11**

(1)   From 53 Fed. Register, No. 168, 33317, August 30, 1988.

(2)   This lack of disposal options for hazardous wastes may pose even further problems for the states. Section 104 (c) (9) of the Comprehensive Environmental Response, Compensation and Liability Act (CERCLA) as amended by Section 104K of SARA requires each state to assure the Administration of the U.S. EPA by October 17, 1989 that adequate capacity will exist to manage all hazardous waste generated in the state over the next twenty years. Each state is required to submit a Capacity Assurance Plan. Failure to submit an approved assurance plan will prohibit the state from receiving Superfund money until such time that a plan is submitted and approved. (See 53 Federal Register 33618, August 31, 1988.)

(3)   Spaulding, M., K. Jayko and W. Knauss, "Hindcast of Medical Waste Trajectories in Southern New England Waters", Applied Science Assoc. Inc., Narragansett, Rhode Island, July 27, 1988.

(4)   See Special Considerations Section. Intensive recycling may conflict with those municipal refuse incinerators using refuse derived fuel. Plastics for instance, are a preferred fuel for such facilities due to their ability to sustain elevated incinerator temperatures.

(5)   Note:   The state of New Jersey created the Hazardous Waste Facilities Siting Commission in 1981. This autonomous body was to have selected two sites within the state to build two hazardous waste incinerators. These sites according to the initial legislation were to be located in "virgin", sparsely populated, areas, later termed "green fields". As opposition in these areas increased the concept of using brown fields (i.e. older industrial areas) has developed. Much to the Commission's chagrin, however, residents in these areas have also expressed their violent opposition to such

incinerators.  At present, two sites have been tentatively selected.  It is interesting to note that the Commission must provide $50,000 to the "host community" to review and/or refute the Commission's selection.

(6)    It is interesting to note that the U.S. Army Toxic and Hazardous Materials Agency in Aberdeen, Maryland is examining composting (a non-destructive method) as an alternative to incinerating explosives contaminated soils.  A pilot study on 16 explosive contaminated lagoons at the Louisiana Army Ammunition Plant indicates some initial success in reducing levels of both TNT and HMX explosives.  World Wastes, November, 1988, Page 19.

(7)    From American Public Works Assoc., "Municipal Refuse Disposal", Chicago, Illinois, 1970, Page 93.

(8)    See earlier discussion of the 1965 federal Solid Waste Disposal Act and the 1970 Resource Recovery Act.

(9)    The term sanitary landfill was first coined in the 1930's when it was used to describe the "cut and cover" or "trench" method of landfilling used in Fresno, California.

(10)   From 53 Fed. Register, 33316, 33317, August 30, 1988.

(11)   See also Chapter 5 on Land Treatment Systems.

(12)   From 53 Fed. Register, 33318, August 30, 1988.

(13)   From 53 Fed. Register, August 30, 1988.

(14)   See Federal Aviation Administration Order 5200.5.

(15)   It is interesting to note that at least one county government, Collier County, Florida intends to remove decomposed garbage from some of its older landfill cells to use as cover material for new trash.  The county will then refill the excavated cells with new garbage thus extending the landfills useful life.  From Public Works, October, 1988, page 44.

(16)   Some old landfills in Arizona (a state with an annual rainfall of 10" to 20") were recently excavated as part of a University of Arizona research program.  Due to the state's relatively low amount of annual rainfall most of the waste uncovered was relatively unaffected by burial.

Newspapers, for instance, which were more than 20 years old were still legible. Organic wastes such as cardboard and food wastes were in a well preserved condition (Ref. 11.28).

(17)  Clay or clay-soil liners should be a minimum of 2 to 3 feet thick with a permeability of 1 x $10^{-7}$ cm/sec.

(18)  Synthetic liners should be a minimum of 30 mil in thickness if not reinforced and a minimum of 36 mil if reinforced.

(19)  From Cadwallader, Mark and Erin Dixon, "Liner Materials for Waste Containment", Pollution Engineering, July, 1988, page 70.

(20)  Some experts urge caution on the use of geotextiles claiming that information on their long term performance is lacking. They fear overburden pressure could decrease in-plane permeability resulting in an interference with leachate flow. They recommend instead a continuance of the minimum six inch thick layer of sand (classified as SW or SQ by USCS) above and below the liner.

(21)  Some of the nation's more arid regions continue to argue that such collection systems are unnecessary due to their negative precipitation-infiltration rates.

(22)  From "The Solid Waste Forum", Public Works, November, 1984, page 46.

(23)  The LEL of a gas is the lowest percent, by volume, of that gas in a mixture of explosive gases, that will propagate a flame in air at $25^{\circ}$C and atmospheric pressure at sea level.

(24)  The author believes that this criterion should be applied at the actual physical boundary of the landfill in those instances where the landfill boundary and the property boundary do not coincide. This would provide a greater margin of safety.

(25)  From "The Solid Waste Forum", Public Works, March, 1985, page 54.

(26)  From Stearns, Robert P. and Gaylen S. Petroyan, "Active Systems for Landfill Gas Control", Public Works, April, 1982, page 42.

(27)  From Federal Register, 40 CFR, Parts 257 and 258, August 30, 1988.

(28)    Excess amounts of methane can be particularly damaging to vegetation when it displaces oxygen in the root zone.

(29)    In extremely arid regions where perennial vegetation is not expected to be significant, this topsoil requirement may be deleted.

(30)    In this process, shredded organic matter was placed into a fully enclosed, partially mechanized digester that dropped the material from one level to another thereby encouraging aerobic digestion. it was intended to produce compost in 28 days.

(31)    If the moisture content exceeds 60 percent, the compost becomes more compact thereby reducing the amount of air (oxygen) present. This can lead to the development of anaerobic conditions. Likewise, if the moisture content is too low the temperature of the composting mass is lowered thereby extending the time of degradation.

(32)    As a general rule if the windrow composting method is used a digestion period of 9 to 12 days will be required if the C to N ratio is 20; 10 to 16 days if the C to N ratio is 30 to 50; and 21 days if the C to N ratio is 78.

(33)    It should be noted that the measurement of the pH of the compost can also be used to evaluate the progress of digestion. The initial pH of digesting municipal refuse for instance ranges between 5.0 and 7.0. In the first 2 to 3 days of composting the pH will drop to 5.0 or less and then will rise to about 8.5 for the remainder of the digestion process if kept aerobic. If the digestion process turns anaerobic the pH will drop to about 4.5.

(34)    Leaves 4 to 6 months old generally.

**REFERENCES**

**CHAPTER 11**

11.1    United States Environmental Protection Agency, "The Solid Waste Dilemma: An Agenda for Action", EPA/530-SW-88-052, September, 1988.

11.2    United States Environmental Protection Agency, "Characterization of Municipal Solid Waste in the United States", 1960 to 2000 (Update, 1988), EPA/530-SW-88-033, March, 1988.

11.3    Fessitore, Joseph L. and Frank L. Cross, "Incineration of Hospital Infectious Waste", Pollution Engineering, November, 1988, page 83.

11.4    Del Bello, Alfred B., "Understanding Medical Waste", Pollution Engineering, November, 1980, page 26.

11.5    Pollution Engineering, "Burn or Not to Burn: The Hospital's Modern Day Dilemma", November, 1988, page 97.

11.6    Michaels, Abraham, "Household Hazardous Waste", Public Works, December, 1988, page 98.

11.7    United States Environmental Protection Agency, "Process Design Manual, Land Application of Municipal Sludge", EPA 625/1-83-016, October, 1983.

11.8    Brunner, Calvin R., "Incineration Systems, Selection and Design", Van Nostrand Reinhold Company, New York, 1984.

11.9    Pavoni, Joseph L., John E. Heer Jr., D. Joseph Hagerty, "Handbook of Solid Waste Disposal--Materials and Energy Recovery", Van Nostrand Reinhold Co., New York, 1975.

11.10   Goldstein, Jerome, "Garbage As You Like It", Rodale Books Inc., Penna., 1969.

11.11   Wilson, David Gordon, "Handbook of Solid Waste Management", Van Nostrand Reinhold Company, New York, 1977.

11.12   American Public Works Assoc., Institute for Solid Wastes, "Municipal Refuse Disposal", Public Administration Service, Chicago, Illinois, 1970.

11.13   United States Environmental Protection Agency, "Project Summary--Resistance of Flexible Membrane Liners to Chemical and Wastes", EPA/600/S2-85/127, January, 1986.

11.14   United States Environmental Protection Agency, "Design and Construction of Covers for Solid Waste Landfills", EPA-600/2-79-165, August, 1979.

11.15   United States Environmental Protection Agency, "Handbook--Remedial Action at Waste Disposal Sites", EPA-625/6-82-006, June, 1982.

11.16   Stearns, Robert P. and Gaylen S. Petroyan, "Active Systems For Landfill Gas Control", Public Works, April, 1982.

11.17   Connelly, Thomas L., "Exploding Methane Migration Myths", Public Works, April, 1983.

11.18   Brown, K.W. and D.C. Anderson, "The Case for Above Ground Landfills", Pollution Engineering, November, 1983.

11.19   de Bertoldi, Marco, Franco Zucconi and M. Civili, "Temperature, Pathogen Control and Product Quality", Biocycle, February, 1988.

11.20   Biocycle, "Options for Municipal Leaf Composting", October, 1988, page 38.

11.21   Goldstein, Nora, "Enclosures For Compost Projects--To Cover or Not to Cover", Biocycle, February, 1988.

11.22   Logsdon, Gene, "Composting Industrial Wastes Solves Disposal Problems", Biocycle, May/June, 1988.

11.23   J.G. Press Inc., "The Bio-Cycle Guide to In-Vessel Composting", March, 1986.

11.24   Ernst, A.A., "30 Years of Refuse/Sludge Composting", Biocycle, July, 1988.

11.25   Mathur, S.P. and J.Y. Daigle, J.L. Brooks, M. Levesque, J. Arsenault, "Composting Seafood Wastes", Biocycle, September, 1988.

11.26   Fiedler, David M., "Collecting/Composting Leaves in Michigan", Biocycle, September, 1988.

11.27  United States Environmental Protection Agency, "Process Design Manual, Sludge Treatment and Disposal", EPA 625/1-79-011, September, 1979.

11.28  Michaels, Abraham, "The Solid Waste Forum", Public Works Magazine, July 1989, page 47-48.

11.29  United States Environmental Protection Agency, "Process Design Manual, Municipal Sludge Landfills", EPA-625/1-78-010 SW-705, October, 1978.

11.30  Federal Register, Vol. 53, No. 168, Tuesday, August 30, 1988.

11.31  Pollution Engineering, March, 1984, page 25.

11.32  Brunner, C., Design of Sewage Sludge Incineration Systems, Noyes, Park Ridge, N.J., 1980.

11.33  Darcey, Sue, The Management of World Wastes, "States Fight to Ban Out-of-State Wastes", March, 1990, p. 36.

11.34  United States Environmental Protection Agency, "Municipal Waste Combustion Multipollutant Study, Shutdown/Startup Emission Test Report", EMB Report No. 87-MIN-04A, Vol. 1, September, 1988.

11.35  United States Environmental Protection Agency, "Municipal Waste Combustion Multipollutant Study, Appendix D of Background Information Document: Emission Measurement and Continuous Monitoring", September, 1989.

11.36  United States Environmental Protection Agency, "Municipal Waste Combustion Multipollutant Study, Summary Report", EMB Report No. 86-MIN-03A, September, 1988.

# 12

# Roadways

The public road system in the United States is a complex mixture of municipal, county, state and interstate highways, many of which were constructed with little, if any, consideration for the impact on the human and natural environment--at least up until 1969. It was in that year that the National Environmental Policy Act forced the commencement of such considerations on all federally funded projects, including the nation's federally funded interstate highway system. Eventually, any roadway that received federal funding in part or whole was included. As a result, the environmental impact statement became an integral part of the overall highway design process.

Thousands, perhaps tens of thousands of environmental impact statements (and sometimes statements of Negative Declaration)[1] have been prepared for federally funded roadway projects since then. Have they improved the quality of the environment along our highway corridors? The answer is yes, to the extent that these roadways now include mitigating measures which probably would not have been included or ever considered prior to 1969. The construction of such things as noise barriers, overpasses for wildlife, storm water retention basins and replacement wetlands are all a direct result of this federal initiative. However, roadways, new and old, continue to have negative impacts on the human and natural environment. In this chapter, we shall discuss several that sometimes escape the attention of even the most adroit land planners.

## ROADWAY NOISE POLLUTION

Noise, or airborne sound, is caused by a rapid fluctuation in air pressure called sound waves. The loudness of that sound is related to the size of those fluctuations and their frequency of occurrence. Surrounded, as we are by a constant envelope of air, we are continually bombarded by a wide variety of sounds whose loudness can be measured by a unit of measure called an "A" weighted decibel which is often abbreviated as dBA. The A-weighted decibel approximates the way an average person hears sounds.

The A-weighted decibel scale begins at zero. This represents the faintest sound that can be heard by humans with very good hearing. As shown in Figure 12.1 the dBA scale increases in value as the perceived noise levels increase. Based on many tests of large numbers of people, a sound level of 70 is twice as loud to the listener as a level of 60.

The level of highway traffic noise depends on three things: (1) the volume of traffic, (2) the speed of the traffic, and (3) the number of trucks in the traffic flow (Ref. 12.1). Generally, the loudness of traffic noise is increased by heavier traffic volumes, higher speeds and greater numbers of trucks. Vehicle noise is a combination of the noises produced by the engine, exhaust and tires. The loudness of traffic noise can also be increased by defective mufflers. Any condition such as steep grades that causes heavy laboring of motor vehicle engines also increases noise levels. Highway traffic noise is rarely constant and varies with the number, type and speed of the motor vehicles. For this reason statistical descriptors are almost always used to describe varying traffic noise levels. The two most common statistical descriptors are $L_{10}$ and $L_{eq}$. $L_{10}$ is the sound level that is exceeded 10 percent of the time. The $L_{eq}$ is the constant, average sound level, which over a period of time, contains the same amount of sound energy as the varying levels of the traffic noise. The $L_{eq}$ for typical traffic conditions is usually about 3dBA less than the $L_{10}$ for the same conditions.

Highway traffic noise often interferes with conversations and disrupts sleep. A 1975 survey by the U.S. Bureau of Census identified street noise as the most annoying neighborhood problem (Ref. 12.2).

Figure 12.1: Some common noise sources and their relationship to the dBA noise scale.

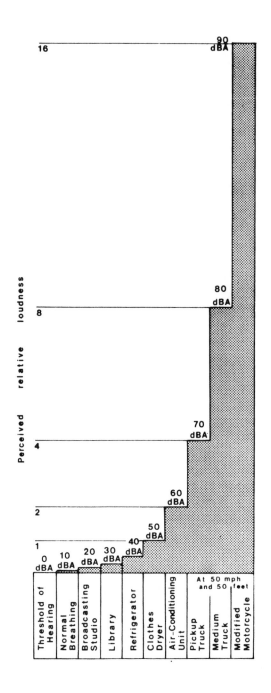

Source: Ref. 12.1

Today nearly all new major highway design includes some evaluation of potential noise pollution. Included in such evaluations are measurements of existing noise levels along the proposed route as well as predictions of sound levels if the project is constructed. If the predicted levels exceed the recommended level based on land use (Table 12.1) then mitigating measures are to be incorporated into the design. Unfortunately, mitigating measures are not included for any vacant or undeveloped land located along the proposed alignment. As a result several hundred thousand miles of existing highways in this country are bordered by vacant, potentially developable land for which no noise mitigation measures have been considered or implemented. It was anticipated that local land planners would consider noise impacts in these areas either during local master plan development or during the approval process for individual land development applicants.

It appears that our expectations of local government in this regard may have been too high.

Generally, roadway noise pollution has not been a major consideration in the local land subdivision process particularly for roadways in residential subdivisions where vehicle speeds are lower and heavy truck traffic is absent. As a result planners, perhaps because of their preoccupation with other environmental considerations, often overlook noise pollution entirely during the subdivision review process. This, in spite of the fact that some of the proposed lots may adjoin a major interstate highway, turnpike, or parkway. It is only after homes have been built and occupied that planners are suddenly reminded, often by a mob of angry new home owners, that there are sources of roadway noise pollution other than the internal roads of a subdivision.

Once homes have been constructed remedies are limited and are almost always structural in nature. They include primarily 1) the installation of noise barriers, or 2) the sealing of highway facing windows and the installation of air conditioning units. Planting rows of vegetation may be attempted but unless such vegetation is 200 feet or more in width and extremely dense it will generally be ineffective.

Table 12.1: Noise Abatement Criteria for Highway Design

Hourly A-Weighted Sound Level - decibels (dBA) 1/

| Activity Category | Leq(h) | $L_{10}$(h) | Description of Activity Category |
|---|---|---|---|
| A | 57 (Exterior) | 60 (Exterior) | Lands on which serenity and quiet are of extraordinary significance and serve an important public need and where the preservation of those qualities is essential if the area is to continue to serve its intended purpose. |
| B | 67 (Exterior) | 70 (Exterior) | Picnic areas, recreation areas, playgrounds, active sports areas, parks, residences, motels, hotels, schools, churches, libraries, and hospitals. |
| C | 72 (Exterior) | 75 (Exterior) | Developed lands, properties, or activities not included in Categories A or B above. |
| D | -- | -- | Undeveloped lands. |
| E | 52 (Interior) | 55 (Interior) | Residences, motels, hotels, public meeting rooms, schools, churches, libraries, hospitals, and auditoriums. |

1/Either $L_{10}$(h) or Leq(h) (but not both) may be used on a project.

Source: Ref. 12.3

## Noise Barriers

Noise barriers are solid obstructions built between the highway and the affected homes. These barriers can reduce noise levels by 10 to 15 decibels, cutting the loudness of traffic noise in half. They are generally constructed of soil (earthen berms), concrete, steel, timber or a combination of any of the above (Figures 12.2 and 12.3). Earth berms while more natural looking require a substantial amount of land. Walls require considerably less space but are usually limited to heights of 25 feet or less for structural and aesthetic reasons.

Noise barriers do have limitations. For a noise barrier to work, it must be high enough and long enough to block the view of the road. Noise barriers do little good for homes on a hillside overlooking a road or for buildings which rise above the barrier. In addition, any openings in these barriers to accommodate driveway connections or intersecting streets destroy their effectiveness.

## Preventative Planning

It is much easier and generally less expensive to incorporate noise prevention and mitigation measures into the land use planning and approval process than it is to retrofit an already constructed site. As with current highway design practices all land developers should be required to measure ambient noise levels in those parts of the subdivision contiguous to a major highway. If the measured noise levels exceed the $L_{10}$ or $L_{eq}$ values shown in Table 12.1 then mitigating measures must be incorporated into any proposed design plans.

Preventative planning techniques available to lessen the acoustical impact on residential development from highway noise include:

1) Placing as much distance as possible between the noise source (highway) and the receptors. This is usually accomplished by requiring large buffer zones (usually open space) between the highway and any residences (Figure 12.4).

2) Placing noise-compatible activities such as parking lots, open space and commercial

Figure 12.2: This wooden noise barrier is approaching the recommended maximum height for such structures.

Figure 12.3:  This all-concrete noise barrier offers a pleasing alternative to wood.

Figure 12.4: Open space can be left as a buffer zone between residences and a highway.

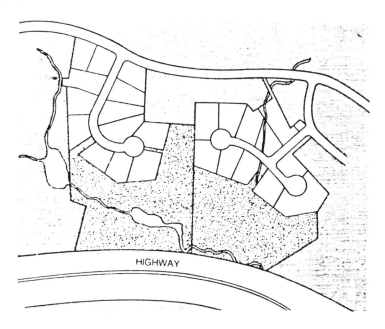

Figure 12.5: Using non-residential buildings as noise buffers.

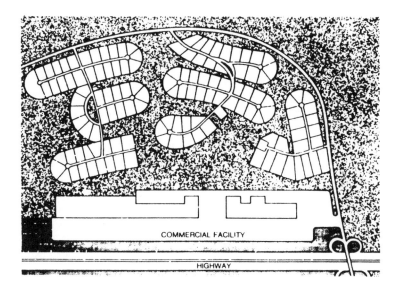

Source for Figures 12.4 and 12.5: Ref. 12.1

facilities between the noise source and the most sensitive receptor (Figure 12.5).

3) Using buildings as barriers.

4) Orienting noise-sensitive buildings to face away from the noise source.

5) Incorporating noise reducing concepts into architectural design. This can include placing windows, balconies and courtyards on the sides of buildings away from the noise source.

6) Incorporating noise-absorbing material into the walls of new buildings, and

7) Constructing noise barriers as previously discussed or providing wide and dense vegetative buffers.

## CUL-DE-SACS AND SNOW REMOVAL

It would be rare to find a contemporary residential subdivision (especially in rural or suburban areas) that does not have at least one cul-de-sac (a circular dead-end turnaround) included in its roadway layout. It is a useful civil engineering design practice that enables developers to create lots that might not otherwise be possible. Because these cul-de-sacs have been so widely used in residential subdivision road designs, land planners have come to accept them almost without reservation--an attitude that frustrates some municipal road departments who must eventually assume responsibility for their maintenance. In areas of the United States where snowfall is common in winter, these cul-de-sacs are a bane to municipal snow removal crews. Unlike normal width roadways, these 40 to 60 foot wide circular turnarounds required as much as a half hour of additional time to clean them completely of snow. This is due to the relative closeness of driveways around the cul-de-sac's circumference and the placement of shrubbery and mailboxes-- all of which contribute to the general lack of space to stockpile the removed snow. As a consequence, snow must often be removed from the cul-de-sac by truck. Trying to shortcut removal by simply making a single pass with a snow plow around the outside edge of the cul-de-sac often results in numerous complaints from subdivision residents who insist that their cul-de-sac be completely cleaned. The deeper the snowfall, the more critical this removal becomes.

## ROADWAY SALTING

This proliferation of suburban roadways into rural America frequently produces impacts that are rarely considered in the local land subdivision approval process. The application of deicing salts to a large number of these roadways is just such an impact.

The American motorist's demand for speed and convenience even under the most adverse weather conditions has led to a dramatic increase in the quantity of salts applied to U.S. roadways for ice and snow removal. The two salts most commonly used are sodium chloride (rock salt) and calcium chloride. Both salts melt ice and snow by dissolving in water and lowering the freezing point of the resultant brine solution. When combined with the heat of the sun and pressure of highway traffic enough ice and snow is melted so that plows can clear the road.

Once the brine has accomplished its initial objective its presence becomes a significant but often ignored threat to the environment and human health. Whatever brine does not seep into the soil and ultimately the ground water along the roadway's edge eventually flows to adjacent watercourses.. These uncontrolled releases of saltwater produce a myriad of negative impacts, not the least of which is the contamination of ground water. Hundreds of shallow, hand dug wells and an equal number of public supply wells in shallow aquifers located near heavily salted roadways have been contaminated with excessive levels of salt. Salt concentrations in highway runoff can be extremely high. Chloride concentrations in direct highway runoff for instance have been measured as high as 10,250 ppm in Wisconsin and 11,000 to 25,000 ppm in Chicago (Ref. 12.4). The recommended maximum level of chloride in a potable water supply is 250 ppm. Sodium concentrations in potable water supplies are even more critical with a recommended maximum level of 50 ppm. Sodium poses a danger to persons subject to heart disease and hypertension as well as those on low salt diets due to complications of pregnancy or obesity.

Where roadways pass through the uppermost reaches of a watershed, as many now do, streams are generally smaller in size with lesser volumes of water. Consequently, the amount of water available to dilute the incoming brine solution is generally less. Concentrations of salt, therefore, may remain at the thousands parts per million level for longer distances in the stream. Such high levels can disrupt the osmotic balance in the gills of fish and cause either their direct death or increase their susceptibility to secondary infections.

Excessive amounts of deicing salts can also damage the structure of soils. Salts in solution yield the positive ions of sodium and calcium and the negative ions of chloride. The chloride ions are of little consequence to soil structure. The sodium ions, however, can displace calcium in the soil. When sodium ions occupy 15 percent of the total amount of binding sites in the soil the soil structure begins to deteriorate. Permeability and water-holding capacity decrease sharply. In addition the capability of the soil to retain further sodium ions is significantly diminished. Sodium, therefore, will tend to filter rather quickly through the altered soil structure and into the ground water.

Increased salinity of the soil is also toxic to plants. Highly saline soil water will inhibit the water intake of plants, as well as suppress the uptake of potassium, calcium and magnesium. The first symptom of salt damage is growth depression, followed by the death of the shoots (browning) and eventually complete death of the plant. The damaging effects of salt, however, may not become apparent for several years. Generally, trees and woody shrubs are more susceptible to salt damage than herbaceous plants (Ref. 12.4). Trees such as maples, pines, hemlock and American elm show an increased sensitivity.

In those communities where deicing salts are regularly applied in wintertime it is imperative that the annual salt burden be determined for all surface watercourses expected to receive storm water runoff from any proposed roadway. Claims by consultants that all pollutants will be effectively removed by storm water detention basins are untrue. Dissolved materials such as deicing salts pass through such structures relatively unaffected, unless of course the storm water is recharged through the ground. Then the burden is shifted to the soil and the ground water. Here, as we discussed previously, some treatment by sodium cation exchange can occur.

An annual average concentration of deicing salt for any point on a surface stream may be calculated as follows[2]:

1)    Select the point on the stream to be evaluated.

2)    Delineate the drainage area of the watershed upstream of this selected point.

3)    Determine the lane miles of existing and proposed roadway expected to contribute storm water runoff to the point.

4)    Estimate the tonnage of salt to be spread on these roadway(s). This information may be obtained from the local or state highway department and will be in tons per lane mile.

5)    Estimate the average annual runoff from the delineated watershed including that expected from the proposed roadways.

6)    If runoff is determined in inches, the annual average concentration may be estimated as follows:

$$\frac{\text{Tons of salt per lane mile} \times \text{number of lane miles}}{\text{Inches of runoff} \times \text{drainage area in square miles}} \times 13.79 = \begin{array}{l}\text{Concentration} \\ \text{of salt as} \\ \text{NaCL in} \\ \text{mg/liter}\end{array}$$

If the concentration of chloride only is desired, use a factor of 8.37 instead of 13.79 in the above equation.

7)    If runoff is in average cubic feet per second per square mile then the following calculation should be used:

$$\frac{\text{Tons of salt per lane mile x number of lane miles}}{\text{Runoff in cubic feet per second per square mile x drainage area in square miles}} \quad x \quad 1.02 \quad = \quad \begin{array}{l}\text{Concentration}\\\text{of salt as}\\\text{NaCL in}\\\text{mg/liter}\end{array}$$

If the concentration of chloride only is desired, use a factor of 0.61 instead of 1.02 in the prior equation.

## ROADWAY AIR POLLUTION

The nation's air contains a variety of pollutants ranging from minute particles of solid and liquid material to noxious and potentially toxic gases. The sources of these materials are equally as varied with some coming from such natural sources as volcanic eruptions and dust storms while others are produced by man-made sources such as industrial smokestacks and motor vehicles. It is this latter source that is of interest to us in this chapter.

The primary pollutants from motor vehicles are carbon monoxide (CO), hydrocarbons (HC), oxides of nitrogen ($NO_x$), oxides of sulfur ($SO_x$) and particulates of lead and carbon soot. Motor vehicles are the nation's major source of CO emissions (over 90 percent). Some significant secondary pollutants are also produced when HC and $NO_x$ are released into the atmosphere. Here in the presence of sunlight (ultraviolet radiation) new chemicals are formed. They include ozone, peroxyacetyl nitrate (PAN) and peroxybenzoyl nitrate (PBN), all of which can have serious impacts on human health and vegetation. Ozone for instance, a strong oxidant made up of three oxygen atoms, can cause chest pain, coughing, nausea in just a few hours of exposure at levels as low as 0.12 ppm. Similar deleterious effects can also be found in plants exposed to persistent levels of ozone. Yields of soybeans, wheat and corn have been reduced by as much as 30 percent when repeatedly exposed to low levels of ozone. The limiting factor in the formation of secondary pollutants is the amount of HC emitted.

The less hydrocarbons emitted, the less secondary pollutants will be formed.  Since 1975 light duty vehicles (passenger cars and trucks) have been equipped with air pollution control devices (primarily catalytic converters) to reduce the emissions of HC and CO.  While the amounts of HC and CO have been reduced these catalytic converters are contributing new pollutants--principally a sulfuric acid mist.

The consideration of air pollution impacts from motor vehicles on local roadways has been notably absent in the local land subdivision review and approval process as well as in the development of municipal master plans.  The general consensus has been that such considerations are relevant only to the largest state and federal highways where higher traffic volumes produce a nearly constant linear source of pollution.  While the consideration of air pollution impacts are unquestionably a significant part of the design of major highways it should not be overlooked during the approval process for local subdivisions and site plans either.

Similar to the concerns addressed in the noise pollution section of this chapter any proposed residential building lots expected to adjoin an existing major highway need some additional evaluation.  Ambient measurements of CO, HC and $NO_x$ need to be taken within the projected building area of each lot to determine if current air quality exceeds the U.S. Department of Transportation Standards shown in Table 12.2.  Such sampling should include both morning and evening peak traffic periods and ideally seasonal variations as well.  Such sampling, if done correctly, could require sampling for up to one year--a proposition few developers would greet with enthusiasm.  In lieu of this extensive sampling some municipalities have accepted as little as a single 24 hour day of sampling or up to a month or more.  Site microclimatology will play a key role in this decision.  Such things as the prevailing wind patterns (wind rose) may indicate that pollution will be consistently blown away from the expected receptors (houses) (Figure 12.6).

Where ambient levels of air pollutants are found to exceed the standards it then becomes necessary to either eliminate these lots and create open space in

Table 12.2:  Maximum Allowable Air Pollution Levels Used in Highway Design

**Carbon Monoxide**

**9 PPM** - for an 8 hour period.

**35 PPM** - for a 1 hour period.

**Nitrogen Oxides**

**0.05 PPM** - annual average

**Hydrocarbons**

**0.24 PPM** - 3 hour average (6-9 a.m.)

Table 12.3:  Potential Health Effects of Some Selected Air Pollutants

**Carbon Monoxide**

**15 to 50 PPM\*** - some loss of judgment and vision

**100 PPM** - headaches occur more frequently.  Levels
greater than 300 PPM can cause collapse
and death.

**Nitrogen Oxides**

**2 to 10 PPM** - acute injury to plants

**10 to 40 PPM** - possible chronic pulmonary fibrosis
and emphysema.

\*   PPM = parts per million

Figure 12.6: Dispersion of pollutants from a highway source in a deep wide valley.

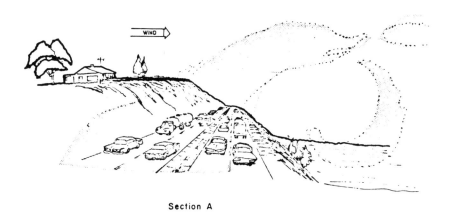

Section A

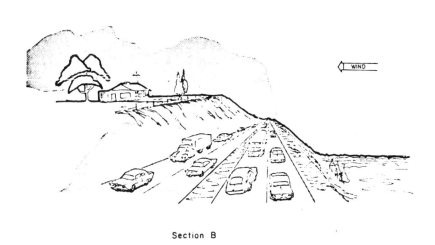

Section B

Source: Ref. 12.6

their place or mandate the location of any proposed dwelling as far away from the highway as possible.


## FOOTNOTES

### CHAPTER 12

(1)   The Statement of Negative Declaration has been largely replaced by the FONSI or Finding of No Significant Impact and the Categorical Exclusion (CE). Regardless of the terminology, all of the above documents generally declare that after consideration of all potential environmental impacts it is the belief of the preparing agency that no negative impacts will occur and, therefore, a complete environmental impact statement (E.I.S.) will not be necessary.

(2)   From Ref. 12.5.


## REFERENCES

### CHAPTER 12

12.1   United States Department of Transportation, Federal Highway Administration, "Highway Traffic Noise", September, 1980.

12.2   United States Department of Transportation, Federal Highway Administration, "Highway Traffic Noise and Future Land Development Can Be Compatible", 1979.

12.3   United States Department of Transportation, Federal Highway Administration, Federal-Aid Highway Program Manual, Vol. 7, Chapter 7, Section 3, "Procedures For Abatement of Highway Traffic Noise and Construction Noise", August 9, 1982.

12.4   McConnell, H. Hugh and Jennifer Lewis, "...Add Salt to Taste", Environment, November, 1972.

12.5    United States Geological Survey, "Effect of Deicing Chemicals on Surface and Ground Water", prepared for the Massachusetts Dept. of Public Works, Research and Materials Division, Interim Report USGS-MDPW-003, December, 1973.

12.6    United States Department of Transportation, Federal Highway Administration, "Meteorology and Its Influence on the Dispersion of Pollutants From Highway Line Sources", Air Quality Manual, Vol. 1, FHWA-RD-72-33, April, 1972.

12.7    United States Geological Survey and the Massachusetts Department of Public Works, "Hydrologic Effects of Highway-Deicing Chemicals in Massachusetts", Open-File Report 81-209, 1981.

# 13

# Coastal Areas

## COASTAL DEGRADATION--A WORLDWIDE PROBLEM

The nation's coastal areas and estuaries have not been spared from environmental degradation. While the extent of degradation varies in intensity with geographical location, the symptoms are strikingly similar:

1) Eutrophication with recurring, massive blooms of algae.
2) Events of hypoxia resulting in the death of fish, shellfish and crustaceans.
3) Loss of submerged vegetation.
4) Accumulation of toxics in fish and shellfish rendering them unfit for human consumption (Table 13.1).
5) Accumulation of toxics in estuarine and coastal sediments to levels in some instances that qualify for Superfund cleanup.
6) Abnormalities, reproductive failure and deaths of commercially important food fish.
7) An increase in the number of bathing beach closures due to contaminated water or presence of large amounts of floating solid wastes (Figure 13.1)
8) Accelerated erosion of beaches.

The United States has no monopoly on such symptoms, either. Similar occurrences are becoming unsettlingly routine in marine waters and coastal areas in nearly every part of the world. A number of beaches along the Black Sea and the Baltic Sea in the Soviet Union, for instance, have been closed due to bacterial contamination. Usually red, Norwegian lobsters caught

Table 13.1: Primary Pollutants Associated with Fishing Advisories and Bans[a]

| Pollutant | No. of States |
|---|---|
| PCBs | 15 |
| Mercury | 11 |
| Chlordane | 10 |
| Pesticides (unspecified) | 7 |
| Dioxin | 7 |
| DDT | 5 |
| Metals (unspecified) | 5 |
| Organics (unspecified) | 3 |
| Dieldrin | 3 |

(a) Includes waters other than coastal and marine

Source: Ref. 13.24

Figure 13.1

off the coast of Denmark have turned black and biologists have found seals in the Baltic Sea starving to death because their mouths and fins were so badly deformed. Japan's Inland Sea is plagued with hundreds of toxic red tides annually. Remote beaches on Mexico's Yucatan Peninsula are littered with plastics and tires, and even the remotest islands in the Hawaiian chain are littered with tons of plastics and other solid wastes.

While engineers, scientists, politicians, developers, planners and citizens carry on eloquent but recondite debates about causes and seek to foist the blame upon each other, the fact of the matter is, we have simply overwhelmed these areas with our presence and by our actions.

The nation's estuaries and coastal beaches combine to form an environmentally significant transition zone between two of the world's major mediums-- land and sea. Estuaries, for instance, remove a portion of the sediment, nutrients and other solids carried off the land by streams and rivers. They, likewise, provide nurseries and spawning grounds for a variety of sport and commercial fish and shellfish (Table 13.2). In addition, they provide habitat for wildlife and serve as buffers for controlling some of the erosional impacts of the outgoing freshwater and incoming saltwater. Coastal beaches, on the other hand, serve primarily as shock absorbers against the unrelenting undulations of the sea. They are flattened or raised, widened (prograded) or narrowed (eroded) and in some instances have become islands unto themselves separated from and paralleling the mainland. These we call barrier islands.

The general estuarine, biophysical regions of the United States are shown on Figure 13.2. All estuaries function similarly, however, their morphology and other characteristics may vary considerably (Figure 13.3). This is the result of such factors as freshwater flows, coastline slope and configuration resulting from tectonic influences, ocean currents, rates of sedimentation, climate and tides. A brief description of each estuarine region in the United States is shown in Table 13.3.

Man's attraction to these areas has been relentless, first for the fishery resources, then ultimately for commerce, manufacturing, habitation and recre-

Table 13.2:  Estuarine Dependence of Important Sport and Commercial Fish

| Biophysical region | Sports fish: Type of dependence | | |
| --- | --- | --- | --- |
| | Permanent residence | Passage zone | Nursery zone |
| North Atlantic | Croaker, Atlantic mackerel, bluefish. | Atlantic salmon, shad | Striped bass. |
| Middle Atlantic | Croaker, drums, Atlantic mackerel, spot, bluefish. | Shad | Do. |
| Chesapeake | Crabs, croaker, drums, spot, bluefish. | ....do | Do. |
| South Atlantic | Crabs, croaker, drums, spotted sea trout, spot, bluefish. | ..do | Do. |
| Caribbean | Spotted sea trout, spot, bluefish. | | |
| Gulf of Mexico | Crabs, croaker, drums, spotted sea trout, spot, bluefish. | Shad | Striped bass. |
| Pacific Southwest | Abalone, rockfish, barracuda. | | |
| Pacific Northwest | Abalone, rockfish. | Salmon (chum, coho, king, red). | Pink salmon. |
| Alaska | Crabs. | ..do | Do. |
| Pacific Islands | Barracuda. | | |

| Biophysical region | Commercial fish: Type of dependence | | |
| --- | --- | --- | --- |
| | Permanent residence | Passage zone | Nursery zone |
| North Atlantic | Oysters, clams, croaker, flatfish. | Atlantic salmon, eel. | Menhaden, lobsters. |
| Middle Atlantic | ....do | Eel. | Menhaden. |
| Chesapeake | Oysters, clams, crabs, croaker, flatfish. | ....do. | Do. |
| South Atlantic | Oysters, crabs, croaker, flatfish. | do. | Shrimp, menhaden. |
| Caribbean | Flatfish. | | Lobsters. |
| Gulf of Mexico | Oysters, crabs, croaker, flatfish. | | Shrimp, menhaden. |
| Pacific Southwest | Clams, abalone, flatfish. | | |
| Pacific Northwest | Oysters, abalone, crabs, flatfish. | Salmon (chum, coho, king, red). | Pink salmon. |
| Alaska | Crabs, flatfish. | ....do. | Shrimp, pink salmon. |
| Pacific islands | Oysters, flatfish. | | Lobsters. |

Source:  Ref. 13.1

Figure 13.2:  Biophysical regions of the United States.

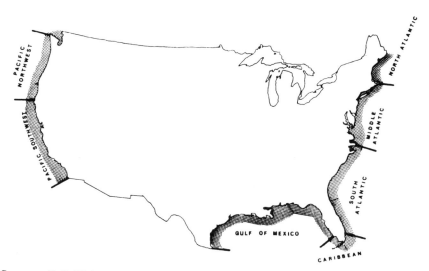

Source:  Ref. 13.1

Figure 13.3: Morphological classification of estuaries and estuarine zones.

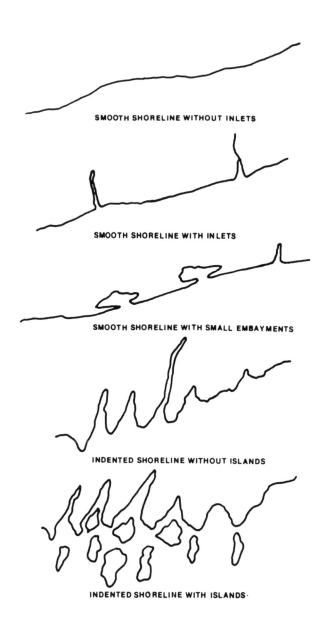

SMOOTH SHORELINE WITHOUT INLETS

SMOOTH SHORELINE WITH INLETS

SMOOTH SHORELINE WITH SMALL EMBAYMENTS

INDENTED SHORELINE WITHOUT ISLANDS

INDENTED SHORELINE WITH ISLANDS·

(continued)

Figure 13.3: (continued)

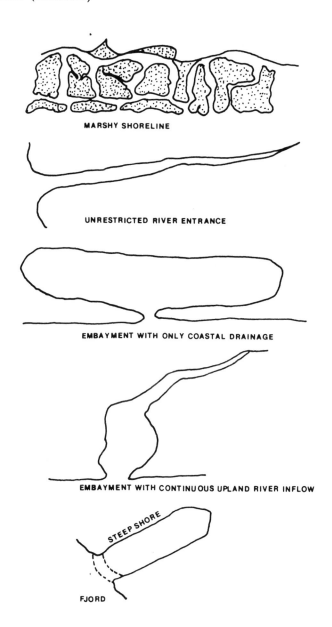

MARSHY SHORELINE

UNRESTRICTED RIVER ENTRANCE

EMBAYMENT WITH ONLY COASTAL DRAINAGE

EMBAYMENT WITH CONTINUOUS UPLAND RIVER INFLOW

STEEP SHORE

FJORD

Source:  Ref. 13.1

Table 13.3:  Characteristic Natural Estuarine Zone Circulation and Water Quality Conditions

| Biophysical region | Smooth shoreline (1) | Indented shoreline (2) | Marshy shoreline (3) | Unrestricted river entrance (4) | Embayment, coastal drainage only (5) | Embayment, continuous upland river flow (6) | Fjord (7) |
|---|---|---|---|---|---|---|---|
| North Atlantic | Deep near shore, oceanic water, longshore currents, some suspended sand and clay. | Deep near shore, oceanic water, erratic tidal currents; eddies and tidal pools. | Strong currents in many small channels through marsh, some turbidity, high oxygen. | Highly stratified, some turbidity, high oxygen, temperatures warmer in summer, colder in winter than ocean. | Little turbidity, water of oceanic character; strong tidal currents through inlets. | Little turbidity, high oxygen, may be stratified, upper layer fresh, with temperatures warmer in summer; colder in winter than the ocean. | |
| Middle Atlantic | Oceanic water, longshore currents; suspended mud, clay, silt. | Generally shallow, suspended mud and sand, oceanic water. | Moderate currents in well-defined channels, high dissolved oxyganic material, little turbidity, high oxygen. | Moderate stratification, suspended mud and silt, high oxygen, strong currents. | Generally shallow, small tides, clear water with lowered salinity, high oxygen. | Variable stratification, suspended mud and silt, high oxygen, small amounts of organic material. | |
| Chesapeake | Longshore tidal currents highly variable salinities, small amounts of organic material. | Moderate tidal currents, highly variable salinities, some turbidity. | Poorly defined channels, small currents, dissolved organic material, moderately fluctuating oxygen. | Moderate stratification, suspended mud and silt, high oxygen, strong currents. | Generally shallow, small tides, clear water with lowered salinity, high oxygen. | Variable stratification, suspended mud and silt, high oxygen, small amounts of organic material. | |
| South Atlantic | Primarily tidal and wave induced currents, oceanic water with mud, clay and silt. | Moderate tidal currents, highly variable salinities, some turbidity. | Small currents, high dissolved organics, highly variable oxygen, sometimes low, high temperatures. | Strong stratification, high suspended mud and clay, strong currents, dissolved organics, moderate oxygen. | Some color, small currents, generally shallow, high dissolved organics, highly fluctuating oxygen. | Slight and variable stratification, river water cooler than ocean, slight color, some oxygen fluctuation. | |
| Caribbean | Clear ocean water, gentle currents, warm temperatures throughout the year. | Clear ocean water, gentle currents, eddies, warmer than ocean. | High dissolved organics, color, suspended mud, very small currents, hot. | Slightly turbid, strong currents, river cooler than ocean water. | Very small currents, generally shallow, quite warm, clear ocean water. | Slightly turbid, eddying currents, slight stratification, high oxygen. | |

(continued)

Table 13.3: (continued)

| Biophysical region | Smooth shoreline (1) | Indented shoreline (2) | Marshy shoreline (3) | Unrestricted river entrance (4) | Embayment, coastal drainage only (5) | Embayment, continuous upland river flow (6) | Fjord (7) |
|---|---|---|---|---|---|---|---|
| Gulf of Mexico | Clear, generally warm ocean water, long-shore currents. | Very small currents, ocean water with slight turbidity, warmer than ocean. | High dissolved organics, color, very small currents, slightly turbid, very warm. | Slightly turbid, strong currents, river cooler than ocean water. | Very small currents except in inlet, shallow, warm, slight turbidity from sand and silt, highly fluctuating oxygen. | Slight and variable stratification, river water cooler than ocean some oxygen fluctuation. | |
| Southwest Pacific | Strong wave action, cool oceanic water, some silt and clay turbidity. | Moderate suspended solids, erratic currents, high oxygen, cool. | High suspended solids, erratic tidal currents, warmer than ocean and rivers. | Strong stratification, offshore bar formation, cool, high oxygen. | Some suspended silt, erratic currents, cool, high oxygen. | Moderate to strong, stratification, high suspended silt, strong currents, high oxygen, cool. | |
| Northwest Pacific | Strong wave action, cold ocean water, some silt and clay turbidity. | Moderate suspended solids, erratic currents, high oxygen, cold. | High suspended solids, erratic tidal currents, warmer than ocean and rivers. | Strong stratification, offshore bar formation, cold, high oxygen. | Some suspended silt, erratic currents, cold, high oxygen. | Moderate to strong stratification, high suspended silt, strong currents, high oxygen, cold. | |
| Alaska | Very cold oceanic water, usually ice, salinities slightly depressed. | Very cold oceanic water, overlain by some fresh water, high oxygen. | Very cold water, variable salinity, much fine silt, debris from freezing. | Strong currents, high suspended solids frequently glacial in origin, very cold. | Very cold oceanic water, much ice, surface layer of fresh water, high oxygen. | High turbidity with glacial debris, seasonal freezeups, strong currents during runoffs. | Stagnant below silt depth, very little oxygen, high salinity, hydrogen sulfide. |
| Pacific Islands | Clear warm ocean water, strong wave action. | Clear ocean water, gentle currents, eddies, warmer than ocean. | High dissolved organics, color, suspended mud, very small currents, hot. | Slightly turbid, strong currents, river cooler than ocean water. | Very small currents, generally shallow, quite warm, clear ocean water. | Slightly turbid, eddying currents, slight stratification, high oxygen. | |

References: (1) U.S. Coast and Geodetic Survey, Coast Pilots. Tidal Current Tables. (2) U.S. Army Project and Study Reports. (3) FWPCA Reports and unpublished data. (4) National Estuarine Inventory.

Source: Ref. 13.1

ation.  Unfortunately, little of this human intrusion
was planned with any real sensitivity to maintaining
the vitality or function of these special ecosystems.
Few Americans have a real grasp of the magnitude of
man's influence in these estuarine-coastal zones--a
shortcoming we will shortly rectify.

## MAJOR CONTRIBUTORS TO DEGRADATION

There are three major contributors to the degradation of these areas.  They are physical alteration,
direct waste disposal and non-point source pollution.
Let us examine each source in more detail.

### Physical Alteration

The physical alteration of the nation's estuaries and coastal beaches has in many instances been the
precursor of the degradation caused by direct waste
disposal and non-point source pollution.  Consequently, it is appropriate that we make it the initial
topic of our discussion.

Some of the most common alterations of the estuarine areas have been the ditching and filling (often
with refuse) of tidal marsh areas, and the subsequent
erection of dwellings, commercial buildings and large
industrial complexes.  As a consequence, we have lost
significant amounts of fish and wildlife habitat, some
of it critical to the world's commercial fisheries.
In addition, thousands of acres of valuable floodwater
storage areas have been lost, as well.  Ditching has
not been limited to the tidal marsh area, either.
Large canals have also been dug into the inland or bay
side of many barrier islands to provide waterfront
lots and/or boat docking facilities.  Unfortunately,
such wholesale ditching has lowered ground water
tables and caused infiltration of salt or brackish
water into underground aquifers.  Homes subsequently
built along these new waterways often fill them with
septic tank seepage and excess nutrients resulting in
anoxic or hypoxic conditions and fish kills.  When
some of these canal diggers carried their projects
nearly to the oceanside of an island, subsequent
storms often found them to be the path of least resistance for surging floodwaters and islands have been
severed in half by erosion as a result.

The nation's coastal beaches, by far the greatest attraction of the estuarine zone, have been similarly altered.    A  typical  cross-section  of  an unaltered  mid-Atlantic  coastal  beach  is  shown  in Figure 13.4.  You will note the prominence of the sand dune.  These coastal dunes serve as barriers to storm-wave overwash and flooding and further serve as reservoirs of sand for the replenishment of beaches ravaged by  storms  or  lost  by  attrition  from  longshore currents.

In our desire to have unobstructed views of, and be as close as possible to the pounding surf, we have either  removed  these  dunes  completely  or  placed dwellings on top of them.  Some have even been foolhardy enough to place their structures completely in front of the dune line.  Once these dunes have been removed, storm waves can surge inland virtually unimpeded.  As a result, soils inland of the dune line are eroded, structures are undermined, and beaches can no longer receive an allotment of sand to replace that lost in the storm.  It is this last consequence that causes  great  consternation  amongst  surfside  inhabitants.   The  very  beach  that  attracted  them  in  the first place, begins to slowly erode away.  The result is the placement of more man-made structures and further  interference  with  the  natural  cycle  of  the coastal environment.  Groins, which are short walls built perpendicular to the shoreline (Figure 13.5) are often one of the first structures to be built.  They are intended to trap whatever sand is flowing with the longshore currents.  While effective, one of the drawbacks is that sand builds up on the upstream side and erodes  on  the  downstream  side,  thereby  producing  a beach of discontinuous widths.  In addition, groins are only effective where there is sufficient supply of sand moving along with the longshore currents.

A second  structure  often  used  to  replace  an excavated dune is a bulkhead--a short vertical wall designed to retain or prevent sliding of the land (Figure  13.6).   Bulkheads  are  very  susceptible  to storm damage and are frequently undermined.  When bulkheads fail, the structure of last resort has been a seawall.  These are generally massive and expensive structures (Figure 13.7) with some serious side effects.   Such  massive  structures,  usually  built  of stone or concrete, reflect wave energy to any remain-

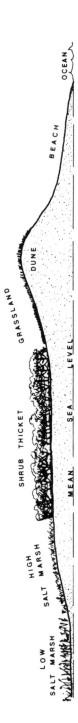

Figure 13.4: Typical cross-section of a natural mid-Atlantic coastal beach on a barrier island.

Figure 13.5:  Typical groin construction and placement.

Figure 13.6:  A vertical bulkhead showing an advanced stage of undermining and partial collapse.

Figure 13.7: Coastal towns slowly become fortresses as seawalls rise against an adversary that will, in the end, eventually prevail.

Responding to the unrelenting assault of the ocean, sea walls are widened and heightened. Owners of beach front dwellings find themselves forced to add to the height of their residences. The ones shown here are already several stories high.

ing beach, thereby, hastening its erosion. This, in turn, steepens the offshore bottom profile which increases the storm wave energy striking the wall. As a result, seawalls weaken and must be made larger and larger. Eventually, seawalls may be so high and large that it will be impossible to see the surf zone. How much simpler it would have been to place our structures well behind the dune line, forgo the need to view the pounding surf from a sun deck and walk the several hundred feet or so to the top of the dunes.

The final physical alteration, we will discuss here, is very often the least noticed. This is ground subsidence caused by the withdrawal of fresh water and petroleum from subterranean formations in coastal areas. As a consequence of these withdrawals, layers of sand and clay compress and any land above it sinks lower. If we continue to extract such materials, it is possible that the coastline may recede inland as much as 200 feet in the next 50 years, particularly in areas morphologically similar to the coastline of the northeastern United States. In some points along the Gulf Coast, where the land is flatter, particularly in Florida, this inland intrusion could reach as far as 500 feet.

## Direct Waste Disposal

Direct waste disposal is the intentional release of wastes either through direct dumping or through pipeline discharges. Wastes dumped directly, and often with approval of regulatory agencies, include dredged materials, municipal sewage sludge and industrial wastes. About 180 million wet metric tons (WMT) of dredged spoils are disposed of each year in the nation's marine waters (Table 13.4) accounting for about 80 to 90 percent of the total volume of materials dumped in these waters (Ref. 13.3). Two-thirds of that amount is dumped in estuaries. The composition of dredged material varies from site to site. In certain areas, sediments are contaminated, sometimes heavily, with metals and organic chemicals originating from direct industrial or municipal discharges and non-point sources.

The dumping of sewage sludge originating from municipal sewage treatment plants has risen steadily (Figure 13.8) from 2.5 million wet metric tons in 1958

Table 13.4: Quantities of Dredged Material Disposed of Annually in the Marine
Waters of the United States (mmt/yr)

| Coastal region | Average quantities disposed of annually | | | |
| | Estuaries | 0 to 3 miles offshore[b] | Over 3 miles offshore[b] | Total |
| --- | --- | --- | --- | --- |
| Northern Pacific | 5.4 | 10.0 | 0.3 | 15.7 (9) |
| Southern Pacific | 8.4 | 4.3 | 2.4 | 15.1 (8) |
| Gulf of Mexico | 91.3 | 16.4 | 10.9 | 118.6 (66) |
| Southern Atlantic | 6.1 | 1.5 | 10.6 | 18.2 (10) |
| Northern Atlantic | 4.0 | 2.2 | 6.3 | 12.5 (7) |
| Total | 115.2 (64) | 34.4 (19) | 30.5 (17) | 180.1 (100) |

[a]Data were obtained from each U.S. Army Corps of Engineers District Office in the form of an annual average; data were not obtained for individual years. The period over which the data are averaged varies from one district to the next, but generally includes most of the 1970s and early 1980s. Units are millions of metric tons per year (mmt/yr); numbers in parentheses are the percent of the total.
[b]The distinction between "0 to 3 miles offshore" and "over 3 miles offshore" was used by the Corps to classify its data, based on the statutory definition of the territorial sea. This division does not, however, correspond exactly to the division between coastal and open ocean waters used by OTA: some open ocean waters may be included in the "0 to 3 miles offshore" category, and some coastal waters may be included in the "over 3 miles offshore" category

Source:   Ref. 13.3

Figure 13.8: Amounts of industrial waste and municipal sewage sludge dumped in all marine waters, 1973-1985.

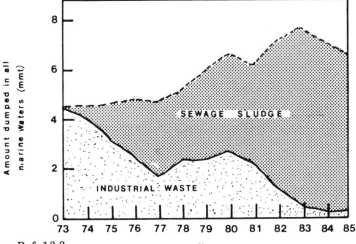

Source: Ref. 13.3

Figure 13.9: Location of current municipal sewage sludge and industrial waste dumpsites in the northern Atlantic Ocean.

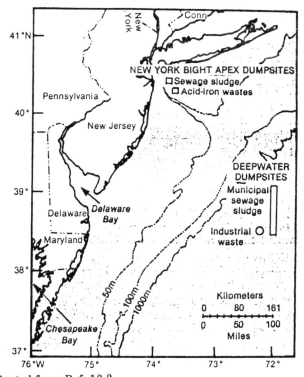

Source: Adapted from Ref. 13.3

to 7.5 million wet metric tons in 1983 (Ref. 13.3)
thanks largely to the absence of available land disposal options and the continued upgrading of our
municipal sewage treatment plants to secondary treatment. Nationally, most sewage sludge has been dumped
either at the Northeast's Deepwater Municipal Sludge
site off of Delaware Bay [1,2] (Figure 13.9) or at the
12 mile Sewage Sludge Dump Site [3] located in the New
York Bight. Recent federal legislation (see Federal
Clean Water Act) requires all sludge dumping in marine
waters to be phased out by 1991--a deadline that has
already been extended several times. Whether this one
will be met remains to be seen. It is interesting to
note that the dumping of industrial wastes (largely
acid or alkaline liquids) has decreased substantially
over the last decade decreasing from a high of 4.6
million wet metric tons originating from 300 firms in
1973 to a current level of about 200,000 wet metric
tons from three firms. Most of this is dumped at the
Deepwater Industrial Waste Site located about 10
nautical miles west of the Deepwater Municipal Sludge
Site.

The second component of direct dumping is
pipeline discharges. Over 1,300 major industrial
facilities (excluding power plants) and about 600 publicly owned treatment works (POTW's) discharge
directly into estuarine and coastal waters (Table
13.5). Ninety-nine percent of these industrial facilities and 89 percent of these POTW's discharge into
estuaries. Over half of these major industrial and
municipal outfalls are located in the nation's northern Atlantic region and the western Gulf of Mexico.
These estuarine discharging POTW's represent only
about 3.5 percent of the total 15,500 POTW's nationwide, yet they account for 1/4 of the total amount of
municipal wastewater discharged nationally. On an
annual basis this amounts to 2.3 trillion gallons (2
trillion gallons into estuaries and 0.3 trillion gallons into coastal waters). About 2/5 of this effluent
receives less than secondary treatment. In addition,
a small number of POTW's still discharge sewage sludge
through their outfalls.

The ultimate fate of these directly discharged
materials is dependent upon a variety of processes.
When a waste enters marine or estuarine waters through
dumping or discharge, it immediately begins to mix

Table 13.5: Number of Municipal and Major Industrial Facilities Discharging
Directly Into the Marine Waters of the United States

| Coastal States | Number of dischargers[a] | | | | | |
|---|---|---|---|---|---|---|
| | Municipal | | Major industrial | | Total | |
| **Northern Atlantic region:** | | | | | | |
| Maine | 38 | (3)[b] | 35 | | 73 | (3) |
| New Hampshire | 2 | | 4 | | 6 | |
| Massachusetts | 20 | (1) | 20 | | 40 | (1) |
| Rhode Island | 8 | (2) | 24 | | 32 | (2) |
| Connecticut | 22 | | 75 | | 97 | |
| New York | 47 | (1) | 29 | | 76 | (1) |
| New Jersey | 48 | (12) | 129 | (2) | 177 | (14) |
| Pennsylvania | 9 | | 33 | | 42 | |
| Delaware | 4 | | 30 | | 34 | |
| Maryland | 34 | (1) | 120 | | 154 | (1) |
| Virginia | 11 | (4) | 76 | | 87 | (4) |
| District of Columbia | 1 | | 1 | | 2 | |
| Total | 244 | (24) | 576 | (2) | 820 | (26) |
| **Southern Atlantic region:** | | | | | | |
| North Carolina | 10 | (1) | 41 | | 51 | (1) |
| South Carolina | 11 | | 22 | | 33 | |
| Georgia | 4 | | 26 | | 30 | |
| Florida (Atlantic) | 34 | (10) | 24 | (1) | 58 | (11) |
| Total | 59 | (11) | 113 | (1) | 172 | (12) |
| **Gulf of Mexico region:** | | | | | | |
| Florida (Gulf) | 22 | (5) | 17 | (1) | 39 | (6) |
| Alabama | 6 | | 29 | | 35 | |
| Mississippi | 6 | | 30 | | 36 | |
| Louisiana | 27 | (1) | 79 | | 106 | (1) |
| Texas | 52 | | 192 | (1) | 244 | (1) |
| Total | 113 | (6) | 347 | (2) | 460 | (8) |
| **California and Hawaii:** | | | | | | |
| California | 50 | (18) | 112 | (5) | 162 | (23) |
| Hawaii | 13 | (4) | ? | | 13 | (4) |
| Total | 63 | (22) | 112 | (5) | 175 | (27) |
| **Northern Pacific region:** | | | | | | |
| Oregon | 17 | (1) | 40 | (5) | 57 | (6) |
| Washington | 51 | | 144 | | 195 | |
| Alaska | 31 | (5) | ? | | 31 | (5) |
| Total | 99 | (6) | 184 | (5) | 283 | (11) |
| **Total United States** | 578 | (69) | 1,332 | (15) | 1,910 | (84) |

[a]Municipal category includes all municipal facilities. Industrial category includes those industrial facilities (excluding steam electric plants) discharging more than 10,000 gallons per day. The most recent available data pertain to dischargers as of 1982 or earlier.
[b]Numbers in parentheses indicate discharges directly into coastal waters. All remaining discharges are into estuarine waters.

Source: Ref. 13.3

with the water to form a broadening plume. As a result, the waste is diluted substantially often by a factor of 5000 or more. This initial dilution takes place within the first few hours after disposal and lasts until the waste particles either stop moving vertically in the water column or reach bottom. The vertical movement of these wastes particles depends on the particle's bulk density and the presence of pycnoclines. Waste particles that are more dense than the surrounding water will sink. Less dense particles will rise. Particles can, therefore, accumulate either along the bottom or along a pycnocline. Any that accumulate along a pycnocline eventually end up on the bottom through other processes, although they may have been transported a considerable distance from their point of origin by then (Figure 13.10). This physical transport is usually accomplished either by currents or mixing or a combination of both. There are two general categories of currents: permanent and transient. Permanent currents occur in both coastal and open ocean waters. The Gulf Stream off the east coast and the weaker California current off the west coast are classic examples. Portions of these currents called eddy rings or jets, can meander off the main current and alter the general direction of flow of the waste material. Transient currents, on the other hand, occur over shorter distances and time periods. They can be found in all marine environments including estuaries and are caused generally by winds, tides and waves. Their effect on the transport of waste materials is most pronounced when the materials are dumped or discharged near shore. Such currents are known to move sediment such as dredged material along and into the shoreline.

Mixing occurs when two masses of water with different densities intermingle along their common boundary. This mixing can be enhanced by wave action and tides and the rate of mixing varies with different marine environments. For instance, complete mixing of bottom and surface waters in the open ocean may take centuries while in estuaries mixing occurs almost immediately (Figure 13.11).

Another less commonly known form of transport is biological transport. This occurs when marine organisms ingest waste particles and carry them to other areas. Plankton and various pelagic fishes are exam-

Figure 13.10:  General fate of effluent discharged into marine waters.

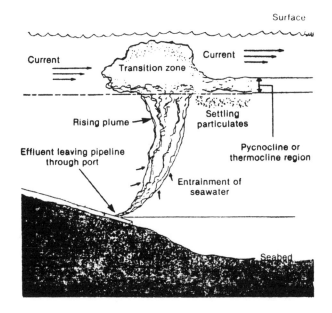

Source:  Ref. 13.3

Figure 13.11:  Typical mixing pattern in the estuarine zone.

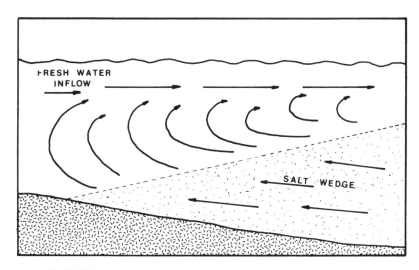

Source:  Ref. 13.1

ples of such organisms. Currents and mixing can transport and disperse wastes in the marine environment for hundreds of kilometers. We are foolish, therefore, to continue to believe that singular discharge pipes and isolated dump sites do not have widespread impacts.

## Non-Point Source Pollution

Non-point sources are now considered significant contributors of pollutants to estuaries and marine waters (Figure 13.12). Some major sources of non-point source pollution include:

1.  Runoff from storm sewers that originate in urban areas, industrial sites and farmland, as well as a mixture of storm water and sewage from combined storm and sanitary sewerage systems (Figures 13.13 and 13.14).
2.  Precipitation.
3.  Atmospheric deposition.
4.  Underground transport through aquifers from landfills or septic systems.

Detailed data, quantifying and characterizing the complete impact of all of these sources have not been sufficiently developed. However, their contribution to the degradation of estuarine waters is considered to be significant (Table 13.6). Some studies of runoff from storm sewers and combined storm and sewage outfalls (CSO's) indicate that such water is a significant source of fecal coliform bacteria and suspended solids.

Extensive amounts of paving and sidewalks normally found in densely developed, urban areas contribute additional pollutants such as oil and grease and lead and chromium, while runoff from agricultural areas carries large quantities of pesticides and herbicides.

## RELEVANT FEDERAL LEGISLATION

There are two major federal statutes that directly regulate waste disposal in the nation's coastal and marine environments. They are the Marine Protection Research and Sanctuaries Act (MPRSA) and the Clean Water Act (CWA), both of which were enacted

Figure 13.12: Nonpoint source parameters most widely reported.

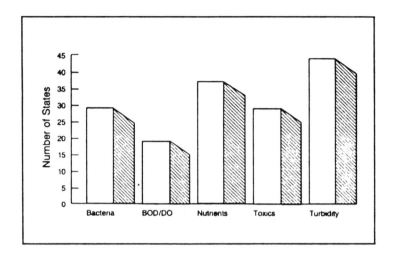

Figure 13.13: Nonpoint sources reported as major causes of use impairments.

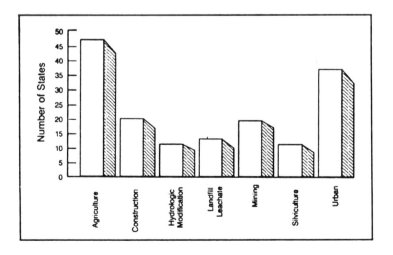

Source for Figures 13.12 and 13.13: Ref. 13.3

Figure 13.14:　Typical combined sewer collection network during a storm.

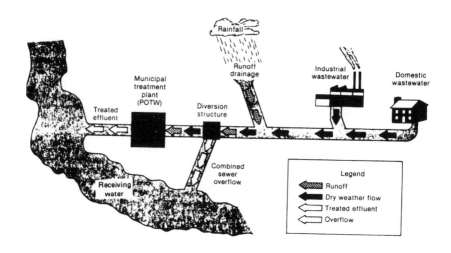

Source:　Ref. 13.3

Table 13.6:　Relative Contribution of Pollutants (in percent) by Major Sources in Coastal Hydrologic Units, Circa 1977–1981

| Region and source | BOD | TSS | TKN | TP | CD | CR | CU | PB | AS | FE | HG | ZN | OIL | CHL HCS | FEC COL |
|---|---|---|---|---|---|---|---|---|---|---|---|---|---|---|---|
| **Northern Pacific:** | | | | | | | | | | | | | | | |
| Industrial[b] | 34 | <1 | 11 | 4 | 98 | 7 | 35 | 46 | 96 | 3 | 86 | 40 | 29 | 99 | 0 |
| Municipal | 27 | <1 | 27 | 23 | <1 | 1 | 1 | 2 | 2 | 2 | 0 | 4 | 16 | <1 | 4 |
| Nonpoint | 40 | 99 | 62 | 73 | 2 | 92 | 63 | 53 | 2 | 95 | 14 | 57 | 55 | <1 | 96 |
| **Southern Pacific:** | | | | | | | | | | | | | | | |
| Industrial | 3 | <1 | 3 | <1 | 81 | 2 | <1 | 6 | 99 | 6 | 88 | 16 | 10 | 90 | 0 |
| Municipal | 55 | 1 | 31 | 21 | 3 | 8 | 7 | 7 | <1 | 4 | 9 | 6 | 58 | 9 | 34 |
| Nonpoint | 43 | 99 | 67 | 78 | 16 | 90 | 91 | 87 | <1 | 90 | 3 | 78 | 33 | <1 | 66 |
| **Gulf of Mexico:** | | | | | | | | | | | | | | | |
| Industrial | 34 | <1 | 31 | 32 | 95 | 39 | 45 | 45 | 100 | 18 | 93 | 47 | 53 | 97 | 0 |
| Municipal | 26 | <1 | 32 | 34 | <1 | 4 | 6 | 2 | <1 | <1 | 3 | 5 | 9 | 3 | 16 |
| Nonpoint | 40 | 99 | 37 | 34 | 4 | 57 | 49 | 53 | <1 | 82 | 3 | 49 | 39 | <1 | 84 |
| **Southern Atlantic:** | | | | | | | | | | | | | | | |
| Industrial | 28 | <1 | 10 | 14 | 73 | 58 | 27 | 12 | 100 | 7 | 89 | 25 | 8 | 98 | <1 |
| Municipal | 35 | <1 | 54 | 73 | 1 | 8 | 9 | 3 | 0 | 1 | 6 | 7 | 13 | 1 | 10 |
| Nonpoint | 37 | 99 | 36 | 13 | 26 | 34 | 64 | 85 | <1 | 92 | 6 | 68 | 79 | <1 | 90 |
| **Northern Atlantic:** | | | | | | | | | | | | | | | |
| Industrial | 6 | <1 | 8 | 7 | 84 | 15 | 18 | 16 | 100 | 21 | 92 | 25 | 5 | 93 | <1 |
| Municipal | 73 | 3 | 74 | 76 | 2 | 32 | 23 | 13 | <1 | 10 | 5 | 19 | 46 | 7 | 12 |
| Nonpoint | 21 | 97 | 17 | 18 | 13 | 52 | 60 | 72 | <1 | 69 | 3 | 56 | 49 | <1 | 88 |
| **Total U.S. coastal:** | | | | | | | | | | | | | | | |
| Industrial | 11 | <1 | 9 | 5 | 89 | 15 | 18 | 20 | 100 | 13 | 91 | 25 | 11 | 94 | <1 |
| Municipal | 56 | 1 | 46 | 36 | 1 | 13 | 11 | 8 | <1 | 5 | 5 | 5 | 9 | 41 | 6 | 16 |
| Nonpoint | 34 | 99 | 45 | 59 | 10 | 72 | 71 | 73 | <1 | 82 | 3 | 66 | 47 | <1 | 84 |

KEY: BOD—Biochemical oxygen demand CD—Cadmium AS—Arsenic OIL—Oil and grease
TSS—Total suspended solids CR—Chromium FE—Iron CHL HCS—Chlorinated hydrocarbons
TKN—Total Kjeldahl nitrogen CU—Copper HG—Mercury FEC COL—Fecal coliform bacteria
TP—Total phosphorus PB—Lead ZN—Zinc

[a]Information regarding contribution of pollutants is aggregated for all maritime hydrologic units in each region. Hydrologic units are designated by the U.S. Geological Survey and represent natural and human-made drainage areas. Only pollutants that first enter surface waters in maritime hydrologic units (i.e., directly adjacent to marine waters) are included. Pollutants originating in upstream hydrologic units and flowing into the maritime units considered here are excluded, although in some instances the upstream units contribute a sizable portion or even a majority of the pollutants entering coastal waters.

[b]The "industrial" category includes powerplants.

Source:　Ref. 13.3

in 1972 (Figure 13.15).   Each has been subsequently
amended since then, with the latter statute being last
amended by the Water Quality Act of 1987.   The major
provisions of the MPRSA and the CWA are shown in Table
13.7.   Other federal statutes[4] that may also impact
directly or indirectly on the quality of the nation's
coastal and estuarine environments include:

1)   The Clean Air Act Amendments (CAA) of 1977
(42 U.S.C. 7401 et. seq.) have directly resulted in
the generation of large amounts of air pollution
control wastes, (fly ash, flue-gas desulfurization
sludges and other air pollution control sludges) which
have been proposed for marine disposal at various
times.   These wastes are generated by the air pollution control equipment installed to comply with
national emission and air quality standards for sta-
tionary sources of air pollutants.

2)   The Coastal Zone Management Act (CZMA) of
1972 (16 U.S.C. 1451 et. seq.) provides Federal grants
to States to develop Coastal Zone Management Plans
that balance the pressure for economic development and
the need for environmental protection.   EPA cannot
issue a permit for an activity affecting land or water
use in a coastal zone until it has certified that the
activity does not violate a State's management plan.
Through the National Estuarine Sanctuary Program, the
act authorizes 50 percent matching grants to States to
acquire and manage estuaries for research and educational purposes.   Amendments to CZMA in 1980 state
that management policies should protect coastal natural resources (including estuaries, beaches and fish
and wildlife and their habitat) and encourage area
management plans for estuaries, bays and harbors.

3)   The Comprehensive Environmental Response,
Compensation and Liability Act (CERCLA) of 1980, (42
U.S.C. 9601 et. seq.) better known as Superfund, was
enacted to provide emergency response and cleanup
capabilities for chemical spills and releases from
hazardous waste treatment, storage and disposal facilities.   Its primary impact on marine waste disposal
involves:   1) the identification of large numbers of
hazardous waste sites in the coastal zone, with
potential for movement of waste pollutants into marine
waters; 2) the suggestion that some wastes generated
by remedial action at Superfund sites possibly be dis-

Figure 13.15: Jurisdictional boundaries of the Federal Clean Water Act and the Marine Protection, Research, and Sanctuaries Act in marine waters.

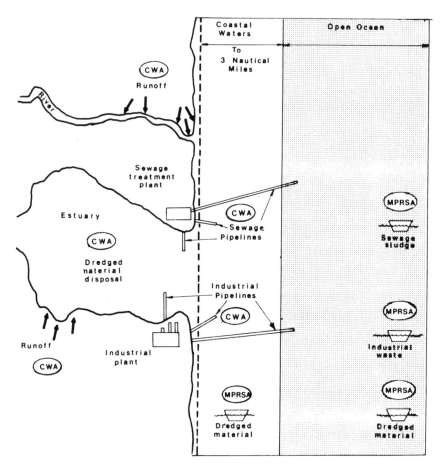

CWA - Clean Water Act, formally known as the Federal Water Pollution Control Act

MPRSA - Marine Protection, Research, and Sanctuaries Act

Dumping beyond the inner boundary of the territorial sea is covered by MPRSA (CWA covers dumping within the territorial sea in principle, but is preempted by MPRSA.        Estuarine dumping falls under CWA

Pipelines (wherever they are located) are covered by CWA

Source:  Adapted from Ref. 13.3

Table 13.7:  Major Provisions of the Marine Protection, Research and Sanctuaries Act and the Clean Water Act

| Statute and section | Purpose |
|---|---|
| **Marine Protection Research, and Sanctuaries Act:** | |
| Sec. 101 | Prohibits, unless authorized by permit, the transportation of wastes for dumping and/or the dumping of wastes into the territorial seas or the contiguous zones. |
| Sec. 102 | Authorizes EPA  to issue permits for dumping of nondredged materials into the contiguous zone and beyond as long as the materials will not "unreasonably degrade" public health or the marine environment, following criteria specified in statute or established by the Administrator. |
| Sec. 103 | Authorizes Corps of Engineers to issue permits for dumping dredged material, applying EPA's environmental impact criteria to ensure action will not unreasonably degrade human health or the marine environment. |
| Sec. 104 | Specifies permit conditions for waste transported for dumping or to be dumped, issued by EPA or the Coast Guard. |
| Sec. 107 | Authorizes EPA and Corps of Engineers to use the resources of other agencies, and instructs the Coast Guard to conduct surveillance and other appropriate enforcement activities as necessary to prevent unlawful transportation of material for dumping or unlawful dumping. |
| **Clean Water Act:** | |
| Sec. 104(n) | Directs EPA to establish national estuaries programs to prevent and control pollution; to conduct and promote studies of health effects of estuarine pollution. |
| Sec. 104(q) | Establishes a national clearinghouse for the collection and dissemination of information developed on small sewage flows and alternative treatment technologies. |
| Sec. 201, 202, 204 | Specifies sewage treatment construction grants program eligibility and Federal share of cost. |
| Sec. 208 | Authorizes a process for States and regional agencies to establish comprehensive planning for point and nonpoint source pollution. |
| Sec. 301 | Directs States to establish and periodically revise water quality standards[c] for all navigable waters; effluent limitations for point sources requiring BPT should be achieved by July 1, 1977; timetable for achievement of BAT and other standards set. Compliance deadlines for publicly owned treatment works (POTWs) to achieve secondary treatment also set. |
| Sec. 301(h) | Authorizes waivers for POTWs in coastal municipalities from secondary treatment for effluent discharged into marine waters if criteria to protect the marine ecosystem can be met. |
| Sec. 301(k) | Allows industrial dischargers to receive a compliance extension from BAT requirements until July 1, 1987, for installation of an innovative technology, if it will achieve the same or greater effluent reduction than BAT at a significantly lower cost. |
| Sec. 302 | Allows EPA to establish additional water quality-based limitations once BAT is established, if necessary to attain or maintain fishable/swimmable water quality (for toxics, the NRDC v. EPA consent decree sets terms). |
| Sec. 303 | Requires States to adopt and periodically revise water quality standards; if they determine that technology-based standards are not sufficient to meet water quality standards, they must establish total maximum daily loads and waste load allocations, and incorporate more stringent effluent limitations into Sec. 402 permits. |
| Sec. 303(e) | Requires States to establish water quality management plans for watershed basins, to provide for adequate implementation of water quality standards by basin to control nonpoint pollution; Section 208 areawide plans must be consistent with these plans. |
| Sec. 304 | Requires EPA to establish and periodically revise water quality criteria to reflect the most recent scientific knowledge about the effects and fate of pollutants, and to maintain the chemical, physical, and biological integrity of navigable waters, groundwater, and ocean waters and establish guidelines for effluent limitations. |
| Sec. 304(b) | Outlines factors to be considered when assessing BPT and BAT to set effluent limitation guidelines, including accounting for "non-water quality impact," age of equipment, etc. |
| Sec. 305(b) | Sets State water quality reporting requirements. |
| Sec. 306 | Sets new source performance standards for a list of categories of sources. |
| Sec. 307 | Requires EPA to issue categorical pretreatment standards for new and existing indirect sources; POTWs required to adopt and implement local pretreatment programs, toxic effluent limitation standards must be set according to the best available technology economically achievable. |
| Sec. 308 | Requires owners or operators of point sources to maintain records and monitoring equipment, do sampling, and provide such information or any additional information. |
| Sec. 309 | Gives enforcement powers primarily to State authorities. Civil penalties, however, and misdemeanor sanctions can be issued by EPA in U.S. district courts for violation of the act, including permit conditions or limitations; EPA also is authorized to issue criminal penalties for violations of Sections 301, 302, 306, 307, and 308. EPA may take enforcement action for violations of Section 307(d) which introduce toxic pollutants into POTWs. |
| Sec. 402 | Establishes National Pollutant Discharge Elimination System (NPDES), authorizing EPA Administrator to issue a permit for the discharge of any pollutant(s) to navigable waters that will meet requirements of Sections 301, 302, 306, 307 and other relevant sections; States can assume administrative responsibility of the permit program. |
| Sec. 403 | Directs EPA to establish Ocean Discharge Criteria as guidelines for permit issuance for discharge into territorial seas, the contiguous zone, and open ocean. |
| Sec. 404 | Directs Secretary of the Army to issue permits for dredged or fill material; EPA must establish criteria comparable to Section 403(c) criteria for dredged and fill material discharges into navigable waters at specified disposal sites. |
| Sec. 405 | Requires EPA to issue sludge use and disposal regulations for POTWs. |
| Sec. 504 | Grants emergency powers to Administrator to assist in abating pollutant releases, establishes a contingency fund, and requires Administrator to prepare and publish a contingency plan to respond to such emergencies. |
| Sec. 505 | Citizen suit provision allows citizens to bring civil action in district court against any person in violation of an effluent standard or limitation of an order by the Administrator for failing to perform a nondiscretionary act. |

Source:  Ref. 13.3

posed in the ocean; and 3) provisions regarding the liability of ocean incineration vessels.

4)    The Endangered Species Act (ESA) of 1973 (16 U.S.C. 1531 et. seq.) requires all Federal agencies and their permittees and licensees to ensure that their actions are not likely to jeopardize the existence of an endangered or threatened species or result in the destruction or adverse modification of critical habitats of such species.  If an activity might affect an endangered or threatened species, the Federal agency must obtain a biological opinion from the U.S. Fish and Wildlife Service or the National Marine Fisheries Service about the potential effects.  These opinions can be an integral part, for example, of the site designation process for marine dumping activities.

5)    The National Environmental Policy Act (NEPA) of 1970 (42 U.S.C. 4321 et. seq.) requires that an Environmental Impact Statement (EIS) be prepared for all proposed legislation and all major Federal actions that could significantly affect the quality of the human environment.  Under NEPA, EPA is exempted from this provision, but it voluntarily prepared an EIS when it proposed revisions to the ocean dumping regulations and criteria in 1977 and it prepares EIS's for site designations.

6)    The National Ocean Pollution Planning Act (NOPPA) of 1978 (33 U.S.C. 1701 et. seq.) directs the National Oceanic and Atmospheric Administration to coordinate the ocean pollution research and monitoring that is conducted by various Federal Agencies and to establish Federal priorities in marine research.

7)    The Port and Tanker Safety Act of 1978 (33 U.S.C. 1221 et. seq.) regulates the operation of ships within U.S. ports and waterways to promote navigational and vessel safety and protect the marine environment.  Regulations for operations at waterfront facilities include requirements for handling, storage, loading and movement of dangerous materials.

8)    The Safe Drinking Water Act (SDWA) of 1974 (42 U.S.C. 300 (f) et. seq.) mandated the development of primary and secondary drinking water standards. Compliance with these standards has resulted in the use of water treatment processes that generate sludges requiring disposal.  Some of these wastes could be disposed of in marine waters.

9) <u>The Toxic Substances Control Act (TSCA) of 1976</u> (15 U.S.C. 2601 et. seq.) regulates the manufacture, processing, distribution, use and disposal of chemical substances that present significant risks to human health or the environment. EPA is authorized to gather information concerning the toxicity of chemicals, mandate additional testing and research where necessary and assess the extent of risk. TSCA affects marine waste disposal primarily through its regulations on the disposal of wastes contaminated with PCB's.

10) <u>The Low Level Radioactive Waste Policy Amendments Act of 1986</u> (42 U.S.C. 2021 (b) et. seq.) places responsibility with the States for managing commercial low-level radioactive waste (LLW) generated within their borders. States may enter into compacts with other States to establish and operate regional disposal sites. The original deadline for establishing operational sites was January 1, 1986, but it has been extended to December 31, 1992. Currently, there are only three operational disposal sites in the country. Although MPRSA effectively bans marine disposal of LLW, the difficulty of establishing future land-based disposal sites could lead to reconsideration of the marine disposal option.

11) <u>The National Flood Insurance Act of 1968</u> (P.L. 90-448) as amended by the Flood Disaster Protection Act of 1973 (P.L. 92-234) encourages communities to provide land-use planning that will minimize property damage in flood-prone areas like barrier beaches. Also, it provides an opportunity for property owners to purchase flood insurance that is generally not available from private insurance companies.

12) <u>The Resource Conservation and Recovery Act (RCRA) of 1976</u> (42 U.S.C. 6901 et. seq.) defines and lists hazardous wastes and controls their generation, transport, treatment, storage and disposal. Some hazardous wastes currently enter marine waters from disposal sites located on land but near estuaries or coastal waters, or as part of "indirect" industrial discharges into municipal treatment plants that subsequently discharge into these waters. The Domestic Sewage Exemption of RCRA, for example, allows legal "indirect" discharges of some hazardous wastes into municipal treatment plants.

13)   The 1984 Hazardous and Solid Waste Amendments to RCRA dramatically increased the scope and complexity of the RCRA program and represented an attempt by Congress to discourage most land-based disposal methods for managing hazardous wastes. These new provisions could lead to the consideration of marine disposal of hazardous wastes.

14)   The Coastal Barrier Resources Act of 1982 is designed to minimize loss of human life, expenditure of federal revenues and damages to natural resources by restricting future federal financial assistance that encourages development on coastal barrier islands (Ref. 13.2).

15)   The Ocean Dumping Ban Act of 1988, Title I of the Marine Protection, Research and Sanctuaries Act. It prohibits further dumping of sewage sludge by August 14, 1989, unless the dumper has received a permit and has entered into an agreement with the U.S. EPA and the State in which the dumper is located that includes a schedule to end the dumping. The Act makes it unlawful to ocean dispose of sewage sludge after December 31, 1991.

## VARIOUS RELEVANT WATERBODY MANAGEMENT PROGRAMS

Numerous programs have been initiated at Federal, State and local levels to address water quality problems in a number of the nation's estuaries and coastal waterbodies (Ref. 13.3). These include:

### The Chesapeake Bay Program

The primary purpose of the Chesapeake Bay Program is to develop a comprehensive understanding of the Bay's ecosystem. It is a combined State and Federal effort initiated by the U.S. EPA in response to legislation passed by Congress in 1976. Specifically, EPA was directed to assess and make recommendations on how to improve water quality management in the Bay, to coordinate all research in the Bay and to establish a system of data collection and analysis. The study of Chesapeake Bay was authorized for 5 years, but was twice extended by a year and was completed in 1987 at a cost of nearly $30 million.

The study focused on the Bay's 10 most critical water quality problems, three of which were studied intensively: nutrient enrichment, toxic substances and the decline of submerged aquatic vegetation (SAV). The findings documented a historical decline in living resources in the Bay and indicated the need for better management. As a result, the Chesapeake Executive Council was established to facilitate the implementation of coordinated plans for the improvement and protection of the Chesapeake Bay estuarine system.

## The Puget Sound Water Quality Authority

The Puget Sound Water Quality Authority was established by the Washington State Legislature in 1983 and authorized in 1985 to develop a comprehensive management plan for Puget Sound and its related waterways. The first Puget Sound Water Quality Management Plan was adopted in late 1986. The plan focuses on protecting Puget Sound from toxic pollutants and pathogens and on the control of non-point source pollution. It emphasizes (land planners, please note) a lead role for local governments in identifying and controlling non-point sources.

## The Gulf Waste Disposal Authority

The Gulf Coast Waste Disposal Authority is a three-county unit of local government established by the Texas legislature in 1969 to abate point source pollution in the heavily industrialized Houston Ship Channel and Galveston Bay. The authority is active in pollution control financing and owns and operates four industrial wastewater treatment facilities. These facilities treat and dispose of liquid wastes from over 40 industrial plants.

## The Southern California Coastal Water Research Project

The Southern California Coastal Water Research Project is responsible for researching

and monitoring the effects of municipal wastewater discharges on marine life. The project
publishes a report on its research every two
years. It is sponsored by the sanitation districts of Orange County and Los Angeles County
and the cities of Oxnard, San Diego and Los
Angeles. The goal of the research is to develop
predictive models that would help determine what
levels of wastewater treatment are needed to
protect marine life.

## The San Francisco Bay Regional Water Quality Control Board

The San Francisco Bay Regional Water
Quality Control Board is one of the major agencies involved in managing the Bay's waters. It
operates independent of, but is responsible to
the State of California Water Resources Control
Board. It is primarily an enforcement agency
and is responsible for monitoring sewage outfalls and other point source discharges. It has
no authority to control impacts caused by pollutants carried by the Sacramento and San Joaquin
Rivers. The Regional Board is active in planning, reviewing and amending the Basin Plan for
the Bay area and in reviewing water quality
standards.

## The National Estuary Program

The National Estuary Program (NEP) was
created within the U.S. EPA in 1985 to oversee
the implementation efforts in the Great Lakes
and Chesapeake Bay and to initiate comprehensive
programs in other estuaries in the United
States. Programs are underway in Puget Sound,
Long Island Sound, Buzzards Bay, Narragansett
Bay, San Francisco Bay and Albemarle-Pamlico
Sounds. The NEP uses existing authorities under
Section 104 of the Clean Water Act. The objective of NEP studies is to characterize the conditions and trends in each system and develop an
integrated management program to maintain and
restore the estuary.

## The National Marine Pollution Program

The National Marine Pollution Program was established to coordinate the 11 departments and agencies that are engaged in research or monitoring related to marine pollution. This program was mandated under the National Ocean Pollution Planning Act of 1978. In 1985 the program issued a four year plan that recommended a greater Federal emphasis on:

1) Resource-oriented monitoring to provide national assessments of the status and trends in environmental quality.

2) Better coordination of monitoring efforts.

3) Research and monitoring programs related to municipal and industrial effluents.

4) Research and monitoring on nutrients and pathogens.

## The Northeast Monitoring Program

The Northeast Monitoring Program (NEMP) monitors waters from the Gulf of Maine to North Carolina's Cape Hatteras. NEMP was established in 1979 by the National Oceanic and Atmospheric Administration (NOAA) to monitor physical, chemical and biological variables of estuarine and coastal waters over long periods.

## REMEDIATION--AN OVERVIEW

The remediation of the nation's coastal areas and estuaries and the prevention of their further deterioration will not be quickly forthcoming unless we make the following items a part of our remediation agenda:

1. First, we need a change in the prevailing national philosophy toward the environment. At present, there are two major philosophical attitudes in the United States. The first, and currently the least influential views the coastal areas and estuaries as having unique properties and functions that deserve stringent protection. This perspective is frequently referred to as the Protectionist Philosophy. The second philosophical point of view and cur-

rently the most popular with the nation's legislators and regulatory agencies is the Managerial Philosophy. This philosophy views the coastal areas and estuaries as renewable and capable of being used in many ways including for waste disposal. Many factors must be weighed, however, in determining appropriate uses, including human health considerations, technological feasibility, economic costs and availability of other options. For legislators these factors also include political repercussions. The selected uses under this latter philosophy may not always be the most environmentally protective. The primary distinction between these two philosophies is that the Managerial Philosophy is a human-centered approach which views the nation's environment as a resource for society's benefit. The Protectionist Philosophy, on the other hand, places primary emphasis on the environment, treating anthropological concerns as peripheral issues.

Regulatory schemes based primarily upon the Managerial Philosophy are best suited for a "perfect world" situation. In the case of the federal Clean Water Act, for example, this is a world where NPDES permitted facilities (see Chapter 2) never exceed their permit limits, where sewage treatment plants never have bypasses of raw sewage and where permit writers have full knowledge of the assimilative capacity of the receiving waterbody before they set permit limitations. Recognizing the shortcomings of this approach in the "real world", complex enforcement programs have been included to encourage adherence to prescribed numerical limitations or other conditions. Unfortunately, a persistent criticism of the nation's environmental agenda generally has been the lack of rigorous enforcement.

There is no doubt that the estuarine and coastal environment would be better served by a shift in the national philosophy toward the Protectionist Philosophy.

2.   Local land use ordinances in the coastal and estuarine areas must be revised to reflect the sovereignty of natural systems.

3.   Inland sources of pollution must be controlled.

In 1988 a Blue Ribbon Panel comprised of state scientists, attorneys, local and state legislators, educators and industrial representatives was convened

at the request of the New Jersey Department of
Environmental Protection to examine the underlying
causes of that state's unsettling coastal zone problems that manifested themselves with a particular
vengeance in the summer of 1987. Their final report
(Ref. 13.21) embodied recommendations that are particularly noteworthy for their Protectionist Philosophy
orientation (perhaps there is hope after all) as well
as their recognition of the intricacy of environmental
protection in our modern society. Because of this we
list them here in their entirety:
"The Blue Ribbon Panel recommends the following:

o  Development

1.  Density: The density of development on the
    coast should be lowered. The following land
    types must be protected: beach fronts,
    shorelines, wetlands, river banks, flood
    plains and shallows. Protection should also
    be given to beach, port, harbor and water-
    front properties needed to maintain tradi-
    tional water dependent use of and access to
    the coastal environment.

2.  Future Development: Future development must
    be limited to areas presently providing ade-
    quate infrastructure, utilities and ser-
    vices. New growth should not occur in
    undeveloped, environmentally sensitive
    areas.

3.  Land Use: Land uses that do not degrade the
    coastal environment should be allowed over
    those that do. Land uses that improve the
    coastal environment should be allowed over
    both. 'Limits-to-Growth' concepts must be
    introduced into coastal land use decision
    making as well as planning and zoning.

4.  Local Zoning: If it is absolutely necessary
    to develop in the coastal zone, local zoning
    ordinances must be sensitive to environmen-
    tal values and strictly enforced to control
    urban sprawl. The Panel recommends that
    planning and development be done to achieve
    the best ratio of developed acreage to open,
    natural lands. Such ratio should be estab-
    lished that is fully protective of the
    environment.

5.   <u>Federal and State Laws</u>:   The Panel recommends that the Coastal Area Facility Review Act (CAFRA) and the Coastal Barrier Resources Act (CBRA) be reviewed, strengthened and expanded."
Author's Special Note:
New Jersey's Coastal Area Facilities Review Act (CAFRA) enacted in 1973 is typical of the emasculated environmental legislation produced in a state dominated by a managerial environmental philosophy. A key provision of CAFRA exempted from its jurisdiction all residential housing developments of less than 25 units. Prior to the bill's enactment, residential projects of a hundred or more units were commonplace in the coastal zone. After the act most projects in the coastal zone were submitted and built in blocks of 24--won't that provide interesting speculation for future archaeologists. The most irritating aspect of this bill, however, is the fact that in the 16 years that this law was on the books the state's DEP never aggressively lobbied for an amendment to eliminate this pre-planned less than 25 unit loophole.

6.   <u>Government Expenditures and Subsidies</u>:   The Panel recommends that direct or indirect subsidies that encourage development in sensitive coastal areas be eliminated. These would include Federally guaranteed flood and storm insurance and other coastal subsidy programs.

7.   <u>Runoff</u>:   Direct runoff of contaminant and debris laden waters into coastal waters should be prohibited and prevented. Effective retention basins, filters, screens and similar devices must be used to control such materials.

8.   <u>Pollution Control</u>:   The DEP should work toward improving public systems designed and implemented to control pollution caused by land development. The major cost of these improvements should be born by the municipalities that made the original land use decisions.

9.   Development Impacts:   The DEP should work
     with municipalities to develop and enforce
     steps that will lessen the impact of devel-
     opment on water quality (e.g., porous
     paving, better storm water control).   If mu-
     nicipalities do not comply, the Commissioner
     of the DEP should be empowered to create
     bans on construction until compliance is
     attained.

o  Water Quality
   1.   Combined Sewer Outfalls:   The DEP must work
        with the appropriate agencies and industry
        to reduce or eliminate the problems caused
        by outfalls that combine sewer and storm
        water runoff, particularly in the New
        Jersey/New York metropolitan area, through a
        structured, long-term, state/local capital
        improvement program.
   2.   Sewage Sludge Disposal:   The Panel recom-
        mends that the DEP immediately consider the
        development of alternatives to ocean dumping
        a high priority program.   The Department,
        with support from the legislature, must ad-
        dress the following questions:   How can
        industrial wastes be pretreated to avoid the
        accumulation of toxic chemicals in the
        sewage sludge?   Can the air quality of the
        state be maintained with the addition of an
        incineration program for sewage sludge?
        What are the possibilities of ground water
        contamination from the land application of
        sludge?   What are the effects of sewage
        sludge disposal on the marine life at the
        106-mile site?   and, If sludges are clean
        enough to dispose of on land, can we safely
        use ocean disposal?   Unless the DEP answers
        these questions there will be no basis to
        make a decision about the future disposal of
        sludge.   The Panel recommends that by the
        end of 1989 the DEP must present the Gover-
        nor and the citizens of the state with a
        plan, validated by an advisory panel, to
        implement alternatives to ocean disposal.
        This plan should begin operation on some
        percentage of the sludge (20%?) in 1990.
        The alternatives must be scaled up to handle

all of the sludge during 1991 and thereafter. If the deadlines are not met, a surcharge should be levied and increased as the
time of noncompliance increases. The surcharge should be used to support the
development of alternatives.

3.    Industrial Waste Pretreatment: The Panel
recommends that the State address the problems of continued release of toxic chemicals
into the coastal environment by improving
research, monitoring, permitting and
enforcement components of its existing
industrial waste pretreatment program. The
improvement should begin with a joint planning effort between the State and private
sector leaders. A large sewage authority in
the State (e.g. Passaic Valley Sewerage
Commission) should be chosen as a case study
for the planning effort and as the first
beneficiary of the plan. Strict time tables
must be set and kept for the planning and
implementation phases of the program. This
program must be coordinated with the sewage
sludge plan noted above.

4.    Contaminated Dredged Material: The Port
Authority of New York and New Jersey must
take a vigorous leadership role in the planning of future dredged material disposal
facilities and sites. Alternatives to present ocean disposal practices exist. The
Port Authority should take the lead in
developing strategies to finance new disposal methodologies (e.g. borrow pit disposal and sanitary landfill cover). Highest
priority should be given to alternatives
that can accommodate highly contaminated
sediments. EPA should implement new
national ocean dumping criteria as soon as
possible. In the interim, the EPA, the Army
Corps of Engineers, other appropriate
federal agencies, New York State Department
of Conservation and New Jersey Department of
Environmental Protection should update
regional guidance and testing protocols to
ensure that ocean dumping is being strictly
regulated. Attention should be given to the

restoration of the entire Mud Dump Site. Restoration should include covering or capping of the entire site with a layer of sand to create habitat with similar grain-size characteristics to that of the adjacent Continental Shelf. The covering will also isolate contaminated sediments from living marine resources. The Panel further recommends prohibiting the practice of placing back-barrier muds on ocean beaches.

5. Monitoring: The DEP has an extensive water quality monitoring program. Consideration should be given to expanding the number of variables monitored and to developing a more statistically sound sampling program. Monitoring should be conducted to detect "unreasonable degradation" and should be sufficiently rigorous to withstand the tests of courts as well as to allow for greater control over pollutant sources. Water quality monitoring variables should include nutrients, salinity, temperature, dissolved oxygen, biochemical oxygen demand (BOD), turbidity, fecal coliform and other appropriate microorganisms. Sediments should be monitored for selected metals, nutrients, total organic carbon (TOC), grain size, biological toxicity, chlorinated hydrocarbons and polynuclear aromatic hydrocarbons (PAHs). Biological effects must be measured to ensure that sensitive organisms are protected. Research is needed to investigate and develop new and improved indicators of water quality.

Steps associated with the improved water quality monitoring program include:

a) Developing a data management system that includes adequate quality assurance/quality control.

b) Reviewing the quality of existing monitoring data and the interpretation and use of that data. It is critical that DEP address external charges that their data are not credible. A science advisory panel should be formed to help with this task.

c)   Developing monitoring criteria for unreasonable degradation that are agreed
to, published, periodically reviewed,
updated and enforced.  The State should
take the initiative and not depend upon
federal standards.

6.   Shellfish Monitoring:  The DEP must continue
to improve its monitoring programs for toxic
chemicals in marine organisms and for
microorganisms in shellfish.  This would
include a surf clam monitoring program to
assess the effects of contaminant inputs on
this species and provide a mechanism to
evaluate changes in water quality over time.
This new program and the present monitoring
program should have a statistically sound
design.  The measured variables should
include metals, chlorinated hydrocarbons and
a public health microbial indicator in addition to fecal coliform counts.  Such a
program would have public health management
significance, lead to resource improvement,
and add to public confidence in seafood.

Because of the critical status of shellfish beds to decisions about coastal development within CAFRA, DEP should raise the
priority and funding level of its shellfish
resource survey.  By seeking change in the
legislation, DEP would strengthen its ability to collect adequate fees from applicants
for CAFRA permits, increasing user contributions for the expenses of CAFRA reviews and
shellfish surveying.

o  Ocean Health

1.   Sanitary Facilities:  County and State
health officials must require that sanitation (toilet) facilities be located at
beaches in sufficient numbers to handle
beach users.  If current enforcement authority is lacking, new legislation should be
introduced.  Since capital costs may prohibit immediate implementation, a fund
should be established which municipalities
can 'tap' now and repay later.  DEP should
take the lead in this endeavor, in cooperation with the Division of Travel and Tourism

in the Department of Commerce, Energy and
Economic Development and with the Department
of Health (DOH).

2.  Beach Fees:  Beach fees should be regulated
to ensure that they are reasonable and that
all revenues are used for beach purposes
only.  Fees should be used to repay loans to
the State for construction of sanitation
facilities and other public health purposes.

3.  Crowded Beaches:  DEP in cooperation with
the counties, DOH and EPA should continue
beach monitoring efforts for bacteria and
other indicator organisms.  As part of the
ongoing Ocean Health Study, sampling should
be conducted to investigate water quality
during crowded bathing conditions.  The
methods should be subject to the scrutiny
and approval of an advisory panel that is
independent of state government.

4.  Bather-to-Bather Contamination:  The data
generated in the Ocean Health Study in 1987
indicates that, in some cases, coastal
bathing waters may not have sufficient
flushing or turnover.  This suggests that
bathers may be exposed to significant
numbers of bacteria generated from the
swimmers themselves.  Therefore, the Panel
recommends that persons engaging in contact
water sports should not be crowded into
limited bathing areas.  Though life saving
should remain of paramount importance, lifeguards should not "corral" bathers into
small areas, particularly during crowded
holiday conditions.  Additional lifeguards
may be necessary to keep watch over wider
areas.  Guidance and/or standards should be
developed to assist lifeguards.

5.  Education Programs:  Lifeguards, beach
administrators and beach cleaning personnel
should be educated by State officials (DEP
and DOH) prior to the bathing season concerning public health and water quality issues.  The educational program should
include information on the water quality
monitoring program, including what the
sample numbers mean and information on

plankton blooms and other environmental events. In addition, the program should include actions that the lifeguards can take (e.g. dispersal of swimmers) to ensure safe pleasant bathing.

6.   Vessel Sanitary Pumpout Stations:  The DEP must require an increase in the number of vessel sanitary pumpout stations in the backbays and estuaries (presently, only 11 are available). In addition, the Department must clarify and more strictly enforce all the Federal and State guidelines for disposal of sanitary wastes by vessels.

7.   Sewage Treatment Plants:  The DEP must work with the legislators and sewerage authorities to improve operation and maintenance of sewage treatment plants.  The DEP should investigate financial incentives or disincentives to facilitate this goal.

o   Floatables

1.   Public Works Activities:  The DEP must continue to develop and enforce statewide standards for municipal public works activities such as drain cleaning and street sweeping. All storm drains must have trash collectors installed and maintained.  Substantial fines should be mandated when noncompliance is demonstrated.

2.   Solid Waste:  The DEP and the Attorney General must continue to work with the State of New York and New York City to improve solid waste handling practices at transfer points within N.Y.C.  It is particularly important to improve trash transfer and transporting in the Hudson-Raritan estuary.  Materials must be prevented from escaping to marine waters from the Fresh Kills landfill and similar systems.

3.   Littering Laws:  The DEP and the Attorney General must work toward increasing penalties, and strengthening and enforcing the littering laws.  Laws should also include sanctions against authorities, towns, cities, institutions, businesses and individuals which allow trash to accumulate without proper retention mechanisms.

Littering laws should extend to all New Jersey activities and operations, including vessels.

4. Litter Prevention Program: The DEP should continue to develop a comprehensive litter prevention program for shore areas. The program should include the following:

   a) An education and public awareness program should be established to teach the public that litter is unsightly, costly and possibly harmful, and that their actions can make a difference.

   b) The DEP or local agencies should provide sufficient litter containers and trash receptacles in the right places. It should encourage nonplastic litter bags in every car and boat.

   c) Litter control grants (under strict regulations and guidelines) should be made available to towns, cities and counties.

   d) Volunteers should be organized at every level including volunteer chair-persons at the state and local levels. The Water Watch program may be the appropriate vehicle or model. Volunteer groups should be formed in every community and full community involvement encouraged. Groups might be eligible for litter control grants under appropriate guidelines and with matching funds.

   e) Manufacturers of products found as wastes in the ocean and beach should be encouraged to effectively instruct users in the proper disposal of their products.

5. Packaging Alternatives: Manufacturers of packaging and floatable products should be encouraged to develop alternatives.

6. Recycling: The DEP should continue to support an aggressive recycling program for glass, cans and plastics.

7. Deposit Bill: The majority of the Panel believes that having a container deposit bill will reduce the amount of litter on the

beaches.  The DEP should provide information on this issue in order to implement legislative action.

o    Fisheries Resources

1.    Bioaccumulation:  Some shellfish resources in condemned areas and some species of fish currently contain chemicals at levels that merit serious concern.  Previous studies have indicated that early life stages and reproductive tissue seem to be most sensitive to pollutants.  Thus, bioaccumulation could potentially lead to a significant reduction in fishery-stock recruitment.  The DEP must commit sufficient financial resources for research in this area to determine the most effective strategy for preventing both potential public health risks and resource degradation.

2.    Communication:  It is recommended that risk assessment data be developed that would allow State spokespersons to make responsible statements to not only protect the public from contaminated seafood but also the fishing industry from the effects of unfounded public fears about contaminated seafood.

o    Enforcement

1.    Environmental and Conservation Laws:  The Panel recommends that a thorough review of all State of New Jersey environmental and conservation laws be undertaken.  As appropriate, laws and regulations should be strengthened and provisions made for more stringent interpretation, stiffer penalties (including imprisonment) and timely judicial processing of violations.  In addition, regulatory agencies should receive increased support for adequate enforcement authority and capabilities.

As an example:

o Conservation laws should mandate legally appropriate minimum fines or imprisonment, with immediate seizure of vessels, equipment and gear used in an offense at sea or in harbor.  Repeat offenses may result in revocation of operators and/or wildlife and fishing licenses, for one or more years.

2. Enforcement Machinery: The DEP must increase its enforcement capabilities to improve levels of compliance with laws designed to protect marine fish and shellfish, the property rights of shellfish bed leaseholders and public health. It must also increase enforcement activity with respect to polluters of the marine environment.

o  Education

1. Ocean Issues Campaign: Building upon the Environmental Education Act of 1971, the DEP should initiate a program with the Department of Education to educate citizens about ocean issue. This program should include grade schools and high schools, as well as people who use marinas, beaches, boardwalks and fishing piers. Its message should be that the ocean is not so big that it cannot be polluted and that everyone must and can help. This should be a well thought-out campaign with a slogan, logo, graphics, a spokesperson, etc. This is separate from and should not be confused with campaigns to promote tourism.

2. Public Participation: The Panel recommends that the DEP increase its efforts to involve the public in attempts to improve and protect the coastal environment. Grant support should be made available to new programs, as well as, existing ones (e.g. Coast Watch, Water Watch, Navesink River Project, Coast Guard Auxiliary) aimed at reducing non-point source pollution. The Office of Communications in the DEP, with the assistance of other appropriate divisions, should develop guidelines and administer the grants.

Care should be taken to accurately inform the public as soon and as thoroughly as possible of actions, events and findings that concern the coastal environment. Where risks to health or economics are concerned, the public should be involved from the start in deliberations about assessment and management of such risks.

o   Coastal Activities Coordinator and Ocean Advocate

1.   Coordinator:   The DEP should institute a
position in the Commissioner's office for a
Coastal Activities Coordinator.   This person
would work within the department with the
directors and assistant directors of appropriate divisions (Water Resources; Coastal
Resources; Science and Research; and Fish,
Game and Wildlife) as well as with the
Office of Communications to coordinate and
oversee activities concerning the coastal
zone.   All decisions involving coastal environment issues must include input from the
Coordinator.   This person would serve as a
focus for other agencies, the legislature
and the scientific and environmental communities.   The Coordinator would be responsible for producing an annual "State of the
Oceans" report for the Commissioner and
Governor.   This report would delineate "hot
spots", identify major issues, and summarize
progress in resolving issues concerning the
coast.   The report would be scientific in
content but with appropriate summaries for
the lay person.   An appropriate Science
Advisory Board would review the report prior
to its release.

The Coordinator would also be responsible
for preparing an annual agenda for ocean
research which would assist the DEP to
achieve its mission.   The Coordinator would
elicit inputs for this research program from
DEP staff and from an independent Science
Advisory Board.   The Coordinator would oversee the implementation of this program and
would provide an annual summary report of
the research accomplishments.

2.   Ocean Advocate:   There should be consideration given to the establishment of the
Office of the Ocean Advocate to promote the
concept that while the oceans are vast,
coastal regions can be polluted and degraded, and that society can stop this pollution and degradation.   In short, the

advocate should be responsible for developing an ethic on coastal zone usage."

## RECOMMENDED REMEDIES FOR THE MAJOR
## CONTRIBUTORS OF DEGRADATION

### Physical Alterations

1.    Local land use ordinances must be redrafted to
reflect the fact that coastal beaches are in a
dynamic equilibrium and that all man-made construction in these areas reduces their natural
flexibility.  They should be considered zones of
change and therefore the placement of permanent
structures would be incompatible (Figure 13.16).
The removal of sand dunes or the placement of
any structures, including dwellings, over or in
front of the natural dune line should be prohibited.

2.    The development of currently vacant barrier
islands should be prohibited and their acquisi-
tions for public use considered.

3.    Subsidized flood insurance coverage should be
eliminated for all of the nation's coastal
beaches and estuaries at or below an elevation
of 10 feet mean sea level.

4.    To minimize further impacts on coastal or estuarine areas already heavily developed, replacement construction should not be permitted at
densities that exceed existing levels.  For example, 2 seasonal cottages could not be replaced
on the same land area by a 10 unit condominium.
This will help maintain a status quo situation
in which no additional traffic or resultant air
pollution is generated and where potable water
consumption is stabilized.

5.    The further filling of estuarine wetlands should
be prohibited and recognition given to the fact
that these areas act as both buffer and filter
for pollutants being carried to the world's
oceans.  Their role as part of an international,
natural treatment system and as nurseries for
many important food fishes is to be preserved.

Figure 13.16: Construction in the coastal zone is never without hazard. This sanitary sewer manhole once sat safely above the high tide line on man-made land. The forces of coastal erosion, however, are relentless and it is only a matter of time before this structure likewise succumbs.

## Direct Waste Disposal

In spite of the present shortcomings of the NPDES permit system, this program will remain the primary regulator of point source discharges to the nation's estuaries and coastal waters. While certain statutory changes, such as mandating nutrient removal and other tertiary treatment for all publicly owned treatment plants, could improve the quality of the nation's estuaries and coastal waters, it is unlikely that they would be easily enacted. Marked improvements in water quality could just as well be achieved by correcting what have been identified as glaring deficiencies of the NPDES program. These include:

1.   Lax state and federal enforcement programs that allow permit violations to continue for months sometimes even years, and,
2.   Inadequate permit compliance monitoring sampling programs. A state like New Jersey, for example, conducts compliance monitoring samplings at less than 5% of its total 1,600 dischargers per year.

## Non-Point Source Pollution

The greatest single contribution the nation could make to control current and future non-point sources of pollution in the coastal areas would be to either reduce the density of development in these areas or prohibit further development altogether. Whether the nation has the courage to take such bold steps remains to be seen. In the meantime, if we assume that development will proceed at its usual unimpeded pace, then we must look at environmental protection alternatives that will accommodate this continued development. We cannot, however, accept business as usual alternatives. For instance, storm water detention and retention basins (see Chapter 2) and storm drains in general receiving runoff from paved surfaces in developed areas must be integrated with a post runoff treatment system (see Item 3, Special Considerations this chapter) or a regular pre-precipitation pavement and storm sewer cleaning program (see Item 4, Special Considerations this chapter). One enterprising inventor has devised a system to capture contaminated storm water run off using floating reinforced plastic curtains located at the end of storm water pipes. Called the Dunkers Flow Balancing Method after its inventor, Swedish engineer Karl Dunkers, it has been used in Sweden to stop urban runoff into freshwater lakes. The first application of this method on American soil took place in 1988 when a trial system was installed on a storm drain on Fresh Creek in Brooklyn, New York (Figs. 13.16A, 13.16B). It will undergo an 18 month evaluation to determine its effectiveness.

No program to control coastal and estuarine pollution can be considered complete without incorporating pollution control programs and environmentally sensitive land use planning in America's inland areas. Many of these practices we have already discussed in earlier chapters and include, watershed protection, soil erosion and sediment control, and a more restricted application of pesticides, herbicides and fertilizers on farmland. Regarding this latter item even the individual home owner can make a significant contribution in this respect by restricting his/her personal use of fertilizers, herbicides and pesticides on home lawns. Land use regulations mandating the retention or replacement of indigenous vegetation (especially in woodland areas) and discouraging the creation of new lawn areas should be encouraged. In fact, there is great merit in forbidding the use of these chemical additives completely (for individual households) or at least limiting their application to once a year and then by permit only. Anyone wishing to apply these chemicals more than once a year would have to pay a surcharge.

Clustered development should be encouraged where possible to preserve existing site conditions. This will minimize the amount of roadways, parking areas and other impervious surfaces that trap, then later release, pollutants during rainfall. A side benefit of this clustering is the reduction in future maintenance costs due to the reduction in the overall amount of improvements, including roadways, sidewalks, parking lots, storm sewers and sanitary sewers and lawn areas.

Figure 13.16A:   A freshwater application of the Dunkers Flow Balancing Method.

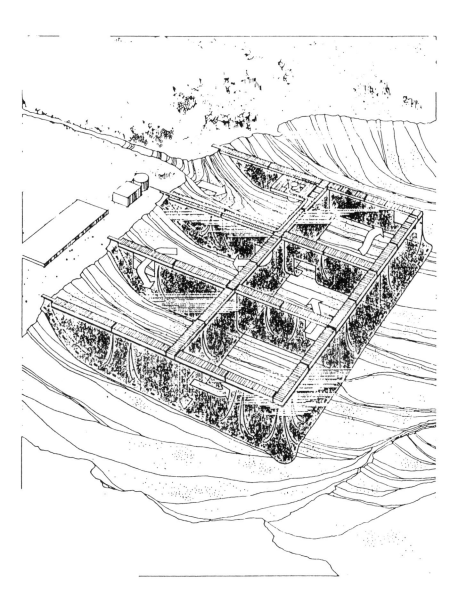

Sketch courtesy of: CM Towers, Inc., P.O. Box 1166, West Caldwell, NJ 07006

Figure 13.16B: This flow balancing method prototype facility was installed on Fresh Creek in the City of Brooklyn, New York during the summer of 1988. The Fresh Creek facility is the first and only facility installed on a tidal estuary and for the purpose of containing sewer runoff.

Photo courtesy of: CM Towers, Inc., P.O. Box 1166, West Caldwell, NJ 07006

## SPECIAL CONSIDERATIONS--COASTAL AREAS

### 1.    Marine Floatables

Major strandings of various solid wastes are occurring with increasing frequency on American beaches and estuaries. These wastes generally referred to as "floatables" are comprised of an enormous variety of materials (Table 13.8) that float in the water until carried to land by the wind or water. In some instances, these materials may be repeatedly stranded, refloated and stranded again at numerous locations. The predominant causes of these refloatations are full moon high tides, hurricanes and other storms.

Sources of these floatables are pandemic and include:

a)    Debris discarded directly on beaches or from commercial or recreational vessels.

b)    Overflow from sewage treatment plants.

c)    Flows from combined sanitary sewers and storm drains.

d)    Barge transfer and transport of municipal and industrial waste including sewage sludge and garbage.

e)    Decay or demolition of wharfs and pilings and old boats and barges.

f)    Storm water runoff carried by streams and rivers.

g)    Illegal discharges and dumping.

h)    Ocean burning of wood.

During these repeated movements and strandings these materials may cause damage to boats, interfere with navigation and various recreational uses (such as bathing, surfing or water skiing) or threaten public health with diseases. This was exactly the situation that confronted a number of states along the nation's northeast coastline in 1986, 1987 and 1988--particularly the states of New Jersey and New York. Sludge greaseballs, plastics and large timbers (along with slimy masses of dead algae) began washing up on beaches in quantities that were noticeably larger than the small amounts usually found. While unpleasant in appearance (Figure 13.17) these materials posed minimal health risks although some large timbers floating in the surf on one of New Jersey's beaches allegedly

Table 13.8: Typical Items in Various Floatable Categories

| | |
|---|---|
| **Glass:** | Bottles, jars |
| **Kitchen Wastes:** | Preparation refuse, uneaten food, bones |
| **Metals:** | Beer and soda cans, other cans, wire |
| **Paper:** | Cardboard boxes, bags, plates, newspapers |
| **Plastics:** | Household containers, tampon applicators, bags, 6-pack wrappers, toys, condoms |
| **Styrofoam:** | Various packaging and household products |
| **Wood:** | Lumber, driftwood, crates |

Source:   Adapted from N.J. Department of Environmental Protection, "New Jersey Floatables Study: Possible Sources, Transport, and Beach Survey Results," November 1987.

Figure 13.17: Stranded marine floatables—an all too common sight on American beaches.

Figure 13.18: Typical medical waste found washing up on the shores of the nation's east coast.

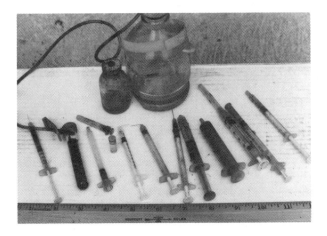

injured two small children playing at the water's edge.

It was the appearance of medical waste (hypodermic needles, intravenous fluid bags, catheters, body parts and blood filled vials) in 1987, however, that finally incited public indignation. Although the quantities washed up were small, when compared to the total amount of other floatables, the public could easily identify with the health hazards posed by such items (Figure 13.18). New Jersey and New York were not well prepared to investigate the sources of such floatables nor was the federal government for that matter. There were few laws dealing specifically with the disposal or tracking of medical waste. Tracking down the sources was difficult if not impossible. Most of the material had no labels or their identifying marks were obscured due to exposure to the sun and salt water. Some materials may have been floating in the ocean or estuaries for months. While other eastern seaboard states besides New Jersey and New York had washups of medical waste floatables (Connecticut, Massachusetts, Maryland, Rhode Island and North Carolina in particular) it was New Jersey's responses to the marine floatables problem that is worth examining in detail.

As the nation's most densely populated state, New Jersey has 127 miles of coastal beaches that support an annual $7 to $8 billion tourist industry. Following the 1986 washups the New Jersey Department of Environmental protection organized three beach surveys for the spring of 1987 to obtain data on the distribution and characterization of floatables along its beaches. Fifteen sites were sampled (Figure 13.19) under three different sets of meteorological and oceanographic conditions: average weather and tides, after a storm, and after a spring high tide (Ref. 13.5). These 3 time periods were selected to see if wet weather or high tides affected the amount of floatables found on the beaches. The sampling was conducted by county and local health department personnel with assistance from the New Jersey Marine Sciences Consortium. The most abundant material found was wood followed by plastics. A summary of the total number of floatables for each of the 3 sampling events is shown in Table 13.9. The following conclusions were drawn from this study (Ref. 13.5):

**Figure 13.19:** Locations of beaches sampled during the New Jersey Department of Environmental Protection's 1987 beach survey.

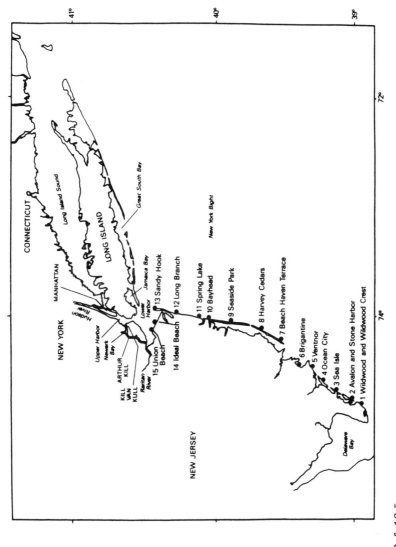

Source:  Ref. 13.5

Table 13.9:  Total Floatables Found on Each Beach During the N.J. DEP Survey

| Beach | Oceanographic/meteorological conditions | | | Mean |
|---|---|---|---|---|
| | Average | High tide | Rain | |
| Wildwood and Wildwood Crest | 248 | 73 | 355 | 225 |
| Avalon and Stone Harbor | 104 | 29 | 223 | 119 |
| Sea Isle | 93 | 178 | 78 | 116 |
| Ocean City | 100 | 34 | 47 | 60 |
| Ventnor | 32 | 82 | 116 | 77 |
| Brigantine | 48 | 131 | 107 | 95 |
| Beach Haven Terrace | 23 | 0 | 66 | 30 |
| Harvey Cedars | 79 | 0 | 42 | 40 |
| Seaside Park | 83 | 109 | 160 | 117 |
| Bayhead | 60 | 91 | 67 | 73 |
| Spring Lake | 72 | 95 | 67 | 78 |
| Long Branch | 602 | 1,198 | 789 | 863 |
| Sandy Hook | 3,040 | 3,051 | 3,089 | 3,060 |
| Ideal Beach | 4,064 | 4,018 | 4,026 | 4,036 |
| Union Beach | 1,556 | 4,099 | 2,451 | 2,702 |
| Total | 10,204 | 13,188 | 11,683 | 11,692 |

Source:  Ref. 13.5

1.  Floatables come from a variety of sources and very few items can be tied specifically to one source.

2.  The sources of floatables are apparently concentrated where activities associated with population centers are concentrated—in the greater New York metropolitan area (including northern New Jersey).

3.  The current patterns in the Hudson-Raritan Estuary and the New York Bight are variable and are affected by winds and tides in such a way that almost any potential floatable source could result in material being deposited on New Jersey beaches.

4.  While individual floatables strandings can affect the entire New Jersey coast, the northernmost New Jersey beaches appear to be most affected by floatable items derived from the Hudson-Raritan Estuary. Similarly, southernmost beaches appear to be affected more by locally generated beach litter and recreational boating litter than by floatables released in the greater New York metropolitan area.

During the fall of 1987 and the spring of 1988 the N.J. Department of Environmental Protection (NJDEP) in collaboration with the U.S. Environmental Protection Agency, Region II continued its study of the floatables problem with the use of released drifters. The purpose of the study was to determine whether floating materials actually released in the Hudson-Raritan Estuary would be deposited on New York and New Jersey beaches. Drifter design posed some problems for the investigators since it had to meet three criteria. First, the drifters needed to float near or on the surface of the water to allow transport by surface currents (thereby simulating floatable debris). Secondly, the drifters needed to be readily identifiable and contain reporting instructions for persons finding them. Lastly, the drifters could not cause any significant adverse effects on marine fish, birds or mammals. The drifter design finally selected consisted of a 500 ml plastic bottle sealed with a screw cap and waterproof tape (Figure 13.20). Placed inside each drifter was a numbered, color coded card that requested specific information. In addition, a

Figure 13.20: Diagram of drifter bottle and return form used during the N.J. DEP drifter study.

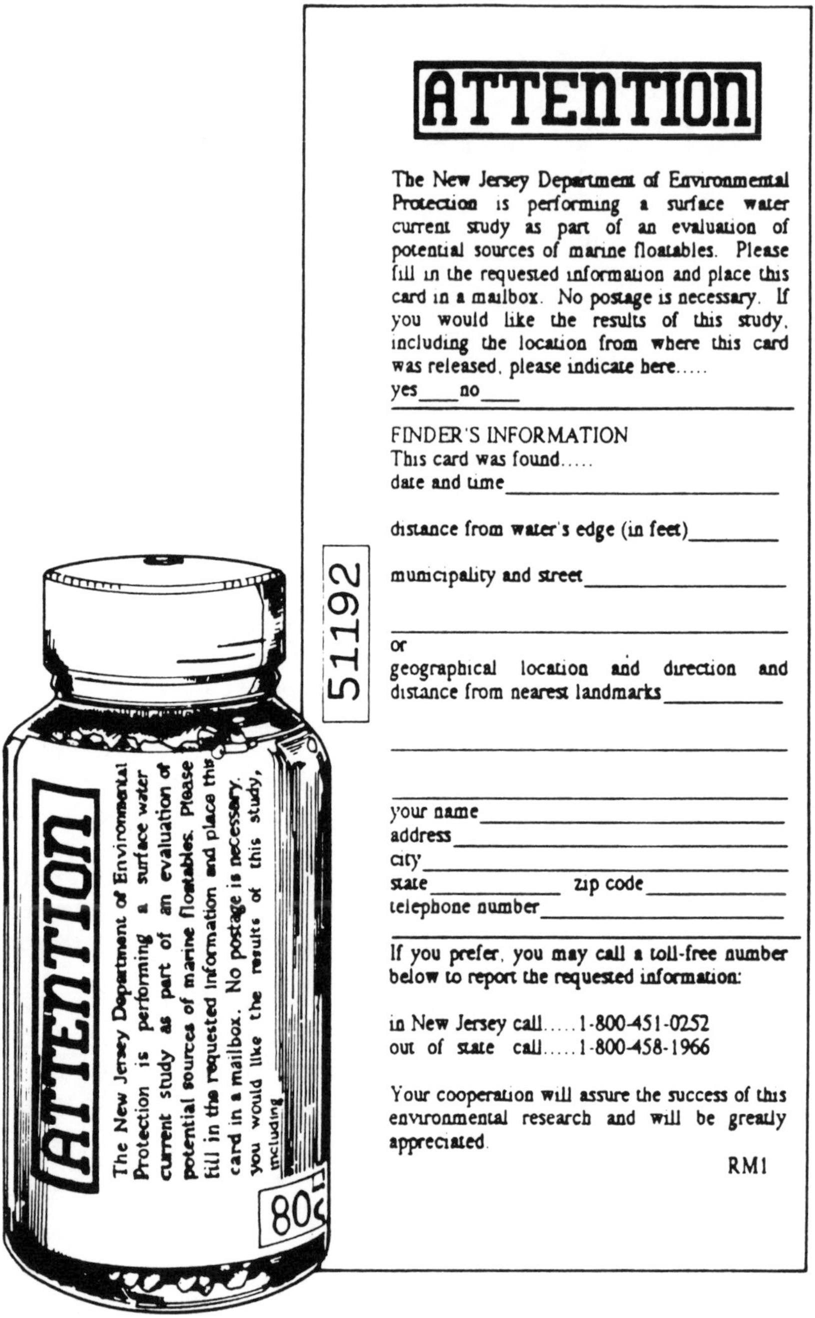

# ATTENTION

The New Jersey Department of Environmental Protection is performing a surface water current study as part of an evaluation of potential sources of marine floatables. Please fill in the requested information and place this card in a mailbox. No postage is necessary. If you would like the results of this study, including the location from where this card was released, please indicate here.....
yes____ no____

FINDER'S INFORMATION
This card was found.....
date and time_____

distance from water's edge (in feet)_____

municipality and street_____

_____
or
geographical location and direction and distance from nearest landmarks_____

_____

_____
your name_____
address_____
city_____
state_____ zip code_____
telephone number_____

If you prefer, you may call a toll-free number below to report the requested information:

in New Jersey call.....1-800-451-0252
out of state call.....1-800-458-1966

Your cooperation will assure the success of this environmental research and will be greatly appreciated.

RM1

Source: Ref. 13.6

sufficient amount of pea gravel was added to keep the bottle floating about 2 inches above the water surface. A blank protective card was inserted between the color coded card and the gravel to reduce damage caused by abrasion.

Three releases were made at 6 sites in the Hudson-Raritan Estuary on September 30 and October 1, 1987 and February 3, 1988. See Figure 13.21 for the location of each release site and Table 13.10 for the number of drifters released. The number of bottles returned/reported is shown on Table 13.11. [5]

In April, 1988 the NJDEP instituted daily helicopter surveillance patrols of the state's beaches as well as the Hudson-Raritan Estuary area in order to keep track of floatables or other materials that could threaten recreation. Several toll free numbers were also set up to provide the public with daily information and predictions on beach conditions based on the helicopter observations. Slicks of floatables were common and some attempts were made to capture several of those spotted by the helicopter. This was done by using two commercial fishing vessels in tandem hauling commercial fishing nets (Figure 13.22). The first attempt was unsuccessful due to undersized floats on the nets which allowed much of the collected material to escape when retrieved. The nets were subsequently modified and in 1989 private fishing vessels were again deployed to collect slicks of floatables. At present the costs of collection are averaging about $300 per cubic yard or $1,000 per ton--a figure that may doom future contracts.

In early 1989 the NJDEP, with the approval of the Governor and in cooperation with the state's Department of Corrections (DOC) commenced a massive cleanup of potential floatables along some 66 miles of its shoreline in the Hudson River-Raritan Bay Estuary. This project nicknamed "Operation Clean Shores" utilized inmate labor (Figure 13.23) supplied by New Jersey's penal institutions. Due to problems with access (as can be expected in a heavily urbanized area) only 16.5 miles of shoreline were able to be cleaned by June, 1989. The amount of material collected, however, was phenomenal. Approximately 5,000,000 pounds or over 8,100 cubic yards of potential floatables were removed from the shoreline at a cost of about $25 per cubic yard. The low cost of

Figure 13.21: Locations of the six release sites used during the New Jersey drifter study.

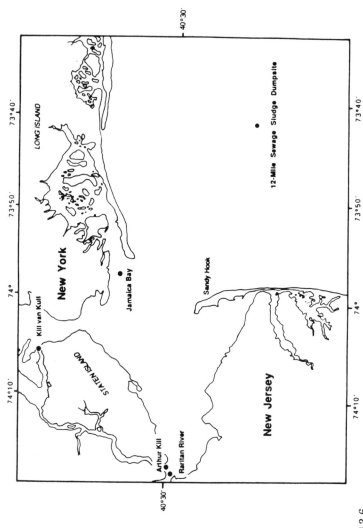

Source: Ref. 13.6

Table 13.10:  Locations of Drifter Release Sites, Dates of Releases, and
Number of Drifters Released

| Name | Location | Release 1 | | Release 2 | |
|------|----------|------|--------|------|--------|
| | | Date | Number | Date | Number |
| Raritan River | At confluence of Raritan River and Raritan Bay | 10/1/87 | 400 | | |
| Raritan River | Midway between Arthur Kill and Raritan River in Raritan Bay | | | 2/3/88 | 400 |
| Arthur Kill | At confluence of Arthur Kill and Raritan River | 10/1/87 | 400 | | |
| Jamaica Bay | At confluence of Jamaica Bay and the Hudson River | 10/1/87 | 400 | 2/3/88 | 400 |
| Kill van Kull | At confluence of Kill van Kull and Hudson River | 10/1/87 | 400 | 2/3/88 | 400 |
| 12-Mile Site | At 12-Mile Sewage Sludge Dumpsite | 9/30/87 | 900 | | |

Source:  Ref. 13.6

Table 13.11:  Numbers of Drifters Released and Returned from Each Release
Site for each Release

| Release Site | Release 1 | | | Release 2 | | |
|--------------|-----------------------|----------------------|--------------|-----------------------|----------------------|--------------|
| | Numbers Released | Numbers Returned | % Returned | Numbers Released | Numbers Returned | % Returned |
| 12-Mile Site | 900 | 0 | 0.0 | | | |
| Arthur Kill | 400 | 364 | 91.0 | | | |
| Jamaica Bay | 400 | 72 | 18.0 | 400 | 109 | 27.3 |
| Kill van Kull | 400 | 144 | 36.0 | 400 | 50 | 12.5 |
| Raritan River | 400 | 323 | 80.8 | 400 | 20 | 5.0 |
| Total for Release | 2500 | 903 | 36.1 | 1200 | 179 | 14.9 |
| Overall Total | 3700 | 1082 | 29.2 | (Sum of both releases) | | |

Source:  Ref. 13.6

Figure 13.22: The cost-effectiveness of using private boats and retrofitted fishing nets to capture marine floatables needs further evaluation. Initial results are not promising.

Figure 13.23: Using minimum security inmates to clean stranded floatables from New Jersey's shorelines has been an unqualified success.

this land based cleanup is the direct result of the utilization of prison inmate labor (Ref. 13.4).

Other states have conducted similar cleanups of their shoreline, however, they have been of a much shorter duration utilizing volunteers for labor. The state of Florida, for example, held a statewide coastal cleanup in September, 1988 in concert with the states of Texas, Louisiana, Mississippi and Alabama. A total of 10,676 Florida volunteers cleaned 914.5 miles of the state's beaches and estuaries and removed 192.7 tons of debris. Nearly 50 percent of the 36 cleanup sites reported finding some medical waste and an unbelievable 279 miles of monofilament fishing line was also recovered. The state of Louisiana held a similar cleanup in 1987 with 3,300 volunteers participating to collect 200 tons of debris (Ref. 13.7).

You can expect such large scale beach cleanups to become routine occurrences for all coastal states in the decade ahead. The ready availability of citizen volunteers and a large potential prison labor force will make this a practical method for tackling the floatables problem. Certain precautions must be taken, however, to provide the collectors with instructions on how to deal with medical or other hazardous waste.

In 1988 the New York State Department of Environmental Conservation (DEC) undertook its own investigation of the sources of floatables in the Hudson River-Raritan Bay Estuary, confining its inquiry largely to sources within its own jurisdiction. A report was completed in December, 1988 (Ref. 13.20) and it identified a number of sources that deserve our attention. In addition, the report provides a further glimpse of the future that awaits those coastal states that continue to allow poorly regulated and dense development along shoreline areas. And lastly, it dramatizes the inherent weakness of environmental safeguards when applied in the real world. Consequently, we list the following excerpts from the report:

"A.   Fresh Kills Landfill

A large volume of medical-related waste is generated in New York City everyday. New York City has 75 hospitals with a total of 35,000 beds. Assuming 100% occupancy, 315 tons of waste would be generated per day, of which 63 tons would be infectious waste.

It is evident from the investigations conducted that some of this medical-related waste is being placed into the solid waste stream of New York City. Hospitals are sometimes negligent in sorting their waste; doctors, laboratories, nursing homes and clinics sometimes negligently mix medical related waste with trash; intentional dumping of regulated medical waste occurs to avoid the expense of disposal; and out-patients like diabetics and other intravenous users dispose of their waste in the trash.

As an example, it is clear that some hospitals do not always comply with New York City Local Law 57, which requires that infectious waste and non-infectious medical waste be removed from the New York City solid waste stream. DOS investigators have conservatively estimated that one-third of all the garbage trucks picking up trash from hospitals also carry regulated medical waste, such as syringes, which the hospitals have improperly mixed with trash. This estimate is based upon reports of DOS investigators who stopped trucks and inspected their loads. Since 1985 over 50 New York City hospitals have been fined--many more than once--for not properly removing their medical-related waste from the solid waste stream...Obviously some of the medical-related waste that is not discovered reaches the Marine Transfer Stations (MTSs) and Fresh Kills landfill as part of the solid waste stream. Thus, when trash spills into the water at Fresh Kills or the MTSs, some of that trash may be medical-related waste.

As stated above, NYCDEP acknowledges placing syringes into the solid waste stream. Their procedures for removing floatables from sewage treatment plants alone place between 25,000 and 36,000

syringes per year into the solid waste
stream. The primary source of these
syringes appears to be diabetics who
inject themselves with insulin. It is
estimated that New York City has between
60,000 and 112,000 Type One diabetics who
must inject themselves at least daily with
insulin. The insulin syringe is used once
and then discarded in either the trash or
the toilet.

Another indication that medical-
related waste reaches Fresh Kills is that
during the investigation, tugboat operators stated that red and blue-tagged
"medical wastes" were observed on the
barges being brought to the Fresh Kills
landfill...According to DOS procedures,
these barges can only be unloaded by the
electric crane (unfortunately this crane
is inefficient and it allows debris to
escape when it digs close to the coaming.
In addition, the DOS employees hose down
the equipment and barges, thus washing any
leftover hospital wastes into the water).

Garbage from Fresh Kills escapes
into the water in a variety of ways.
Barges loaded with garbage are brought to
the Fresh Kills landfill where they enter
a boom-lock system and are unloaded. The
off-loading of the barges results in
spillage into the water. Han-Padron Associates staff have observed the presence of
overloaded buckets and debris falling into
the water and have reported considerable
debris in the water occurring during
approximately one-third of their observations. At times the debris has been
described as "slicks" which contain
debris-filled garbage bags.

The amount of subsurface debris also
cannot be underestimated. As indicated in
the Han-Padron Associates Report for 16
May to 31 May, 1988, skimmer boats have to
be dry docked periodically to clear away
debris caught in their propellers. BECI
investigators observing the Fresh Kills

landfill by boat have also noted considerable debris at least six feet deep in the channel. This subsurface floatable debris can escape and eventually reach the beaches...

The landfill operation uses a boom/skimmer boat approach to prevent waste from escaping the landfill area and to remove any waste that falls into the water. There are three locks (or booms) in operation at the landfill: the main boom which stretches across the inlet leading to the unloading pads; the range boom which is approximately half the distance between the main and third booms and stretches across the Isle of Meadows (and island located directly across from the landfill) and the Fresh Kills landfill; and the third boom which is across the mouth of the channel that leads to the Arthur Kill between the Fresh Kills landfill and the Isle of Meadows. The three booms are fitted with skirts to catch debris floating both on the water surface and below it. There is also a permanent boom across the northern channel of the Fresh Kills landfill and the Isle of Meadows.

A tugboat propelling garbage-laden barges is supposed to enter the first lock only after a skimmer boat has removed the floating debris from the surrounding waters. The third boom is supposed to be closed behind the barge before the range boom is opened. The main boom should be opened after the range boom is closed, and the main boom should be closed before the barges are unloaded. At least three skimmer boats should be in operation at all times.

The evidence substantiates that this system has not succeeded in preventing garbage from escaping the Fresh Kills area. Until approximately mid-May, only two booms were in operation--the main boom and the range boom. (The range boom was located in a slightly different location prior to the addition of the third boom and a 15-foot skirt was added to the new range boom about the same time the third boom was installed.) The booms are often left open without coordination with tugboat movement.

The opening and closing of the booms also permits the escape of floatables. Skimmer boats which pick up water-borne debris do not always operate in front of and/or alongside the booms before opening. Booms are occasionally opened the wrong way, thus releasing floatables trapped by the booms. The boom skirts are apparently never cleaned or repaired. Invariably the submerged debris trapped by the skirts is moved by the opening of the booms towards the shoreline where it will escape if the booms are improperly moored. Assuming the floatation sections of the boom are damaged from time to time by tugboats or runaway barges, one could safely expect the skirt to also be in need of occasional repair. Surprisingly, none of the monitors were aware of repairs made to the boom skirt.

Observations by monitors of the skimmer boats indicate that the boats were often operated at high speeds and were driven in an erratic fashion. This appears to have an impact on the effectiveness of skimming, driving floatables to the subsurface rather than removing them from the water. There are other problems: for example, the skimmer boat is in the water but does not operate until the monitor appears dockside, or the skimmer boat operator watches television on the boat rather than looking for floating debris.

B.    Sewage Discharges

New York's sewers are supposed to operate as follows: most sewage generated in New York City travels by sewer to one of the 14 water pollution control plants. There are three types of sewers: sanitary

sewers which carry household and industrial waste, storm sewers which carry rain
and surface water runoff and combined sewers which carry both waste and water
runoff.

There are over 540 combined sewer
overflow points within the New York City
sewer system. Approximately 70% to 80% of
all New York City sewers are the combined
type. During periods of rainfall greater
than .25 inches per day, treatment plants
are by-passed and sewage is discharged
directly into the New York Harbor complex.

When operating properly, a sewage
treatment plant removes pollutants from
wastewater. After screens remove floatables, the wastewater is subjected to biological     treatment     and     disinfection.
Floatables removed from the screens are
collected and transported to the Fresh
Kills landfill. The effluent is then discharged into the waters of the State.
Sludge produced by these processes currently is disposed of at sea (106 miles
offshore). Approximately 1.7 billion gallons of sewage per day are presently
treated by New York City.

As noted above, rainfall of .25
inches, or at times less, causes the New
York City sewage treatment plants to be
bypassed. When a bypass occurs, any
floatables which would have been screened
out of the sewage are discharged into the
New York Harbor complex. This is in addition to the approximately 6.8 million gallons of untreated sewage discharged into
the New York Harbor complex everyday under
dry weather conditions. When a treatment
plant fails or malfunctions, there is no
backup system. On such occasions, raw
sewage is often discharged directly into
the water.

C.    Marine Transfer Stations (MTS)
There are nine municipal marine
transfer stations run by DOS in New York
City; eight are currently in operation.

There is one privately owned MTS. The New York City Department of Sanitation (DOS) provides off-loading service from truck to barge for both DOS and private carters at the MTS. The barges are supposed to be loaded in a systematic fashion so as to avoid spillage, and the barges are supposed to be loaded to a maximum height of 8 feet above the barge coaming (i.e. side walls) to avoid spillage while the barges are in motion. In addition, DOS is supposed to use overhead sprinklers, dip nets and booms to minimize spillage. Tugboats take 2 to 4 barges to the Fresh Kills landfill where they are off-loaded. Approximately 14,000 tons of garbage per day are taken by barge to the Fresh Kills landfill.

In fact, spillage to water occurs during the loading of barges. The loading process creates an updraft which releases lighter articles of garbage into the air. Eventually they fall into the water. Overloading of barges is another cause of spillage. Interviews with DOS employees indicate that it was common, until the middle of July 1988, for barges to be loaded to capacity, thus causing debris to fall into the water. The monitors of Han-Padron Associates noted that the overloading of barges occurred during about 10% of their observations. It appears that loading practices commonly cause some spillage at the MTS. Overloading, however, causes major spills.

Once garbage enters the water at the MTS's, it is not likely to be removed. The water sprinkler systems are rarely operational and when they are, they are rarely effective. Dip nets simply are not used on a regular basis. The only boom is located at the MTS next to the southwest incinerator, and this boom is ineffective. Materials collected by the boom are not removed and the boom is opened from dockside thus releasing anything captured.

In short, garbage indisputably spills into the water at the MTSs and is not removed. In fact, independent monitors as well as BECI investigators have observed currents carrying debris out of the loading slips and into the New York Harbor complex.

Debris is blown off the barges by the winds and dislodged by the movement of the tugboats while barges are being taken to the Fresh Kills landfill.

Cleaning the barges is another source of spillage. Periodically, each barge is removed from service and cleaned. During use, a layer of debris left at the bottom of the barge serves as a cushion for loading and off-loading. When a barge is cleaned, the floor is swept clean and hosed down. At least until recently it has been the practice to release the debris-laden waste cleaning water directly into the water surrounding the barges.

D.    Pleasure and Commercial Boats

Another source of floatables for the washups of 1988 is waste disposal from commercial vessels. However, very little of the identifiable waste from the 1988 washups has been traced to boaters.

In 1987, approximately 5,000-6,000 vessels over 1,600 gross tonnage arrived at the Port of New York. Little data are known about the sewage, garbage and refuse generated by commercial vessels. There have been two incidents in other states where beach closings were caused by commercial and military shipping debris. In North Carolina, United States Navy vessels discharged debris offshore, which caused approximately 50 miles of beach to close.

The United States Coast Guard (USCG) estimates that over 50,000 pleasure boats can be found in New York waters, and each boater produces approximately 1-1/2 pounds of garbage each day on the water. Inevitably some of this waste will be tossed overboard.

E.    Illegal Disposal

According to the New York Times, there are over 250,000 intravenous drug addicts in New York City. Traditionally, drug addicts do not dispose of their syringes until they are inoperable. However, possession of drug paraphernalia is illegal, and conceivably these syringes will be discarded when an addict is confronted with law enforcement personnel. The most logical point on the street is the storm drain. One can expect some disposal into the water from CSOs that are connected to the storm drains. The same logic would apply to drug users who frequent the beaches at night. The most obvious manner of disposal is to drop the syringes on the beach. From the beach, the syringes can either enter the water or be collected with washed up materials the following morning."

## 2.    Human Pathogens in Marine Waters

Some of the wastes disposed of in the marine environment carry with them a variety of human pathogens--microorganisms capable of inducing human disease (Table 13.12). In the United States the most important waste-borne agents of human disease are viruses and bacteria. The ability of bacteria and viruses to infect humans depends upon a variety of factors, one of which is the minimum effective dose. As few as 10 to 100 bacteria or a single virus are capable of inducing infection and illness. Bacteria and viruses have a strong affinity for particulate matter--a characteristic that increases the risk of exposure two ways. First, their survival is enhanced by the protection provided by the particle and secondly, these same particles can collect numerous viruses and bacteria on their surfaces. Hence, a single ingested particle could provide a large dose of microorganisms.

A growing amount of evidence indicates that some bacteria may persist in the marine environment for many months and remain noncultureable but virulent (Ref.'s 13.3, 13.8, 13.9). In one instance, the agent responsible for an outbreak of cholera along the Gulf Coast of Texas appears to have persisted in coastal waters for at least 5 years (Ref. 13.3, 13.10).

## Table 13.12:  Potential Human Pathogens in Domestic Sewage and Sludge Disposed in Marine Waters

**Viruses**--Over 100 different types of human intestinal viruses are present in sewage and have the potential to be spread through contamination of water or food sources.  Enteric Hepatitis A virus, which induces the potentially life-threatening liver disease infectious hepatitis, probably is the greatest threat to public health; this virus has been demonstrated to have been spread as a result of sewage contamination.  Many outbreaks of viral gastroenteritis linked to contaminated coastal waters or shellfish are reported each year in the United States.

**Bacteria**--A wide range of bacteria are found in sewage, including many known human pathogens that can cause food poisoning, typhoid fever, strep throat, dysentery, and other intestinal fevers.  A number of other human pathogens have been identified as natural members of the marine bacterial community as well as constituents of sewage, including the causative agents of cholera and acute bacterial gastroenteritis.  Several commonly known pathogenic bacteria found in sewage are Salmonella, Streptococcus, and Campylobacter.

Bacteria that are resistant to antibiotics are readily identified in sewage-contaminated waters.  Because such bacteria can pass their resistance characteristics to other bacteria, the potential exists for humans to contract serious bacterial infections that would be difficult to treat.  Pathogenic antibiotic-resistant bacteria can survive and have been detected in marine environments, but there is as yet no strong correlation between their occurrence and the disposal of sewage wastes.  Nor has their actual public health significance been established.

**Protozoa**--Protozoa are single-celled organisms, including such species as Giardia and Entamoeba (causes of giardiasis and amoebic dysentery, respectively), that are typically found in sewage in the form of cysts.  Cysts are inactive and extremely resistant to environmental damage.  Upon ingestion by a host organism, the cysts are activated and can undergo maturation and reproduction to form additional cysts that are excreted in the feces.

**Helminths**--Helminths include various parasitic flatworms and are transmitted through ingestion of cysts present in inadequately cooked meat; infection leads to disease and excretion of eggs in the feces.  Helminths are responsible for trichinosis, schistosomiasis, diarrhea; and various liver disorders.

**Nematodes**--Nematodes are parasitic roundworms and hookworms, the most common of which is Ascaris.  They are capable of forming cysts or eggs, which are excreted in the feces and require activation outside of an organism in order to initiate infection; hence, they are not capable of direct human-to-human spreading.

**Fungi**--These are generally present in significant quantity in sewage material only when given the opportunity to grow during treatment or storage.  Large numbers are commonly found in composting sludge.  Reproduction involves the formation of spores, the inhalation of which provides the most direct pathways to human infection.  Pathogenic fungi include Actinomyces and Candida.

**Cyanobacteria and Eucaryotic Algae**--Certain types of marine bacteria and algae, whose growth appears to be generally stimulated by nutrients, produce toxins that can be subsequently concentrated by shellfish.  Eating toxin-contaminated shellfish can lead to diarrhea, muscular paralysis, neurological damage, and liver disorders.  Shellfishing is usually restricted during "blooms" of toxin-producing algae, called red or green tides.  In addition, skin and respiratory problems have been reported from swimming in such waters.  Limited evidence exists that links blooms in the New York Bight to nutrients or metals introduced through waste disposal.

Source:  Ref. 13.3

Because of their affinity for particulate matter concentrations of enteroviruses may be 10 to 10,000 times more abundant in coastal sediments than in the overlying water.  Subsequent disturbances of these sediments by dredging or other means could increase local concentrations and availability of human pathogens.  While the outbreaks of serious human dis-eases attributable to direct contact with polluted water has diminished substantially in the U.S., water borne gastrointestinal illnesses have conversely shown an increase (Ref. 13.3, 13.11, 13.12).  However, much of these latter illnesses still go unreported because of their relatively short duration and rather benign nature.

The disposal of wastes in the marine environment also has the potential to significantly alter the com-position of indigenous marine microbial communities (Ref. 13.13).  One study of a pharmaceutical waste dump off the coast of Puerto Rico (Ref. 13.8) used from 1972 to 1981 reveals the drastic change in the types of bacteria that can occur as a result of waste dumping.  Where a wide variety of marine microor-ganisms once existed, one bacterial genus, Vibrio, be-came the dominant microorganism.  Vibrios occur natu-rally in marine waters and are particularly responsive to an influx of nutrient-rich water--the very condi-tion that prevails in much of the nation's estuaries and near coastal waters influenced by man.  Vibrios, once a small part of the marine microbial community, now dominate large areas of our coastal waters.  A 1988 advisory released by the University of Florida (Ref. 13.14) warns of the potential dangers of Vibrio vulnificus infections which can be transmitted to humans through open wounds in contact with seawater or through the consumption of improperly cooked or raw shellfish.  The massive die-off of bottle-nose dol-phins along the eastern Atlantic coastline in 1987 and 1988 (see Chapter 1) was in part attributable to Vibrio infections.

The safety of the nation's coastal and estuarine waters for recreation is of paramount concern.  Rou-tine sampling of marine waters for indicator organisms is a common practice in many coastal states.  Fecal coliform bacteria are the most frequently used indica-tor organisms with a maximum standard of 200 fecal co-liform colonies per 100 milliliters of water (written

200/100 ml).   This   standard   is   based   upon   research
done   in   the   1950's   by   Albert   H.   Stevenson   who   visited
beaches   on   Lake   Michigan,   the   Ohio   River   and   Long   Island   (Ref.   13.15).     He   measured   various   types   of
bacteria   in   the   water   including   fecal   coliform   and
simultaneously   surveyed   swimmers   to   count   how   many   got
sick.     He   subsequently   developed   a   mathematical   equation   linking   overall   bacteria   counts   to   health   complaints.     In   1968   the   U.S.   Department   of   the   Interior
mathematically   translated   his   data   into   a   fecal   coliform   count.     Their   derived   maximum   permissible   level
was   400   fecal   coliforms   per   100   ml   of   water.     At   this
level   they   expected   swimmers   to   suffer   noticeable   excess   health   effects.     Taking   a   conservative   approach
they   halved   this   figure   to   get   the   present   standard   of
200   per   100   ml.     In   1972   a   panel   of   the   National
Academy   of   Sciences   and   the   National   Academy   of   Engineers   recommended   against   the   adoption   of   this   standard   due   to   the   lack   of   supporting   evidence   (Ref.
13.16).     In   1976   the   U.S.   EPA   adopted it   as   a   national
standard   despite   this   recommendation[6].

In   all   likelihood   the   use   of   the   200/100   ml
fecal   coliform   standard   significantly   underestimates
the   true   number   of   viable   pathogens   entering   the
marine   environment.

3.     **Chlorinating Storm Sewer Outfalls That Discharge
Near Ocean Recreational Beaches**

In   July   of   1989   several   local   merchants   in   the
New   Jersey   shore   community   of   Wildwood   Crest   received
civil   summonses   for   dumping   chlorine   (considered   a
deleterious   substance   under   state   law)   into   the   ocean
at   one   of   the   municipalities   recreational   beaches.
Their   actions   were   apparently   motivated   by   several
days   of   recreational   beach   closures   due   to   high   fecal
coliform   counts   in   the   near-shore   waters   of   the
Atlantic   Ocean.     While   the   hundred   pounds   or   so   of
chlorine   they   added   apparently   did   no   identifiable
harm,   the   incident   prompted   some   adjacent   resort   communities   to   openly   admit   that   they   had   been   routinely
adding   chlorine   to   storm   water   outfalls   that   discharged   near   ocean   recreational   beaches.     Most   often
it   was   applied   in   the   form   of   cakes   or   granules
wrapped   in   porous   cloth   and   thrown   or   suspended   in
manholes   or   catch   basins.     When   queried   the   state's
Department   of   Environmental   Protection   (DEP)   indicated
it   was   aware   that   such   unregulated   disinfection   was

occurring. A spokesman for that department indicated that in marine waters this was an acceptable practice but that it was discouraged for discharges to fresh water.

Environmental groups took issue with the DEP's position indicating that allowing such disinfection masked the severity of non-point source pollution from storm water outfalls and created a false sense of security amongst the recreating public--both valid complaints.

The chlorination of storm water and combined storm water and sanitary sewer outfalls can be of some benefit in reducing the level of fecal coliform bacteria (considered a pollutant of significance from non-point sources). However, it is not to be considered a long term or even interim solution for non-point source pollution without some significant pre-planning. Items that need to be considered prior to implementation include:

a)   Unlike many of the present crude methods of chlorine addition, the application of chlorine needs to be carefully dosed and metered, using either a liquid or gaseous chlorine addition system. In those storm water systems that constantly discharge due to infiltrating ground water dosing should likewise be constant. For those storm sewers that discharge only during rainfall, sensors would be required to activate the dosing system to add the precalculated amount of chlorine depending upon the depth (i.e. volume) of storm water passing the chlorinating unit. The preferred method of dosing is to collect the "first flush" of storm water (considered to be the most contaminated) in a separate reservoir, holding tank or pond and chlorinating it while it was being held. Dosages of 25 mg/l of chlorine gas have successfully reduced fecal coliform and fecal strep bacteria to 200 colonies per 100 ml in two minutes (Ref. 13.23).

b)   The concentration of chlorine at the discharge end of the outfall pipe should be regularly measured to assure that the resultant levels of chlorine pose no danger to those who might come in direct contact with the discharged water.

c)   All storm water outfalls receiving chlorination should have a NPDES permit and be regularly inspected and maintained by a qualified (preferably licensed) operator.

The reader should be aware that chlorination could have negative impacts, one of which is the formation of trihalomethane organic chemicals[7] as a result of interactions between the added chlorine and naturally occurring organic material in the storm water. We do not at this time recommend chlorination for outfalls discharging to bodies of fresh water.

4.   **Combined Sewers--Interim Solutions**

The decision to build singular sewer systems to jointly convey sanitary sewage and storm water runoff from urban areas has lost much, if not all, of its original appeal--thanks to decades of hindsight. When originally conceived few designers had any real inkling as to the magnitude and density which America's urban areas would grow. One of the major problems with combined sewers is the buildup of wastewater solids along the bottom of the pipes. This is the result of installing pipes that are much larger than needed for dry weather sanitary sewage flows but necessary for the conveyance of intermittent rainfall runoff which can be up to 1,000 times larger in volume. Dry weather flow velocities are typically inadequate to keep settleable sewage solids in suspension and as a consequence they accumulate along the bottom of the pipes. During heavy rainfall, most of these solids are flushed from the sewer lines and if the hydraulic capacity of the sewer lines is exceeded these solids are carried directly to surface waterbodies with no treatment. An intense rainstorm lasting two hours after four days of antecedent dry weather may wash the equivalent of a full day's flow of raw sewage into the receiving water (Ref. 13.22). This "shock" type pollution loading can do substantial harm to water quality.

In some instances thanks to unrestrained and ill-planned growth, dry weather wastewater flows may occupy nearly all of the hydraulic capacity previously reserved for rainfall. As a consequence some wastewater may be discharging to nearby waterbodies round the clock.  Pipeline deterioration and collapse due to lack of maintenance may also be the cause of structural obstructions that likewise divert wastewater continuously to surface waterbodies.  Similarly, the unregulated expansion of the contributing storm sewer system to service the greatly enlarged urban area may have also created a situation where there is insufficient hydraulic capacity to accommodate all of the rainwater runoff.  This again leads to extended periods of bypassing.

The costs to implement the preferred controls (i.e. sewer separation and enlargement) are mind-boggling with estimates ranging from several hundred billion to nearly a trillion dollars.  Needless to say, amounts of this magnitude have forced legislators and regulatory agencies to seek other less costly alternatives.  These options are generally non-structural in nature.

One option that has been closely studied and successfully implemented is dry weather sewer line flushing.  The purpose of sewer line flushing is to either scour and transport deposited solids to the downstream sewage treatment plant or to displace solids deposited in the upper reaches of large collection systems closer to the system outlet (i.e. treatment plant).  The intent is to reduce the amount of deposited solids that may be resuspended and discharged untreated during rainfall and to decrease the time the solids are transported within the collection system.  Moving these materials closer to the treatment works reduces their susceptibility to premature overflows during rainfall and increases their chances of being captured by either wet weather storage facilities or the treatment plant.

Sewer flushing may be accomplished through either one of the following methods:

a)    By the introduction of water from a tanker truck capable of delivering several hundred gallons of water at a minimum rate of 0.50 c.f.s. into accessible manholes (Figure 13.24); or

b)    By the manual blockage of incoming sewer lines to a manhole resulting in a backup of sewage followed by a sudden release to produce a "flush wave" (Figure 13.25).  To reduce labor costs this process may be ac-

Figure 13.14:  Representative external source flush injection methods.

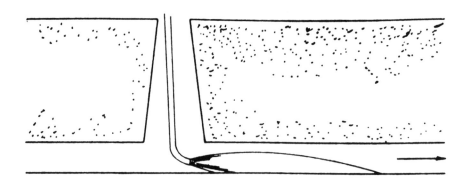

A.   Low Rate Flush (Gravity Feed) (Volume High or Low).

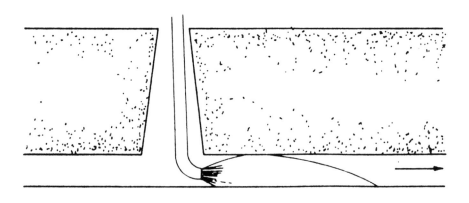

B.   High Rate Flush (Pressurized Feed) (Volume High or Low).

Source: Ref. 13.22

**Figure 13.25:** Representative backup and release flush methods.

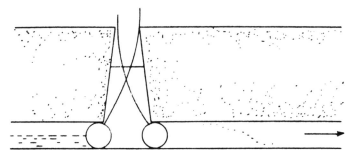

A.    Manhole Surcharge - Rapid Release (External Source)

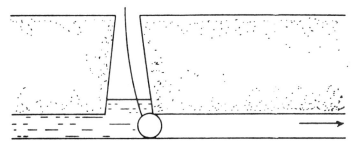

B.    Back and Release - Collection Pipe and Manhole (Higher Head)

C.    Backup and Release - Collection Pipe

Source: Ref. 13.22

complished by the installation of an automated sewer flushing module (Figure 13.26)
that is pre-programmed to produce such
flush waves at intervals of 6 to 72 hours.
The final selection of the method to be used
would of course be determined by actual field conditions.

**Figure 13.26:** Automated sewer flushing module.

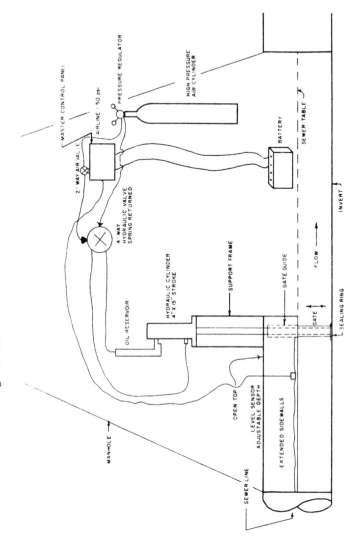

Source: Ref. 13.22

**FOOTNOTES**

**CHAPTER 13**

(1)      These sites are located beyond our adminis-
tratively defined estuarine zone, however, it
is possible that some impact upon the estuarine zones could occur from these areas.  See
discussion on the New Jersey Floatables Study
under Special Considerations in this chapter.

(2)      The Deepwater Municipal Sludge Site occu-
pies an area of approximately 100 square nautical miles.  It is located approximately 120
nautical miles southeast of Ambrose Light,
New York, and 115 nautical miles from
Atlantic City, New Jersey, in water depths
ranging from 2,250 to 2,750 meters (49 FR
19005-19012, May 4, 1984).

The Deepwater Industrial Waste Site occu-
pies an area of approximately 30 square nautical miles.  It is located approximately 125
nautical miles southeast of Ambrose Light,
New York, and 105 nautical miles from
Atlantic City, New Jersey, in water depths
ranging from 2,250 to 2,750 meters (49 FR
19005-19012, May 4, 1984).

(3)      This site has now been shut down.

(4)      Source:  Ref. 13.3.

(5)      During the fall of 1988 after the initial
results on the drifter releases had been
tabulated and reported additional drifters
were found at some rather unexpected locations.   Some of the drifter bottles had
reached Nantucket Island and Martha's Vine-
yard off the coasts of Rhode Island and
Massachusetts and 2 bottles had floated to
the shores of Bermuda.  Most astonishing was
the discovery of a drifter bottle on the
shores of the eastern Azores off the coast of
Portugal.   For the author who was directly
involved in these studies these discoveries
reaffirmed the close interrelationship
between the earth's major ecosystems.

(6)      The U.S. EPA has recently adopted enterococci
as a microbiological indicator of water qual-

ity for marine recreational waters (Ref. 13.17). That standard is 35 enterococci/100 ml.

(7)     Some of the most common trihalomethane compounds include chloroform, bromodichloromethane, dibromochloromethane and bromoform.

## REFERENCES

## CHAPTER 13

13.1   United States Department of the Interior, "The National Estuarine Pollution Study--Report of the Secretary of the Interior to the United States Congress", U.S. Government Printing Office, Senate Document 91-58, 1970.

13.2   Nordstrom, Karl F., et. al., "Living With the New Jersey Shore", Duke University Press, 1986.

13.3   United States Congress, Office of Technology Assessment, "Wastes in Marine Environments", OTA-0-334, Washington, D.C., U.S. Government Printing Office, April, 1987.

13.4   Honachefsky, William B., "Cleaning Up Coastal Beaches Utilizing Inmate Labor--A Case Study", (as yet unpublished report).

13.5   N.J. Department of Environmental Protection, "New Jersey Floatables Study: Possible Sources, Transport and Beach Survey Results", November, 1987 (Prepared by Science Applications International Corp./Battelle Ocean Studies).

13.6   N.J. Department of Environmental Protection, "New Jersey Floatables Study: Drifter Study Results", August, 1988 (prepared by Science Applications International Corporation/ Battelle Ocean Sciences.

13.7   Louisiana Department of Natural Resources, "Louisiana Coastlines", August, 1988.

13.8   Grimes, D.J., "Human Health Impacts of Waste Constituents, Pathogens and Antibiotic--and Heavy Metal Resistant Bacteria", Office of Technology Assessment, College Park, MD, 1986.

13.9   Morita, R.Y., "Starvation and Miniaturization of Heterotrophs, With Special Emphasis on Maintenance of the Starved Viable State in Bacteria in Their Natural Environments", M. Fletcher and G.D. Floodgate editors, Academic Press Inc., 1985, page 111-130.

13.10   Blake, P.A., et. al., "Cholera--A Possible Endemic Focus in the United States", The New England Journal of Medicine, 302:305-309, February 7, 1980.

13.11   Centers for Disease Control, "Water Related Disease Outbreaks in the United States--1980", Morbidity and Mortality Weekly Report, 30:623-634, 1982.

13.12   Goyal, S.M., "Viral Pollution of the Marine Environment", CRC Critical Reviews in Environmental Control, 14(1):1-32, 1984.

13.13   Geraci, J.R., "Interim Report on the Investigation of Mass Mortality of Bottlenose Dolphins", U.S. Marine Mammal Commission, May, 1988.

13.14   University of Florida, Food Science and Human Nutrition Department, "Vibrio Vulnificus, An Advisory Note", Gainesville, Florida, 1988.

13.15   Stevenson, Albert H., "Studies of Bathing Water Quality and Health", American Journal of Public Health, May 1953.

13.16   Birnbaum, Shira, "Testing Method Under Scrutiny by Officials", Courier Post, New Jersey, August 30, 1988.

13.17   United States Environmental Protection Agency, Office of Research and Development and Office of Waste Regulations and Standards, "Bacterial Ambient Water Quality Criteria For Marine and Freshwater Recreational Waters", EPA-440/5-84-002, Washington, D.C., January, 1986.

13.18   Toufexis, Anastasia, "The Dirty Seas", Time Magazine, August, 1988.

13.19   Komar, Paul D., "Beach Processes and Sedimentation", Prentice Hall Inc., New Jersey, 1976.

13.20   New York Department of Environmental Conservation, "Investigation: Sources of Beach Washups in 1988", December, 1988.

13.21   State of New Jersey, Report by the Blue Ribbon Panel on Ocean Incidents--1987, "The State of the Ocean", May, 1988.

13.22   United States Environmental Protection Agency, "Dry-Weather Deposition and Flushing for Combined Sewer Overflow Pollution Control", EPA-600/2-79-133, August, 1979.

13.23    United States Environmental Protection Agency, "Disinfection/Treatment of Combined Sewer Overflows", Syracuse, New York, EPA-600/2-79-134, August, 1979.

13.24    United States Environmental Protection Agency, "National Water Quality Inventory, 1986 Report to Congress", EPA-440/4-87-008, November, 1987.

# 14

# Wildlife

## AN UNDERVALUED RESOURCE

In spite of the fact that 141 million Americans participate in wildlife related activities, the nation's wildlife resources continue to be assigned an extremely low, nearly non-existent value in the land planning process, especially at the local level. The number of projects denied nationally by local planning boards due to adverse impacts on wildlife or its habitat could be counted on two hands or less. Yet municipal ordinances calling for an assessment of the impact on wildlife as a result of man's "land improvements" are abundant. Why then does wildlife continue to receive such feeble recognition in the land planning process? The answer is complex. One of the major reasons, however, is that many land planners (and many Americans as well) are unfamiliar with the life cycles or habitat requirements for wildlife in their immediate geographical area. As a consequence, they must rely heavily on documentation provided by "experts" as required by their ordinances.

Unfortunately, many of these required assessments, usually in the form of environmental impact statements (E.I.S's) leave much to be desired. They are not only often poorly written but often lack both substance and accuracy.

The following excerpts from 6 environmental impact statements, submitted as part of some recent land subdivision projects will give you some idea of the quality of information being provided to municipal land planners.

443

E.I.S. Example 1

"The existing wildlife on this site seems to be of a visiting nature, rather than that of actual inhabitance. Small animals such as rabbits, or birds such as crows would visit the site to satisfy their eating habits only. Because of the barrenness of the land, inhabitance is unlikely."

E.I.S. Example 2

"An area now serving as habitat for many wild animals and birds will no longer be available for this purpose. It is expected that many animals now dwelling at the site will seek new homes."

E.I.S. Example 3

"The species threatened most seriously are those that are most conspicuous and those that are sensitive to changes in land use. Hawks and some owls may be harassed, but will be affected most by the conversion of cultivated and abandoned fields to homesites, roads, lawns and parks with a resultant reduction in rodent population."

E.I.S. Example 4

"The proposed construction on this site will force most of the wildlife to leave the immediate area, although the proposed use should allow most of the wildlife to live adjacent to this site in the proposed County Park."

E.I.S. Example 5

"There will be a substantial reduction in wildlife visiting the area."

E.I.S. Example 6

"No known wildlife systems will be destroyed other than rodents."

Your attention is directed in particular to E.I.S. Examples 2 and 4 which indicate that wildlife will somehow find new habitat in adjoining land areas--a convenient assumption but one that is more fiction than fact. Wildlife will move when forced too, but whether it can survive that displacement is dependent upon two major variables--availability and suitability of the new habitat. Rarely are these two items adequately discussed in contemporary environmental impact statements. In some parts of America our wildlife have simply run out of places to move, yet this "Displacement Myth" persists and few land planners have risen to challenge it's validity.

## HISTORICAL SIGNIFICANCE AND CONFLICTS IN USES

America was not particularly kind to its wildlife resources as a young nation. You will recall in Chapter 1 that we discussed the three phases of the American environmental movement. It was largely during the first phase and the early part of the second phase that America's wildlife resources suffered the most severe damage with some species like the passenger pigeon and the heath hen extirpated completely and the American bison (buffalo) coming dangerously close to sharing a similar fate. In the case of the passenger pigeon, habitat loss, a prime concern for today's wildlife, is believed to have been a significant contributing factor although market hunting certainly assisted in the process. The American bison on the other hand was the unwitting victim of a national scheme to subdue certain hostile American Indian cultures.

I do not mean to imply that the nation was totally indifferent to the fate of its wildlife resources during these time periods. Some of the nation's earliest laws pertaining to wildlife even predate the establishment of an American nation. In 1667, for example, Connecticut prohibited the export of game across its borders and in 1738 Virginia banned the harvest of doe deer (Ref. 14.1). One of the most significant laws, however, was passed by Rhode Island in 1846 when that state established the first seasonal regulations to protect waterfowl from spring shooting. The concept of daily bag limits was established by Iowa in 1878 and by 1900 twelve additional states had set up similar regulations. That same year Congress passed the Lacey Act which finally outlawed market hunting and by 1910 thirty-three states were obtaining revenue for wildlife protection and restoration from the sale of hunting licenses.

The preponderance of hunting related regulations amongst these early American wildlife laws reflects the dominant uses that Americans have traditionally

Table 14.1:  Some American Wildlife Population Changes

| Species | 1935 Population | 1985 Population |
|---|---|---|
| Bison | 12,800 | 65,000 |
| North American Elk | 225,000 | 500,000 |
| Pronghorn Antelope | 40,000 | 750,000 |
| Trumpeter Swan | 73 | 10,000 |
| White-tailed Deer | 5,000,000 | 14,700,000 |
| Whooping Crane | 29 | 138 |
| Wild Turkey | 31,250 | 2,500,000 |

Source:  Ref. 14.2

Figure 14.1: This suburban-dwelling raccoon has lost much of its natural fear of humans and boldly shares dinner with this trio of domestic cats. These mammals, however, may harbor the deadly rabies virus and therefore pose a serious threat to these cats and their owner.

made of their wildlife resources--namely sustenance and sport hunting. Today these two uses are under some intense criticism especially from a number of groups that have formed within the past decade and who strongly oppose the hunting or killing of any wildlife. Obviously this rigid opposition to harvesting wildlife is an irritation to America's 50 million plus hunters, fishermen and trappers who, for over 50 years have contributed enormous sums of money[1] (some $3 billion since the 1920's) (Ref. 14.1) to assist in the restoration and management of American wildlife. While these contributions have admittedly been self-serving all wildlife species, game and non-game alike, have been beneficiaries. Indeed many species have made major comebacks (Table 14.1).

The contention by some of these anti-hunting groups that legalized sport hunting and fishing are wildlife's greatest threat is an irksome allegation even to environmental planners. I can assure you that legitimate sport hunting and fishing will not result in the demise of America's wildlife. Unfortunately, such a groundless accusation not only lessens the chances for a much needed collaboration of all groups concerned with the welfare of the nation's wildlife but worst of all it diverts the focus of the nation away from the real threat to wildlife survival--habitat loss.

Man and wildlife have been and continue to be competitors for the same habitat--a contest in which wildlife has rarely, if ever, been the victor. When woodland and trees are cut and removed, some existing and potential wildlife nesting sites and food sources are likewise removed. The filling of wetlands and the placement of concrete and asphalt pavement and buildings result in similar losses. Even agriculture contributes to these losses when brushy hedgerows between fields that serve as corridors for the passage of wildlife and nesting and resting areas are plowed under. Habitat loss beyond American borders is even more threatening. For instance, nearly half of all species of birds that live in North America are songbirds (Order Passeriformes). In order to survive each year they must migrate to warmer climates. Twenty-two of our 32 flycatchers and 47 of our 52 warblers migrate to Mexico and areas further south and remain there from October to March (Ref. 14.3). Nine of our

11 vireos, 4 of our 5 native orioles and all four of our tanagers overwinter in Mexico, the Caribbean Islands and Central America. The estimate of birds that migrate south from North America each fall is around 5 billion.

Why so many of our North American birds migrate is the result of evolutionary selection. Most of America's insect eating birds are descendants of species that lived year round in the tropics (Ref. 14.3). As these ancestral species expanded their ranges northward into the temperate zones they found insect prey that was underutilized during the summer months. Consequently, some species learned to seasonally exploit this habitat and slowly pushed further and further northward from their original ranges. With the advent of winter, however, they always returned. Although Americans consider many of them "their" birds they are in reality tropical birds that only visit here to breed amidst an abundant supply of insect food. Today when these migrants return to their winter habitat there is no guarantee that the habitat they left will still be there. The Nature Conservancy has estimated that 74,000 acres of tropical forest are cleared worldwide every day (Ref. 14.3). That equates to a loss of 40,000 square miles annually--an area equivalent in size to the state of Virginia. Much of this tropical deforestation results from general dependence of many of the economies of these regions on exports to the United States. America's enormous consumption rate of coffee, bananas and beef requires the clearing of vast tracts of forest to keep up with the demand.

Not all of America's wildlife suffer as a result of habitat loss or change. Mammals such as the white-tailed deer (Odocoileus virginianus), the raccoon (Procyon lotor), and the western coyote (Canis latrans) for example, have adapted remarkably well to life in America's suburbs and cities. Such adaptations, however, can pose special dangers to humans as familiarity encourages the illusion that such animals have been at least partially domesticated.

A large population of urban and suburban dwelling raccoons now survives quite well in all parts of the United States thanks to an abundance of garbage cans, bird feeders, human handouts (Figure 14.1) and thousands of miles of underground highways called storm drains.   In California, the western coyote has adapted so well to the urban lifestyle that it can be found boldly walking city streets in broad daylight. In this instance, however, the coyotes supplement their normal garbage can fare with pet cats and dogs-- a trait that has alienated a once tolerant human population.   An ongoing rabies epidemic involving urban raccoon populations in at least 6 Middle-Atlantic states has fortunately brought many Americans back to reality as have the 16 coyote attacks on humans in Los Angeles County, California since 1975 (Ref. 14.5).   Seven of these attacks, by the way, were on children under the age of 5, one of which, in 1981, was fatal to a 3 year old girl.

Even the supposedly timid white-tailed deer[2], the frequent object of many anti-hunting protests, can be more than just a nuisance.   Take the case of Princeton Township, New Jersey where the discharge of firearms has been banned since 1972 as the result of one careless person who shot from a roadway and hit the picture window of a township home.   This 16 square mile suburb now harbors a population of between 700 and 900 deer and after a decade and a half of uncontrolled population growth their cuteness has worn thin.   The township, as of 1987, was averaging over 160 deer-car collisions a year resulting in many human injuries and an extensive amount of automobile damage.

## A NEW APPROACH

Americans are fortunate in that a significant number of wildlife refuges have already been set aside across the nation.   These actions, however, have fostered the additional myth that wildlife can be stockpiled forever in such areas.   They cannot.   Concentrating wildlife into such confined habitats (literally oases in some areas of dense human development) can be lethal to wildlife.   When these areas exceed their carrying capacity, death can result from starvation, stress or waste induced diseases (Figure 14.2).   Wildlife need areas in which they may move about freely to breed, nurse and hide their young, rest undisturbed and feed.   Despite an abundance of provisions for the consideration of wildlife in local land use regulations, few if any, planning boards ever deny subdivision or site plan approval over concerns

Figure 14.2: This whitetailed buck, his body covered with tumors, is a product of overcrowding on a wildlife refuge where hunting was forbidden.

Photo by:  John Mihatov

Figure 14.3: Harrassed by pet dogs these whitetailed deer were seeking refuge in less densely populated woodlands. Lacking safe corridors for their passage, five of a herd of twelve were struck and killed by a passing truck.

for the needs of wildlife. The now common practice of requiring donations of land from developers for open space preservation is no guarantee that such areas will be suitable for the needs of wildlife. In many instances such parcels are often discontinuous, isolated vestiges of habitat surrounded by human development. We need to start thinking in terms of greenways and corridors through which both wildlife and man may pass undisturbed (Figure 14.3). In Chapter 2 you will recall we spoke of the need to protect watersheds and recommended the establishment of buffer zones along even the tiniest of tributaries and wetlands. Such buffers can, with some modifications, provide a valuable network to link habitats as well as provide corridors for the unimpeded movement of wildlife. In fact, wildlife use stream corridors more than any other type of habitat (Ref. 14.8, 14.9). Modifications to these buffer zones generally include the expansion of the buffer widths beyond that necessary to accommodate runoff flows and the filtering of pollutants. This writer generally concurs with a recommendation for a buffer width of 100 to 300 feet on both sides of the streambed (Ref. 14.9) as a minimum. The final selection of buffer width will of course depend upon the wildlife species you are trying to assist. Other modifications include the connection of these corridors to feeding, resting and breeding areas. Figure 14.4 demonstrates the pleasing and extremely functional design that results from the use of such stream corridor networks. If we are successful in implementing such concepts it will be comforting to know that there will always be secluded mountaintops and ravines along which both stag and man may freely ramble, safe for a moment at least from a way of life that sometimes has little time for either.

## SPECIAL CONSIDERATIONS--WILDLIFE

### 1.    Canadian Geese

The Canadian Goose (Branta canadensis) has become an unwelcome resident in many parts of the United States. Like many other migratory birds, Canadian geese normally migrate south in the fall in search of habitat and food to sustain them during the winter months. Not all of these geese, however, are completing this annual trek. They are instead taking

Figure 14.4: Linking delineated stream and wetland buffer zones to each other as well as with areas of critical wildlife habitat provides a more realistic opportunity to preserve the nation's wildlife.

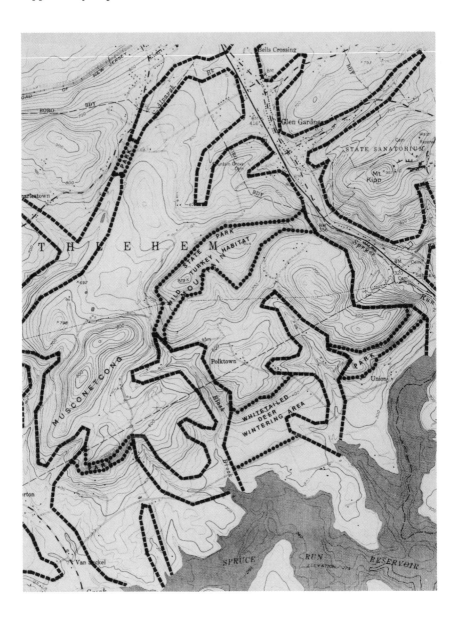

up residence in city parks, landscaped corporate and industrial campuses, golf courses and even wastewater treatment ponds. Once there, they are further enticed to stay by handouts of feed from well intentioned individuals who are unwittingly interfering with the geese's normal migration pattern. While the food offered is sometimes comprised of nutritious small grains like corn or wheat, most often, especially in an urban development, it is "junk food", which includes potato chips, popcorn or bread. These items may be filling but are far from nutritious. With regular feeding the geese will stop seeking out natural, more nutritious foods and become dependent upon these handouts. As a result large numbers of birds will begin to congregate, producing overcrowded conditions. Crowding produces stress which when combined with cold weather and nutritionally deficient food increases their susceptibility to life threatening diseases like avian cholera, duck plague and avian botulism.

In addition, such concentrated numbers of birds produce unsightly accumulations of feathers and feces, the latter of which can pollute nearby waterbodies. Lawn areas will be trampled and often overgrazed to the point that soil erosion begins to occur. And finally the noise that they make may be particularly irritating as the geese communicate amongst themselves, even at night.

Land planners may be able to assist in discouraging these geese from deviating from their normal migratory patterns by requiring certain items to be included as part of subdivision or site plan approval. These are especially important wherever artificial lakes, pond or wastewater treatment ponds are being considered. These inhibitory practices include:

1. The placement of a low fence (minimum of 2 feet high) or other solid barrier along the entire perimeter of the pond or lake to restrict access to adjacent lawn areas. Geese tend to walk to and from water to feed.
2. The installation of rock rip-rap or vertical wood bulkhead around the periphery of the pond to further retard goose movement.
3. The prohibition of small islands or peninsulas in ponds because they provide ideal nesting sites for geese.

In addition to the above practices, the municipality can make a significant contribution to this effort by enacting a strict ordinance prohibiting the feeding of geese and other migratory waterfowl. A copy of such an ordinance can be found in Appendix 1, Part F.

In this last regard it is interesting to note that one New Jersey town passed such an ordinance but was nonplussed to find that it did not apply to a major state park and lake located within its municipal boundaries. Strong-headed residents defiantly continued feeding the geese from state owned lands. The best solution, therefore, may be to have a statewide prohibition with some provision for an emergency permit system in the event of an unusual situation where emergency feeding might become necessary.

### FOOTNOTES

### CHAPTER 14

(1)   This has been largely in the form of hunting and fishing license fees, excise taxes on firearms, ammunition and archery equipment (taxes which sportsmen strongly endorsed and helped enact) and contributions to private, non-profit conservation organizations such as the National Wildlife Federation, Ducks Unlimited and the Izaak Walton League. Current contributions from these same sources are about $225 million annually (Ref. 14.1).

(2)   At times even the white-tailed deer can be physically dangerous to humans, especially during the rut or breeding season. Persons have been gored severely by bucks that were considered to be tame.

## REFERENCES

## CHAPTER 14

14.1   National Shooting Sports Foundation, Inc., "The Hunter and Conservation", Riverside, Connecticut, 1982.

14.2   National Wildlife Federation, "National Wildlife", April/May, 1986, pg. 15.

14.3   Connor, Jack, "Empty Skies", Arrowsmith Magazine, August, 1988.

14.4   Lawren, Bill, "Something to Sing About", National Wildlife Magazine, December/January, 1989.

14.5   Harrison, George, "Nature/Urban Coyotes", Sports Afield Magazine, November, 1987.

14.6   Lipske, Mike, "Night Stalker", National Wildlife Magazine, June/July, 1985.

14.7   Ferrara, Jerry L., "Why Vultures Make Good Neighbors", National Wildlife Magazine, June/July, 1987.

14.8   Odum, E.P., "Ecological Importance of the Riparian Zone", Proceedings of the National Symposium on Strategies for Protection of Floodplain Wetlands and Other Riparian Ecosystems, U.S. Forest Service, General Technical Report WO-12, 1979.

14.9   New Jersey Office of State Planning, "The N.J. Freshwater Wetlands Protection Act as it Relates to Stream Corridor Buffer Considerations in the State Development and Redevelopment Plan", January 11, 1988, by Rogers, Golden and Halpern, Phila., PA.

14.10  National Wildlife Federation, "National Wildlife", August/September, 1987, pg. 25.

# 15

# Miscellaneous Environmental Problems

## BURIED STUMPS AND OTHER WOOD PRODUCTS

A common practice among builders, particularly in large subdivisions, is to bury tree stumps and other wood construction debris in pits on some or all of the lots. In the normally damp conditions of the subsoil, all of these materials will decay[1]--with lumber sometimes decomposing in as little as 5 or 6 years. Tree stumps will generally persist much longer, varying from 10 to 20 years. As these materials decay, voids form and the adjoining soil may eventually collapse into them. Stumps in particular pose a special danger, especially if they are buried at a shallow depth. It is a fact that some home owners and their guests have been injured when these voids suddenly opened to or near the ground surface to form pits as much as five feet deep and several feet wide. In some instances, lawns have been literally pockmarked with these subsidence pits. Most of the anxious home owners were unable to determine the cause until an expert showed them the residual fibers of wood around the edges of the holes.

Burying of such materials should be discouraged, and their recycling encouraged. Some states now require wood stumps to go to stump grinding facilities where they are reduced to wood chips and sold as mulch or compost. Untreated lumber may likewise be chipped into mulch or compost. If this option is not available, such materials should be buried at least 8 feet below the ground surface except for pressure treated lumber which should be taken to a secured landfill.

## STONEWALL REMOVAL

A unique industry has recently developed in some rural and suburban areas of the northeastern United States precipitated by a heavy demand for landscaping materials, in this case, natural rock.  As a result, enterprising suppliers have begun to remove miles of stonewalls (Figure 15.1) that once marked the perimeters of early American agricultural fields and roadways.   Unfortunately, these same stonewalls also denote property boundaries, with older deed descriptions frequently referencing corners and junctions of these very same walls.  If such wholesale removal is allowed to continue, hundreds of property corners referenced in thousands of recorded deeds may be lost forever.  The impact on local land surveying practices could be devastating, particularly in very rural areas.  Consequently, the prohibition of such wholesale removals is recommended.

## HOUSEBOATS

A new form of residential housing has begun to appear along portions of the coastal and inland waterways of the United States.  These units have been described both as floating homes and houseboats with the latter terminology probably being a more accurate description.  This, of course, was the intent of the original innovator of this unique type of dwelling.  As mobile vessels, these homes, much to the chagrin of local authorities, are frequently exempt from municipal property taxes, even though they may be permanently moored at a marina.  In order to maintain the status of a vessel, each unit is equipped with an outboard motor (Figure 15.2) that the developers have demonstrated, in the more quiescent bays at least, will adequately propel and maneuver the houseboat to the satisfaction of the U.S. Coast Guard--the regulative authority over such vessels.  Whether these boat-dwellings could survive the rigors of the open sea is still in question.

These floating homes come in various sizes and shapes with some sized to easily accommodate a family of four on a permanent basis given adequate electrical, water and sewer hookups.  These hookups are most practical, however, only when anchored at a semi-per-

Figure 15.1: This mountaintop stonewall may be a vital reference point for land surveyors. Its removal could make future land surveying activities more lengthy and more costly.

Figure 15.2:  Houseboat at dock. Note outboard engine at rear of structure.

Figure 15.3: Several outlets for grey water discharges can be seen along the left side of this docked houseboat.

Photos by:  Ed Markowski.

manent mooring such as a marina.  It is this practice
of mooring and the provision of adequate sewage that
have proven to be the nemesis of the owners and developers of such housing.  When adjoining, land-based
residents discover that children from these boat-homes
are attending local schools at no cost (thanks to an
exemption from municipal taxes) animosity quickly
develops.  Local officials are soon besieged with
petitions from their dry-land constituents calling for
some action to end this decidedly unfair tax advantage.  In most cases, little can be done.

     Frustrated with the unsuccessful attempts to
deal with the problem, municipal officials have
requested state and federal environmental agencies to
investigate the manner in which these floating homes
dispose of their sewage.  They are further nonplussed
to find out that as a declared marine vessel, these
units may discharge their "grey water" (i.e. water
from sinks, tubs, showers) directly into the waterways
without treatment (Figure 15.3).  Wastewater from
toilets (black water) however, must be retained and
disposed of in an approved manner.  The discharge of
grey water from true maritime vessels on the open sea
is believed to have a negligible impact on water quality.  That same grey water, however, discharged from a
moored vessel in a still lagoon, will ultimately produce sheens of soap and grease, odors, discoloration
and ultimately complaints from nearby residents, boat
owners and fishermen.

     If you are fascinated by this new concept in
housing, I urge you to proceed with caution, keeping
in mind some of the problems outlined above.  Legislative changes and decisions from pending court cases
will most certainly affect some of the current
attributes of this type of housing.

## INSECT VECTORS

### Lyme Disease

     The tranquil countryside that lures Americans to
the nation's sprawling suburbs may harbor hazards
rarely considered in the land planning process.  Take
the case of a young mother in the Connecticut suburb
of East Haddam, who was plagued by tiny bugs that
attached themselves to her clothing and skin whenever

she jogged along the narrow pathways of a nearby community open space area. Despite her splendid aerobic regimen, her health began to deteriorate less than a year after moving to East Haddam. Chronic headaches, dizziness, fatigue and tingling sensations in the extremities finally forced her to consult a physician. His diagnosis--Lyme disease, most likely caused by the tiny insects that sometimes dotted her legs and arms following each workout. These tiny bugs were actually blood sucking ticks of the species Ixodes dammini[2], also known as deer ticks, barely the size of the period at the end of this sentence. Deer ticks were, of course, only the carrier, the actual agent of infection was the bacterium, Borrellia burgdorferi, a corkscrew shaped spirochete.

Lyme disease, named for a town in Connecticut where it was first recognized, is spreading rapidly and is now the most frequently diagnosed tick-transmitted illness in the United States (Ref. 15.1). It is often referred to as the "suburban disease", since it usually occurs in neighborhoods close to woodlands and overgrown fields where deer, mice and other mammals proliferate (see Chapter 14). Incidences of Lyme disease have been reported in 24 states, however, 90% of all cases have occurred in the following states: New Jersey, New York, Connecticut, Rhode Island, Massachusetts, Wisconsin and Minnesota. Clusters of cases have also occurred in Georgia, Texas, California, Oregon, Utah and Nevada. The federal Center for Disease Control (CDC) in Atlanta listed 7,400 confirmed cases nationwide in 1989[3]--a dramatic increase from the 1498 in 1984 and the 599 in 1983.

Lyme disease usually occurs in three stages (Ref. 15.1). The first is characterized by the appearance of an erythema chronicum migrans (ECM) rash, (a migrating red rash) which develops from 2 to 30 days after an individual is bitten. The rash is frequently accompanied by profound fatigue, fever, chills, backache and headache. In 25 to 50 percent of all cases, secondary lesions appear at various sites on the body.

The second stage is usually marked by neurological complications and migratory musculoskeletal pain. About five percent of the patients develop cardiac difficulties lasting from 3 days to 6 weeks. These patients may experience palpitations, dizziness and shortness of breath associated with irregular electrical impulses to the heart. Some may require temporary pacemakers.

The third stage usually involves the onset of arthritis. Joint problems characteristic of rheumatoid arthritis occur in approximately 60% of the patients who have received no treatment.

The nervous system may be involved at all stages of the disease. Live spirochetes have been detected in the cerebrospinal fluid and brain tissue of diagnosed Lyme disease patients. Neurological symptoms can include meningitis, inflamed nerve roots in the neck and Bell's palsy, which affects facial muscles.

Fortunately, Lyme disease, if detected early, can be successfully treated with antibiotics such as penicillin, tetracycline and erythromycin.

The life cycle of the deer tick, I. dammini, is of particular interest to land planners for reasons we shall shortly see. The life cycle of this arthropod normally spans two years. Eggs are deposited in the spring and hatch into free living larvae a month later. During this first summer, the larvae feed once (usually for a period of two days) on the blood of a host and then enter a resting stage coincident with cooler weather. The following spring, the larvae molt and enter a second immature stage called the nymphal stage. It, again, attaches itself to an animal host to feed for three to four days. The most frequently utilized host at these two immature stages is the white footed mouse (Peromyscus leocopus) although other small mammals such as raccoons, skunks and rabbits may also be used. At the end of the second summer, nymphs molt into the adult stage. They are then usually found in the brush about one meter above the ground where they can attach themselves to larger mammals. The adults feed on a variety of hosts including man, but in the northeastern United States, they are found predominantly on the white-tailed deer, Odocoileus virginianus.

It is on these larger hosts that the ticks mate. The males dies soon after mating and only the females overwinter.

Some of you, I'm sure, are already asking why we have spent so much time on an insect induced disease that, while important, seems far removed from the main topic of this book, land planning. The answer, of course, is that this disease and its vector have been abetted, although somewhat subtly, by some of our present land use planning practices. For instance, our zoning ordinances by their indirect encouragement of the abandonment of agriculture, are partly responsible. No planted crop can provide the cash return that a "crop" of residential building lots can. As a result, large tracts of cultivated farmland have fallen idle and during the lengthy subdivision process often revert to a brushy, successional field environment conducive to the proliferation of such mammals as the white-tailed deer, raccoon, skunk and the white-footed mouse, and, of course, the deer tick. By the time the land is finally carved into individual building lots, the deer tick population will have been well established. In addition, our consistent failure to set aside realistic amounts of wildlife habitat has produced abnormal concentrations and distributions of wildlife. This is further compounded by the enactment of ordinances prohibiting trapping and the discharge of firearms which prevent subsequent hunting and harvesting of this surplus wildlife, thereby creating a valhalla for the deer tick.

While quick cure schemes calling for the elimination of all white-tailed deer would be helpful in curtailing the spread of this disease, the idea would, I'm sure, be greeted with little enthusiasm from sportsmen and anti-hunting groups, alike. No, the long term solution lies with the nation's land planners. Municipal land use master plans or state land use master plans, if necessary, must be retrofitted with provisions to provide a network of wildlife habitat as outlined in Chapter 14. This will keep wildlife dispersed and at the same time allow ample areas in which trapping and hunting may be safely undertaken when necessary to control overpopulation. The deer tick will then have to look a lot farther to find a mammalian host.

In the meantime, however, new devices like that developed by Dr. Andrew Spielman, the scientist who first identified the species of tick that causes Lyme disease, may help. His device called Damminix consists of a cardboard tube packed with insecticide treated cotton balls. The tube when placed in mouse infested areas encourages the mice to take the cotton balls back to their nest sites where they use it for a lining. The insecticide bearing cotton then kills the nymphal ticks residing in the mice's fur (Ref. 15.2).

Deer ticks are only one of a number of disease carrying insects that are rarely, if ever, considered in the land planning process. Mosquitoes, for example, of which there are numerous species, and which are prevalent in wetlands or other low lying areas receive virtually no attention during the consideration of residential subdivisions in these areas. Yet, they can also transmit some rather nasty and potentially lethal viruses such as Eastern and St. Louis encephalitis. As with Lyme disease, there is a similar link to wildlife intermediaries--in this case migratory waterfowl and the ring-necked pheasant.

## FOOTNOTES

### CHAPTER 15

(1)    Arid climates are generally an exception.
(2)    Ixodes dammini is generally found in the northeastern and midwestern sections of the United States. Other transmitters include Ixodes pacificus (found along the pacific coast area) and Amblyomma americanum, the Lone Star tick.
(3)    Personal conversation with C.D.C. Lyme disease center in Fort Collins, Colorado.

## REFERENCES

### CHAPTER 15

15.1    Habicht, Gail S. and Gregory Beck and Jorge Benach, "Lyme Disease", Scientific American, Vol. 257, Number 1, July, 1987.
15.2    Insight, "Simple Device Kills Disease-Ridden Ticks", Dec. 1988, p.56.

# 16

# Radon

## THE DISCOVERY

In the fall of 1984 when Stanley Watras, a construction engineer from Boyerton, Pennsylvania, began setting off radiation detectors on his way into work at the Limerick Nuclear Power Plant, both he and his employer were perplexed. When plant technicians eventually tested his home, however, the mystery was quickly solved. Radioactive radon gas had been seeping into his basement from naturally occurring uranium in the granite bedrock located beneath his home. Prior to this, few Americans had ever heard of radon. Today it is a priority concern on many state and federal environmental protection agency agendas.

## RADON AND ITS SOURCE

In simplest terms radon is a naturally occurring, odorless, colorless, radioactive gas found in the soils and rocks[1] that make up the earth's crust. Because radon is a gas it can, under the proper conditions, travel considerable distances through the soil.

Technically radon is identified as radon-222. It is derived from the emission of alpha radiation during the decay of radium-226, which is part of the natural decay chain of uranium-238. Uranium-238 is one of three isotopes (i.e. uranium-234, uranium-235, uranium-238) comprising natural uranium and it is generally the most prevalent. The relative occurrences of uranium-234, uranium-235 and uranium-238 are 0.006, 0.72 and 99.27%, respectively (Ref. 16.5). Similarly, there are isotopes of radon other than radon-222. These include radon-219 which is derived from the decay of uranium-235 and radon-220 derived

from the decay of thorium-232.  Of these 3 radon isotopes radon-222 is the longest lived.  Hence, the combination  of  uranium-238  dominance  and  radon-222
longevity makes radon-222 the isotope of most interest
to engineers, scientists, planners and health officials.

Radon-222 also eventually decays by releasing
alpha radiation and is transformed into polonium-218,
which decays further into lead-214 which is followed
by further steps of decay into bismuth-214, polonium-
214, lead-210 and the final stable, non-radioactive
end product, lead-206.  These intermediate decay products are referred to collectively as radon progeny or
radon daughters.  Radon is the only gaseous member of
this decay chain.  The above process occurs continually wherever uranium bearing material is found.  The
half life of radon is about four days (3.82 days
exactly) which simply means that in that time period
one half of the quantity of radon will have decayed
into one or more of its progeny.  The half life of
uranium-238 in comparison is 4.5 billion years.
Uranium-238 decay product concentrations in soil are
known to range from 0.6 picocuries (pCi) per gram
(considered background) to hundreds of pCi per gram in
uranium rich areas.  A picocurie is one-trillionth of
a curie$^{(2)}$ --a common measure of radioactivity.

## RADON ENTRY ROUTES INTO DWELLINGS

The most common entry routes for radon gas into
the house are shown in Figure 16.1.

## TESTING METHODOLOGIES

The  radon  measuring  device  most  familiar  to
Americans is the charcoal adsorption canister.  Because of its low cost and simplicity it has become the
most commonly used do-it-yourself device for initial
home screenings.  These units are about the size of a
shoe polish tin (3 inches in diameter and 3/4 of an
inch thick), although some now come packaged as rectangular envelopes.  They are sent sealed and upon
receipt are opened and left exposed to the indoor
atmosphere for 2 to 7 days.  At the end of that time
the charcoal has captured all it can and the canister
is resealed and sent to a laboratory for analysis.

Figure 16.1:  Major radon entry routes into detached houses.

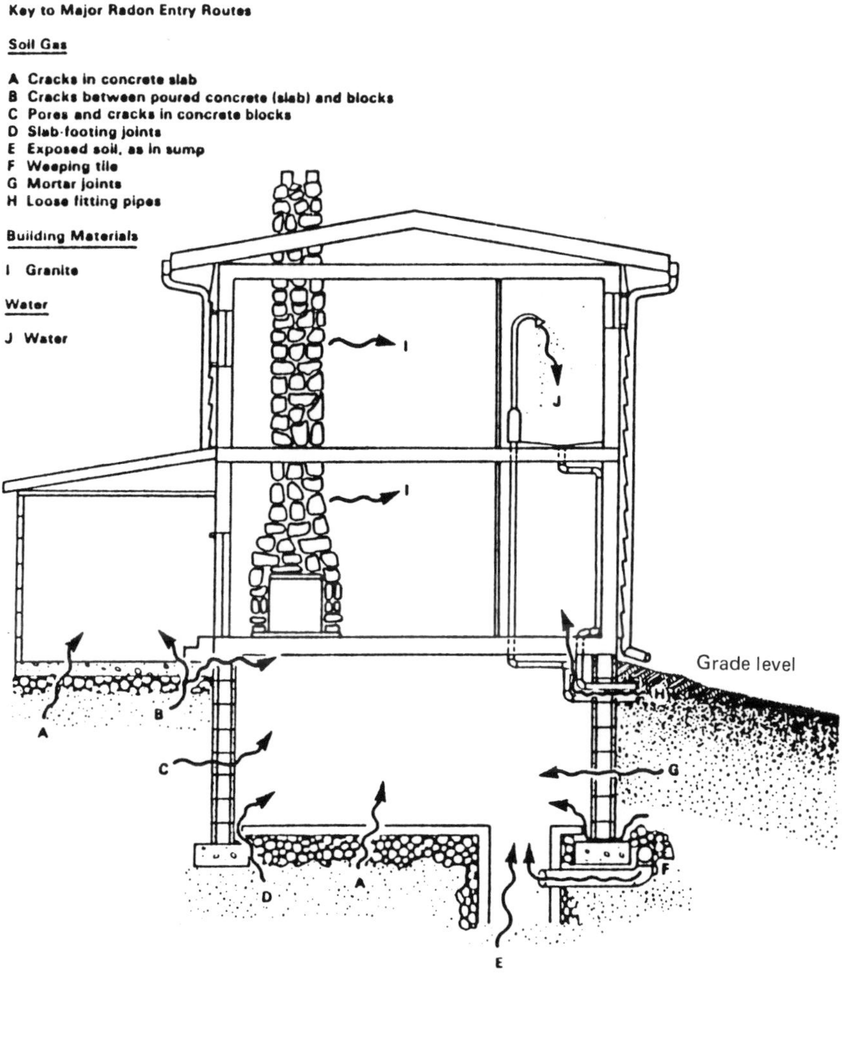

**Key to Major Radon Entry Routes**

Soil Gas

A  Cracks in concrete slab
B  Cracks between poured concrete (slab) and blocks
C  Pores and cracks in concrete blocks
D  Slab-footing joints
E  Exposed soil, as in sump
F  Weeping tile
G  Mortar joints
H  Loose fitting pipes

Building Materials

I  Granite

Water

J  Water

Grade level

Source:  Ref. 16.1

These charcoal units measure radon progeny rather than the parent element. Another device, the alpha track detector (ATD) is less commonly used but offers the advantage of a larger time averaged measurement. The ATD is a small cylinder, the diameter of a silver dollar and about one inch high. For a screening measurement the ATD is left in place from 30 to 90 days, but can be left in place for up to a year to determine the annual exposure. After the period of exposure the ATD is read under a microscope to determine concentration levels. These ATD's measure only radon and not any of its progeny.

Radon may also be measured directly at the site with instruments such as the Continuous Working Level Monitor (CWLM) and the Radon Progeny Integrating Sampling Unit (RPISU). These instruments have recommended sampling times of 24 hours and 72 hours respectively. These latter testing methods are generally more expensive.

**EFFECTS ON MAN**

Wherever radon gas and its progeny remain in the soil and rock or are liberated directly to the atmosphere their presence has little health significance. Outdoor concentrations of radon gas are reported to average around 0.25 pCi per liter. Where there is extensive mineralization this increases the concentration to around 0.75 pCi per liter (Ref. 16.1). When trapped within well insulated homes, however, radon and its progeny accumulate and can attach themselves to dust particles or tobacco smoke which if inhaled damage lung cells[3] and lead to cancer. The U.S. EPA recommends that mitigation measures be initiated when indoor atmospheric radon levels reach 4 pCi per liter.

Much of our knowledge about the health effects of radon and its progeny is based upon analyses of the effects of high exposure to radon on underground miners (Ref. 16.1, 16.7). Relevant findings from such health studies emphasize that:

    1)    There is no doubt that sufficient doses of radon and its progeny can produce lung cancer in humans.

    2)    An excess incidence of cancer has been associated with exposures that were 2 to 3 orders of magnitude (100 to 1,000 times)

greater than those found in normal indoor environments.

It is generally believed that radon and its progeny are responsible for most of the lung cancer risk to the general nonsmoking public.

A commonly used unit for the measurement of radon risk that was developed as a result of these earlier studies on miners is the "working level" (WL). The working level is defined as the quantity of short-lived progeny that will result in $1.3 \times 10^5$ MeV of potential alpha energy per liter of air. Exposures are measured in working level months (WLM); e.g. an exposure to 1 WL for 1 working month (173 hours) is 1 WLM (Ref. 16.1). Figure 16.2 provides further information on radon risk evaluation.

## MITIGATING THE PROBLEM

Specific radon reduction methods are shown on Figures 16.3 through 16.12. Generally, all of these methods fall into one of the following categories:

1) Sealing and closing of all pores, voids, open joints and exposed earth that permit gaseous radon to enter the house.

2) Reversing the predominant direction of soil transmitted radon gas flow so that air movement is from the house to the soil and outside air.

Tables 16.1 and 16.2 provide a summary of all techniques indicating their relative effectiveness and estimated costs (Ref. 16.1, 16.15).

Two additional methods of radon mitigation for the home include:

1) Avoiding use of water supplies containing radon or removing radon from potable water (see discussion below).

2) Avoiding the use of building materials such as granite that may contain radium.

## RADON IN GROUND WATER AND POTABLE WATER SUPPLIES

While much of the present emphasis is on indoor atmospheric concentrations of radon, we should not forget that radon can also invade ground water and ultimately potable water supply wells. In Maine, for instance, radon-222 in ground water has been a recog-

Figure 16.2:  Radon risk evaluation chart.

| pCi/l | WL | Estimated number of lung cancer deaths due to radon exposure (out of 1000) | Comparable exposure levels | | Comparable risk |
|-------|-----|-----------|-----------------|---|----------------|
| 200 | 1 | 440—770 | 1000 times average outdoor level | | More than 60 times non-smoker risk |
| | | | | | 4 pack-a-day smoker |
| 100 | 0.5 | 270—630 | 100 times average indoor level | | |
| | | | | | 20,000 chest x-rays per year |
| 40 | 0.2 | 120—380 | | | |
| | | | 100 times average outdoor level | | 2 pack-a-day smoker |
| 20 | 0.1 | 60—210 | | | |
| | | | | | 1 pack-a-day smoker |
| 10 | 0.05 | 30—120 | 10 times average indoor level | | |
| | | | | | 5 times non-smoker risk |
| 4 | 0.02 | 13—50 | | | |
| | | | 10 times average outdoor level | | 200 chest x-rays per year |
| 2 | 0.01 | 7—30 | | | |
| | | | | | Non-smoker risk of dying from lung cancer |
| 1 | 0.005 | 3—13 | Average indoor level | | |
| 0.2 | 0.001 | 1—3 | Average outdoor level | | 20 chest x-rays per year |

Source:  Ref. 16.2

Figure 16.3:  Drain tile ventilation where tile drains to sump.

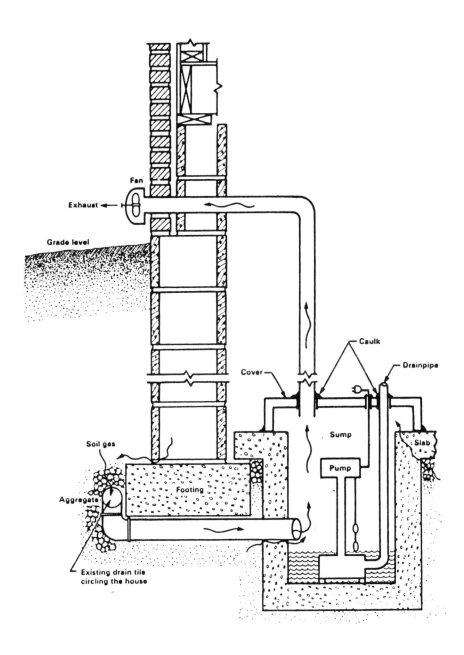

Source:  Ref. 16.1

Figure 16.4:  Drain tile ventilation where tile drains to soakaway.

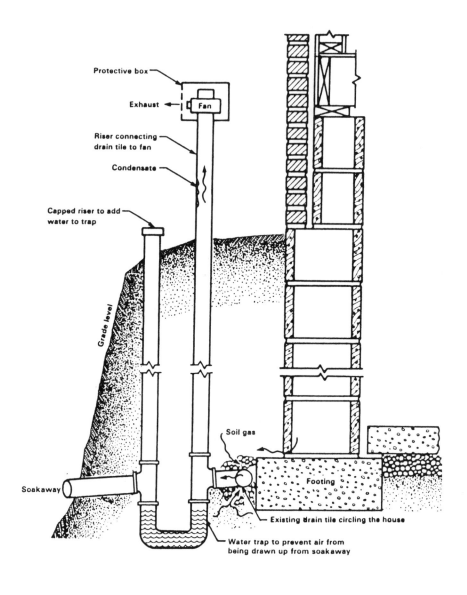

Source:  Ref. 16.1

Figure 16.5: Wall ventilation with individual suction points in each wall.

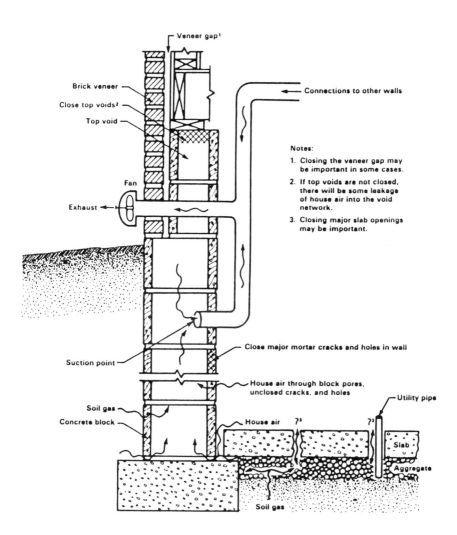

Source: Ref. 16.1

Figure 16.6: Closing the top void in a block foundation when a fair amount of the void is exposed.

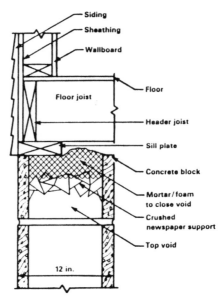

Figure 16.7: Closing the top void in a block foundation when an exterior brick veneer is present.

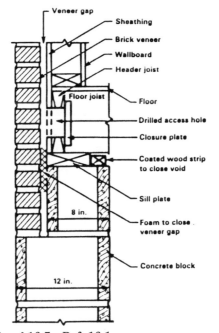

Source for Figures 16.6 and 16.7:  Ref. 16.1

Figure 16.8:  Wall ventilation with baseboard duct.

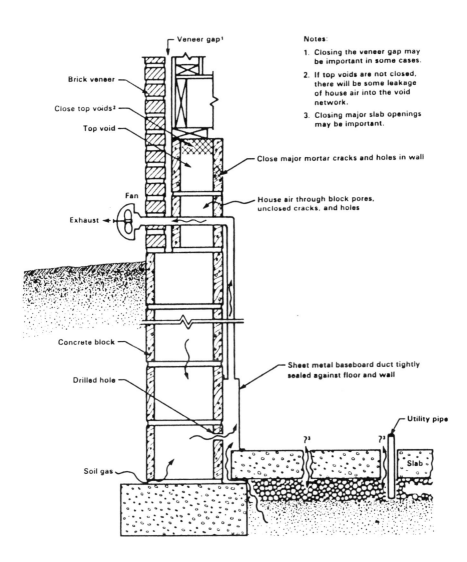

Notes:

1. Closing the veneer gap may be important in some cases.

2. If top voids are not closed, there will be some leakage of house air into the void network.

3. Closing major slab openings may be important.

Source:  Ref. 16.1

Figure 16.9: Sub-slab ventilation using individual suction point approach.

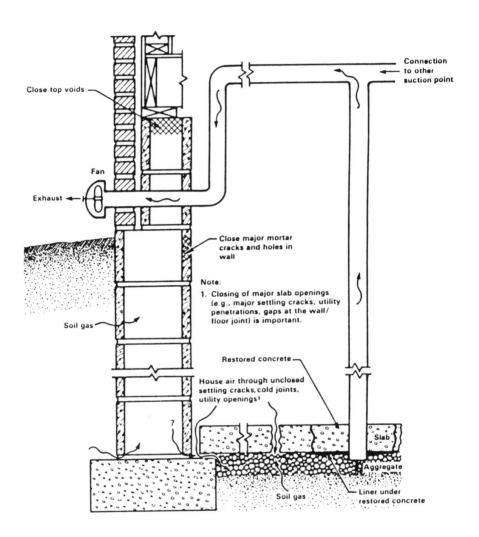

Figure 16.10: Sub-slab ventilation using individual suction point approach with optional horizontal run under slab.

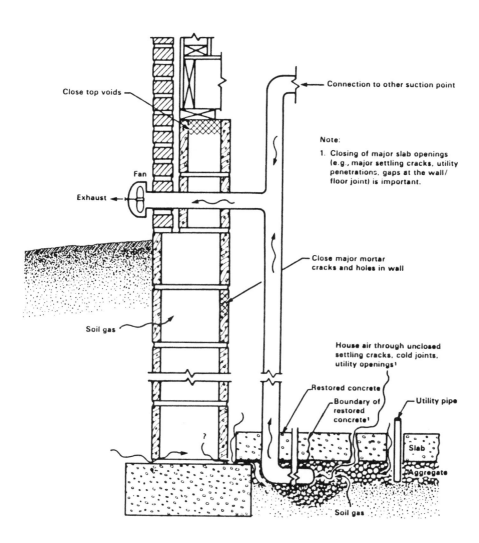

Connection to other suction point

Close top voids

Note:

1. Closing of major slab openings (e.g., major settling cracks, utility penetrations, gaps at the wall/floor joint) is important.

Fan

Exhaust

Close major mortar cracks and holes in wall

Soil gas

House air through unclosed settling cracks, cold joints, utility openings[1]

Restored concrete

Boundary of restored concrete[1]

Utility pipe

Slab

Aggregate

Soil gas

Source: Ref. 16.1

Figure 16.11: Sub-slab piping network suggested for new house construction (from Central Mortgage and Housing Corp. of Canada).

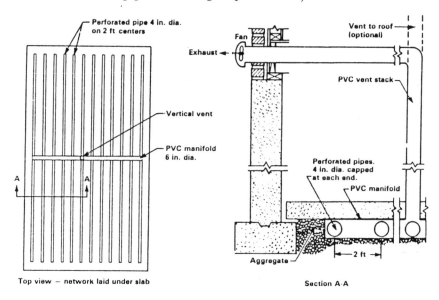

Top view — network laid under slab          Section A-A

Figure 16.12: Sub-slab piping network around perimeter of slab.

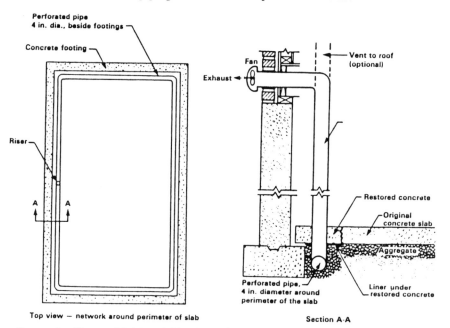

Top view — network around perimeter of slab          Section A-A

Source for Figures 16.11 and 16.12:  Ref. 16.1

Table 16.1: Summary of Radon Reduction Techniques

| Method | Principle of Operation | House Types Applicable | Estimated Annual Avg. Concentration Reduction, % | Confidence in Effectiveness | Operating Conditions and Applicability | Estimated Installation and Annual Operating costs |
|---|---|---|---|---|---|---|
| Natural ventilation | Air exchange causing replacement and dilution of indoor air with outdoor air by uniformly opening windows and vents | All[a] | 90[b] | Moderate | Open windows and air vents uniformly around house. Air exchange rates up to 2 ach may be attained. May require energy and comfort penalties and/or loss of living space use | No installation cost. Operating costs for additional heating are estimated to range up to a 3.4-fold increase from normal (0.25 ach) ventilation conditions[c] |
| Forced air ventilation | Air exchange causing replacement and dilution of indoor air with outdoor air by the use of fans located in windows or vent openings | All | 90[b] | Moderate | Continuous operation of a central fan with fresh air makeup, window fans, or local exhaust fans. Forced air ventilation can be used to increase air exchange rates rates up to 2 ach. May require energy and comfort penalties and/or loss of living space use | Installation costs range up to $150. Operating costs range up to $100 for fan energy and up to a 3.4-fold increase in normal (0.25 ach) heating energy costs[c] |
| Forced air ventilation with heat recovery | Air exchange causing replacement and dilution of indoor air with outdoor air by the use of a fan powered ventilation system | All | 96[d] | Moderate to high | Continuous operation of units rated at 25-240 cubic feet per minute (cfm). Air exchange increased from 0.25 to 2 ach. In cold climates units can recover up to 70% of heat that would be lost through house ventilation without heat recovery | Installation costs range from $400 to $1500 for 25-240 cfm units. Operating costs range up to $100 for fan energy plus up to 1.4-fold increase in heating costs assuming a 70% efficient heat recovery[c] |
| Active avoidance of house depressurization | Provide clean makeup air to household appliances which exhaust or consume indoor air | All | 0-10[e] | Moderate[f] | Provide outside makeup air to appliances such as furnaces, fireplaces, clothes dryers, and room exhaust fans | Installation costs of small dampered duct work should be minimal. Operating benefits may result from using outdoor air for combustion sources |

(continued)

## Table 16.1: (continued)

| Method | Principle of Operation | House Types Applicable | Estimated Annual Avg. Concentration Reduction,% | Confidence in Effectiveness | Operating Conditions and Applicability | Estimated Installation and Annual Operating costs |
|---|---|---|---|---|---|---|
| Sealing major radon sources | Use gas-proof barriers to close off and exhaust ventilate sources of soil-gas-borne radon | All | Local exhaust of the source may produce significant house-wide reductions | Extremely case specific | Areas of major soil-gas entry such as cold rooms, exposed earth, sumps, or basement drains may be sealed and ventilated by exhausting collected air to the outside | Most jobs could be accomplished for less than $100<br><br>Operating costs for a small fan would be minimal |
| Sealing radon entry routes | Use gas-proof sealants to prevent soil-gas-borne radon entry | All | 30-90 | Extremely case Specific | All noticeable interior cracks, cold joints, openings around services, and pores in basement walls and floors should be sealed with appropriate materials | Installation costs range between $300 and $600 |
| Drain tile soil ventilation | Continuously collect, dilute, and exhaust soil-gas-borne radon from the footing perimeter of houses | BB PCB S | Up to 98 | Moderate[g] | Continuous collection of soil-gas-borne radon using a 160 cfm fan to exhaust a perimeter drain tile<br><br>Applicable to houses with a complete perimeter footing level drain tile system and with no interior block walls resting on sub-slab footings | Installation cost is $1200 by contractor<br><br>Operating costs are $15 for fan energy and up to $125 for supplemental heating |
| Active ventilation of hollow-block basement walls | Continually collect, dilute, and exhaust soil-gas-borne radon from hollow-block basement walls | BB | Up to 99 + | Moderate to high | Continuous collection of soil-gas-borne radon using one 250 cfm fan to exhaust all hollow-block perimeter basement walls<br><br>Baseboard wall collection and exhaust system used in houses with French (channel) drains | Installation costs for a single suction and exhaust point system is $2500 (contractor installed in unfinished basement)<br><br>Installation cost for a baseboard wall collection system is $5000 (contractor installed in unfinished basement)<br><br>Operating costs are $15 for fan energy and up to $125 for supplemental heating |

(continued)

## Table 16.1:  (continued)

| Method | Principle of Operation | House Types Applicable | Estimated Annual Avg. Concentration Reduction,% | Confidence in Effectiveness | Operating Conditions and Applicability | Estimated Installation and Annual Operating costs |
|---|---|---|---|---|---|---|
| Sub-slab soil ventilation | Continually collect and exhaust soil-gas-borne radon from the aggregate or soil under the concrete slab | BB PCB S | 80-90, as high as 99 in some cases | Moderate to high | Continuous collection of soil-gas-borne radon using one fan (~ 100 cfm, ≥ 0.4 in.; H$_2$O suction) to exhaust aggregate or soil under slab<br><br>For individual suction point approach, roughly one suction point per 500 sq ft of slab area<br><br>Piping network under slab is another approach, might permit adequate ventilation without power-driven fan | Installation cost for individual suction point approach is about $2000 (contractor installed)<br><br>Installation costs for retrofit sub-slab piping network would be over $5000 (contractor installed)<br><br>Operating costs are $15 for fan energy (if used) and up to $125 for supplemental heating |

[a] BB (Block basement) houses with hollow-block (concrete block or cinder block) basement or partial basement, finished or unfinished
PCB (Poured concrete basement) houses with full or partial, finished or unfinished poured-concrete walls
C (Crawl space) houses built on a crawl space
S (Slab, or slab-on-grade) houses built on concrete slabs.

[b] Field studies have validated the calculated effectiveness of fourfold to eightfold increases in air exchange rates to produce up to 90 percent reductions in indoor radon.

[c] Operating costs are ascribed to increases in heating costs based on ventilating at 2 ach the randon source level; as an example, the basement with 1) no supplementary heating or 2) supplementary heating to the comfort range. It is assumed the basement requires 40 percent of the heating load and if not heated would through leakage still increase whole house energy requirements by 20 percent. Operating costs are based on fan sizes needed to produce up to 2 ach of a 30x30x8 ft (7200 cu ft) basement or an eightfold increase in ventilation rate.

[d] Recent radon mitigation studies of 10 inlet/outlet balanced mechanical ventilation systems have reported radon reduction up to 96 percent in basements. These studies indicate air exchange rates were increased from 0.25 to 1.3 ach.

[e] This estimate assumes that depressurizing appliances (i.e., local exhaust fans, clothes dryers, furnaces, and fireplaces) are used no more than 20 percent of the time over a year. This suggests that during the heating season use of furnaces and fireplaces with provision of makeup air may reduce indoor radon levels by up to 50 percent.

[f] Studies indicate that significant entry of soil-gas-borne radon is induced by pressure differences between the soil and indoor environment. Specific radon entry effects of specific pressurization and depressurization are also dependent on source strengths, soil conditions, the completeness of house sealing against radon, and baseline house ventilation rates.

[g] Ongoing studies indicate that where a house's drain tile collection system is complete (i.e., it goes around the whole house perimeter) and the house has no interior hollow-block walls resting on sub-slab footings, high radon entry reduction can be achieved.

Source:  Ref. 16.1

## Table 16.2: Summary of Results from Radon Mitigation Tests in 40 Eastern Pennsylvania Houses

| House No. | Substructure Type | Final Mitigation | Mean Radon Levels, pCi/L | | Reduction, % |
|---|---|---|---|---|---|
| | | | Before | After | |
| 1 | Block basement | Wall and sub-slab pressurization (baseboard duct) | 161 | 5 | 97 |
| 2 | Block basement | Wall and sub-slab pressurization (baseboard duct and carbon filter on well water) | 238 | 3 | 99 |
| 3 | Block basement | Wall and sub-slab suction | 1205 | 5 | 99 |
| 4 | Block basement | Sub-slab suction | 20 | 3 | 86 |
| 5 | Block basement | Wall pressurization | 110 | 5 | 95 |
| 6 | Block basement | Sub-slab suction | 60 | 5 | 92 |
| 7 | Block basement | Sub-slab and wall suction | 402 | 4 | 99 |
| 8 | Block basement | Wall suction | 88 | 6 | 93 |
| 9 | Block basement | Wall & Sub-slab pressurization (baseboard duct over French drain) | 360 | 7 | 98 |
| 10 | Block basement | Drain tile suction (exterior) | 209 | 7 | 97 |
| 11 | Block basement | Wall & sub-slab suction (baseboard duct over French drain) | 60 | 21 | 65 |
| 12 | Block basement | Drain tile suction (exterior) | 11 | 3 | 75 |
| 13 | Block basement | Drain tile suction (exterior) | 94 | 2 | 98 |
| 14 | Block basement | Wall suction | 61 | 1 | 98 |
| 15 | Block basement | Drain tile suction (exterior) | 18 | 1 | 98 |
| 16 | Block basement & paved crawl space | Wall suction | 240 | 4 | 98 |
| 17 | Block basement | HRV[a] | 60 | 38 | 37 |
| 18 | Block basement | HRV | 2 | 1 | 50 |
| 19 | Block Basement | Wall Suction | 35 | 11 | 68 |
| 20 | Block basement & paved crawl space | Sub-slab & wall suction, & suction on interior drain tiles in crawl space | 282 | 4 | 99 |
| 21 | Block Basement | Sub-slab suction | 111 | 3 | 97 |
| 22 | Poured concrete basement & slab on grade | Sub-slab suction (basement & slab) | 34 | 9 | 74 |

[a]Heat recovery ventilator

(continued)

Table 16.2:  (continued)

| House No. | Substructure Type | Final Mitigation | Mean Radon Levels, pCi/L | | Reduction, % |
|---|---|---|---|---|---|
| | | | Before | After | |
| 23 | Poured concrete basement & slab on grade | Sub-slab suction (basement & slab) | 95 | 3 | 97 |
| 24 | Poured concrete basement | Sub-slab suction | 44 | 3 | 93 |
| 25 | Poured concrete basement | Sub-slab suction | 148 | 8 | 93 |
| 26 | Block Basement | Drain tile suction (exterior) | 89 | 1 | 99 |
| 27 | Block Basement | Drain tile suction (exterior) | 42 | 3 | 93 |
| 28 | Block Basement | HRV | 16 | 10 | 38 |
| 29 | Block basement & unpaved crawl space | Drain tile suction (interior, sump) & crawl space liner/vent | 47 | 2 | 96 |
| 30 | Block Basement | Carbon filter on well water | 29 | 5 | 83 |
| 31 | Block Basement | Sub-Slab suction | 485 | 4 | 99 |
| 32 | Block Basement | Sub-Slab suction | 6 | 1 | 80 |
| 33 | Poured concrete basement | Sub-Slab suction | 84 | 5 | 94 |
| 34 | Poured concrete basement | Sub-Slab suction | 696 | 5 | 99 |
| 35 | Poured concrete basement | Sub-Slab suction | 164 | 1 | 99 |
| 36 | Poured concrete basement & slab on grade | Sub-slab suction (basement & slab) | 142 | 2 | 99 |
| 37 | Poured concrete basement & slab on grade | Sub-slab suction (basement only) | 19 | 1 | 97 |
| 38 | Block Basement | Sub-Slab suction | 375 | 5 | 99 |
| 39 | Block Basement | Sub-Slab suction | 24 | 2 | 93 |
| 40 | Poured concrete basement | Sub-Slab suction | 113 | 3 | 97 |

Source:  Ref. 16.15

nized problem since the late 1950's (Ref. 16.8). Here the greatest levels of radon have been found in water pumped from igneous plutons, which contain water with a median radon level of 6550 pCi per liter. Water from granites containing the micas muscovite and biotite have especially high levels of radon. The Maine Department of Human Services suggested limit for radon in water is 20,000 pCi per liter. The U.S. EPA is presently considering a maximum concentration limit for radon in water in the range of 100 to 40,000 pCi per liter. The 40,000 pCi/liter concentration is estimated to be the equivalent of 4 pCi/liter in air. Radon can be released to the air during cooking (boiling water) or when aerated through a shower head, faucet or flushing toilet. This release of radon to the indoor atmosphere, rather than direct ingestion, is the primary reason we are concerned about high levels of radon in incoming potable water.

## MEASURING RADON IN WATER

There are at present 3 methods to measure radon concentrations in water. They are:
1)   The emanation-Lucas cell method
2)   The liquid scintillation method
3)   The alpha track detector (ATD) method
Of these 3 methods the most frequently used is liquid scintillation.

## REMOVAL OF RADON FROM WATER SUPPLIES

Radon-222 can be effectively removed from potable water supplies and ground water by either aeration or filtering through granulated activated carbon (GAC).

Some recent studies dramatize the advantages and disadvantages of these two removal technologies. Kinner, et. al., for example (Ref. 16.10) in their study of two small flow (5,000-15,000 GPD), mobile home community water supply systems in New Hampshire made the following observation:
"Though the GAC units successfully removed radon from the community's drinking water, several areas of concern have developed. Gamma exposure rates rose rapidly over the first several days of op-

eration.    Background rates in the community range from 0.04-0.075 mR/hr.    The
rates at the surface of the first filter
have remained in the range of 41-46 mR/hr
at the top and middle and 24-34 mR/hr at
the bottom.    Rates for the second filter
have remained at 10-19 mR/hr (top), 9-14
mR/hr (middle) and 6-11 mR/hr (bottom).
Though the exposure rates decrease sharply
with increasing distance from the filters,
they are still high when compared with
background levels.    At a location of 5
feet from the first filter the rate was 4-
5 mR/hr.    To date, personnel have not been
exposed to significant levels of gamma
radiation.    Special design considerations
may be required to insure that operators
and residents are not exposed to increased
health and safety problems as a result of
gamma emissions from the GAC.

The affinity of the GAC for uranium
has caused concern over the disposal of
the carbon.    Depending on a state's specific standards for disposal of radioactive wastes, this could lead to a significant problem in the final disposal of the
GAC.

Theoretically, the units should
operate for very long periods without need
for disposal because radon decays so
rapidly, constantly allowing the carbon to
regenerate naturally.    In cases where the
carbon must be replaced due to fouling or
clogging by iron, manganese, particulate
TOC or microorganisms, disposal may be a
concern."

Lowry and Lowry (Ref. 16.11) in their research
on radon removal for household and other small community water supplies concluded:

"It has been previously documented that
aeration is a feasible method for $^{222}$Rn
removal in point of entry applications.
However, a number of factors have kept it
from becoming as popular as GAC treatment.
These include:

1. Aeration is performed at atmospheric pressure and, therefore, requires re-pressurization of the water supply.

2. The initial cost of aeration systems designed for $^{222}$Rn removal is relatively high, partly due to the re-pressurization requirement. The installed cost ranges from $2,500 to over $4,000, compared to approximately $650 to $1,200 for GAC.

3. Several of the aeration methods are limited in their removal, as compared to GAC. A novel but costly spray aeration system was developed by the Maine Department of Human Services, Division of Health Engineering and six units are operating in the field. These units achieve 90 to 95 percent removal but may be limited to wells containing only 10,000 to 20,000 pCi/L if, for example, the future USEPA MCL for $^{222}$Rn is set at 1,000 pCi/L. While the MCL would apply only to public water supplies, the real $^{222}$Rn issue is in private household supplies for which lending institutions are already requiring removal in many cases. These institutions tend to use MCL's as guidance for private supplies.

    Packed-tower aeration systems are confined by the ceiling height in the cellar or living area and therefore are limited to about 90 percent removal. A multi-staged diffused bubble aeration system developed for organics removal has been tested on a supply that contains 250,000 pCi/L and removed $^{222}$Rn to below detection for virtually 100 percent removal; however, even a less expensive version designed specifically for $^{222}$Rn removal would cost approximately two to three times a GAC unit.

4.      Aeration methods require significant
        O&M compared to GAC, which increases
        the cost differential over the long
        term."

Finally, Dixon and Lee (Ref. 16.12) of the
American Water Work Service Co. in their 1986 and 1987
survey of larger public water supply systems conclude:

"GAC can effectively reduce radon concentrations in drinking water supplies to
very low levels, however, the amount of
contact time within the carbon bed,
required to do so would be prohibitive to
many water utilities from an operational
and economical standpoint."

By their calculations a 99 percent reduction in
radon concentration would require 130 minutes of
detention or contact time with the GAC. They are
therefore more optimistic about aeration and further
conclude that:

"Aeration is very effective in the removal
of radon from drinking water. Packed
tower aerators achieved greater than 95%
reduction in radon concentrations and conventional tray aerators achieved greater
than 75% reduction in radon concentrations."

For the moment at least it would appear that GAC
is the preferred system for removal of radon from
individual households and small community supplies.
Aeration, on the other hand, seems to be the preferred
technology for the larger public water supply systems.

**PREVENTATIVE PLANNING**

Radon is but another environmental problem that
can best be handled through the process of preventative planning. State, county and local land planners,
therefore, can play a major role in bringing this
problem under fairly rapid control. The following
systematic approach is recommended:

1)      First those areas having a high probabil-
        ity that they contain radon emitters must
        be delineated. This can be accomplished
        by using existing geological survey maps
        and other geological studies supplemented

with the collection of new soil and rock core samples as necessary.

2)    Once these potential high risk areas have been delineated, radon screening tests of homes and potable water supply wells already constructed in these areas should be undertaken. This information will help confirm the accuracy of the original delineations.

3)    Lastly these delineated areas should be incorporated into a map and used during the land subdivision approval process. All dwellings or commercial use buildings proposed to be constructed in these areas would then be required to have a radon prevention or mitigation system incorporated into their design.

The federal Indoor Radon Abatement Act of 1988 will be of some assistance in regard to the above recommended procedures. The bill authorizes $10 million per year in state grants for fiscal years 1989 through 1991 to assist states in establishing radon programs, conducting surveys and conducting demonstration mitigation projects. The act also provides the following:

1)    $3 million to provide technical assistance grants to the states.

2)    $1 million for the U.S. EPA to conduct a radon study in schools.

3)    $500,000 for diagnostic and remedial activities in schools.

4)    $1.5 million to establish proficiency programs for radon testing and mitigation companies.

5)    $1 million for grants to universities to establish regional radon training centers.

Finally, the act also requires federal departments and agencies to conduct radon studies in their buildings and mandates the U.S. EPA to develop model construction standards. Planners are advised to keep abreast of all reports resulting from this grant program.

## RADON SCREENING STUDIES OF SIGNIFICANCE

As of November, 1988 the U.S. EPA had tested indoor radon levels in homes in 17 states. Based on these surveys the agency predicts that over 3 million houses in these states will have indoor radon levels greater than 4 pCi per liter (Ref. 16.14).  The 17 states surveyed so far include:  Alabama, Arizona, Colorado, Connecticut, Indiana, Kansas, Kentucky, Massachusets, Michigan, Minnesota, Missouri, North Dakota, Pennsylvania,   Rhode  Island, Tennessee, Wisconsin and Wyoming.  Homes in the following 7 additional states will be surveyed in 1989:  Alaska, Iowa, Maine, New Mexico, Ohio, Vermont and West Virginia. All of these surveys are done in the winter months when homes are less vented in order to obtain measurements when radon levels should be at their highest.

One state not included in this list but certainly deserving of some attention is the state of New Jersey.  With large portions of the northern part of the state underlain by the notorious, uranium rich, granite formation known as the Reading Prong, the state has already produced the highest levels of indoor atmospheric radon ever recorded (3500 pCi/liter). This is even higher than the levels found in Stanley Watras's home.  Fortunately, the state's Department of Environmental Protection has already conducted a statewide radon screening program.  Based on this screening study the N.J. DEP estimates that 1.9 million dwellings are at risk and in need of radon testing as soon as possible.  Table 16.3 shows the average household radon readings for each of New Jersey's 21 counties.

Table 16.3:  Average Indoor Atmospheric Radon Levels in New Jersey
by County[a]

| | COUNTY | Measurement (in pCi/liter) |
|---|---|---|
| 1. | Atlantic | 0.75 |
| 2. | Bergen | 1.85 |
| 3. | Burlington | 1.80 |
| 4. | Camden | 2.27 |
| 5. | Cape May | 1.43 |
| 6. | Cumberland | 1.79 |
| 7. | Essex | 1.23 |
| 8. | Glouchester | 3.12 |
| 9. | Hudson | 2.47 |
| 10. | Hunterdon | 6.88 |
| 11. | Mercer | 4.46 |
| 12. | Middlesex | 2.16 |
| 13. | Monmouth | 2.68 |
| 14. | Morris | 5.13 |
| 15. | Ocean | 0.95 |
| 16. | Passaic | 3.67 |
| 17. | Salem | 2.49 |
| 18. | Somerset | 5.20 |
| 19. | Sussex | 6.47 |
| 20. | Union | 2.32 |
| 21. | Warren | 11.83 |

(a) Based on 6,000 homes surveyed statewide

Source:  Adapted from Ref. 16.13

**FOOTNOTES**

**CHAPTER 16**

(1)   Radon can be found in high concentrations in soil and rocks containing uranium.   These include granite, shale, phosphorite sediments, phosphate and pitchblende.

(2)   One curie of radiation is equivalent to that level of radiation emitted from one gram of radium-226, that is 3.700 x $10^{10}$ disintegrations per second.

(3)   When radon and its progeny decay, alpha radiation is emitted.   These particles are positively charged, high energy units consisting of two protons and two neutrons.   They move relatively slowly and travel only a short distance.   They can be blocked by a sheet of paper over even a layer of dead skin cells on the human body. However, if an alpha particle is inhaled it can disrupt the electrical balance of atoms in the lung cells causing disorganization and breakdown.

## REFERENCES

## CHAPTER 16

16.1   United States Environmental Protection Agency, "Radon Reduction Techniques for Detached Houses, Technical Guidance", EPA/625/5-86/019, June, 1986.

16.2   United States Environmental Protection Agency and U.S. Department of Health and Human Services, "A Citizen's Guide to Radon--What It is and What to do About It", OPA-86-004, August, 1986.

16.3   United States Environmental Protection Agency, "Radon Reduction Methods--A Home-owner's Guide", OPA-86-005, August, 1986.

16.4   American Water Works Association, "Radionuclides in Drinking Water-AWWA Seminar Proceedings", American Water Works Assoc., Colorado, 1987.

16.5   Graves, Barbara, "Radon in Ground Water," Lewis Publishers, Inc., Michigan, 1987.

16.6   Cobb, Charles E. Jr., "Living With Radiation", National Geographic, Vol. 175, No. 4, April, 1989.

16.7   National Academy of Sciences, "Indoor Pollutants", National Academy Press, Washington, D.C., 1981.

16.8   United States Geological Survey, "National Water Summary--1986", Water Supply, 2325, 1988.

16.9   Association of New Jersey Environmental Commissions, "Chemicals in the News", New Jersey Hazardous Waste News, Nov./Dec., 1985.

16.10  Kinner, N.E. and C.E. Lessard, G.S. Schell, et. al., "Radon Removal from Small Community Public Water Supplies Using Activated Carbon and Low Technology/Low Cost Techniques", AWWA Seminar Proceedings, American Water Works Association, Colorado, 1987.

16.11  Lowry, Jerry D. and Sylvia B. Lowry, "Modeling Point-of-Entry Radon Removal by GAC", AWWA Seminar Proceedings, American Water Works Association, Colorado, 1987.

16.12    Dixon, Kevin L. and Ramon G. Lee, "Radon Survey of the American Water Works System", NWWA Conference Proceedings, Lewis Publishers, Michigan, 1987.

16.13    Johnson, Tom, "State Releases Results of Home Radon Study", Newark Star Ledger, September 11, 1987, page 1.

16.14    Pollution Engineering, "EPA and Assistant Surgeon General Call For Radon Home Testing", November, 1988, p. 64.

16.15    United States Environmental Protection Agency, Arthur G. Scott, "Installation and Testing of Indoor Radon Reduction Techniques in 40 Eastern Pennsylvania Houses", EPA/600/58-88/002, Feb. 1988.

# Glossary

**Acre feet** - An engineering term used to denote a volume 1 acre in area and 1 foot in depth containing approximately 326,000 gallons.

**Absorption** - The process by which one substance is taken into and included within another substance, as the absorption of water by soil or nutrients by plants.

**Adsorption** - The increased concentration of molecules or ions at a surface, including exchangeable cations and anions on soil particles.

**Aerosol** - A suspension of fine, solid or liquid particles in air or gas.

**Aggrade** - The alteration of a stream or beach caused by the deposition of sediment or sand.

**Ammonium nitrogen** - A tenuous form of nitrogen ($NH_4^+$) produced by the interaction of ammonia ($NH_3$) with water as in the equation: $NH_3 + H_2O \rightleftharpoons NH_4^+ + OH^-$. At pH levels above 7, the equilibrium is displaced to the left; at levels below a pH of 7 the ammonium ion is predominant.

**Anaerobe** - A microorganism that can live or grow where there is no free oxygen. Anaerobes get oxygen by the decomposition of compounds containing it.

**Anoxic** - A condition in which there is total deprivation of oxygen.

**Application rate** - The rate at which a liquid is dosed to the land usually as inches/hour or feet/year.

**Aquifer** - A body of permeable rock or soil in which water can be stored temporarily as it moves downgradient under the influence of gravity.

**Berm** - A rounded mound generally composed of soil.

**Biodegradable** - Capable of being readily decomposed by biological means especially by bacteria.

**B.O.D.** - Acronym for Biochemical Oxygen Demand which is a measurement to determine the approximate quantity of oxygen necessary to biologically stabilize the organic matter present in a unit volume of water or wastewater. As the amount of organic waste increases, more

494

oxygen is required, resulting in a higher B.O.D.

**Brackish** - A mixture of fresh and salt water generally found in estuaries.

**CAA** - An acronym for the federal 1977 Clean Air Act and amendments.

**CAFRA** - Refers to New Jersey's Coastal Area Facilities Review Act of 1973.

**Capillarity** - The property of exerting or having capillary action--a force that is the result of adhesion, cohesion and surface tension in liquids which are in contact with solids.

**CBRA** - An acronym for the 1982 federal Coastal Barrier Resources Act.

**CERCLA** - An acronym for the 1980 federal Comprehensive Environmental Response, Compensation and Liability Act and subsequent amendments.

**C.O.D.** - An acronym for Chemical Oxygen Demand--a test similar to the B.O.D. measurement. However, here the oxygen equivalent of the organic matter is measured by using a strong chemical oxidizing agent in an acidic medium. Because more compounds can be chemically oxidized than can be biologically oxidized the C.O.D. of waste is generally higher than the B.O.D.

**Condensate** - A product of condensing as in the reduction of a gas to a liquid.

**CSO** - An acronym for Combined Sewer Outfall, a term used to describe the subsurface collection system designed to carry both sanitary wastewater and stormwater.

**CWA** - Refers to the 1972 federal Clean Water Act and subsequent amendments.

**CZMA** - The federal Coastal Zone Management Act of 1972.

**Decibel** - A numerical expression of the relative loudness of sound.

**Denitrification** - The biochemical reduction of nitrate or nitrite nitrogen to gaseous, molecular nitrogen or an oxide of nitrogen.

**Design storm** - A statistically developed storm of a certain return frequency (say once every 100 years) used to estimate the amount of rainfall that would occur for a certain duration.

**Disinfection** - The process of destroying harmful bacteria, viruses and other similar microorganisms.

**Dredged spoils** - Material removed from the bottom of the ocean or in estuaries during the process of channel widening or deepening. Often found to contain high concentrations of hazardous materials.

**E.I.S.** - Acronym for an Environmental Impact Statement.

**Electrical conductivity** - The ability of a material to transmit electrical current.

**E.S.A.** - Acronym for the federal Endangered Species Act of 1973.

**Eutrophication** - The addition of excess nutrients (i.e. nitrogen and phosphorus) and organic matter to waterbodies that accelerates their normal process of aging. Such waterbodies are noted for their thick plant growth, lack of oxygen in summer and warmer temperatures.

**Evapotranspiration** - The combined loss of water from a given area and during a specified period of time, by evaporation from the soil surface, snow or intercepted precipitation and by the transpiration and building of tissue by plants.

**Fecal coliform bacteria** - Any of a number of types of bacteria of the genus Escherichia coli found in the intestinal tract of warm-blooded animals and excreted with feces.

**Fixation** - A combination of physical and chemical mechanisms in the soil that act to retain wastewater constituents within the soil, including adsorption, precipitation and ion exchange.

**Flocculant** - A collection of discrete particles that have formed together to become one mass.

**Floodplain** - The relatively flat area adjoining the channel of a natural stream which has been or may be hereafter covered by floodwater.

**Floodway** - The part of the floodplain that carries the major portion of the flood flow. Water depths are greater and velocities are higher and consequently it is considered a high energy zone.

**Flood fringe area** - The part of the floodplain that adjoins the floodway and is inundated to a lesser degree. Velocities are not as high as in the floodway and consequently this area is considered a lower energy zone.

**Foliation** - A parallel or nearly parallel structure in metamorphic rocks along which the rock tends to break into flakes or thin slabs.

**FONSI** - Acronym for Finding Of No Significant Impact. Used during evaluation of environmental impacts especially for highway projects.

**Fugitive dust** - Dust released to the atmosphere by the disturbance of a dry soil surface or any other dry, finely divided material such as compost.

**Geological mottling** - This is mottling or discoloration caused by other than soil saturation. Some sandy soils, for instance, have uniform grey colors because there are no surface coatings on the sand grains.

**Geology** - The science dealing with the physical nature and history of the earth, including the structure and development of its crust, the composition of its interior and individual rock types.

**Ground water contour map** - A map depicting the surface contours of the top of the ground water table with lines connecting points of equal elevation. This map provides information as to which way the ground water table is sloping.

**HHW** - Household hazardous waste.

**Homeostasis** - Internal stability brought about by coordinated responses that automatically compensate for environmental changes.

**HSWA** - The 1984 federal Hazardous and Solid Waste Amendments to RCRA.

**Hydrated** - Formed by the chemical combination with water.

**Hydrology** - The science dealing with the waters of the earth and their distribution and movement.

**Hypoxia** - A condition in which the level of oxygen has been reduced.

**Infiltration** - A term used to describe the flow of ground water through cracks or leaks or seepage into subsurface pipelines.

**Induration** - The process of hardening.

**Inflow** - The flow of surface water directly into a subsurface pipeline either through an opened manhole or other large void.

**LCS** - Acronym for leachate collection system.

**LEL** - Acronym for lower explosive limit. See footnote 23 of Chapter 11.

**Lithosphere** - The crust and upper part of the Earth's mantle.

**LLW** - Low level radioactive waste.

**MCL** - Maximum contaminant level as used in the federal Safe Drinking Water Act.

**Mesophilic bacteria** - Bacteria whose optimum growth temperature falls in the range of $20^{\circ}$C to $40^{\circ}$C.

**Metamorphic rock** - A rock formed within the Earth's crust by the transformation of pre-existing rock as a result of high temperature, high pressure or both.

**Methane** - A colorless, odorless, flammable gas ($CH_4$) usually formed by the decomposition of vegetable matter.

**Methemoglobinemia** - An illness in which excess nitrates, usually from drinking water, displace oxygen carried by the blood's hemoglobin. Particularly dangerous to infants and young children.

**MPRSA** - Acronym for the the federal Marine Protection Research and Sanctuaries of 1972.

**MSWLF** - Denotes a municipal solid waste landfill.

**NEMP** - The Northeast Monitoring Program established in 1979 by the National Oceanic and Atmospheric Administration (NOAA) to monitor coastal and estuarine waters from the Gulf of Maine to North Carolina.

**NEPA** - Acronym for the federal National Environmental Policy Act of 1970.

**Nitrates** - A stable form of nitrogen denoted by the symbol $NO_3$.

**Nitrified effluent** - An effluent in which most nitrogen compounds have been converted to their stable nitrate form.

**Nitrite nitrogen** - An intermediate and generally short-lived form of nitrogen formed as ammonia begins to breakdown to nitrate nitrogen. Designated by the symbol $NO_2$.

**NOAA -**    Acronym for the federal National Oceanic and Atmospheric Administration.

**NOPPA -**    Acronym for the federal National Ocean Pollution Planning Act of 1978.

**Oxidant -** An oxidizing agent.

**PAH -**    Acronym for polynuclear aromatic hydro-carbons.

**PAN -**    Acronym for chemical compound peroxyacetyl nitrate formed by the interaction of nitrogen oxides and sunlight.

**Parasite -** A plant or animal that lives on or in another organism and from which it derives sustenance or protection without benefitting its host.

**Pathogen -** Any microorganism or virus that can cause disease.

**PBN -**    Acronym for chemical compound peroxybenzoyl nitrate formed by the interaction of nitrogen oxides and sunlight.

**PCB's -**    Polychlorinated biphenyls formerly used exten-sively as heat transfer liquids especially in transformers.   Now banned in the United States.

**Pelagic -** Of the open sea.

**pH -**    A measurement of the degree of acidity or alkalinity of a solution. Expressed as the logarithm of the reciprocal of the hydrogen ion concentration in gram equivalents per liter of solution.   pH values from 0 to 7 indicate acidity.   pH values from 7 to 14 indicate alkalinity.

**Plastic limit -** The moisture content at which a soil changes from a semi-solid to a plastic state.

**POTW -**    Acronym for a publicly owned treatment works.

**Putrescible -** Susceptible to decay and decomposition accompanied by disagreeable or foul odors.

**Pycnocline -** That portion of the water column of the ocean where density increases rapidly with depth.

**RCRA -**    Denotes the federal Resource Conservation and Recovery Act of 1976 and subsequent amendments.

**Rip-rap -** Broken rock of various sizes placed along waterways or other areas to retard soil erosion.

**Riser pipe** - A vertical pipe placed over the horizontal outlet pipe of a sediment basin to slow the outflow of storm water and in doing so trap silt and sediment.

**Sanitary sewage** - Wastewater discharged from residences and from commercial and institutional sources. Also called domestic wastewater.

**Sanitary sewer** - A system of subsurface pipes used to convey domestic and industrial wastewater.

**Saprophytes** - Organisms that live on dead organic matter.

**SAV** - Submerged aquatic vegetation.

**SDWA** - Acronym for the federal Safe Drinking Water Act of 1974 and amendments.

**Seasonal high water table** - The highest elevation during some part of the year to which the water table rises toward the surface of the land.

**Seasonal high water table mottling** - Zones or patches of grey in the soil profile produced by the bacterial oxidation and removal of iron and manganese compounds.

**Soil log** - An excavated pit used to examine the various horizons of a soil profile.

**Specific gravity** - The ratio of the weight or mass of a given volume of liquid or solid to water.

**Spirochetes** - Spiral or corkscrew shaped bacteria that are able to flex and wriggle while moving about.

**SQG** - A small quantity generator of hazardous waste producing more than 100 kilograms of hazardous waste but less than 1000 kilograms in a calendar month.

**STORET** - This is a computerized data base maintained by the U.S. EPA for the storage and retrival of parametric data relating to the quality of waterways of the United States.

**Storm drain** - Subsurface collection pipes used to convey storm water runoff.

**Terrestrial** - Growing or living on the land or in the soil.

**TDS** - An acronym for total dissolved solids which are the organic and inorganic molecules and ions that are present in true solution in water.

**Thermophilic bacteria** - These are bacteria whose optimum growth takes place above $45^{\circ}$ or $50^{\circ}$C. Some may even grow above $65^{\circ}$C.

**TOC** –      An acronym for total organic carbon. This is a measurement of the organic matter present in water. The measured TOC values will usually be slightly less than the actual amount present in the sample.

**TSCA** –     Acronym for the federal Toxic Substances Control Act of 1976 and amendments.

**Turbidity** – This is a measure of the light-transmitting properties of water. It is used to indicate the quality of waste discharges and natural waters with respect to colloidal matter.

**Venturi** – A short tube with a constricted, throat-like passage that increases the velocity and lowers the pressure of a fluid or gas that passes through it.

**Voltage** – The difference in electrical potential, measured in volts.

**Watershed** – The area drained by a river or river system.

# Appendix 1

# Part A

# SEVEN TIER GROWTH MANAGEMENT SYSTEM FOR NEW JERSEY'S STATE DEVELOPMENT AND REDEVELOPMENT PLAN

GROWTH AREAS

**Redeveloping Cities and Suburbs--Tier One:** The primary goal here is revitalization. Decline in these locations needs to be stemmed. Public and private investment should be directed to revitalize downtowns and main streets, stabilize neighborhoods, and improve infrastructure systems.

**Stable Cities and Suburbs--Tier Two:** The primary goal here is to sustain some growth while preserving community character and the quality of life. Infill development and redevelopment need to be prudently balanced with infrastructure and public service requirements.

**Suburban and Rural Towns--Tier Three:** The primary goal here is to concentrate development which would otherwise spread into surrounding suburban, rural and environmentally sensitive areas. This can be accomplished through infill, redevelopment, and fringe development that augments residential and employment choices while preserving the rural character within limited growth areas.

**Suburbanizing Areas--Tier Four:** The primary goal here is to develop corridor centers and residential uses along with related commercial activities while optimizing use of existing or planned infrastructure. In tis way, these areas will absorb projected growth thereby relieving some development pressures on nearby rural, agricultural, and environmentally sensitive areas.

LIMITED GROWTH AREAS

**Future Suburbanizing Areas--Tier Five:** The primary goal here is to manage growth within rural, sparsely developed areas which do not yet contain adequate infrastructure to support intensive development. It is likely that these areas will be designated growth areas by the year 2010. Consequently, the general land development pattern will be at low intensities with development concentrated into corridor centers, villages, and rural clusters to allow for efficient land use and infrastructure extensions in the future.

**Agricultural Areas--Tier 6:** The primary goal here is to ensure that new development is supportive of agriculture. These areas are currently farmed or have a high potential to support agricultural activity. Within this tier, non-agricultural development is encouraged to concentrate into corridor centers, towns, and villages. Development supporting or related to agriculture is encouraged.

**Environmentally Sensitive Areas--Tier 7:** The primary goal here is to create a development pattern characterized by low-density residential, low-intensity recreation, and undeveloped open space compatible with the natural resources found in these locations. This tier includes pristine watersheds, reservoir watersheds and habitats of endangered or threatened species. Protecting such resources requires maintaining low intensities of development. Moreover, the infrastructure and services needed to support intensive development are generally absent from and should not be extended into these areas.

Source:  New Jersey Draft Preliminary State Development and Redevelopment Plan, N.J. State Planning Commission, January, 1988.

STATE OF NEW JERSEY "ENVIRONMENTAL CLEAN-UP
RESPONSIBILITY ACT"

[OFFICIAL COPY REPRINT]
ASSEMBLY COMMITTEE SUBSTITUTE FOR
## ASSEMBLY, No. 1231

# STATE OF NEW JERSEY

ADOPTED MAY 26, 1983

An Act concerning the proper closure of certain industrial establishments *[and]* *, providing penalties for improper closure
of these establishments, amending P. L. 1976, c. 141,* supplementing Title 13 of the Revised Statutes*, and making an
appropriation*.

1   Be it enacted by the Senate and General Assembly of the State
2   of New Jersey:
1     1. This act shall be known and may be cited as the "Environ-
2   mental Cleanup Responsibility Act."
1     2. The Legislature finds and declares that the generation,
2   handling, storage and disposal of hazardous substances and wastes
3   poses an inherent danger of exposing the citizens, property and
4   natural resources of this State to substantial risk of harm or
5   degradation; that the closing of operations and the transfer of
6   real property utilized for the generation, handling, storage and
7   disposal of hazardous substances and waste should be conducted
8   in a rational and orderly way, so as to mitigate potential risks;
9   and that it is necessary to impose a precondition on any closure or
10  transfer of these operations by requiring the adequate preparation
11  and implementation of acceptable cleanup procedures therefor.
1     3. As used in this act:
2     a. "Cleanup plan" means a plan for the cleanup of industrial
3   establishments, approved by the department, which may include a
4   description of the location, types and quantities of hazardous sub-
5   stances and wastes that will remain on the premises; a description
6   of the types and location of storage vessels, surface impoundments,
7   or secured landfills containing hazardous substances and wastes;

EXPLANATION—Matter enclosed in bold-faced brackets [thus] in the above bill
is not enacted and is intended to be omitted in the law.
Matter printed in italics thus is new matter.
Matter enclosed in asterisks or stars has been adopted as follows:
*—Senate committee amendments adopted June 23, 1983.

2

8   recommendations regarding the most practicable method of clean-
9   up; and a cost estimate of the cleanup plan.
10   The department, upon a finding that the evaluation of a site for
11   cleanup purposes necessitates additional information, may require
12   graphic and narrative descriptions of geographic and hydrogeo-
13   logic characteristics of the industrial establishment and evaluation
14   of all residual soil, groundwater, and surface water contamination.
15   b. "Closing, terminating or transferring operations" means the
16   cessation of all operations which involve the generation, manu-
17   facture, refining, transportation, treatment, storage, handling or
18   disposal of hazardous substances and wastes, or any temporary
19   cessation for a period of not less than two years, or any other trans-
20   action or proceeding through which an industrial establishment be-
21   comes nonoperational for health or safety reasons or undergoes
22   change in ownership, except for corporate reorganization not sub-
23   stantially affecting thte ownership of the industrial establishment,
24   including but not limited to sale of stock in the form of a statutory
25   merger or consolidation, sale of the controlling share of the assets,
26   the conveyance of the real property, dissolution of corporate iden-
27   tity, financial reorganization and initiation of bankruptcy proceed-
28   ings;
29   c. "Department" means the Department of Environmental
30   Protection;
31   d. "Hazardous substances" means those elements and com-
32   pounds, including petroleum products, which are defined as such
33   by the department, after public hearing, and which shall be con-
34   sistent to the maximum extent possible with, and which shall
35   include, the list of hazardous substances adopted by the Environ-
36   mental Protection Agency pursuant to Section 311 of the "Federal
37   Water Pollution Control Act Amendments of 1972" (33 U. S. C.
38   § 1321) and the list of toxic pollutants designated by Congress or
39   the Environmental Protection Agency pursuant to Section 307 of
40   that act (33 U. S. C. § 1317); except that sewage and sewage sludge
41   shall not be considered as hazardous substances for the purposes
42   of this act;
43   e. "Hazardous waste" means any amount of any waste sub-
44   stances required to be reported to the Department of Environ-
45   mental Protection on the special waste manifest pursuant to
46   N. J. A. C. 7:26-7.4, or as otherwise provided by law.
47   f. "Industrial establishment" means any place of business en-
48   gaged in operations which involve the generation, manufacture,
49   refining, transportation, treatment, storage, handling, or disposal
50   of hazardous substances or wastes on-site, above or below ground,

3

51  having a Standard Industrial Classification number within 22–39
52  inclusive, 46–49 inclusive, 51 or 76 as designated in the Standard
53  Industrial Classifications manual prepared by the Office of Manage-
54  ment and Budget in the Executive Office of the President of the
55  United States. Those facilities or parts of facilities subject to
56  operational closure and post-closure maintenance requirements
57  pursuant to the "Solid Waste Management Act," P. L. 1970, c. 39
58  (C. 13:1E–1 et seq.), the "Major Hazardous Waste Facilities Siting
59  Act," P. L. 1981, c. 279 (C. 13:1E–49 et seq.) or the "Solid Waste
60  Disposal Act" (42 U. S. C. § 6901 et seq.), or any establishment
61  engaged in the production or distribution of agricultural com-
62  modities, shall not be considered industrial establishments for the
63  purposes of this act. The department may, pursuant to the "Ad-
64  ministrative Procedure Act," P. L. 1968, c. 410 (C. 52:14B–1
65  et seq.), exempt certain sub-groups *or classes of operations within
66  those sub-groups* within the Standard Industrial Classification
67  major group numbers listed in this subsection upon a finding that
68  the operation of the industrial establishment does not pose a risk
68A  to public health and safety.

69      g. "Negative declaration" means a written declaration, sub-
70  mitted by an industrial establishment and approved by the depart-
71  ment, that there has been no discharge of hazardous substances or
72  wastes on the site, or that any such discharge has been cleaned up
73  in accordance with procedures approved by the department, and
74  there remain no hazardous substances or wastes at the site of the
75  industrial establishment.

1      4. a. The owner or operator of an industrial establishment plan-
2  ning to close operations shall:
3      (1) Notify the department in writing, no more than five days
4  subsequent to public release, of its decision to close operations;
5      (2) Upon closing operations, or 60 days subsequent to public
6  release of its decision to close or transfer operations, whichever
7  is later, the owner or operator shall submit a negative declaration
8  or a copy of a cleanup plan to the department for approval and
9  a surety bond or other financial security for approval by the de-
10  partment guaranteeing performance of the cleanup in an amount
11  equal to the cost estimate for the cleanup plan.
12      b. The owner or operator of an industrial establishment plan-
13  ning to sell or transfer operations shall:
14      (1) Notify the department in writing within five days of the
15  execution of an agreement of sale or any option to purchase;
16      (2) Submit within *[30]* *60* days prior to transfer of title a
17  negative declaration to the department for approval, or within

4

18   *[30]* *60* days prior to transfer of title, attach a copy of any
19   cleanup plan to the contract or agreement of sale or any option to
20   purchase which may be entered into with respect to the transfer
21   of operations. In the event that any sale or transfer agreements or
22   options have been executed prior to the sumission of the plan to
23   the department, the cleanup plan shall be transmitted, by certified
24   mail, prior to the transfer of operations, to all parties to any trans-
25   action concerning the transfer of operations, including purchasers,
26   bankruptcy trustees, mortgagees, sureties, and financers;

27      (3) Obtain, upon approval of the cleanup plan by the depart-
28   ment, a surety bond or other financial security approved by the
29   department guaranteeing performance of the cleanup plan in an
30   amount equal to the cost estimate for the cleanup plan.

31      c. The cleanup plan and detoxification of the site shall be im-
32   plemented by the owner or operator, provided that the purchaser,
33   transferee, mortgagee or other party to the transfer may assume
34   that responsibility pursuant to the provisions of this act.

1      5. a. The department shall, pursuant to the "Adiminstrative
2    Procedure Act," P. L. 1968, c. 410 (C. 52:14B-1 et seq.), adopt
3    rules and regulations establishing: minimum standards for soil,
4    groundwater and surface water quality necessary for the detoxi-
5    fication of the site of an industrial establishment, including build-
6    ings and equipment, to ensure that the potential for harm to public
7    health and safety is minimized to the maximum extent practicable,
8    taking into consideration the location of the site and surrounding
9    ambient conditions; criteria necessary for the evaluation and ap-
10   proval of cleanup plans; a fee schedule, as necessary, reflecting
11   the actual costs associated with the review of negative declarations
12   and cleanup plans; and any other provisions or procedures neces-
13   sary to implement this act. *Until the minimum standards described
13A  herein are adopted, the department shall review, approve or dis-
13B  approve negative declarations and cleanup plans on a case by case
13C  basis.*

14      b. The department shall, within 45 days of submission, approve
15   the negative declaration, or inform the industrial establishment
16   that a cleanup plan shall be submitted.

17      c. The department shall, in accordance with the schedule con-
18   tained in an approved cleanup plan, inspect the premises to deter-
19   mine conformance with the minimum standards for soil, ground-
20   water and surface water quality and shall certify that the cleanup
21   plan has been executed and that the site has been detoxified.

1      6. a. The provisions of any law, rule or regulation to the con-
2    trary notwithstanding, the transferring of an industrial establish-

5

3 ment is contingent on the implementation of the provisions of this
4 act.

5    b. If the premises of the industrial establishment would be sub-
6 ject to substantially the same use by the purchaser, transferee,
7 mortgagee or other party to the transfer, and upon written certi-
8 fication thereto and approval by the department thereof, the im-
9 plementation of a cleanup plan and the detoxification of the site
10 may be deferred until the use changes or until the purchaser,
11 transferee, mortgagee or other party to the transfer closes, termi-
12 nates or transfers operations.

13    (1) Within 60 days of receiving notice of the sale or realty
14 transfer and the certification that the industrial establishment
15 would be subject to substantially the same use, the department
16 shall approve, conditionally approve, or deny the certification;

17    (2) Upon approval of the certification, the implementation of a
18 cleanup plan and detoxification of the site shall be deferred.

19    (3) Upon denial of the certification, the cleanup plan and de-
20 toxification of the site shall be implemented pursuant to the pro-
21 visions of this act.

22    c. The authority to defer implementation of the cleanup plan
23 set forth in subsection b. of this section shall not be construed to
24 limit, restrict, or prohibit the department from directing site
25 cleanup under any other statute, rule, or regulation, but shall be
26 solely applicable to the obligations of the owner or operator of an
27 industrial establishment, pursuant to the provisions of this act,
28 nor shall any other provisions of this act be construed to limit,
29 restrict, or prohibit the department from directing site cleanup
30 under any other statute, rule, or regulation.

1    7. No obligations imposed by this act shall constitute a lien or
2 claim which may be limited or discharged in a bankruptcy pro-
3 ceeding. All obligations imposed by this act shall constitute con-
4 tinuing regulatory obligations imposed by the State.

1    8. a. Failure of the transferor to comply with any of the pro-
2 visions of this act is grounds for voiding the sale or transfer of
3 an industrial establishment or any real property utilized in con-
4 nection therewith by the transferee entitles the transferee to re-
5 cover damages from the transferor, and renders the owner or oper-
6 ator of the industrial establishment strictly liable, without regard
7 to fault, for all cleanup and removal costs and for all direct and
8 indirect damages resulting from the failure to implement the
9 cleanup plan.

10    b. Failure to submit a negative declaration, or cleanup plan
11 pursuant to the provisions of section 4 of this act is grounds for

6

12 voiding the sale by the department.

13　c. Any person who knowingly gives or causes to be given any
14 false information or who fails to comply with the provisions of this
15 act is liable for a penalty of not more than $25,000.00 for each
16 offense. If the violation is of a continuing nature, each day during
17 which it continues shall constitute an additional and separate
18 offense. Penalties shall be collected in a civil action by a summary
19 proceeding under "the penalty enforcement law" (N. J. S. 2A:58-1
20 et seq.). Any officer or management official of an industrial estab-
21 lishment who knowingly directs or authorizes the violation of any
22 provisions of this act shall be personally liable for the penalties
23 established in this subsection.

1　*9. Section 16 of P. L. 1976, c. 141 (C. 58:10-23.11o) is amended
2 to read as follows:

3　16. Moneys in the New Jersey Spill Compensation Fund shall be
4 disbursed by the administrator for the following purposes and no
5 others:

6　(1) Costs incurred under section 7 of this act;

7　(2) Damages as defined in section 8 of this act;

8　(3) Such sums as may be necessary for research on the preven-
9 tion and the effects of spills of hazardous substances on the marine
10 environment and on the development of improved cleanup and
11 removal operations as may be appropriated by the Legislature;
12 provided, however, that such sums shall not exceed the amount of
13 interest which is credited to the fund;

14　(4) Such sums as may be necessary for the boards, general ad-
15 ministration of the fund, equipment and personnel costs of the
16 department and any other State agency related to the enforcement
17 of this act as may be appropriated by the Legislature;

18　(5) Such sums as may be appropriated by the Legislature for
19 research and demonstration programs concerning the causes and
20 abatement of ocean pollution; provided, however, that such sums
21 shall not exceed the amount of interest which is credited to the fund.

22　*(6) Such sums as may be requested by the commissioner, up to a
23 limit of $400,000.00 per year, to cover the costs associated with the
24 administration of the "Environmental Cleanup Responsibility Act,"
25 P. L. 198 , c.　(C.　) (now pending before the
26 Legislature as Assembly Committee Substitute for Assembly Bill
27 No. 1231 of 1982.).*

28　The Treasurer may invest and reinvest any moneys in said fund
29 in legal obligations of the United States, this State or any of its
30 political subdivisions. Any income or interest derived from such
31 investment shall be included in the fund.*

7

1    *10. There is appropriated to the Department of Environmental
2    Protection from the New Jersey Spill Compensation Fund created
3    pursuant to P. L. 1976, c. 141 (C. 58:10-23.11 et seq.) the sum of
4    $400,000.00.*

1    *[9.]* *11.* This act shall take effect immediately but shall re-
2    main inoperative for 120 days, but the department may take antici-
3    patory action by developing regulations prior to the effective date
4    of this act.

———————

# Part B

# MODEL STORMWATER MANAGEMENT ORDINANCE

TOWNSHIP OF CLINTON
HUNTERDON COUNTY, NEW JERSEY
SURFACE WATER MANAGEMENT
ORDINANCE

ARTICLE I
Intent, Purposes and Definitions

113-1.  General intent.
    The general intent of this chapter is to manage the increased rate and velocity of surface water runoff created by alterations in the ground cover and natural runoff patterns.

113-2.  Purposes.
    To protect the public health, safety and welfare of the citizens of Clinton Township and the surrounding communities, this chapter is deemed necessary and essential in order to:
A.  Maintain the adequacy of natural stream channels and prevent accelerated bank erosion by controlling the rate and velocity of runoff discharge to these watercourses so as to avoid increasing the frequency of the bank-full stage.
B.  Prevent degradation of the stream biota caused by excessive flushing and sedimentation.
C.  Prevent degradation of stream water quality due to impairment of the stream's biological function.
D.  Enhance the quality of nonpoint runoff by water retention measures.
E.  Preserve present adequacy of culverts and bridges by reducing artificially induced flood peaks.
F.  Reduce public expenditures for replacement or repair of public facilities resulting from artificially induced flood peaks.
G.  Prevent damages to life and property from flooding resulting from excessive rates and velocities of runoff.
H.  Prevent the degradation of property by enhancing the environmental character of the streams of the township.

113-3.  Definitions.
    As used in this chapter, the following terms shall have the meanings indicated:

APPLICANT--Any person, partnership, corporation or public agency requesting permission to engage in land disturbance activity, construction or development.

APPROVED PLAN--A plan to control surface water runoff, approved by the Municipal Planning Board.

CHANNEL--A watercourse with a definite bed and banks which confine and conduct continuously or intermittently flowing water.

CONSERVATION EASEMENT--An agreement or convenant, attached to deed, dedicating land to permanent open space and prohibiting all land or vegetation disturbance, each agreement to be entered into between the applicant and the municipality.

CRITICAL AREA--Any area which should not be disturbed by uses incompatible with the paramount public interest in the management of surface water runoff and attendant environmental damage.  Examples of critical impact areas include but are not limited to lakes, ponds, floodplains and flood hazard areas, designated stream corridors, steep slopes, highly erodible soils, swamps, marshes, bogs, identified aquifer recharge and discharge areas and heavily wooded areas.

DRAINAGEWAY--Any watercourse, trench, ditch, depression or other hollow space in the ground, natural or artificial, which collects or disperses surface water from land.

LAND DISTURBANCE--Any activity involving the clearing, grading, transporting or filling of land, or any other activity which alters topography or vegetative cover.

MAPS--Those listed by title on page 10 of the Regional Stormwater Management Study and Plan and maps of the Clinton Township Natural Resource Inventory series referenced by number, all available for reference in the Planning Board office.

MEADOW OF GOOD HYDROLOGIC CONDITION--As defined by sheet RTSC--Engineer 200, Sheet 1 of two (2) published by USDA-SCS February 1970.

NATURAL DRAINAGE FLOW--The topographical pattern or system of drainage of surface water runoff from a particular site, including the various drainageways and watercourses which carry surface water only during periods of heavy rains, storms or floods.

NONPOINT RUNOFF--Surface water entering a channel from no definable discharge source.

REGIONAL STORMWATER MANAGEMENT STUDY AND PLAN--The Regional Stormwater Management Study and Plan--Clinton Township, Lebanon Borough, Readington Township, published August 1974, sponsored by the Municipal Environmental Commission using local funds plus Ford Foundation grant and matching funds.

SEASONAL HIGH GROUNDWATER TABLE--As depicted on Map 4 of the Clinton Township Natural Resource Inventory series.

SURFACE WATER--All water produced by rain, flood, drainage, springs and seeps flowing over the land or contained within a natural or artificial watercourse.

SURFACE WATER MANAGEMENT PLAN--A plan consistent with the purposes and policies of this chapter which fully indicates necessary land treatment measures and techniques, including a schedule for implementation and maintenance.

SURFACE WATER RUNOFF--Any overland flow of water across the ground surface.

SURFACE WATER RUNOFF DAMAGE--All damage or harm to property values, land, vegetation and water supplies which results or is likely to result when the dispersion of surface water, typical of land in a meadow of good hydrologic condition, is increased in rate, velocity or quantity.   Such damage or harm includes but is not limited to flooding, soil erosion, siltation and other pollution of watercourses and diminished recharge of groundwater.

WATERCOURSE--All rivers, streams, brooks, waterways, lakes, ponds, marshes, swamps, bogs and other bodies of water, natural or artificial, public or private, which are contained within, flow through or border on Clinton Township or any portion thereof.

WATERSHED--An area of surface water runoff related to a point of concentration as shown on the map, Watersheds Overlay, of the Regional Stormwater Management Study and Plan or Map No. 9 of the Clinton Township Natural Resource Inventory series.

**ARTICLE II**
**Administration**

113-4.  Applicability.
A.  A surface water management plan shall be approved by the Planning Board prior to site plan approval, special exception, zoning variance, issuance of building permit, preliminary subdivision approval or any other land disturbance activity.
B.  No person shall engage in any land disturbance activity on any property within the township without having submitted a surface water runoff management plan, together with erosion and sediment control plans, to the Planning Board and obtained approval of such plan and a permit or written waiver of necessity from the Planning Board, except as exempt in 113-6 below.

**113-5. Waiver of plan.**
The Planning Board may waive the need for a surface water management plan upon recommendation of the Municipal Engineer, certifying there is no appreciable increase anticipated in rate or velocity of runoff based on plans submitted.

**113-6. Exemptions.**
The following are exempt from the provisions of this chapter:
A. Plans for development where the vegetative cover of the land will not be disturbed.
B. Agricultural use of lands when operated in accordance with farm conservation practices approved by the local Soil Conservation District.
C. General improvements to single lots containing existing single-family residences, such as but not limited to paving of existing driveways, landscaping improvements, construction of swimming pools which will not discharge water beyond the property containing the swimming pool, gardening.

**113-7. Planning Board approval.**
A. The surface water management plan or any major revision shall be approved by the Planning Board in the manner and form and according to the regulations hereafter set forth.
B. The Planning Board, on approving said surface water management plan, may impose lawful conditions or requirements designated or specified on or in connection therewith. These conditions and requirements shall be provided and maintained as a condition to the establishment, maintenance and continuance of any use or occupancy of any structure or land.

**113-8. Minor amendments.**
Minor revisions to a surface water management plan may be approved by the Municipal Engineer, who shall notify the Planning Board of the nature and reason for the change.

**113-9. Enforcement.**
If at any time the Municipal Engineer finds existing conditions not as stated in the applicant's approved plan, the Planning Board or its designated agent shall, by certified mail, return receipt requested, order cessation of all work and seek to enjoin the violation or take such steps looking to the enforcement of the plan as may be lawful.

ARTICLE III
Standards

**113-10. General standards.**
In the preparation of a surface water management plan, the following general principles shall be adhered to:
A. The rate and velocity of runoff from the site following completion of the planned development shall not exceed that which would prevail under total coverage in a meadow of good hydrologic condition, as defined by Soil Conservation Service standards, or previous cover, whichever produces the least amount of runoff.
B. Maximum use shall be made of presently existing surface water runoff control devices, mechanisms or areas, such as existing berms, terraces, grass waterways, favorable hydrologic soils, swamps, swales, watercourses, woodlands, floodplains, as well as any proposed retention structures.
C. Whenever practicable and when permitted by the Zoning Ordinance, cluster development should be used if it will reduce the total area of impervious surfaces and preserve the open space and topographic features critical to surface water management.
D. Evaluation shall be made of the nature of the subwatershed(s) of which the site is a part, the receiving stream channel capacities and point of concentration structure as shown on the Base Map Showing Roads, Streams, Culverts and Bridges, Clinton Township, and Appendix B of the Regional Stormwater Management Study and Plan.
E. Surface water runoff shall generally not be transferred from one watershed to another.
F. The plan shall coordinate with the soil erosion and sediment control plan and, where applicable, relate to other environmental protection ordinances in force.
G. To the greatest possible extent, the plan shall avoid the concentration of flow and shall provide for dissipation of velocities at all concentrated discharge points.

H. Reestablishing vegetative cover shall be in accordance with Standards and Specifications for Soil Erosion and Sediment Control in New Jersey, adopted by the Hunterdon County Soil Conservation District, latest edition.

I. Timing for the plan shall establish permanent surface water management measures prior to construction or other land disturbance, including seeding and establishing sod in grass waterways.

113-11. **Design standards.**

For engineering review by the Municipal Engineer, the following standards shall apply:

A. For calculating runoff and controls, either of the following methods may be used in computing runoff: Soil Conservation Service method under USDA or the Rational method.

| Criteria | SCS Method | Rational Method |
|---|---|---|
| Surface conditions | Meadow | Average cultivation or light growth R (0.20 to 0.40) |
| Collection system | 15-year storm | 15-year storm |
| Storage | 100-year storm | 100-year storm |
| Outlet discharge | 10-year storm | 10-year storm |
| Emergency spillway | 100-year storm | 100-year storm |
| Soil type | A,B,C,D as determined by Soil Survey, Hunterdon County, by USDA Soil Conservation Service, issued November 1974 | Loam |
| Maximum velocity at pipe outlets | 4 fps | 4 fps |
| Intensity | SCS method | Water policy rainfall curves |

B. All outfalls are to be designed in a manner to retard velocities at the outfall and provide stream channel protection.

C. When a natural drainage pattern is necessarily intercepted, as by a street, this shall be considered.

D. All structures and land treatment practices shall conform to Standards and Specifications for Soil Erosion and Sediment Control in New Jersey, latest edition, adopted by the Hunterdon County Soil Conservation District.

E. All water-carrying structures and/or retention areas shall be completed and stabilized prior to diversion of water to them.

F. Prior to developing a surface water management plan, there shall be an inventory of the site showing all existing natural and man-made drainage-related features as listed in 113-10B (berms, terraces, grass waterways, favorable hydrologic soils, poorly drained soils, swamps, swales, watercourses, woodlands, floodplains). These shall be incorporated in the plan to the greatest possible extent in accordance with their functional capability.

G. Drainageways and watercourses which normally carry or receive surface water runoff shall not be overloaded with increased runoff, sediment or other pollution resulting from disturbance of soil and vegetation or incident to development, constuction or other activity.

H. Due consideration shall be given to the relationship of the subject property to the natural or established drainage pattern of the subwatershed(s) of which it is a part as shown on the map, Watersheds Overlay, of the Regional Stormwater Management Study and Plan or Map No. 9 of the Clinton Township Natural Resource Inventory.

I. Surface water runoff controls shall be designed to assure that the land in question uses no more than its proportionate watershed share of the natural stream and culvert capacity as set forth in tables, Appendix C, Regional Stormwater Management Study and Plan, or as ascertained by field measurements.

J. Innovative surface water runoff control and recharge devices may be proposed, such as rooftop storage, dry wells, cisterns, roof drain infiltration trenches, provided they are accompanied by detailed engineering plans and performance capabilities.

## ARTICLE IV
### Data Required

**113-12. Submission of plan; contents.**

A. The applicant shall submit to the Planning Board a separate surface water management plan for any proposed lot, subdivision, cluster development, site plan review, special exception, zoning variance or any land disturbance activity except as exempted in 113-5 and 113-6.  This plan shall be coordinated with the soil erosion and sediment control plan required by the Soil Erosion and Sediment Control Ordinance.

B. Such plan shall contain:

  (1) Lot and block numbers of the site as shown on the current Tax Map of the township.

  (2) Name and address of the owner of the land.

  (3) Size of subwatershed and location of the site within the subwatershed(s) as shown on the map, Watersheds Overlay, of the Regional Stormwater Management Study and Plan or Map No. 9 of the Clinton Township Natural Resource Inventory.

  (4) Location, description and quantification of significant natural and man-made features on and surrounding the site, including topography, all impervious surfaces, soil and drainage characteristics, with particular attention to the location and description of presently existing surface water runoff control devices, mechanisms or areas, swamps, floodplains, swales, woods and vegetation, steep slopes and other features critical to the purposes of this chapter.

  (5) Size of nearest culvert or bridge downstream of discharge area, profiles and cross section of stream channel upstream of that structure as inventoried on the base Map Showing Roads, Streams, Culverts and Bridges, Clinton Township, and Appendix B of the Regional Stormwater Management Study and Plan, as well as profiles and cross sections of stream channel at all points of proposed surface water discharge from the site as required by the Municipal Engineer.

  (6) Location, description and quantification of proposed changes to the site, whether of a permanent or temporary nature, with particular attention to impervious surfaces and interception of presently dispersed flow which may impact upon the capacity of the soil, vegetative cover and drainageways to absorb, retard, contain or control surface water runoff.

  (7) Designation of critical or other areas to be left undisturbed shown in sufficient detail to be accurately marked on the land.

  (8) Computation of the total surface water runoff before, during and after the disturbance of land and/or construction of impervious surfaces.

  (9) Proposed measures for surface water management.

  (10) A schedule of the sequence of installation of the surface water management plan, related to the starting and completion dates of the project.

  (11) Proposed maintenance schedule for all surface water management structures, stipulating current maintenance, continued maintenance and responsibility for same.

  (12) All proposed revisions of required data as well as such additional data as the Planning Board may require.

## ARTICLE V
### Review and Approval

**113-13. Factors for consideration.**

    Surface water management plans shall be reviewed by the Planning Board with the advice and assistance of the Municipal Engineer, the Soil Conservation District and the Environmental Commission.  The Planning Board's consideration of applications shall be guided by but not limited to the following factors:

A. The suitability of the applicant's proposed surface water management measures, devices and planning techniques, whether involving on-site or off-site measures or some combination thereof in respect to the total surface water runoff, velocities and rates of discharge which the applicant's proposed construction or land disturbance may generate.

B. Existing topography, present vegetation and hydrologic soil factors as shown on the map, hydrologic Soils, USDA Soil Conservation Service, of the Regional Stormwater

Management Study and Plan or Map No. 7 of the Clinton Township Natural Resource Inventory series, subject to field verification.

C. Groundwater recharge and discharge areas and wet soils as shown on the map, Alluvial, Poorly Drained and Somewhat Poorly Drained Soils, USDA Soil Conservation Service, of the Regional Stormwater Management Study and Plan, subject to field verification.

D. Seasonal high groundwater table as shown on Map No. 4 of the Clinton Township Natural Resource Inventory series.

E. The design storm.

F. Natural drainage flow and pattern throughout the subwatershed(s) affected by the plan.

G. Land uses in both the immediate vicinity and the surrounding drainage region.

H. Any other applicable or relevant environmental and resource protection ordinances, statutes and regulations.

**113-14.  Requisites for approval.**

The Planning Board shall approve the surface water management plan and issue the necessary permit only after it determines that the proposed land disturbance or construction will generate no surface water runoff that will not be managed in accordance with the standards of this chapter.

**113-15.  Denial of permit.**

If the proposed land disturbance or construction will generate surface water runoff which will not be managed in accordance with the standards of this chapter and which will be detrimental to the public health, safety and general welfare in light of the paramount public interest in the prevention of conditions which may result in surface water runoff damage and environmental degradation, the Board shall deny the permit and shall clearly and concisely state the reasons.

### ARTICLE VI
### Implementation, Penalties and Appeals

**113-16.  Implementation.**

A. Limit of contract.  Critical impact areas and other areas to be left undisturbed shall be physically marked with survey stakes or protected with temporary snow fence prior to any land disturbance.

B. Timing.  The Planning Board shall require the construction and/or installation of surface water management improvements in accordance with the schedule of sequence of installation as approved.

C. Bonding.  The Planning Board shall provide for the posting of performance guaranties and maintenance bonds when necessary.

D. Inspection.

(1) The applicant shall bear full and final responsibility for the installation and construction of all required surface water runoff control measures according to the provisions of his approved plan and this chapter.  The Municipal Engineer shall inspect the site during its preparation and development and certify that all surface water management measures have been constructed in accordance with the provisions of the applicant's approved plan under this chapter.

(2) During the twelve (12) months subsequent to the date of completion, the Engineer periodically shall inspect the site to ascertain that the provisions of the applicant's approved plan are complied with, including limit of contract for areas to be left undisturbed.  The Engineer shall give the applicant, upon request, a certificate indicating the date on which the required measures were completed and/or accepted.

E. Review fees and inspection fees.  For the review and inspection required by this chapter, the fees that must be paid to the Planning Board Clerk are as follows:

(1) Review fee for a single lot is fifty dollars ($50.).

(2) Inspection fee for a single lot is twenty-five dollars ($25.).

(3) Where a minor subdivision is involved, the review fee is twenty-five dollars ($25.) per lot and the inspection fee is ten dollars ($10.) per lot.

(4) For major subdivisions and site plan review, the fees are included in the application fees set forth in the pertinent ordinances.

F. Maintenance.

    (1)  At the time of approval of the plan, responsibility for continued maintenance of surface water runoff control structures and measures shall be stipulated and properly recorded.

    (2)  The township shall retain the right to enter and make repairs and improvements where necessary to ensure that all control measures as well as areas dedicated to surface water retention or groundwater recharge are adequately maintained and preserved. The township may charge the property owner for the costs of these services if such maintenance is his responsibility.

**113-17. Violations and penalties.**

A. Any person violating any of the provisions of this chapter shall, upon conviction thereof, be subject to a fine of not more than five hundred dollars ($500.) or ninety (90) days in jail, or both, and any costs of correcting said violation.

B. Continued violation or violations not corrected within fourteen (14) days of receipt of written notice shall be subject to a fine of up to five hundred dollars ($500.) for each day of such continuing violation.

C. The Planning Board may extend the designated time of correction due to adverse weather conditions.

**113-18. Appeals.**

Appeals from decisions under this chapter may be made to the municipal governing body in writing within ten (10) days from the date of such decision. The appellant shall be entitled to a hearing before the municipal governing body within thirty (30) days from date of appeal.

Source: Township of Clinton, Hunterdon County, New Jersey

## MODEL ORDINANCES FOR:  STEEP SLOPES, SOIL EROSION AND
## SEDIMENT CONTROL, TREE PROTECTION

II

### STEEP SLOPES

Building upon and otherwise disturbing steep slopes is a major contributor to flooding at lower levels.  Until development arrives, steep land is usually covered with trees or brush. The soils are too thin and erodable to warrant other use. Steep slopes should be highly restricted or eliminated from gross acreage for lots.  Not doing so results in massive and continuing public and private costs for flood damage, erosion and stream destruction.

Note:

To regulate steep slopes separately from the zoning ordinance, Sections 2, 3 and 4 should be reworded as to headings.

If written as a separate ordinance, an appeals provision should be added.

Source: South Branch Watershed Assoc., "Model Environmental Ordinances", Clinton, N.J.

STEEP SLOPE ORDINANCE AS AMENDMENT TO
ZONING ORDINANCE

AN ORDINANCE TO AMEND AN ORDINANCE ENTITLED "THE ZONING ORDINANCE
OF THE _____(Township)_____ OF __(Municipality)__ " TO PROVIDE FOR THE
REGULATION OF DISTURBANCE OF SOIL AND THE CONSTRUCTION OF BUILDINGS
AND STRUCTURES ON CRITICAL SLOPE AREAS.

BE OR ORDAINED by the ___(governing body)___ of the ___(Township)___
of ___(Municipality)___ that the Ordinance entitled "The Zoning Ordinance of the
_____(Township)_____ of ____(Municipality)____ " adopted _____
shall be and is hereby amended as follows:

Section 1. Purpose and Policy. It is hereby found that soil disturbance and the construction of buildings and structures in certain areas of critical slopes within the (Township)
of (Municipality _____ will increase the rate of surface water run-off, soil erosion
and siltation and pollution of streams, as well as the danger of flooding, thereby having the
potential of endangering public and private property and life; that this condition is aggra-
vated by soil disturbance, construction and development on these slopes which creates an
additional hazard to the lives and property of those dwelling on the slopes and below them,
and that the most appropriate method of alleviating such conditions is through regulation of
such soil disturbance, construction and development. It is therefore determined that the
special and paramound public interest in these slopes justifies the regulation of property
located thereon as provided in this ordinance which is in the exercise of the police power
of the minicipality for the protection of the persons and property of its inhabitants and for
the preservation of the public health, safety and general welfare.

Section 2. Said Zoning Ordinance shall be and is hereby amended to add thereto a
new section be designated Section _____ to provide as follows.

" _____ Critical Sopes. Any slopes which have a grade of 12% (twelve
percent) or more identified as "D, E, and F" slopes on the Soil Conservation Service map entitled "Soil Limitations Map" prepared by U. S. Department of agriculture, Soil Conservation Service (New Jersey)."

Section 3. Said Zoning Ordinance shall be and is hereby amended to add to Section
_____ thereof a new paragraph as follows:

"The Soil Limitations Map is hereby adopted as supplement to the Zoning
Map of ___(Municipality)___ . Said map is hereby adopted by reference
and made part of this Ordinance as fully as if entirely set forth herein. A
copy of said map is on file in the office of the Clerk of (Municipality)
and is available for inspection."

Section 4. Said Zoning Ordinance shall be and is hereby amended to add thereto a
new section to be designated Section _____ entitled "General Regulations" to provide
as follows:

"_____General Regulations. The following regulations shall govern development on any Critical Slope in the _____(Township)_____ of ___(Municipality)_____.

a. Permitted Uses. Any of the following uses are permitted on any Critical Slope provided that no building or structure, including swimming pools, vehicular facility (including roads, drives and parking areas), on-lot sewage disposal facility and/or substantial, non-agricultural displacement of soil is required within the critical area as herein defined, provided any such use is permitted in the district in which the premises are located.

1.  Wildlife sanctuary, woodland preserve, aboretum, open spaces.

2.  Game farm, hunting reserve.

3.  Forestry, lumbering and reforestation, in accordance with recognized conservation practices.

4.  Pasture and controlled grazing land.

5.  Recreation use such as park, picnic grove, golf course, hunting club.

6.  Vineyards and/or orchards.

7.  Harvesting of crops according to recognized soil conservation practices.

8.  Utility transmission lines.

9.  Accessory uses customarily incidental to any of the foregoing permitted uses but excluding buildings, vehicular facilities, on-lot sewage facilities and/or non-agricultural displacement of soil, when approved as a special exception by the Board of Adjustment.

b.  Conditional Uses. The following uses shall be permitted on any Critical Slope upon approval by the Board of Adjustment as a special exception.

1.  Front, side and/or rear yards of any use permitted by the primary zoning district, except that an area so utilized shall not be inconsistent with the objectives and standards set forth in the Purpose and Policy of this Ordinance (Section 1.) and with any other pertinent municipal regulations.

Section 4.b.1 (cont.)

Inclusion of Critical Slope areas in order to meet mini-
mun lot area or quard requirements is contingent upon
complying with the objectives and standards set forth
in the Purpose and Policy of this Ordinance (Section
1.) and with any other pertinent municipal regulations.
If such compliance cannot be shown, the land area
within the Critical Slope shall not be calculated for
purposes of determining lot areas or yard requirements.

2.    (Additional uses depending on local conditions)

c.    Issuance of Special Zoning Permit.  The municipal Zoning Board
of Adjustment, in its issuance of a special zoning permit shall
take into consideration the recommendations of a) the minicipal
Environmental Commission and/or b) the _____ County
Soil Conservation District and that such variance can be granted
without substantial detriment to the public good and will not substantially impair the intent and purpose of the zone plan and zoning ordinance and purpose and policy of this ordinance and other
pertinent municipal regulations.

All applications for a Special Zoning Permit within a Critical
Slope shall include site and construction plans for the building,
vehicular facilities, on-lot sewage disposal facilities and/or
non-agricultural displacement of soil.  Such plans shall include
existing slope conditions (two foot contour intervals), existing
vegetation and be of appropriate scale and detail.  Both existing and proposed grades will be shown on both topographic and
profile drawings.  Profile, or cross-sectional, drawings shall be
prepared so as to show clearly the relationship between the proposed buildings, vehicular facilities, on-lot sewage disposal
systems and/or non-agricultural displacement of soil and both
the existing and proposed grades and slopes.  In addition, the
various plans and/or drawings shall include (but not necessarily
be limited to) evidence that:

1.    Proposed buildings (or structures) are of sound engineering design and that footings are designed to
extend to stable soil and/or rock.

2.    Proposed vehicular facilities (including roads, drives
and/or parking areas) shall be so designed that land
clearing and/or regrading will not cause excessive
erosion.

a) Both vertical and horizontal alignments vehicular facilities shall be so designed that hazardous circulation conditions will not be created.

Section 4.c.3 (cont.)

3.   Proposed on-lot sewage disposal facilities shall be
     properly designed and installed in conformance
     with all pertinent health regulations.

4.   Proposed non-agricultural displacement of soil shall
     be for causes consistent with the intent of this ordi-
     nance and that it shall be executed in a manner that
     will not cause excessive erosion or other unstable
     conditions.

5.   Surface run-off of water will not create unstable
     conditions, including erosion, and that appropri-
     ate storm water drainage facilities and/or systems
     will be constructed as deemed necessary.

6.   Downstream flooding will not be increased.

7.   Other conditions necessary to protect the public
     health, safety, and general welfare have been
     considered.

d.   Specifically Prohibited Uses.  Notwithstanding the rule of statutory
construction that any use not permitted is prohibited, for the guidance of
the Board of Adjustment it is deemed appropriate to state that certain
specific uses are so repugnant to the purpose establishing and maintaining
critical slopes that such uses shall not be permitted.  The specifically prohibited uses are:

1.   The filling of steep slope lands and the removal of
     top soil.

2.   Land fill, dump, junk yard, outdoor storage of ve-
     hicles and/or materials.

Section 5.  Municipal Liability.  The granting of a Zoning Permit, a Special Zoning
Permit or approval of a subdivision plan in any Critical Slope shall not constitute a representation, guarantee warranty of any kind by the township or by any official or employee
thereof of the practicability or safety of any structure, use or other plan proposed, and
shall create no liability upon, or a cause of action against such public body, official or
employee for any damage that may result pursuant thereto.

Section 6.  Penalties.  Any person or persons, firm or corporation violating any of the
provisions of this Ordinance shall upon conviction thereof be subject to a fine not to exceed
$200,000 or imprisonment for a period not to exceed ninety (90) days, or both, at the discre-
tion of the Judge imposing the same.  Each and every day that said violation continues shall,
constitute a separate and specific violation.

Section 7. Repealer. All Ordinances inconsistent with the provisions of this Ordinance are hereby repealed to the extent of such inconsistency.

Section 8. Validity. If any part of this Ordinance is found not valid, the remainder of this Ordinance shall remain in effect.

Section 9. Effective Date. This Ordinance shall take effect immediately upon its enactment and publication in the manner provided by law."

Note: Insert Appeals Section if written as a separate ordinance

III

## SOIL EROSION AND SEDIMENT CONTROL

Soil, in mud form, is the single greatest pollutant in the waterways of the United States. Soil erosion and resulting sedimentation in streams and ponds has many negative effects. First, the most fertile soil is being lost; second, the stream bottom becomes coated with a slimy mud, reducing recreation and wildlife values, third, the water becomes clouded and reduces food production; fourth, soil holds pollutants and will concentrate them in the stream bottom. The most visible cost, however, is the bill to clean this sediment out of bridges and culverts and storm sewers so they can carry flood water. Another permanent cost to the municipality is the loss of clean, productive surface water streams.

Note: This ordinance is the product of a joint task force of the Departments of Environmental Protection and Agriculture. It has worked well where adopted with attention to monitoring and enforcement procedures.

MODEL MUNICIPAL LAND DISTURBANCE ORDINANCE
TO CONTROL
SOIL EROSION AND SEDIMENTATION

An ordinance to establish regulations for controlling soil erosion and sedimentation within the municipality of _____.

## ARTICLE I
### TITLE AND PURPOSE

A.    TITLE

This ordinance shall be known as the "_____Municipal Soil Erosion and Sediment Control Ordinance."

B.    PURPOSE

The purpose of this ordinance is to control soil erosion and sediment damages and related environmental damage by requiring   adequate provisions for surface water retention and drainage and for the protection of exposed soil surfaces in order to promote the safety, public health, convenience and general welfare of the community.

## ARTICLE II

A.    RULES APPLYING TO TEXT

For the purpose of this ordinance certain rules or work usage apply to the text as follows:

1.   Words used in the present tense include the future tense; and the singular includes the plural, unless the context clearly indicates the contrary.

2.   The term "shall" is always mandatory and not discretionary: the word "may" is permissive.

3.   The word or term not interpreted or defined by this article shall be used with a meaning of common or standard utilization.

B.    DEFINITIONS

The following definitions shall apply in the interpretation and enforcement of this ordinance, unless otherwise specifically stated:

1.   Applicant- A person, partnership, corporation or public agency requesting permission to engage in land disturbance activity.

2.  Critical Area – A sediment-producing highly erodible or severly eroded area.

3.  Excavation or Cut – Any act by which soil or rock is cut into, dug, quarried, uncovered, removed, displaced or relocated.

4.  Erosion – Detachment and movement of soil or rock fragments by water, wind, ice and gravity.

5.  Erosion and Sediment Control Plan – A plan which fully indicates necessary land treatment measures, including a schedule of the timing for their installation, which will effectively minimize soil erosion and sedimentation. Such measures shall be in accordance with standards as adopted by the State Soil Conservation Committee.

6.  Farm Conservation Plan– A plan which provides for use of land, within its capabilities and treatment, within practical limits, according to chosen use to prevent further deterioration of soil and water resources.

7.  Land – Any ground, soil, or earth including marshes, swamps drainageways and areas not permanently covered by water within the municipality.

8.  Land Disturbance – Any activity involving the clearing, grading, transporting, filling and any other activity which causes land to be exposed to the danger of erosion.

9.  Permit – A certificate issued to perform work under this ordinance.

10. Mulching – The application of plant residue or other suitable materials to the land surface to conserve moisture, hold soil in place, and aid in establishing plant cover.

11. Sediment – Solid material, both mineral and organic, that is in suspension, is being transported, or has been moved from its site of origin by air, water or gravity as a product of erosion.

12. Sediment Basin – A barrier or dam built at suitable locations to retain rock, sand, gravel, silt or other material.

13. Soil – All unconsolidated mineral and organic material of any origin.

14. Soil Conservation District – A governmental subdivision of this State, which encompasses this municipality, organized in accordance with the provisions of Chapter 24, Title 4, N.J.R.S.

15. Site – Any plot, parcel or parcels of land.

16. State Soil Conservation Committee - An agency of the State established in accordance with the provisions of Chapter 24, Title 4, N.J.R.S.

17. Stripping - Any activity which significantly disturbs vegetated or otherwise stabilized soil surface including clearing and grubbing operations.

## ARTICLE III
## PROCEDURE

A. REGULATION

1. No land area shall be disturbed by any person, partnership, corporation, municipal corporation or other public agency within this municipality unless; the applicant has submitted to the building inspector a plan to provide for soil erosion and sediment control for such land area in accordance with the Standards for Erosion and Sediment Control adopted by the New Jersey State Soil Conservation Committee and administered by the _____ _____ Soil Conservation District and such plan has been approved; and a valid land disturbance permit has been issued by the building inspector except as exempted by ARTICLE V.

B. DATA REQUIRED

The applicant must submit a separate soil erosion and sediment control plan for each noncontiguous site. The applicant may consult with the _____ Soil Conservation District in the selection of appropriate erosion and sediment control measures and the development of the plan. Such plan shall contain:

1. Location and description of existing natural and man-made features on and surrounding the site including general topography and soil characteristics and a copy of the soil conservation district soil survey (where available).

2. Location and description of proposed changes to the site.

3. Measures for soil erosion and sediment control which must meet or exceed STANDARDS FOR SOIL EROSION AND SEDIMENT CONTROL adopted by the State Soil Conservation Committee. STANDARDS shall be on file at the offices of the local soil conservation district and the township clerk.

4. A schedule of the sequence of installation of planned erosion and sediment control measures as related to the progress of the project including anticipated starting and completion dates.

5. All proposed revisions of data required shall be submitted for approval.

C.  REVIEW AND APPROVAL

1.  Erosion and sediment control plans submitted with subdivision and site plan applications shall be reviewed by the planning board and approved when in conformance with the Standards for Erosion and Sediment Control. The board may seek the assistance of the _____ Soil Conservation District in the review of such plans and may deem as approved those plans which have been reviewed and determined adequate by the _____ Soil Conservation District.

2.  All other types of soil disturbance not exempted in Article V shall come under the review of the municipal public body or designated municipal officer in accordance with their memorandum of understanding with the _____ Soil Conservation District.

NOTE:  PLANNING BOARD REVIEW POWERS

The planning boards of New Jersey have explicit statutory authority for their subdivision review powers and have judicial support for their site plan review powers.

There are, however, many other land use matters that reflect on planning and physical development of the municipality over which the planning board has no explicit or implicit review powers.

However, pursuant to the provisions of N.J.S.A. 40:55-1.13, the governing body of a municipality may require the planning board to review and report upon any class of matters, (i.e., soil disturbance other than that associated with subdivisions or site plans) prior to final action thereon by any municipal public body or municipal officer.

Therefore, in instances of soil disturbance other than that proposed in subdivision and site plan applications, the initial enforcement jurisdiction may be delegated to the building inspector with the requirement (by ordinance) that he refer the matter to the municipal planning board pursuant to the provisions of 40:55-1.13 for report and recommendation.

C.    REVIEW AND APPROVAL (alternate procedure)

Soil erosion and sediment control plans shall be reviewed by the municipal engineer (or other designated official) and approved when in conformance with the Standards for Erosion and Sediment Control.

The municipal engineer may seek the assistance of the _____ Soil Conservation District in the review of such plans and may deem as approved those plans which have been reviewed and determined adequate by the said district.

ARTICLE IV
PRINCIPLES AND REQUIREMENTS

A.    GENERAL DESIGN PRINCIPLES

Control measures shall apply to all aspects of the proposed land disturbance and shall be in operation during all stages of the disturbance activity. The following principles shall apply to the soil erosion and sediment control plan.

1.    Stripping of vegetation, grading or other soil disturbance shall be done in a manner which will minimize soil erosion.

2.    Whenever feasible, natural vegetation shall be retained and protected.

3.    The extent of the disturbed area and the duration of its exposure shall be kept within practical limits.

4.    Either temporary seeding, mulching or other suitable stabilization measure shall be used to protect exposed critical areas during construction or other land disturbance.

5.    Drainage provisions shall accommodate increased runoff, resulting from modified soil and surface conditions, during and after development or disturbance. Such provisions shall be in addition to all existing requirements.

6.    Water runoff shall be minimized and retained on site wherever possible to facilitate ground water recharge.

7.    Sediment shall be retained on site.

8.    Diversions, sediment basins, and similar required structures shall be installed prior to any on-site grading or disturbance.

B.   MAINTENANCE

All necessary soil erosion and sediment control measures installed under this ordinance, shall be adequately maintained for one year after completion of the approved plan or until such measures are permanently stabilized as determined by the municipal engineer. The municipal engineer shall give the applicant upon request a certificate indicating the date on which the measures called for in the approved plans were completed.

C.   MAINTENANCE BONDS (OPTIONAL)

1.   In the event that certain work required under the permit cannot be performed within the time set for completion and is postponed, a cash bond in form approved by the Borough Attorney in an amount equal to the cost of such uncompleted work may be deposited with the __(designated official)__ to guarantee performance, which amount shall be determined by the Building Inspector. No such permission to postpone work shall exceed six (6) months. Upon failure to complete the work, the __(designated official)__ shall use the money deposited with the Borough to finish the unfinished portion of the work in accordance with the land disturbance permit. Any unused portion of the monies deposited shall be returned to the applicant.

2.   Prior to acceptance of work performed in accordance with the land disturbance permit, the Building Inspector shall obtain a cash maintenance bond from the applicant. Such bond shall be in an amount equal to fifteen (15) percent of the work performed by the applicant as determined by the Buildi ng Inspector. Such bond shall guarantee proper maintenance of the disturbed site for the period required by this Ordinance. Upon expiration of the period of the maintenance bond, the full amount less the costs incurred for maintenance shall be returned to the applicant.

D.   FEES (SUGGESTED SCHEDULE)

1.   Initial application $25.00 per acre or part thereof.

2.   Extension $5.00 per acre or part thereof.

3.   In addition to the above, the applicant shall pay a reasonable engineering inspection fee upon submission of a voucher by the Municipal Engineer. Said fee shall not exceed $500.00.

E.   PENALTIES (SUGGESTED PROVISION)

Any person, partnership, corporation, municipal corporation, public agency or other entity violating any provision of this Ordiance shall be subject to a fine of not more than $500.00 and not more than ninety (90) days in jail, or both. Notwithstanding this provision, the __(municipality)__ may proceed to obtain injunctive or other appropriate relief. In the event that construction is involved, the Building Inspector shall issue a "stop work notice" which shall be effective until recinded. In addition to the penalties herein provided, expenses incurred by the (municipality) in repairing or correcting any such violation shall also be recoverable from the violating party.

### ARTICLE V
### EXEMPTIONS

The following activities are specifically exempt from this Ordinance:

1. Land Disturbance associated with existing one and two family dwellings.

2. Use of land for gardening primarily for home consumption.

3. Agricultural use of lands when operated in accordance with a farm conservation plan approved by the local soil conservation district or when it is determined by the local soil conservation district that such use will not cause excessive erosion and sedimentation.

### ARTICLE VI
### INSPECTION AND ENFORCEMENT

The requirements of this Ordinance shall be enforced by the municipal engineer who shall also inspect or require adequate inspection of the work. If the municiapl engineer finds existing conditions not as stated in the applicant's erosion and sediment control plan he may refuse to approve further work and may require necessary erosion and sediment control measures to be promptly installed and may seek other penalties as provided in Article IV E.

### ARTICLE VII
### APPEALS

Appeals from decisions under this ordinance may be made to the municipal governing body in writing within ten days from the date of such decision. The appellant shall be entitled to a hearing before the municipal governing body within thirty days from the date of appeal. The municipal governing body may thereafter reserve, modify, or affirm the aforesaid decision.

### ARTICLE VIII
### REPEALER

All Ordinances or parts of Ordinances inconsistent with the provisions of this Ordinance are hereby repealed to the extent of such inconsistency.

## ARTICLE X
### VALIDITY AND SEVERABILITY

If any part of this Ordinance is found not valid, the remainder of this Ordinance shall remain in effect.

## ARTICLE XI
### EFFECTIVE DATE

This ordinance shall take effect immediately upon its enactment and publication in the manner provided by law.

V

## TREE PROTECTION

Because trees and forest groupings are most often found in combination with such critical areas as steep slopes and wet soils, and
have profound effect upon the local climate they should be regarded
as being of utmost importance to the public health and welfare, and
therefore protected as a matter of right.  Protecting forest under-
brush is equally important to the forest ecology and builders should
be restricted from trying to make lawn of the forest floor.  None
of these things are assured without an ordinance.

Note:  The list of desirable shrubs and trees may be expanded to include other desirable
native species of your area.

AN ORDINANCE TO REGULATE AND CONTROL
THE REMOVAL OF TREES

BE IT ORDAINED BY THE _____(Governing Body)_____ of the __(Township)__ of
____(Municipality)____ in the County of _____ in the State of
New Jersey:

Section 1.    The _(Governing Body)_ of the __(Township)__ of (Municipality)
does herein decide and find that the indiscriminate, uncontrolled and excess destruction, removal and cutting of trees upon lots and tracts of land within the (Township)
of ____(Minicipality)____ has resulted in creating increased municipal costs
to control drainage and has further caused increased soil erosion, decreased fertility
of soil, increased dust, which has deteriorated property values, and has further rendered the land unfit and unsuitable for their most appropriate use, with the result that
there has been and will result in the future a deterioration of conditions affecting the
health, safety and general well being for the inhabitants of the (Township) of
__(Municipality)__ and has caused the passage of this ordinance to regulate and control the indiscriminate and excessive cutting of trees in the __(Township)__ .

Section 2.    The intent and purpose of this ordinance is to preserve coniferous
and deciduous trees, flowering trees and shrubs growing in a natural state within this
municipality.

Section 3.    Definitions.

(a)    Tree: Any living deciduous tree having a trunk of a diameter greated than
three (3) inches D.B.H. (Diameter Breast High), any living coniferous
tree having a trunk of a diameter greated than four (4) inches D.B.H. or
any living Dogwood (Carnus Florida) or American Holly (Ilex Opaca)
tree having a diameter of one (1) inch or greater D.B.H.

(b)    Shrub: Native Laurel (Kalmia Latifolia) having a root crown of three (3)
inches or greater measured at the soil or surface level.

(c)    Building Lot: A parcel of land upon which a dwelling house has been or,
in accordance with the zoning code or variance, may be erected.

(d)    (optional) Township Forester: A person professionally qualified and appointed by the __(Municipality)__ to supervise the conservation of
trees and soil within the municipality which position is hereinafter created.

(e)    Nursery, Garden Center, Orchard or Tree Farm:  Only such land within the municipality used for horticultural purposes under controlled agricultural conditions.

(f)    Sanitary Landfill:  Land used by the municipality for the disposal of rubbish or garbage.

(g)    Public Right of Way:  Any street or road shown upon a map or plan filed in the _____ County Clerk's Office or on the Official Map of the (Township) of (Municipality).

(h)    Management Plan:  A plan for the management of timbered or forested lands developed by the New Jersey Department of Environmental Protection Brueau of Forestry or similar state or federal agency or Township Forester.

(i)    Permits:  A license issued by the Township Forester (or other designated official) of the (Township) of (Municipality) to remove or destroy trees or shrubs.

Section 4.    No person is to remove or destroy or cause to be removed or destroyed any Tree or Shrub growing in the (Township) of (Municipality) without having obtained a Permit as provided in this section.

Application for Permit with building permit:  Upon application for a building permit for any construction which would cause the removal or destruction, wholly or partially, of any tree or shrub, the applicant shall indicate upon a plot plan, which shall accompany such application, the location of all Trees and Shrubs and shall designate which trees, if any are to be removed or destroyed.  A copy of such Plot Plan shall be filed by the applicant with the Township Forester (or other designated official) and (Municipality) Environmental Commission, which shall review the same and notify the Township Forester of its recommendations.

Said recommendations shall be forwarded to the Township Forester within fifteen (15) days of receipt of same by the Environmental Commission, failing in which the Township Forester may grant or deny a Permit without further delay.  The Building Inspector shall issue a building permit whenever an application and Plot Plan indicates that trees or shrubs shall be destroyed or removed only upon approval by the Township Forester.

Application for Permit without building permit:  Application for a Permit for the removal or destruction of Trees or Shrubs as required under this ordinance where a building permit is not immediately involved shall be made directly to the Township Forester and shall contain the name of the applicant, location of the property and Plot Plan as aforesaid and in addition thereto the purposes for which the application is being made including but not limited to the following:  clearing of land for agricultural use, harvesting timber, fire protection, industrial use, private parks, scenic improvement, hardship or danger to adjacent properties, removal of diseased or damaged trees, transplanting or removal in a growing condition to other locations, installation of utilities or drainage of surface water.  A copy of the plot plan shall be filed with the Environmental Commission, which shall act upon the application as hereinabove stated within fifteen (15) days of the date of receipt and if the Commission shall not have acted upon the application within such period, the Township Forester may grant or deny a Permit without further delay.

Standards:    Prior to the issuance of a Permit by the Township Forester, the lands covered by each application shall be viewed by the Township Forester, who shall inspect the same as to the Trees and Shrubs which are the subject of the application as well as drainage and other physical conditions existing on the property and adjacent property, and the Township Forester shall issue the permit upon finding that the destruction or removal to be permitted would not impair the growth and development of the remaining trees and shrubs on the property of applicant or adjacent properties, would not cause soil erosion, would not impair existing drainage patterns, would not lessen property values in the neighborhood and would not impair substantially the aesthetic values of the area. The Township Forester shall have the authority to affix reasonable conditions to the grant of a Permit hereunder.

Section 5.    Protection of Trees:  No material or temporary soil deposits shall be placed within six (6) feet of any Tree or Shrub.

Section 6.    Excepted from this ordinance shall be:

(a)    Any Tree to be cut for use solely for firewood by owner, provided that this exception shall not apply to any Tree to be cut in connection with the issuance of any permit for soil disturbance, or building.

(b)    Any dead, diseased or other Tree that is likely to endanger life or property.

(c)    If the applicant discloses that no more than five Trees in excess of the size referred to aforesaid in this ordinance are to be removed, cut or destroyed or have been cut, removed or destroyed from any lot or tract of land identified in said application, then and in that event the duly appointed officer may issue a Permit for the removal of said tree or trees.

(d)    Any tree growing on land actually being used as a Nursery, Garden Center, Tree Garm or Orchard.

(e)    Any tree growing on land actually being used as a Sanitary Landfill operation or for surface mining.

(f)    Any Tree growing on a public right of way.

(g)    Any Tree cut or removed in accordance with a Management Plan.

Section 7.    Fees:  Upon the issuance of a Permit, the applicant shall pay to the (Township)    of   (Municipality)    the following fees:

Ten ($10.00) dollars per vacant Building Lot.

Lands other than vacant Building Lots, twenty-five ($25.00) dollars per acre or part thereof.

No fee shall be charged for the removing and transplanting of Trees or Shrubs in living condition to other locations and the removal of diseased or damaged Trees or Shrubs.

Section 8. Appeals: Appeals from decisions under this ordinance may be made to the municipal governing body in writing within ten (10) days from the date of such decision. Said appeal shall be by written notice to the Municipal Clerk. The appellant shall be entitled to a hearing before the __(governing__ __body)__ of the __(township)__ of __(municipality)__ within thirty (30) days from date of appeal. The municipal governing body may thereafter reverse, modify or affirm the aforesaid decision.

Section 9. Penalties: Any person violating any provision of this ordinance shall be subject to a fine not exceeding five hundred ($500.00) dollars, or imprisonment for ninety (90) days, or both, in the discretion of the judge. Each Tree or Shrub destroyed or damaged in violation of this ordinance shall be considered a separate violation.

Section 10. Conditions: Activities conducted under a permit pursuant to this ordinance shall be open to inspection at any time by any agency or agent of the __(township)__ of __(municipality__ .

Section 11. (Optional) There is hereby established the office of Township Forester, whose duties shall be as herein indicated. The annual salary shall be the sun of twenty-five hundred ($2,500.00) dollars (subject to variation by municipality), which shall be paid monthly. The term of office of Township Forester shall be for a period of one (1) year, but in no event shall said term extend more than fourteen (14) days beyond December 31 of any calendar year.

Section 12. Validity: If any part of this ordinance is found not valid, the remainder of this ordinance shall remain in effect.

Section 13. Repealer: All ordinances or parts of ordinances inconsistent with the provisions of this ordinance are hereby repealed to the extent of such inconsistency.

Section 14. Effective Date: This ordinance shall take effect immediately upon its enactment and publication in the manner provided by law.

# Freshwater Wetlands Protection Act

## N. J. S. A. 13:9B-1 et seq.

Printed as public information only by

## Division of Coastal Resources New Jersey Department of Environmental Protection CN 401, Trenton, NJ 08625

For official version, refer to

P. L. 1987, c. 156

# CHAPTER 156

**AN ACT** concerning the regulation of freshwater wetlands, amending P.L. 1977, c. 74, supplementing Title 13 of the Revised Statutes, and making an appropriation.

BE IT ENACTED *by the Senate and General Assembly of the State of New Jersey:*

**C. 13:9B-1**          **Short title.**
1.      This act shall be known and may be cited as the "Freshwater Wetlands Protection Act."

**C. 13:9B-2**          **Findings, declarations.**
2.      The Legislature finds and declares that freshwater wetlands protect and preserve drinking water supplies by serving to purify surface water and groundwater resources; that freshwater wetlands provide a natural means of flood and storm damage protection, and thereby prevent the loss of life and property through the absorption and storage of water during high runoff periods and the reduction of flood crests; that freshwater wetlands serve as a transition zone between dry land and water courses, thereby retarding soil erosion; that freshwater wetlands provide essential breeding, spawning, nesting, and wintering habitats for a major portion of the State's fish and wildlife, including migrating birds, endangered species, and commercially and recreationally important wildlife; and that freshwater wetlands maintain a critical baseflow to surface waters through the gradual release of stored flood waters and groundwater, particularly during drought periods.
    The Legislature further finds and declares that while the State has acted to protect coastal wetlands, it has not, except indirectly, taken equally vigorous action to protect the State's inland waterways and freshwater wetlands; that in order to advance the public interest in a just manner the rights of persons who own or possess real property affected by this act must be fairly recognized and balanced with environmental interests; and that the public benefits arising from the natural functions of freshwater wetlands, and the public harm from freshwater wetland losses, are distinct from and may exceed the private value of wetland areas.

The Legislature therefore determines that in this State, where pressures for commercial and residential development define the pace and pattern of land use, it is in the public interest to establish a program for the systematic review of activities in and around freshwater wetland areas designed to provide predictability in the protection of freshwater wetlands; that it shall be the policy of the State to preserve the purity and integrity of freshwater wetlands from random, unnecessary or undesirable alteration or disturbance; and that to achieve these goals it is important that the State expeditiously assume the freshwater wetlands permit jurisdiction currently exercised by the United States Army Corps of Engineers pursuant to the Federal Act and implementing regulations.

## C. 13:9B-3          Definitions.

3.    As used in this act:

"Bank" means the Wetlands Mitigation Bank established pursuant to section 14 of this act;

"Commissioner" means the Commissioner of the Department of Environmental Protection;

"Council" means the Wetlands Mitigation Council established pursuant to section 14 of this act;

"Department" means the Department of Environmental Protection;

"Environmental commission" means a municipal advisory body created pursuant to P.L. 1968, c. 245 (C. 40:56A-1 et seq.);

"Federal Act" means section 404 of the "Federal Water Pollution Control Act Amendments of 1972" as amended by the "Clean Water Act of 1977" (33 U.S.C. § 1344) and the regulations adopted pursuant thereto;

"Freshwater wetland" means an area that is inundated or saturated by surface water or groundwater at a frequency and duration sufficient to support, and that under normal circumstances does support, a prevalence of vegetation typically adapted for life in saturated soil conditions, commonly known as hydrophytic vegetation; provided, however, that the department, in designating a wetland, shall use the 3-parameter approach (i.e. hydrology, soils and vegetation) enumerated in the April 1, 1987 interim-final draft "Wetland Identification and Delineation Manual" developed by the United States Environmental Protection Agency, and any subsequent amendments thereto;

"Freshwater wetlands permit" means a permit to engage in a regulated activity issued pursuant to this act;

"Hydrophyte" means plant life adapted to growth and reproduction under periodically saturated root zone conditions during at least a portion of the growing season;

"Linear development" means land uses such as roads, drives, railroads, sewerage and stormwater management pipes, gas and water pipelines, electric, telephone and other transmission lines and the rights-of-way therefor, the basic function of which is to connect two points. Linear development shall not mean residential, commercial, office, or industrial buildings;

"Person" means an individual, corporation, partnership, association, the State, municipality, commission or political subdivision of the State or any interstate body;

"Regulated activity" means any of the following activities in a freshwater wetland:

(1)   The removal, excavation, disturbance or dredging of soil, sand, gravel, or aggregate material of any kind;

(2)   The drainage or disturbance of the water level or water table;

(3)   The dumping, discharging or filling with any materials;

(4)   The driving of pilings;

(5)   The placing of obstructions;

(6)   The destruction of plant life which would alter the character of a freshwater wetland, including the cutting of trees;

"Transition area" means an area of land adjacent to a freshwater wetland which minimizes adverse impacts on the wetland or serves as an integral component of the wetlands ecosystem.

## C. 13:9B-4   Exemptions from permit, transition area requirements.

4.   The following are exempt from the requirement of a freshwater wetlands permit and transition area requirements unless the United States Environmental Protection Agency's regulations providing for the delegation to the state of the federal wetlands program conducted pursuant to the Federal Act require a permit for any of these activities, in which case the department shall require a permit for those activities so identified by that agency:

a.   Normal   farming,   silviculture,   and   ranching activities such as plowing, seeding, cultivating, minor drainage, harvesting for the production of food and fiber, or upland soil and water conservation practices; construction or maintenance of farm or stock ponds or irrigation ditches, or the maintenance of drainage ditches; construction or maintenance of farm roads or forest roads constructed and maintained in accordance with best management practices to assure that flow and circulation patterns and chemical and biological characteristics of freshwater wetlands are not impaired and that any adverse effect on the aquatic environment will be minimized;

b.   Normal harvesting of forest products in accordance with a forest management plan approved by the State Forester;

c.   Areas regulated as a coastal wetland pursuant to P.L. 1970, c. 272 (C. 13:9A-1 et seq.);

d.   Projects for which (1) preliminary site plan or subdivision applications have received preliminary approvals from the local authorities pursuant to the "Municipal Land Use Law," P.L. 1975, c. 291 (C. 40:55D-1 et seq.) prior to the effective date of this act, (2) preliminary site plan or subdivision applications have been submitted prior to June 8, 1987, or (3) permit applications have been approved by the U.S. Army Corps of Engineers prior to the effective date of this act, which projects would otherwise be subject to State regulation on or after the effective date of this act, shall be governed only by the Federal Act, and shall not be subject to any additional or inconsistent substantive requirements of this act; provided, however, that upon the expiration of a permit issued pursuant to the Federal Act any application for a renewal thereof shall be made to the appropriate regulatory agency. The department shall not require the establishment of a transition area as a condition of any renewal of a permit issued pursuant to the Federal Act prior to the effective date of this act. Projects not subject to the jurisdiction of the United States Army Corps of Engineers and for which preliminary site or subdivision applications have been approved prior to the effective date of this act shall not require transition areas;

e.   The exemptions in subsections a. and b. of this section shall not apply to any discharge of dredged or fill material into a freshwater wetland incidental to any activity which involves bringing an area of freshwater wetlands into a use to which it was not previously subject, where the flow or circulation patterns of the waters may be impaired, or the reach of the waters is reduced.

## C. 13:9B-5          Permit process.

5. a. The department shall consolidate the processing of wetlands related aspects of other regulatory programs which affect activities in freshwater wetlands, including, but not limited to, sewer extension approvals required pursuant to P.L. 1977, c. 74 (C. 58:10A-1 et seq.), permits required pursuant to P.L. 1973, c. 185 (C. 13:19-1 et seq.), and any permits and approvals required pursuant to P.L. 1977, c. 75 (C. 58:11A-1 et seq.) and P.L. 1962, c. 19 (C. 58:16A-50 et seq.), with the freshwater wetlands permit process established herein so as to provide a timely and coordinated permit process consistent with the Federal Act.

b. Within 60 days after the department receives comment on a complete application for a permit from the United States Environmental Protection Agency, or upon receipt of notice from the United States Environmental Protection Agency that no comment will be forthcoming, the department may hold a public hearing on the application for a permit. If such a hearing is held, it shall be in the county wherein the freshwater wetland is located whenever practicable. The department may issue or deny a permit without a public hearing, unless there is a significant degree of public interest in the application as manifested by written requests for a hearing within 20 days after the publication of notice of the permit application in the bulletin of the department.

c. The department shall issue or deny a permit within 90 days of receipt of comments, or notice that comments will not be forthcoming, from the United States Environmental Protection Agency, or within 180 days of submittal of a complete application, whichever is later. Until the State assumes the implementation of the Federal Act, the department shall issue or deny a permit within 180 days of submittal of a complete application, except as may otherwise be provided by the Federal Act. The department shall review an application for a permit for completeness, and make any necessary requests for further information, within 30 days of receipt of the application for a permit;

provided, however, that this deadline shall not apply to requests for further information made by the department on the basis of comments received from the United States Environmental Protection Agency. If the department issues the permit, the department shall send notice thereof to the applicant. If the department denies, or requests a modification of, the complete permit application, the department shall send notice thereof to the applicant. The department may issue a permit imposing conditions necessary for compliance with this act and the "Water Pollution Control Act," P.L. 1977, c. 74 (C. 58:10A-1 et seq.).

d.    The fees authorized pursuant to sections 8, 9, and 17 of this act shall be dedicated to further the specific purposes of this act.

## C. 13:9B-6    Meadowlands, Pinelands exemptions.

6. a.    Activities in areas under the jurisdiction of the Hackensack Meadowlands Development Commission pursuant to P.L. 1968, c. 404 (C. 13:17-1 et seq.) shall not require a freshwater wetlands permit, or be subject to transition area requirements, except that the discharge of dredged or fill material shall require a permit issued under the provisions of the Federal Act, or under an individual and general permit program administered by the State under the provisions of the Federal Act and applicable State laws.

b.    Activities in areas under the jurisdiction of the Pinelands Commission pursuant to P.L. 1979, c. 111 (C. 13:18A-1 et seq.) shall not require a freshwater wetlands permit, or be subject to transition area requirements established in this act, except that the discharge of dredged or fill material shall require a permit issued under the provisions of the Federal Act, or under an individual and general permit program administered by the State under the provisions of the Federal Act and applicable State laws, provided that the Pinelands Commission may provide for more stringent regulation of activities in and around freshwater wetland areas within its jurisdiction.

## C. 13:9B-7    Classification system.

7.    The department shall develop a system for the classification of freshwater wetlands based upon criteria which distinguish among wetlands of exceptional resource value, intermediate resource value, and ordinary resource value.

a.    Freshwater wetlands of exceptional resource value shall be freshwater wetlands which exhibit any of the following characteristics:

(1)    Those which discharge into FW-1 waters and FW-2 trout production (TP) waters and their tributaries; or

(2)    Those which are present habitats for threatened or endangered species, or those which are documented habitats for threatened or endangered species which remain suitable for breeding, resting, or feeding by these species during the normal period these species would use the habitat. A habitat shall be considered a documented habitat if the department makes a finding that the habitat remains suitable for use by the specific documented threatened and endangered species, based upon information available to it, including but not limited to, information submitted by an applicant for a freshwater wetlands permit. An applicant shall have the opportunity to request the department that a documented habitat not result in the classification of a freshwater wetland as a freshwater wetland of exceptional value if the applicant can demonstrate the loss of one or more requirements of the specific documented threatened or endangered species, including, but not limited to wetlands or overall habitat size, water quality, or vegetation density or diversity.

b.    Freshwater wetlands of ordinary value shall be freshwater wetlands which do not exhibit the characteristics enumerated in subsection a. of this section, and which are certain isolated wetlands, man-made drainage ditches, swales, or detention facilities.

c.    Freshwater wetlands of intermediate resource value shall be all freshwater wetlands not included in subsection a. or b. of this section.

d.    As used in this section "threatened or endangered species" shall be those species identified pursuant to "The Endangered and Nongame Species Conservation Act," P.L. 1973, c. 309 (C. 23:2A-1 et seq.) or which appear on the federal endangered species list, and "FW-1, FW-2, trout production (TP) waters" shall mean those waters delineated as such by the department under regulations adopted pursuant to the "Water Pollution Control Act," P.L. 1977, c. 74 (C. 58:10A-1 et seq.) and the "Water Quality Planning Act," P.L. 1977, c. 75 (C. 58:11A-1 et seq.).

e.    The classification system established in this section shall not restrict the department's authority to require the creation or restoration of freshwater wetlands pursuant to the provisions of section 13 of this act.

## C. 13:9B-8    Letter of interpretation.

8. a.  A person proposing to engage in a regulated activity in a freshwater wetland or in an activity which requires a transition area waiver may, prior to applying for a freshwater wetlands permit or transition area waiver, request from the department a letter of interpretation to establish that the site of the proposed activity is located in a freshwater wetland or transition area.

b.     Within 20 days after receipt of a request for a letter of interpretation, the department may require the submission of any additional information necessary to issue the letter of interpretation.

c.     If no additional information is required, the department shall issue a letter of interpretation within 30 days after receiving the request.

d.     If additional information is required the department shall issue a letter of interpretation within 45 days after receipt of the information.

e.     The department may require an applicant for a letter of interpretation to perform and submit to the department an onsite inspection to determine or verify the general location of the freshwater wetland boundary and the applicable transition area. This inspection shall be subject to approval and verification by the department. If the department determines that onsite inspection by the department is necessary, the department shall make the inspection. If an on-site inspection is required by the department the time specified in this section for issuance of the letter of interpretation shall be extended by 45 days.

f.     If a person requesting the letter has not made a reasonable good faith effort to provide the department with information sufficient to make a determination, the department shall issue a letter of interpretation requiring the application for a freshwater wetlands permit or transition area waiver.

g.     A person applying for a letter of interpretation may also submit a report of an onsite freshwater wetlands delineation and receive within the time specified in this section a letter of interpretation verifying the actual freshwater wetlands and transition area boundaries.

h.     The department may charge a fee not to exceed the costs for reviewing the information submitted, conducting on-site inspections pursuant to subsection e. of this section, and for issuing a letter of interpretation.

i.     Any person who requests a letter of interpretation pursuant to the provisions of this act and does not receive a response from the department within the deadlines imposed in this section shall not be entitled to assume that the site of the proposed activity which was the subject of the request for a letter of interpretation is not in a freshwater wetland. A person who receives a letter of interpretation pursuant to this section shall be entitled to rely on the determination of the department, except as provided in subsection j. of this section.

j.     The department shall transmit to the United States Environmental Protection Agency a copy of any letter of interpretation determining that the site of a proposed regulated activity is not in a freshwater wetland. Any letter of interpretation which determines that the site of a proposed regulated activity is not in a freshwater wetlands shall be subject to review, modification, or revocation by the United States Environmental Protection Agency.

k.     The department shall publish in the bulletin of the department a list indicating the status of each application for a permit submitted to the department pursuant to the provisions of this act.

## C. 13:9B-9          Permit application; conditions for issuance.

9. a. A person proposing to engage in a regulated activity shall apply to the department for a freshwater wetlands permit, for a fee not to exceed the cost of reviewing and processing the application, and on forms and in the manner prescribed by the commissioner pursuant to the "Administrative Procedure Act," P.L. 1968, c. 410 (C. 52:14B-1 et seq.). An agency of the State proposing to engage in a regulated activity shall also apply to the department for a freshwater wetlands permit on forms and in a manner prescribed by the commissioner, but shall not be required to pay a fee therefor. The application shall include the name and address of the applicant, the purpose of the project, the names and addresses of all owners of property adjacent to the proposed project, and at least the following:

(1)    A preliminary site plan or subdivision map of the proposed development activities, or another map of the site if no preliminary site plan or subdivision map exists, and a written description of the proposed regulated activity, the total area to be modified, and the total area of the freshwater wetland potentially affected;

(2)    Verification that a notice has been forwarded to the clerk, environmental commission, and planning board of the municipality in which the proposed regulated activity will occur, the planning board of the county in which the proposed regulated activity will occur, landowners within 200 feet of the site of the proposed regulated activity, and to all persons who requested to be notified of proposed regulated activities, which notice may be filed concurrently with notices required pursuant to P.L. 1975, c. 291 (C. 40:55D-1 et seq.), describing the proposed regulated activity and advising these parties of their opportunity to submit comments thereon to the department;

(3)    Verification that notice of the proposed activity has been published in a newspaper of local circulation;

(4)    A statement detailing any potential adverse environmental effects of the regulated activity and any measures necessary to mitigate those effects, and any information necessary for the department to make a finding pursuant to subsection b. of this section.

b.    The department, after considering the comments of the environmental commission and planning boards of the county and municipality wherein the regulated activity is to take place, federal and State agencies of competent jurisdiction, other affected municipalities and counties, and the general public, shall issue a freshwater wetlands permit only if it finds that the regulated activity:

(1)    Is water-dependent or requires access to the freshwater wetlands as a central element of its basic function, and has no practicable alternative which would not involve a freshwater wetland or which would have a less adverse impact on the aquatic ecosystem, and which would not have other significant adverse environmental consequences, and also complies with the provisions of paragraphs (3)-(9) of this subsection; or

(2)    Is nonwater-dependent and has no practicable alternative as demonstrated pursuant to section 10

of this act, which would not involve a freshwater wetland or which would have a less adverse impact on the aquatic ecosystem, and which would not have other significant adverse environmental consequences;  and

(3)    Will result in minimum feasible alteration or impairment of the aquatic ecosystem including existing contour, vegetation, fish and wildlife resources, and aquatic circulation of the freshwater wetland; and

(4)    Will not jeopardize the continued existence of species listed pursuant to "The Endangered and Nongame Species Conservation Act," P.L. 1973, c. 309 (C. 23:2A-1 et seq.) or which appear on the federal endangered species list, and will not result in the likelihood of the destruction or adverse modification of a habitat which is determined by the Secretary of the United States Department of the Interior or the Secretary of the United States Department of Commerce as appropriate to be a critical habitat under the "Endangered Species Act of 1973," (16 U.S.C. § 1531 et al.); and

(5)    Will not cause or contribute to a violation of any applicable State water quality standard; and

(6)    Will not cause or contribute to a violation of any applicable toxic effluent standard or prohibition imposed pursuant to the "Water Pollution Control Act," P.L. 1977, c. 74 (C. 58:10A-1 et seq.); and

(7)    Will not violate any requirement imposed by the United States government to protect any marine sanctuary designated pursuant to the "Marine Protection, Research and Sanctuaries Act of 1972," (33 U.S.C. § 1401 et al.); and

(8)    Will not cause or contribute to a significant degradation of ground or surface waters; and

(9)    Is in the public interest as determined pursuant to section 11 of this act, is necessary to realize the benefits derived from the activity, and is otherwise lawful.

## C. 13:9B-10          Rebuttable presumption.

10. a. It shall be a rebuttable presumption that there is a practicable alternative to any nonwater-dependent regulated activity that does not involve a freshwater wetland, and that such an alternative to any regulated activity would have less of an impact on the aquatic ecosystem. An alternative shall be practicable if it is available and capable of being carried out after taking into

consideration cost, existing technology, and logistics in light of overall project purposes, and may include an area not owned by the applicant which could reasonably have been or be obtained, utilized, expanded, or managed in order to fulfill the basic purpose of the proposed activity.

b.    In order to rebut the presumption established in subsection a. of this section an applicant for a freshwater wetlands permit must demonstrate the following

(1)    That the basic project purpose cannot reasonably be accomplished utilizing one or more other sites in the general region that would avoid, or result in less, adverse impact on an aquatic ecosystem; and

(2)    That a reduction in the size, scope, configuration, or density of the project as proposed and all alternative designs to that of the project as proposed that would avoid, or result in less, adverse impact on an aquatic ecosystem will not accomplish the basic purpose of the project; and

(3)    That in cases where the applicant has rejected alternatives to the project as proposed due to constraints such as inadequate zoning, infrastructure, or parcel size, the applicant has made reasonable attempts to remove or accommodate such constraints.

c.    In order to rebut the presumption established in subsection a. of this section with respect to wetlands of exceptional resource value, an applicant, in addition to complying with the provisions of subsection b. of this section, must also demonstrate that there is a compelling public need for the proposed activity greater than the need to protect the freshwater wetland that cannot be met by essentially similar projects in the region which are under construction or expansion, or have received the necessary governmental permits and approvals; or that denial of the permit would impose an extraordinary hardship on the part of the applicant brought about by circumstances peculiar to the subject property.

## C. 13:9B-11        Determination of public interest.

11.   In determining whether a proposed regulated activity in any freshwater wetland is in the public interest, the department shall consider the following:

a.   the public interest in preservation of natural resources and the interest of the property owners in reasonable economic development;

b.    the relative extent of the public and private need for the proposed regulated activity;

c.    where there are unresolved conflicts as to resource use, the practicability of using reasonable alternative locations and methods, including mitigation, to accomplish the purpose of the proposed regulated activity;

d.    the extent and permanence of the beneficial or detrimental effects which the proposed regulated activity may have on the public and private uses for which the property is suited;

e.    the quality of the wetland which may be affected and the amount of freshwater wetlands to be disturbed;

f.    the economic value, both public and private, of the proposed regulated activity to the general area; and

g.    the ecological value of the freshwater wetlands and probable impact on public health and fish and wildlife.

## C. 13:9B-12    Accessibility to approved site.

12.    If a freshwater wetlands permit is approved and issued pursuant to the provisions of this act the department shall waive or modify the requirement for a transition area to the extent required to provide access to the site of the approved regulated activity.

## C. 13:9B-13    Mitigation of adverse environmental impacts.

13. a. The department shall require as a condition of a freshwater wetlands permit that all appropriate measures have been carried out to mitigate adverse environmental impacts, restore vegetation, habitats, and land and water features, prevent sedimentation and erosion, minimize the area of freshwater wetland disturbance and insure compliance with the Federal Act and implementing regulations.

b.    The department may require the creation or restoration of an area of freshwater wetlands of equal ecological value to those which will be lost, and shall determine whether the creation or restoration of freshwater wetlands is conducted onsite or offsite.  The department shall accept and evaluate a proposal to create or restore an area of freshwater wetlands only after the department has evaluated the permit application for which the proposal is made, and shall evaluate the proposal to create or restore an area of freshwater wetlands independently of the permit application. The department's evaluation of a proposal to create or restore an area of freshwater wetlands shall be conducted in consultation with the United States Environmental Protection Agency.

c.    If the department determines that the creation or restoration of freshwater wetlands onsite is not feasible, the department, in consultation with the United States Environmental Protection Agency, may consider the option of permitting the creation of freshwater wetlands or the restoration of degraded freshwater wetlands offsite on private property with the restriction on these wetlands of any future development, or the making of a contribution to the Wetlands Mitigation Bank. The contribution shall be equivalent to the lesser of the following costs: (1) purchasing and restoring existing degraded freshwater wetlands, resulting in preservation of freshwater wetlands of equal ecological value to those which are being lost; or (2) purchase of property and the cost of creation of freshwater wetlands of equal ecological value to those which are being lost. The applicant may also donate land as part of the contribution if the Wetlands Mitigation Council determines that the donated land has potential to be a valuable component of the freshwater wetlands ecosystem. The department shall permit the donation of land as a part of the contribution to the Wetlands Mitigation Bank only after determining that all alternatives to the donation are not practicable or feasible.

## C. 13:9B-14          Wetlands Mitigation Bank; council membership.

14. a. There is established in the Executive Branch of State Government the Wetlands Mitigation Bank. For the purpose of complying with Article V, section IV, paragraph 1 of the New Jersey Constitution, the bank is allocated within the Department of Environmental Protection but, notwithstanding this allocation, the bank shall be independent of any supervision or control by the department or the commissioner, or any other officer or employee thereof.

b.    The bank shall be governed by the Wetlands Mitigation Council which shall comprise seven members as follows: the Commissioner of Environmental Protection, who shall serve ex officio; and six members of the general public to be appointed by the Governor with the advice and consent of the Senate, two of whom shall be appointed from persons recommended by recognized building and development organizations; two of whom shall be appointed from persons recommended by recognized environmental and conservation organizations; and two of whom shall be appointed from institutions of higher learning in the State. Each of the members appointed from the general public shall serve for a term of three years and until a successor is appointed and qualified, except that of the members first appointed, two shall serve terms of one year, and two shall serve terms of two years. All vacancies, except those created through the expiration of term, shall be filled for the unexpired term only, and in the same manner, and with a member having the same class, as the original appointment. Each member shall be eligible for reappointment, but may be removed by the Governor for cause.

c.   A majority of the membership of the council shall constitute a quorum for the transaction of council business. Action may be taken and motions and resolutions adopted by the council at any meeting thereof by the affirmative vote of a majority of the full membership of the council.

d.   The Governor shall appoint a chairman from the public members and the council may appoint such other officers as may be necessary. The council may appoint such staff or hire such experts as it may require within the limits of appropriations made for these purposes.

e.   Members of the council shall serve without compensation, but may be reimbursed for expenses necessarily incurred in the discharge of their official duties.

f.   The council may call to its assistance such employees as are necessary and made available to it from any agency or department of the State or its political subdivisions.

g.   The council may adopt, pursuant to the "Administrative Procedure Act," and in consultation with the department, any rules and regulations necessary to carry out its responsibilities.

## C. 13:9B-15   Powers of Wetlands Mitigation Council.

15.   a.   The Wetlands Mitigation Council shall be responsible for disbursements of funds from the bank to finance mitigation projects. The council shall

have the power to purchase land to provide areas for the restoration of degraded freshwater wetlands, and to preserve freshwater wetlands and transition areas determined to be of critical importance in protecting freshwater wetlands. The council shall not engage in the restoration of degraded freshwater wetlands on public lands, except those lands which are acquired by the bank. The council shall assist the department in preparing the portions of the report required pursuant to section 29 of this act which pertain to mitigation.

b. The council may contract with nonprofit organizations, the Division of Fish, Game and Wildlife in the department, the United States Fish and Wildlife Service, and other appropriate agencies to carry out its responsibilities, and may aggregate mitigation actions to achieve economies of scale. Any contract proposed by the council pursuant to this subsection shall be subject to review and approval by the United States Environmental Protection Agency.

c. The council may transfer any funds or lands restricted by deed, easement or other appropriate means to mitigation and freshwater wetlands conservation purposes, to a state or federal conservation agency that consents to the transfer, to expand or provide for:

(1)   Freshwater wetlands preserves;

(2)   Transition areas around existing freshwater wetlands to preserve freshwater wetland quality;

(3)   Future mitigation sites for freshwater wetlands restoration; or

(4)   Research to enhance the practice of mitigation.

## C. 13:9B-16      Transition areas.

16. a. There shall be transition areas adjacent only to freshwater wetlands of exceptional resource value and of intermediate resource value. A transition area shall serve as:

(1)   An ecological transition zone from uplands to freshwater wetlands which is an integral portion of the freshwater wetlands ecosystem, providing temporary refuge for freshwater wetlands fauna during high water episodes, critical habitat for animals dependent upon but not resident in freshwater wetlands, and slight variations of freshwater wetland boundaries over time due to hydrologic or climatologic effects; and

(2)   A sediment and storm water control zone to reduce the impacts of development upon freshwater wetlands and freshwater wetlands species.

b.   The width of the transition area shall be determined by the department as follows:

(1)   No greater than 150 feet nor less than 75 feet for a freshwater wetland of exceptional resource value;

(2)   No greater than 50 feet nor less than 25 feet for a freshwater wetland of intermediate resource value.

c.   The minimum width of a transition area established pursuant to this section may be further reduced consistent with a transition area averaging plan approved under section 18 of this act.

## C. 13:9B-17       Prohibited activities.

17.  a.  The following activities, except for normal property maintenance or minor and temporary disturbances of the transition area resulting from, and necessary for, normal construction activities on land adjacent to the transition area, are prohibited in the transition area, except in accordance with a transition area waiver approved by the department pursuant to section 18 of this act:

(1)   Removal, excavation, or disturbance of the soil;

(2)   Dumping or filling with any materials;

(3)   Erection of structures, except for temporary structures of 150 square feet or less;

(4)   Placement of pavements;

(5)   Destruction of plant life which would alter the existing pattern of vegetation.

b.   A person proposing to engage in an activity prohibited pursuant to subsection a. of this section within 150 feet of a freshwater wetland of exceptional resource value, or within 50 feet of a freshwater wetland of intermediate resource value, shall apply to the department for a transition area waiver, for a fee not to exceed the cost of reviewing and processing the waiver application, and on forms and in the manner prescribed by the commissioner pursuant to the "Administrative Procedure Act," P.L. 1968, c. 410 (C. 52:14B-1 et seq.). An agency of the State proposing to engage in such an activity in a transition area shall also apply to the department for a transition area waiver on forms and in a manner prescribed by the commissioner but shall not be required to pay a fee therefor. The waiver application shall include at least the following:

(1)   A preliminary site plan or subdivision map of the site, or another map of the site if no preliminary site plan or subdivision map exists, containing proposed activities and a written description of the proposed activity, the total areas to be modified, and the total area of the transition area potentially affected; and

(2)   Verification that a notice has been forwarded to the clerk, environmental commission, and planning board of the municipality, and the planning board of the county wherein the activity is to occur, which notice shall describe the activity and advise these instrumentalities of local government of their opportunity to submit comments thereon to the department; and

(3)   A statement detailing any potential adverse environmental effects of the activity on the freshwater wetlands and any measures that may be necessary to mitigate those effects; and

(4)   A transition area averaging plan, if an averaging plan is required in connection with a transition area waiver requested pursuant to section 18 of this act.

c.   At the applicant's option, the maximum transition area distances established in subsection b. of section 16 of this act, or a lesser transition area distance established pursuant to a waiver approved pursuant to section 18 of this act, shall be further reduced, or the transition area adjacent to a portion of a wetlands shall be eliminated, pursuant to a transition area averaging plan submitted by the applicant, provided that the plan is consistent with the provisions of subsection a. of section 16 of this act.

### C. 13:9B-18      Transition area waivers.

18. a. The department shall grant a transition area waiver reducing the size of a transition area to not less than the minimum distance established in subsection b. of section 16 of this act  provided that (1) the proposed activity would have no substantial impact on the adjacent freshwater wetland or (2) the waiver is necessary to avoid a substantial hardship to the applicant caused by circumstances peculiar to the property.  If the proposed activity is the construction of a stormwater management facility having no feasible alternative on-site location or is linear development having no feasible alternative location,  the department  shall

approve a further transition area waiver or elimination of a portion of a transition area as necessary to permit the activity. A transition area waiver approved pursuant to this subsection shall not require transition area averaging to compensate for the reduction of transition area distance or for partial elimination of the transition area.

   b.    The department shall also approve transition area waivers reducing the transition area distances established in subsection b. of section 16 of this act and shall also approve waiver applications eliminating portions of transition areas, provided that the applicant submits a transition area averaging plan. The transition area requirements of this act shall be satisfied if the transition area averaging plan expands a portion of the transition area to compensate, on a square footage basis, for reduction of a transition area distance or for partial elimination of a transition area. The applicant shall have the right to determine the area of transition area reduction or partial elimination provided that the transition area averaging plan will result in a transition area consistent with the provisions of subsection a. of section 16 of this act. If a transition area waiver is approved pursuant to subsection a. of this section, the average transition area required by this subsection shall be based upon the transition area distance established pursuant to subsection a. of this section. If no waiver is approved pursuant to subsection a. of this section, the average transition area shall be based upon the maximum applicable transition area distance provided in subsection b. of section 16 of this act.

   c.    Any other provision of this act to the contrary notwithstanding, the transition area distance from a freshwater wetland of exceptional resource value may be reduced to no less than 75 feet except pursuant to section 12 of this act. A transition area waiver shall be approved pursuant to this subsection only if a transition area distance reduction would have no substantial adverse impact on the adjacent freshwater wetlands or if denial of a transition area waiver would result in extraordinary hardship to the applicant because of circumstances peculiar to the subject property. A transition area waiver approved pursuant to this subsection shall be conditioned on a transition area averaging plan which provides an average transition area of not less than 100 feet.

d.   The department shall issue or deny an application for a transition area waiver within 90 days of submission of a complete application; provided, however, that if the project or activity for which the transition area waiver is requested also involves a regulated activity in a freshwater wetland, or if an application for a permit to conduct a regulated activity in a freshwater wetland adjacent to the transition area for which the transition area waiver is requested is pending before the department, the department shall approve or deny the transition area waiver within the time period set forth for the approval or denial of a permit in subsection c. of section 5 of this act.

## C. 13:9B-19        Consideration for tax purposes.

19.   If the department denies an application for a freshwater wetlands permit, the owner of record of the property affected may request, and the local tax assessor shall provide, that this fact be taken into account when the property is valued, assessed, and taxed for property tax purposes.

## C. 13:9B-20        Administrative hearing.

20.   An applicant for a freshwater wetlands permit issued pursuant to this act may request the commissioner for an administrative hearing on any decision to issue or deny a permit made by the department pursuant to this act. Upon receipt of such a request, the commissioner shall refer the matter to the Office of Administrative Law, which shall assign an administrative law judge to conduct a hearing on the matter in the form of a contested case pursuant to the "Administrative Procedure Act," P.L. 1968, c. 410 (C. 52:14B-1 et seq.). Within 45 days of receipt of the administrative law judge's decision, the commissioner shall affirm, reject, or modify the decision. The commissioner's action shall be considered the final agency action for the purposes of the "Administrative Procedure Act," and shall be subject only to judicial review as provided in the Rules of Court.

## C. 13:9B-21        Remedies for violations.

21. a. Whenever, on the basis of available information, the commissioner finds that a person is in violation of any provision of this act, or any rule or regulation adopted, or permit or order issued, pursuant to this act, the commissioner may:

(1)    Issue an order requiring any such person to comply in accordance with subsection b. of this section; or

(2)    Bring a civil action in accordance with subsection c. of this section;  or

(3)    Levy a civil administrative penalty in accordance with subsection d. of this section; or

(4)    Bring an action for a civil penalty in accordance with subsection e. of this section; or

(5)    Petition the Attorney General to bring a criminal action in accordance with subsection f. of this section.

Recourse to any of the remedies available under this section shall not preclude recourse to any of the other remedies.

b.     Whenever, on the basis of available information, the commissioner finds a person in violation of any provision of this act, or of any rule or regulation adopted, or permit or order issued, pursuant to this act, the commissioner may issue an order: (1) specifying the provision or provisions of this act, or the rule, regulation, permit or order of which he is in violation; (2) citing the action which constituted the violation; (3) requiring compliance with the provision or provisions violated; (4) requiring the restoration of the freshwater wetland or transition area which is the site of the violation; and (5) providing notice to the person of his right to a hearing on the matters contained in the order.

c.     The commissioner is authorized to institute a civil action in Superior Court for appropriate relief from any violation of any provisions of this act, or any rule or regulation adopted, or permit or order issued, pursuant to this act. Such relief may include, singly or in combination:

(1)    A temporary or permanent injunction;

(2)    Assessment of the violator for the costs of any investigation, inspection, or monitoring survey which led to the establishment of the violation, and for the reasonable costs of preparing and bringing legal action under this subsection;

(3)    Assessment of the violator for any costs incurred by the State in removing, correcting, or terminating the adverse effects upon the freshwater wetland resulting from any unauthorized regulated activity for which legal action under this subsection may have been brought;

(4)    Assessment against the violator for compensatory damages for any loss or destruction of wildlife, fish or aquatic life, and for any other actual damages caused by an unauthorized regulated activity. Assessments under this

subsection shall be paid to the State Treasurer, except that compensatory damages shall be paid by specific order of the court to any persons who have been aggrieved by the unauthorized regulated activity;

(5)    A requirement that the violator restore the site of the violation to the maximum extent practicable and feasible.

d.    The commissioner is authorized to assess a civil administrative penalty of not more than $10,000.00 for each violation, and each day during which each violation continues shall constitute an additional, separate, and distinct offense. Any amount assessed under this subsection shall fall within a range established by regulation by the commissioner for violations of similar type, seriousness, and duration. No assessment shall be levied pursuant to this section until after the party has been notified by certified mail or personal service. The notice shall identify the section of the statute, regulation, or order or permit condition violated; recite the facts alleged to constitute a violation; state the amount of the civil penalties to be imposed; and affirm the rights of the alleged violator to a hearing. The ordered party shall have 20 days from receipt of the notice within which to deliver to the commissioner a written request for a hearing. After the hearing and upon finding that a violation has occurred, the commissioner may issue a final order after assessing the amount of the fine specified in the notice. If no hearing is requested, the notice shall become a final order after the expiration of the 20–day period. Payment of the assessment is due when a final order is issued or the notice becomes a final order. The authority to levy an administrative order is in addition to all other enforcement provisions in this act, and the payment of any assessment shall not be deemed to affect the availability of any other enforcement provisions in connection with the violation for which the assessment is levied. Any civil administrative penalty assessed under this section may be compromised by the commissioner upon the posting of a performance bond by the violator, or upon such terms and conditions as the commissioner may establish by regulation.

e.    A person who violates this act, an administrative order issued pursuant to subsection b., or a court order issued pursuant to subsection c., who fails to pay a civil administrative assessment in full pursuant to subsection d., shall be subject, upon order of a court, to a civil penalty not to exceed $10,000.00 per day of such violation, and each day during which the violation continues shall constitute an additional, separate, and distinct offense. Any civil penalty imposed pursuant to this subsection may be collected with costs in a summary proceeding pursuant to "the penalty enforcement law" (N.J.S. 2A:58-1 et seq.). The Superior Court shall have jurisdiction to enforce "the penalty enforcement law" in conjunction with this act.

    f.    A person who willfully or negligently violates this act shall be guilty, upon conviction, of a crime of the fourth degree and shall be subject to a fine of not less than $2,500.00 nor more than $25,000.00 per day of violation. A second offense under this subsection shall subject the violator to a fine of not less than $5,000.00 nor more than $50,000.00 per day of violation. A person who knowingly makes a false statement, representation, or certification in any application, record, or other document filed or required to be maintained under this act, or who falsifies, tampers with or knowingly renders inaccurate, any monitoring device or method required to be maintained pursuant to this act, shall, upon conviction, be subject to a fine of not more than $10,000.00.

    g.    In addition to the penalties prescribed in this section, a notice of violation of this act shall be recorded on the deed of the property wherein the violation occurred, on order of the commissioner, by the clerk or register of deeds and mortgages of the county wherein the affected property is located and with the clerk of the Superior Court and shall remain attached thereto until such time as the violation has been remedied and the commissioner orders the notice of violation removed.

    h.    If the violation is one in which the department has determined that the restoration of the site to its previolation condition would increase the harm to the freshwater wetland or its ecology, the department may issue an "after the fact" permit for the regulated activity that has already occurred; provided that assessment against the violator for costs or damages enumerated in subsection c. of this section has been made, the creation or restoration of freshwater wetlands resources at another site has been required of the violator, an opportunity has been afforded for public hearing and comment, and the reasons for the issuance of the "after the fact" permit are published in the New Jersey Register and in a newspaper of general circulation in the geographical area of the violation. Any person violating an "after the fact" permit issued pursuant to this subsection shall be subject to the provisions of this section.

i.    The burden of proof and degree of knowledge or intent required to establish a violation of this act shall be no greater than the burden of proof or degree of knowledge or intent which the United States Environmental Protection Agency must meet in establishing a violation of the Federal Act or implementing regulations.

j.    The department shall establish and implement a program designed to facilitate public participation in the enforcement of this act which complies with the requirements of the Federal Act and implementing regulations.

k.    The department shall make available without restriction any information obtained or used in the implementation of this act to the United States Environmental Protection Agency upon a request therefor.

l.    The department may require an applicant or permittee to provide any information the department requires to determine compliance with the provisions of this act.

m.    The department shall have the authority to enter any property, facility, premises or site for the purpose of conducting inspections, sampling of soil or water, copying or photocopying documents or records, and for otherwise determining compliance with the provisions of this act.

## C. 13:9B-22    Taking without just compensation.

22. a. Any person having a recorded interest in land affected by a freshwater wetlands permit issued, modified or denied pursuant to the provision of this act may file an action in a court of competent jurisdiction to determine if the issuance, modification or denial of the freshwater wetlands permit constitutes a taking of property without just compensation.

b.    If the court determines that the issuance, modification, or denial of a freshwater wetlands permit by the department pursuant to this act constitutes a taking of property without just compensation, the court shall give the department the option of compensating the property owner for the full amount of the lost value, condemning the affected property pursuant to the provisions of the "Eminent Domain Act of 1971," P.L. 1971, c. 361 (C. 20:3-1 et seq.), or modifying its action or inaction concerning the property so as to minimize the detrimental effect to the value of the property.

## C. 13:9B-23          General permits.

23. a. The department shall consider for adoption as general permits, to the extent practicable and feasible, and to the extent that this adoption is consistent to the maximum extent practicable and feasible with the provisions of this act, all applicable Nationwide Permits which were approved under the Federal Act as of November 13, 1986 by the U.S. Army Corps of Engineers.

b.    The department shall issue a general permit for an activity in a freshwater wetland which is not a surface water tributary system discharging into an inland lake or pond, or a river or stream, and which would not result in the loss or substantial modification of more than one acre of freshwater wetland, provided that this activity will not take place in a freshwater wetland of exceptional resource value. The department shall issue a general permit for a regulated activity in a freshwater wetland located in an area considered a headwater pursuant to the Federal Act if the regulated activity would not result in the loss or substantial modification of more than one acre of a swale or a man-made drainage ditch. The provisions of this subsection shall not apply to any wetlands designated as priority wetlands by the United States Environmental Protection Agency.

c.    The department shall issue additional general permits on a Statewide or regional basis for the following categories of activities, if the department determines, after conducting an environmental analysis and providing public notice and opportunity for a public hearing, that the activities will cause only minimal adverse environmental impacts when performed separately, will have only minimal cumulative adverse impacts on the environment, will cause only minor impacts on freshwater wetlands, will be in conformance with the purposes of this act, and will not violate any provision of the Federal Act:

(1)    Maintenance, reconstruction, or repair of roads or public    utilities    lawfully    existing    prior    to    the

effective date of this act or permitted under this act, provided that such activities do not result in disturbance of additional wetlands upon completion of the activity;

(2)   Maintenance or repair of active irrigation or drainage ditches lawfully existing prior to the effective date of this act or permitted under this act, provided that such activities do not result in disturbance of additional freshwater wetlands upon completion of the activity;

(3)   Appurtenant improvements or additions to residential dwellings lawfully existing prior to the effective date of this act, provided that the improvements or additions require less than a cumulative surface area of 750 square feet of fill and will not result in new alterations to a freshwater wetland outside of the fill area;

(4)   Mosquito management activities determined to be consistent with best mosquito control and freshwater wetlands management practices and for which all appropriate actions to minimize adverse environmental effects have been or shall be taken.

(5)   Activities, as determined by the department, which will have no significant adverse environmental impact on freshwater wetlands, provided that the issuance of a general permit for any such activities is consistent with the provisions of the Federal Act and has been approved by the United States Environmental Protection Agency.

(6)   Regulated activities which have received individual or general permit approval or a finding of no jurisdiction by the U.S. Army Corps of Engineers pursuant to the Federal Act, and which have received a grant waiver pursuant to the "National Environmental Policy Act of 1969" (42 U.S.C. § 4321 et seq.); provided, that upon the expiration of a permit any application for a renewal or modification thereof shall be made to the department.

(7)   State or federally funded roads planned and developed in accordance with the "National Environmental Policy Act of 1969" and the Federal Act, and with Executive Order Number 53, approved October 5, 1973 and for which application has been made prior to the effective date of this act to the United States Army Corps of Engineers for an individual or general permit under the Federal Act; provided that upon expiration of a permit any application for a renewal or modification thereof shall be made to the department, and, provided, further, that the department shall not require transition areas as a condition of the renewal or modification of the permit.

(8) Maintenance and repair of storm water management facilities lawfully constructed prior to the effective date of this act or permitted under this act, provided that these activities do not result in disturbance of additional freshwater wetlands upon completion of the activity.

(9) Maintenance, reconstruction, or repair of buildings or structures lawfully existing prior to the effective date of this act or permitted under this act, provided that these activities do not result in disturbance of additional freshwater wetlands upon completion of the activity.

d. The department may, on the basis of findings with respect to a specific application, modify a general permit issued pursuant to this section by adding special conditions. The department may rescind a general permit and require an application for an individual permit if the commissioner finds that additional permit conditions would not be sufficient and that special circumstances make this action necessary to insure compliance with this act or the Federal Act.

e. The department shall review general permits adopted or authorized pursuant to subsection c. every five years, which review shall include public notice and opportunity for public hearing. Upon this review the department shall either modify, reissue or revoke a general permit. If a general permit is not modified or reissued within five years of publication in the New Jersey Register, it shall automatically expire.

f. The date of publication of the general permits authorized by subsections a. and b. of this section shall be the effective date of this act.

g. A person proposing to engage in an activity covered by a general permit shall provide written notice to the department containing a description of the proposed activity at least 30 working days prior to commencement of work. The department, within 30 days of receipt of this notification, shall notify the person proposing to engage in the activity covered by a general permit as to whether an individual permit is required for the activity.

## C. 13:9B-24          Temporary emergency permit.

24. a. Notwithstanding the provisions of this or any other act to the contrary, the department may issue a temporary emergency freshwater wetlands permit for a regulated activity if:

(1)    An unacceptable threat to life or severe loss of property will occur if an emergency permit is not granted; and

(2)    The anticipated threat or loss may occur before a permit can be issued or modified under the procedures otherwise required by this act and other applicable State law.

b.    The emergency permit shall incorporate, to the greatest extent practicable and feasible but not inconsistent with the emergency situation, the standards and criteria required for non–emergency regulated activities under this act and shall:

(1)    Be limited in duration to the time required to complete the authorized emergency activity, not to exceed 90 days;

(2)    Require the restoration of the freshwater wetland within this 90 day period, except that if more than the 90 days from the issuance of the emergency permit is required to complete restoration, the emergency permit may be extended to complete this restoration.

c.    The emergency permit may be issued orally or in writing, except that if it is issued orally, a written emergency permit shall be issued within five days thereof.

d.    Notice of the issuance of the emergency permit shall be published and public comments received, in accordance with the provisions of the Federal Act, and applicable State law, provided that this notification shall be sent no later than 10 days after issuance of the emergency permit.

e.    The emergency permit may be terminated at any time without process upon a determination by the department that this action is appropriate to protect human health or the environment.

## C. 13:9B–25    Rules, regulations.

25. a. Within 10 months of the enactment of this act, and after a 60 day comment period, the department shall adopt, pursuant to the provisions of the "Administrative Procedure Act," any rules and regulations necessary to implement the provisions of this act. These rules and regulations shall include the general permits which the department will issue pursuant to section 23 of this act.

b.    Within one year of the enactment of this act, the department shall adopt, in consultation with the United States Environmental Protection Agency, a list of vegetative species classified as hydrophytes, as defined in section 3 of this act, indicative of freshwater wetlands and consistent with the geographical regions of the State.

c. The department shall develop a functional, complete, and up to date composite freshwater wetlands map and inventory using the most recent available data, which shall include, but need not be limited to, aerial photographs and soil inventories at a scale suitable for freshwater wetlands regulatory purposes, and shall make appropriate sections of this map and inventory available on a periodic basis to the county clerk or register of deeds and mortgages in each county, as appropriate, and to the clerk of each municipality.

## C. 13:9B-26   Distribution of National Wetlands Inventory maps.

26. The department shall, within 180 days of enactment of this act, forward to the clerk of each municipality copies of the appropriate National Wetlands Inventory maps for the State prepared by the United States Fish and Wildlife Service and direct the clerk to notify the residents of the municipality of the availability for inspection of these maps, by publication in a newspaper of general circulation. The department shall inform the clerk of each municipality that these maps have not been determined to be accurate for the purposes of locating the actual wetlands boundary, and that the department will be preparing a composite freshwater wetlands map and inventory at the specified uniform scale.

## C. 13:9B-27   Assumption of permit jurisdiction.

27. a. The department and the Attorney General shall take all appropriate action to secure the assumption of the permit jurisdiction exercised by the United States Army Corps of Engineers pursuant to the Federal Act. The department shall make an initial application to the United States Environmental Protection Agency for this assumption within one year of enactment of this act, and shall provide the Governor and the Legislature with a schedule therefor and a copy of the application and supporting material forwarded to the federal government.

b. The department shall utilize, to the maximum extent practicable and feasible, forms and procedures for permit applications which are identical to those used by the United States Army Corps of Engineers in issuing permits under the Federal Act.

c.    The department shall seek to conduct the review of an application for a freshwater wetlands permit in conjunction with federal personnel responsible for reviewing an application for a permit under the Federal Act.

d.    It is the intention of the Legislature that the permit process imposed in this act be conducted by the department concurrently with the review conducted by the federal government until such time as the department secures assumption of the permit jurisdiction exercised by the United States Army Corps of Engineers.

### C. 13:9B-28        Public education program.

28.    The department shall, within one year of the effective date of this act, conduct a public education program on the provisions of this act and the rules and regulations adopted pursuant hereto.

### C. 13:9B-29        Report.

29.    The department shall, within two years of the effective date of this act, prepare and submit a report to the Governor, the President of the Senate and the Speaker of the General Assembly, and the Senate Energy and Environment Committee and the Assembly Energy and Natural Resources Committee, or their designated successors.  The report shall describe:

(1)    The success or failure of mitigation measures performed in actual development situations, both within the State and in other states, the nature of the mitigation measures, and the state-of-the-art techniques used for mitigation; and

(2)    Recommendations for legislative or administrative action necessary to ensure the long term protection of freshwater wetlands from damage and degradation resulting from land use activities, pollution, and hydrologic changes which occur in upstream regions of the same watersheds of particular freshwater wetlands.

### C. 13:9B-30        Local regulation preempted.

30.    It is the intent of the Legislature that the program established by this act for the regulation of freshwater wetlands constitute the only program for this regulation in the State except to the extent that these areas are regulated consistent with the provisions of section 6 of this act.  To this end no municipality, county, or political subdivision thereof, shall enact, subsequent to the

effective date of this act, any law, ordinance, or rules or regulations regulating freshwater wetlands, and further, this act, on and subsequent to its effective date, shall supersede any law or ordinance regulating freshwater wetlands enacted prior to the effective date of this act. Between the enactment and effective date of this act, no municipality, county, or political subdivision thereof shall enact any law, ordinance, or rule and regulation requiring a transition area adjacent to a freshwater wetland; provided however, that any such law, ordinance, or rule and regulation adopted prior to the enactment of this act shall be valid until the effective date of this act.

31. Section 5 of P.L. 1977, c. 74 (C. 58:10A-5) is amended to read as follows:

**C. 58:10A-5** **Powers of department.**

5. The department is empowered to:

a. Exercise general supervision of the administration and enforcement of this act and all rules, regulations and orders promulgated hereunder;

b. Assess compliance of a discharger with applicable requirements of State and Federal law pertaining to the control of pollutant discharges and the protection of the environment and, also, to issue certification with respect thereto as required by section 401 of the Federal Act;

c. Assess compliance of a person with applicable requirements of State and federal law pertaining to the control of the discharge of dredged and fill material into the waters of the State and the protection of the environment and, also, to issue, deny, modify, suspend, or revoke permits with respect thereto as required by section 404 of the "Federal Water Pollution Control Act Amendments of 1972," as amended by the "Clean Water Act of 1977," (33 U.S.C. § 1344), and implementing regulations;

d. Advise, consult, and cooperate with other agencies of the State, the federal government, other states and interstate agencies, including the State Soil Conservation Committee, and with affected groups, political subdivisions and industries in furtherance of the purposes of this act;

e. Administer State and federal grants to municipalities, counties and other political subdivisions, or any recipient approved by the commissioner according to terms and conditions approved by him in order to meet the goals and objectives of this act.

32.   Section 6 of P.L. 1977, c. 74 (C. 58:10A-6) is amended to read as follows:

## C. 58:10A-6        NJPDES permits; exemptions.

6. a.   It shall be unlawful for any person to discharge any pollutant, except in conformity with a valid New Jersey Pollutant Discharge Elimination System permit that has been issued by the commissioner pursuant to this act or a valid National Pollution Discharge Elimination System permit issued by the administrator pursuant to the Federal Act, as the case may be.

b.   It shall be unlawful for any person to build, install, modify or operate any facility for the collection, treatment or discharge of any pollutant, except after approval by the department pursuant to regulations adopted by the commissioner.

c.   The commissioner is hereby authorized to grant, deny, modify, suspend, revoke, and reissue NJPDES permits in accordance with this act, and with regulations to be adopted by him.  The commissioner may reissue, with or without modifications, an NJPDES permit duly issued by the federal government as the NJPDES permit required by this act.

d.   The commissioner may, by regulation, exempt the following categories of discharge, in whole or in part, from the requirement of obtaining a permit under this act; provided, however, that an exemption afforded under this section shall not limit the civil or criminal liability of any discharger nor exempt any discharger from approval or permit requirements under any other provision of law:

(1)   Additions of sewage, industrial wastes or other materials into a publicly owned sewage treatment works which is regulated by pretreatment standards;

(2)   Discharges of any pollutant from a marine vessel or other discharges incidental to the normal operation of marine vessels;

(3)   Discharges from septic tanks, or other individual waste disposal systems, sanitary landfills, and other means of land disposal of wastes;

(4)   Discharges of dredged or fill materials into waters for which the State could not be authorized to administer the section 404 program under section 404(g) of the "Federal Water Pollution Control Act Amendments of 1972," as amended by the "Clean Water Act of 1977" (33 U.S.C. § 1344) and implementing regulations;

(5)    Nonpoint source discharges;

(6)    Uncontrolled nonpoint source discharges composed entirely of storm water runoff when these discharges are uncontaminated by any industrial or commercial activity unless these particular storm water runoff discharges have been identified by the administrator or the department as a significant contributor of pollution;

(7)    Discharges conforming to a national contingency plan for removal of oil and hazardous substances, published pursuant to section 311(c)(2) of the Federal Act.

e.    The commissioner shall not issue any permit for:

(1)    The discharge of any radiological, chemical or biological warfare agent or high-level radioactive waste into the waters of this State;

(2)    Any discharge which the United States Secretary of the Army, acting through the Chief of Engineers, finds would substantially impair anchorage or navigation;

(3)    Any discharge to which the administrator has objected in writing pursuant to the Federal Act;

(4)    Any discharge which conflicts with an areawide plan adopted pursuant to law.

f.    A permit under this act shall require the permittee:

(1)    To achieve effluent limitations based upon guidelines or standards established pursuant to the Federal Act or this act, together with such further discharge restrictions and safeguards against unauthorized discharge as may be necessary to meet water quality standards, areawide plans adopted pursuant to law, or other legally applicable requirements;

(2)    Where appropriate, to meet schedules for compliance with the terms of the permit and interim deadlines for progress or reports of progress towards compliance;

(3)    To insure that all discharges are consistent at all times with the terms and conditions of the permit and that no pollutant will be discharged more frequently than authorized or at a level in excess of that which is authorized by the permit;

(4)    To submit application for a new permit in the event of any contemplated facility expansion or process modification that would result in new or increased discharges or, if these would not violate effluent limitations or other restrictions specified in the permit, to notify the commissioner of such new or increased discharges;

(5) To install, use and maintain such monitoring equipment and methods, to sample in accordance with such methods, to maintain and retain such records of information from monitoring activities, and to submit to the commissioner such reports of monitoring results as he may require;

(6) At all times, to maintain in good working order and operate as effectively as possible, any facilities or systems of control installed to achieve compliance with the terms and conditions of the permit.

g. The commissioner shall have a right of entry to all premises in which a discharge source is or might be located or in which monitoring equipment or records required by a permit are kept, for purposes of inspection, sampling, copying or photographing.

h. In addition, any permit issued for a discharge from a municipal treatment works shall require the permittee:

(1) To notify the commissioner in advance of the quality and quantity of all new introductions of pollutants into a facility and of any substantial change in the pollutants introduced into a facility by an existing user of the facility, except for such introductions of nonindustrial pollutants as the commissioner may exempt from this notification requirement when ample capacity remains in the facility to accommodate new inflows. Such notifications shall estimate the effects of such changes on the effluents to be discharged into the facility.

(2) To establish an effective regulatory program, alone or in conjunction with the operators of sewage collection systems, that will assure compliance and monitor progress toward compliance by industrial users of the facilities with user charge and cost recovery requirements of the Federal Act or State law and toxicity standards adopted pursuant to this act and pretreatment standards.

(3) As actual flows to the facility approach design flow or design loading limits, to submit to the commissioner for his approval, a program which the permittee and the persons responsible for building and maintaining the contributory collection system shall pursue in order to prevent overload of the facilities.

i. All owners of municipal treatment works are hereby authorized to prescribe terms and conditions, consistent with applicable State and federal law, upon which pollutants may be introduced into such works, and to exercise the same right of entry, inspection, sampling and copying with respect to users of such works as are vested in the commissioner by this act or by any other provision of State law.

j.    In reviewing permits submitted in compliance with this act and in determining conditions under which such permits may be approved, the commissioner shall encourage the development of comprehensive regional sewerage facilities which serve the needs of the regional community and which conform to the adopted area-wide water quality management plan for that region.

33.    There is appropriated from the General Fund to the department the sum of $60,000.00, which shall be used to undertake and coordinate all activities required to implement the provisions of this act on the effective date of this act.

34.    This act shall take effect one year after enactment, except that section 25, section 26, section 27, section 30 and section 33 shall take effect immediately, and except that the department shall not implement the provisions of sections 16, 17, and 18 until two years after enactment. The department shall take any administrative actions prior to the effective date of this act necessary to implement the provisions of this act on and after the effective date.

Approved July 1, 1987.

## SUMMARY OF REGULATIONS—SURFACE WATER TREATMENT REQUIREMENTS FOR GIARDIA LAMBLIA AND OTHER BIOLOGICAL CONTAMINANTS

JUN 1 9 1989

### Background

°   Proposal was published in Federal Register on November 3, 1987 (52 FR 42178)

°   Notice of Availability, describing new regulatory options, was published in the Federal Register on May 6, 1988 (53 FR 16348).

°   Final rule promulgated June, 1989.

### Maximum Contaminant Level Goals

| | |
|---|---|
| Giardia Lamblia | 0 |
| Viruses | 0 |
| Legionella | 0 |
| Turbidity | none |
| Heterotrophic Plate Count (HPC) | none |

### General Requirements

Coverage: All public water systems using any surface water or ground water under direct influence of surface water must disinfect, and may be required by the State to filter, unless certain water quality source requirements and site specific conditions are met.

Treatment technique requirements are established in lieu of MCLs for Giardia, viruses, heterotrophic plate count bacteria, Legionella and turbidity.

Treatment must achieve at least 99.9 percent removal and/or inactivation of Giardia lamblia cysts and 99.99 percent removal and/or inactivation of viruses.

All systems must be operated by qualified operators as determined by the State.

### Criteria to be Met to Avoid Filtration

Source Water Criteria

Fecal coliform concentration must not exceed 20/100 ml or the total coliform concentration must not exceed 100/100 ml before disinfection in more than ten percent of the measurements for the previous six months, calculated each month.

Minimum sampling frequencies for fecal or total coliform determinations are:

| SYSTEM SIZE (persons) | SAMPLES/WEEK |
|---|---|
| < 501 | 1 |
| 501-3,300 | 2 |
| 3,301-10,000 | 3 |
| 10,001-25,000 | 4 |
| > 25,000 | 5 |

If not already conducted under the above requirements, a coliform test must be made each day that the turbidity exceeds 1 NTU.

Turbidity levels must be measured every four hours by grab sample or continuous monitoring. The turbidity level may not exceed 5 NTU. If the turbidity exceeds 5 NTU, the system must install filtration unless the State determines that the event is unusual or unpredictable, and the event does not occur more than two periods in any one year, or five times in any consecutive ten years. An "event" is one or more consecutive days when at least one turbidity measurement each day exceeds 5 NTU.

Site Specific Conditions

Disinfection

Disinfection must achieve at least a 99.9 and 99.99 percent inactivation of _Giardia_ cysts and viruses, respectively. This must be demonstrated by the system meeting "CT" values in the rule ("CT" is the product of residual concentration (mg/l) and contact time (minutes) measured at peak hourly flow). "C" and "T" must be determined at or prior to the first customer. The total percent inactivation can be calculated based on unlimited disinfectant residual measurements in sequence prior to the first customer. Failure to meet this requirement on more than one day in a month is a violation. Filtration is required if a system has two or more violations in a year, unless the State determines that at least one of these violations was caused by circumstances that were unusual and unpredictable. A third violation in 12 months, regardless of the cause, triggers filtration.

Disinfection systems must a) have redundant components including alternate power supply, automatic alarm and start-up to ensure continuous disinfection of the water during plant operation or b) have automatic shut-off of delivery of water to the distribution system whenever the disinfectant residual is less than 0.2 mg/L, provided that the State determines that a shut-off would not pose a potential health risk to the system.

For systems using chloramines if chlorine is added prior to ammonia, the CT values for achieving 99.9 percent inactivation of Giardia lamblia cysts can also be assumed to achieve 99.99 percent inactivation of viruses. Systems using chloramines and adding ammonia prior to chlorine must demonstrate with on-site studies that they achieve 99.99 percent inactivation of viruses. For systems using disinfectants other than chlorine, the system may demonstrate that other CT values than those specified in the rule, or other disinfection conditions are provided which achieve at least 99.9 and 99.99 percent inactivation of Giardia lamblia and viruses, respectively.

Disinfectant residuals in the distribution system cannot be undetectable in more than five percent of the samples, each month, for any two consecutive months. Samples must be taken at the same frequency as total coliforms under the revised coliform rule. A system may measure for HPC in lieu of disinfectant residual. If the HPC measurement is less than 500 colonies/ml, the site is considered to have a "detectable" residual for compliance purposes. Systems in violation of this requirement must install filtration unless the State determines that the violation is not caused by a deficiency of treatment of the source water. For systems which cannot maintain a residual or practically monitor for HPC, the State can judge whether adequate disinfection is provided in the distribution system.

Systems must maintain a disinfectant residual concentration of at least 0.2 mg/l in the water entering the system, demonstrated by continuous monitoring. If there is a failure in the continuous monitoring, the system may substitute grab sample monitoring every four hours for up to five days. If the disinfectant residual falls below 0.2 mg/l, the system must notify the State as soon as possible but no later than the end of the next business day; notification must include whether or not the residual was restored within four hours. If the residual is not restored to at least 0.2 mg/l within four hours, it is a violation and the system must filter, unless the State determines that the violation was caused by unusual and unpredictable circumstances. Systems serving 3300 people or less can take daily grab samples in lieu of continuous monitoring. Minimum grab sampling frequencies are: 1/day < 501 people; 2/day 501 - 1000 people; 3/day 1001 - 2500 people; 4/day 2501 - 3300 people. If at any time the residual is below 0.2 mg/l, the system must conduct grab sample monitoring every four hours until the residual is restored.

Other Conditions

Systems must maintain a watershed control program which will
minimize the potential for contamination by human enteric
viruses and Giardia lamblia cysts.  Systems must monitor and
control the activities in the watershed that may have an
adverse impact on water.  Systems must demonstrate through
ownership or written agreements with landowners in the
watershed that they are able to limit and control all human
activities that may have an adverse impact upon water
quality.  The watershed control program and disinfection
treatment must be inspected on-site and approved by the
State annually.

Systems must not have had any waterborne disease outbreaks,
or if they have, such systems must have been modified to
prevent another such occurrence, as determined by the State.

Systems must not be out of compliance with the monthly MCL
for total coliforms for any two months in any consecutive 12
month period, unless the State determines that the
violations are not due to treatment deficiency of the source
water.

Systems serving more than 10,000 people must be in
compliance with MCL requirements for total trihalomethanes.

## Criteria for Filtered Systems

Turbidity Monitoring

Turbidity must be measured every four hours by grab sample
or continuous monitoring.  For systems using slow sand
filtration or filtration technologies other than conventional treatment, direct filtration or diatomaceous earth
filtration, the State may reduce the sampling frequency to
once per day.  The State may reduce monitoring to one grab
sample per day for all systems serving less than 500 people.

Turbidity Removal

Conventional filtration or direct filtration water must
achieve a turbidity level in the filtered water at all times
less than 5 NTU and not more than 0.5 NTU in more than five
percent of the measurements taken each month.  The State may
increase the 0.5 NTU limit up to less than 1 NTU in greater
than or equal to 95% of the measurements, without any
demonstration by the system, if it determines that OVERALL
TREATMENT with disinfection achieves at least 99.9 percent
and 99.99 percent removal/inactivation of Giardia cysts and
viruses, respectively.

Slow sand filtration must achieve a turbidity level in the filtered water at all times less than 5 NTU and not more than 1 NTU in more than five percent of the samples taken each month. The turbidity limit of 1 NTU may be increased by the State (but at no time exceed 5 NTU) if it determines that there is no significant interference with disinfection.

Diatomaceous earth filtration must achieve a turbidity level in the filtered water at all times less than 5 NTU and of not more than 1 NTU in more than five percent of the samples taken each month.

Other filtration technologies may be used if the system demonstrates to the State that they achieve at least 99.9 and 99.99 percent removal/inactivation of Giardia lamblia cysts and viruses, respectively, and are approved by the State. Turbidity limits for these technologies are the same as those for slow sand filtration, including the allowance of increasing the turbidity limit of 1 NTU up to 5 NTU, but at no time exceeding 5 NTU upon approval by the State.

## Disinfection Requirements

Disinfection with filtration must achieve at least 99.9 and 99.99 percent removal/inactivation of Giardia cysts and viruses, respectively. The States define the level of disinfection required, depending on technology and source water quality. Guidance on the use of CT values to make these determinations is available in the Guidance Manual. Recommended levels of inactivation are based on expected occurrence levels of Giardia cysts in the source water and the filtration technology in place. Disinfection requirements for point of entry to the distribution system and within the distribution system are the same as for unfiltered systems.

## Analytical Requirements

Except for ozone, testing and sampling must be in accordance with Standard Methods, 16th edition, or methods approved by EPA for total coliforms, fecal coliform, turbidity, disinfectant residuals, temperature, and pH. Residual disinfectant concentrations for ozone must be measured by the Indigo Method or automated methods which are calibrated in reference to the results obtained by the Indigo Method.

## Reporting

All parameters required in the rule must be reported monthly to the State. Unfiltered water systems must also report annually on their watershed control program and on-site inspections.

Compliance

Surface Water Systems

Unfiltered systems must meet monitoring requirements beginning 18 months following promulgation, unless the State has determined that filtration is required. Unfiltered systems must meet the criteria to avoid filtration beginning 30 months following promulgation, unless the State has determined that filtration is required, or they are in violation of a treatment technique requirement. Unfiltered systems must install filtration within 18 months following the failure to meet any one of the criteria to avoid filtration, or within 48 months following promulgation, whichever is later, or they are in violation.

Filtered systems must meet monitoring and performance requirements beginning 48 months following promulgation.

The interim turbidity monitoring and MCL requirements will remain in effect for unfiltered systems until 30 months following promulgation, and for filtered systems until 48 months following promulgation. For systems which prior to 30 months following promulgation the State determines must filter, the interim turbidity requirements will remain in effect until 48 months following promulgation, or until filtration is installed, whichever is later.

Ground Water Systems Under Direct Influence of Surface Water

All systems using ground water under direct influence of surface water must meet the treatment requirements under the SWTR. States must determine which community and non-community ground water systems are under direct influence of surface water within 5 years and 10 years, respectively, following promulgation. Unfiltered systems under the direct influence of surface water must begin monitoring within 6 months following the determination of direct influence unless the State has determined that filtration is required. Systems under direct influence of surface water must begin meeting the criteria to avoid filtration 18 months after the determination of direct influence, unless the State has determined that filtration is required. Unfiltered systems under direct influence of surface water must install filtration within 18 months following the failure to meet any of the criteria to avoid filtration.

Variances

Variances are not applicable.

Exemptions

Exemptions are allowed for the requirement to filter. Systems using surface water must disinfect (i.e., no exemptions). Exemptions are allowed for the level of disinfection required.


Source:  U.S. Environmental Protection Agency.

Thomas H. Kean, *Governor*
Richard T. Dewling, Ph.D., P.E., *Commissioner*

# TWO-PART PUMP TEST FOR EVALUATING THE WATER-SUPPLY CAPABILITIES OF DOMESTIC WELLS

## New Jersey Geological Survey Ground-Water Report Series No. 1

by
Jeffrey L. Hoffman and Robert Canace

Department of Environmental Protection
Division of Water Resources
CN029
Trenton, NJ 08625

John W. Gaston, Jr., P.E., Director
Haig F. Kasabach, State Geologist

1986

## CONTENTS

## ILLUSTRATIONS

## TABLES

# TWO-PART PUMP TEST FOR EVALUATING THE
# WATER-SUPPLY CAPABILITIES OF DOMESTIC WELLS

by
Jeffrey L. Hoffman and Robert Canace

Key terms: aquifer test, water yield, well construction, constant head test
New terms: peak demand rate, peak demand test

## ABSTRACT

An evaluation of the capability of a well drilled in rock to supply domestic needs can be based on predicted household water use patterns and characteristics of the well. Household water demand is estimated using a residency rate of two per bedroom and a usage rate of 100 gallons per capita daily. Water use is assumed to be split between equal morning and evening periods of peak demand. Rate of usage during peak periods is estimated at 3 gallons per minute per bathroom. Duration of the peak demand periods is estimated by dividing total usage during peak demand periods by the average rate of usage during peaks.

Capability of the well to meet peak demand and total daily demand is evaluated through a two-part pump test. The first part, a peak demand test, is a drawdown test to determine whether or not the combined well storage and aquifer contribution to well flow can meet peak needs. The second part, a constant head test, is to determine whether or not flow from the aquifer is sufficient to meet total daily needs. If so it is assumed that the aquifer can meet long-term household needs. Measurements of drawdown during the pump test are used in determining the depth at which the pump should be placed in the well and the necessary depth if the well needs to be deepened to provide additional storage.

The method does not take into account extreme droughts, interfering stresses on the aquifer from other pumping or decreasing efficiency of the well and pump due to aging.

## INTRODUCTION

In response to inquiries by local health agencies and the public concerning domestic well failures, the New Jersey Geological Survey has developed a method to estimate the water supply needs of private homes and to evaluate the adequacy of wells drilled to supply those households. The method consists of a calculation to estimate total daily needs and peak demand needs of a household and a two-part pumping test to determine whether or not a well can meet these needs. The procedure is intended primarily for use in areas of consolidated bedrock (Regions 2 and 3 of the New Jersey water well construction regulations (NJAC 7:10-35), shown in figure 1).

The yield of a well is usually established by pumping the well and measuring the discharge (well flow), in gallons per minute, from the well head. Unfortunately, pump test requirements for domestic water wells have not been standardized. The New Jersey Department of Environmental Protection requires only that "each well be tested for yield and drawdown" (NJAC 7:10-3.58). No testing procedures are stipulated under the code. A variety of local ordinances regulate well testing. Most of these have been established on the basis of experience in the field or other local ordinances. Most have fixed minimum yield requirements regardless of household size for issuance of the certificate of occupancy.

The method outlined in this report is modified from a procedure developed by the Connecticut Well Drillers Association (Hunt, 1978). It supplements the Connecticut approach by providing a systematic method based on anticipated peak and long-term demand to design, perform and evaluate a pump test.

The pump test is divided into two parts. The first part, peak demand test, is to see if the well can meet the predicted water demand of the house during twice-daily periods of peak use. The second part, the constant head test, is to measure the aquifer's ability to transmit sufficient water to the well for the total daily water demand of the household.

The pump test identifies satisfactory wells and, for wells that are not satisfactory, whether the problem lies with insufficient storage or with inability of the aquifer to transmit sufficient

Figure 1. Geologic regions of New Jersey as defined in water well construction regulations. (NJAC 7:10-35)

water to the well. Insufficient storage can be corrected by deepening the well to provide additional storage within the borehole, redrilling to a greater diameter or constructing a surface storage tank. The necessary depth to which the well must be deepened can be estimated using measurements of aquifer yield and drawdown taken during the constant head test. Insufficient aquifer yield may, in some instances, be corrected by developing the well or deepening it to encounter additional water-bearing zones.

Equations, tables and worksheets are provided to assist in the recording and analysis of pump test data. A flow diagram is also provided which shows the evaluation procedures and options should the well fail. Hypothetical examples have been included to demonstrate the method.

It must be recognized that this method is a simplified mathematical approach that cannot take into account many of the physical factors that can influence long-term well performance. These include seasonal fluctuations in ground water availability, extreme drought, permanent dewatering of fracture zones, stresses on the aquifer from nearby wells and reduction of well and pump efficiency due to clogging or wear. These factors may result in failure of wells which were rated as satisfactory at the time of testing.

## BASIS AND BACKGROUND

A method for estimating the water needs of individual households and evaluating domestic wells was developed by the Connecticut Well Drillers Association (Hunt, 1978). The impetus for the development of the Connecticut minimum well formula was an objection to a requirement by the Federal Housing Authority that a home have a well with a yield of 5 gpm (gallons per minute) or more to qualify for loan assistance. The Well Drillers Association, convinced that 5 gpm was an unreasonable across-the-board requirement, demonstrated that smaller yields could be certified as adequate for domestic needs if household demand were taken into account. The Connecticut minimum well formula:

1. estimates peak load (household water demand, in gallons, within each of two daily peak use periods);

2. estimates peak time (the length of peak use periods in minutes);

3. evaluates the capability of the well to meet the peak demand of the household, and;

4. establishes a minimum pump capacity and pump installation depth to assure an adequate supply.

In the Connecticut method, if the well can deliver the peak load within the peak time it is considered to be satisfactory. This determines whether or not the well can supply peak needs. But like many domestic well tests, it does not evaluate the long-term ability of the aquifer to supply the well. Also, for a well which fails the pump test, the Connecticut method does not determine whether the cause of failure is inadequate storage or inability of the aquifer to transmit sufficient water. The two-part pump test presented here uses a peak demand test followed by a constant head test. The constant head test measures the rate at which water can move from the aquifer to the well. This test determines whether or not the well can transmit sufficient water for long-term needs. If the well fails either test, the constant head test provides guidance as to how the well might be modified to meet the household demand. For instance, if the well can meet the total daily demand but not the peak demand, the constant head test results can be used to estimate the depth to which the well would have to be drilled to provide sufficient storage in the bore-hole to meet the peak demand. If flow from the aquifer falls short of total daily demand, then the well must be deepened to encounter additional water-bearing zones or redrilled at another location, but the test cannot provide an estimate of the required well depth.

## ASSUMPTIONS

In order to calculate peak load and peak time, the Connecticut formula and the modification presented here rely on four assumptions concerning water use. The assumptions are conservative. While one or more of the assumptions may not be met by a particular household, there is a sufficient margin of safety that water use estimates will be valid for determining the adequacy of wells for most peak periods for most families. Orndorff (1966) reported that on days when intensive chores are performed the peak demand may be several times the average demand. The assumptions are not valid for these days. The assumptions are:

1. Each person uses 100 gpd (gallons per day).

    Comments: The use of 100 gpd per capita is consistent with *Standards for the Construction of Public Non-Community and Non-Public Water Supplies* (NJAC 7:10-3.32), in which this value is applied as a planning criterion. This is a conservative figure which exceeds most measured values for water consumption. Reported average per capita consumption is approximately half this volume, or about 50 gpd (U. S. Environmental Protection Agency, 1978; Orndorff, 1966, p. 30, table 7). Average per capita consumptions of up to 80 gpd have been reported (Linaweaver and others, 1967, p. 2). In addition, in a detailed study of domestic water use Orndorff (1966) found that though total usage increases with family size, per capita usage decreases. This provides an additional margin of safety for larger families.

2. Two people occupy each bedroom.

    Comments: The average residency rate in the United States in 1980 was reported to be 2.75 residents per dwelling (U.S. Department of Commerce, 1981, table 10). Inasmuch as most new dwellings have two or more bedrooms, an estimate of two people per bedroom is conservative and allows for later addition of bedrooms or higher-than-average occupancy rates.

3. Most daily water usage occurs during two peak periods.

    Comments: In his study of domestic water use Orndorff (1966, p. 23) concluded that "peak demands tend to occur during two particular times of the day." One of these is in the morning and the other in the evening. While most water use occurs during these peaks, there is additional use at other times. Because of water use during off-peak times, it is reasonable that no more than half the total daily water use will occur during a single peak demand period. Differences in habitual water use patterns among families will thus seldom be such that peak water use will be underestimated.

4. Water flows through fixtures at the rate of three gallons per minute per bathroom during peak periods.

    Comments: This is the key assumption in estimating peak time (the duration of peak demand periods). In the Connecticut formula it is assumed that water use during peak time is in large part bathroom use. Orndorff (1966) points out that, although a normal daily peak demand can be determined statistically, water demand of particular households is established by habitual patterns. Bathroom use would be the prime example.

    New Jersey's *Standards for the Construction of Public Non-Community and Non-Public Water Systems* (NJAC 7:10- 3.10, et seq., 1978) require that water systems provide a minimum flow rate of 2 gpm at each plumbing fixture. Orndorff reported average peak demand rates of 1.60 gpm for a subdivision served by on-site wells and 2.29 gpm in homes served by an external water source. The use of 3 gpm provides a margin of safety above measured rates of water usage and the requirements established for domestic water supply systems.

For making calculations, it is assumed that water flows through the fixtures of a half bath at 1.5 gpm.

## THEORY

### Calculation of Household Water Demand

In order to establish conditions for a pump test, it is necessary to quantify the total daily demand and peak demand which will be placed on the well. Total daily demand is a function of the number of residents and per capita usage. Peak demand can be quantified in terms of volume, time and rate. The total volume of water the household will require during each of two daily periods of peak demand is the **peak load**. The average rate of use during peak demand times is the **peak demand rate**. The length of time within which the peak load demand for water will be exerted is the **peak time**.

To quantify the concepts of peak load, peak time and peak demand rate the following assumptions, discussed above, are applied:

1. each person uses 100 gallons per day.

2. two people occupy a bedroom.

3. most daily water usage occurs during two peak periods.

4. water flows through fixtures at the rate of three gallons per minute per bathroom during peak periods.

Under these assumptions equations for peak load, peak time and peak demand rate can be expressed as:

peak load (gallons) =

$$\frac{(no.\ of\ bedrooms)\ (persons/bedroom)\ (gallons/person/day)}{peak\ periods/day}$$

$$= \frac{(no.\ of\ bedrooms)\ (2)\ (100)}{2} = (no.\ of\ bedrooms)\ (100) \quad (1)$$

peak demand rate (gpm) = (gpm/bathroom) (no. of bathrooms)

$$= (3)\ (no.\ of\ bathrooms) \quad (2)$$

peak time (minutes) = $\dfrac{peak\ load\ (gallons)}{peak\ demand\ rate\ (gpm)}$ \quad (3)

As an example, a three bedroom house with two bathrooms will have:

peak load = 3 bedrooms X 100 gallons/bedroom

= 300 gallons

peak demand rate = 3 gpm/bathroom X 2 bathrooms

= 6 gpm

peak time = 300 gallons/6 gpm

= 50.0 minutes

### Pump Test Design

Well flow (discharge) is a combination of water pumped from the standing column of water in a well (the **well storage contribution**) and water flowing into the well from the aquifer

(the **aquifer contribution**). In any well evaluation it is necessary to recognize that the well acts as a water storage area. Water is taken from well storage during peak demand times and gradually replenished from the aquifer during off-peak times.

At the beginning of the two-part pump test, the water in a well is at the static level: the water in the well and aquifer are at the same pressure and there is no net flow into or out of the well. As soon as the pump goes on for the peak demand test, water is removed from the casing and the water level drops. Because the pressure is now lower in the well than in the aquifer, water will flow from the aquifer into the well. Until the water level stops dropping, the discharge pumped from the well includes well storage and aquifer contribution components. In general, though certainly not always, the aquifer contribution will increase as the water level in the well drops.

For the constant head test the water level must be stable or nearly stable, neither dropping nor rising rapidly. The water level may stabilize during the peak demand test due to increase in aquifer contribution, or it may be stabilized by decreasing the pumping rate. When the water level and pumping rate are stable the well is said to be at equilibrium, the pumping to be at a constant head pumping rate and the water to be at a constant head level.

### Peak Demand Test

The peak demand part of the two-part test determines whether or not the volume of water stored in the well plus the volume which will flow from the aquifer to the well during peak time will be sufficient for peak needs. The well is allowed to come to its static level, then is pumped at the peak demand rate for the peak time. A well for which the combined well storage and aquifer contributions are insufficient will fail before the expiration of the peak time.

### Constant Head Test

The constant head test determines whether or not the aquifer contribution will meet long-term needs. An accurate measurement of the aquifer contribution can be made when the well is being pumped under constant head conditions. If a well is being pumped but the water level is not changing, the volume of water stored in the well is not changing. The well storage contribution is therefore zero and all water flowing from the well is coming from the aquifer; the aquifer contribution is equal to the measured pumping rate. If the aquifer contribution rate is less than the flow required for total daily needs (see tables 3 and 4), the well fails this portion of the test.

### Test Results

If the well passes the constant head test but fails the peak demand test, the aquifer can supply enough water on a daily basis, but additional storage is required for peak demand needs. Storage may be provided in either the well or a surface storage tank.

If a well fails the constant head test it will not supply enough water on a daily average to meet household needs regardless of whether or not it passed the peak demand test. Aquifer contribution must be increased by developing, deepening or relocating the well.

### Depth Required for Adequate Storage

If, for a well which has passed the constant head test but failed the peak demand test, storage is to be provided by deepening the well, the necessary well depth can be calculated from results of the constant head test. Two assumptions are necessary. The first is that there is no aquifer contribution until the drawdown in the well reaches the level measured

3

during the constant head test. This is a conservative assumption. There is indeed aquifer contribution before the water level falls this far, but the rate is not measured during the test and is thus unknown. The second assumption is that as soon as the drawdown in the well reaches the level measured during the constant head test, the aquifer contribution begins at the rate measured at the conclusion of the constant head test and does not increase as the drawdown increases. This, also, is a conservative assumption, but is reasonable in that in a bedrock aquifer the most significant water-bearing zones are commonly associated with weathered fractures within several tens of feet of the soil/rock interface. Additional drawdown below the fractured or weathered zone may not induce much more water to flow into the well.

The volume of water which will enter the well from the aquifer during peak demand periods can be estimated using these assumptions. Subtracting this volume from the peak load gives the total volume of storage needed. Conversion of storage volume to additional depth required for drawdown is discussed under *Additional Drawdown for Wells Without Adequate Storage* in the *Implementation* section, below.

### Surface Storage Tanks

The effect of surface storage of water on the well depth required for reliable peak supply can be taken into account by subtracting the available volume in the storage vessel from the peak demand. This gives the volume which must come from the well during each peak demand period. Dividing this lower volume by the peak demand rate gives the length of time the pump must operate during the peak demand period. The calculation of the necessary volume of well storage then proceeds as discussed above. The peak demand test should be carried out using the lower value for pumping time. The surface storage tank must, of course, be refilled, but if the well has passed the constant head test it should be possible to fill between peak demand periods.

Normally a domestic water supply system includes a hydropneumatic tank. A conventional hydropneumatic tank is intended primarily for maintenance of water pressure and contributes little to available water storage. Only that volume of water which drains from the tank before declining pressure causes the well pump to switch on contributes to available storage. This can be referred to as the available storage volume of the tank. An additional tank, dedicated to water storage and equipped as necessary to deliver water to the plumbing system of the house at the required pressure, could conceivably be installed.

### Total Well Depth and Pump Placement

The pump must be placed in a well so as to allow for the maximum drawdown measured during the peak demand test. Pump size must be based on the required water pressure in the plumbing and the anticipated drawdown. Too deep of a pump setting can be as undesirable as too shallow a setting if the pump cannot deliver water at the required pressure. A balance must be struck between the storage advantage of maximum drawdown and ensuring adequate water pressure. This report attempts to prevent well failure caused by setting the pump at too shallow a depth.

### IMPLEMENTATION

Testing of domestic wells is not usually conducted in a manner that establishes long-term performance capability. In fact, estimates of well yield are frequently made during drilling. In particular this is true with air rotary drilling when estimates of yield are based on the quantity of water lifted from the well

with an air compressor. For an accurate test a pump with a discharge rate control is necessary; this might include the use of valves or a throttle on a generator used to power a pump.

An effective pump test must determine whether the well can supply both peak needs and total daily needs. In the method presented here, this is done using a two-part pumping test. The first part is the peak demand test and the second is the constant head test. Both of these must be performed in one continuous testing session. *It is important to note that the well is not considered to be satisfactory until it passes both tests during one continuous session.*

### Peak Demand Test

The peak demand test is a drawdown pumping test to determine if the well can supply the water needed by the household during times of peak water demand. In this test the well is pumped at the peak demand rate for the household for a time equal to or greater than the estimated peak time. If pumping can be maintained at this rate for the peak time, then the well should be able to support the peak needs of the household. If this pumping cannot be maintained, storage in the well is insufficient. The constant head test will provide further information as to the nature of the difficulty.

### Constant Head Test

The constant head test is to determine if flow from the aquifer to the well can replenish water removed from the well during peak demand periods. In this test the pumping rate is adjusted so that the drawdown stabilizes. When the water in the well is at a constant level, one is certain that all of the discharge is coming from the aquifer, none from well storage. If under this constant head pumping condition the aquifer can supply the total daily household needs, then long-term needs can most probably be met. For the purposes of this report a constant head condition exists when the pumping rate is held steady and the water level changes at a rate of less than 6 inches per hour.

### Outline of Testing Procedure

Planning of a pump test and evaluation of the results (figure 2) can be summarized as follows:

1. The well is constructed in accordance with state and local requirements.

2. The peak demand rate and peak time are calculated from the number of bedrooms and bathrooms using equations 2 and 3 or tables 1 and 2. Unless surface storage is to be taken into account, these are the required discharge rate and duration of the peak demand pump test.

3. The peak demand test is performed. For this test the pump should be positioned so as to take full advantage of the available drawdown in the well. The static water level in the well is measured prior to pumping. Then the well is pumped at the peak demand rate for the peak demand time. If pumping cannot be sustained at this rate the well fails the peak demand pump test. The water level at the completion of the peak demand test must be measured accurately for it is used to measure drawdown during the constant head test and, later, used to establish the pump setting.

4. The constant head test is undertaken *immediately* upon completion of the peak demand test regardless of whether or not the well passed the peak demand test.

4

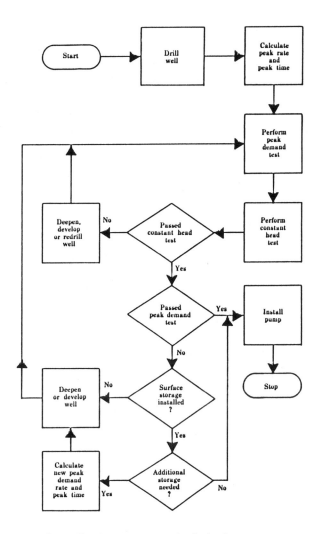

Figure 2. Flow chart of performance and evaluation of pump test.

5

The well is not allowed to recover from the peak demand test. In the constant head test the pumping rate is lessened, if necessary, to a rate at which the drawdown shows insignificant change with time. **A constant head condition** exists if the head changes **at a rate of less than 0.5 feet (6 inches) per hour under a constant pumping rate.** The pumping rate which maintains this condition is termed a **constant head pumping rate.** The water level at the constant head pumping rate must be measured accurately. The difference between this level and the static water level is the **constant head drawdown.** The constant head pumping rate should be as close as possible to the peak demand pumping rate, but the main objective is to achieve a stable water level while pumping. Too low a discharge rate will result in underestimation of the capabilities of the aquifer and may lead to overdeepening of the well.

5. The constant head pumping rate is converted to gallons per day (table 3) and compared to the total daily demand of the household. If the discharge rate during the constant head test will provide the total daily water demand, the well passes the constant head test. If not, it fails.

6. The results of the two tests are evaluated. The four possible outcomes are:

    a.  Peak demand test—Pass
        Constant head test—Pass

        This well can supply the home with enough water. The pump should be installed as discussed under *Total Well Depth and Pump Placement* below.

    b.  Peak demand test—Pass
        Constant head test—Fail

        This well, when full, can supply peak demand needs. However, more water will be withdrawn each day than can be replaced by the aquifer. This well will prove inadequate sooner or later. It must be deepened, developed or redrilled at a new location to provide a larger aquifer contribution. In bedrock aquifers the well may not encounter additional water-bearing zones at greater depth. The aquifer may not provide sufficient supply for domestic needs. If the well is redrilled, developed or deepened it must be retested.

    c.  Peak demand test—Fail
        Constant head test—Pass

        This well lacks sufficient water from combined well storage and aquifer flow to meet peak needs. The aquifer contribution is, however, sufficient to meet total daily needs. More storage must be supplied to satisfy short-term peak demands. This can be accomplished by deepening the well and lowering the pump, redrilling to a greater diameter or adding a water storage tank. If the well is to be deepened, a method for calculating additional required depth is described in *Additional Drawdown for Wells Without Adequate Storage*, below. If a water storage tank is installed see *Surface Storage Tanks*, above. If the well is redrilled or developed it must be retested.

    d.  Peak demand test—Fail
        Constant head test—Fail

This well has neither sufficient storage nor sufficient aquifer contribution to ensure an adequate household supply. It must be deepened to encounter additional water-bearing zones, developed to increase flow to the well or a new well drilled elsewhere. Depending on the success of efforts to increase flow from the aquifer, storage may have to be increased. The well must be retested when any of the above procedures are completed.

### Additional Drawdown for Wells Without Adequate Storage

A well that fails the peak demand test but passes the constant head test can be made satisfactory by providing additional storage. If additional storage is created by deepening the well, the necessary additional volume of storage and corresponding necessary available drawdown are found by calculating as follows:

A.  *Assured volume:* The assured volume of water in the well is the volume in the well between the static water level and the constant head pumping level. Calculate this volume by multiplying the constant head drawdown by the storage capacity per foot of well casing (1.4 gallons per foot for a 6-inch well).

B.  *Assured time:* This is the time it will take to pump the assured volume from the well. It is calculated by dividing the assured volume by the peak demand rate.

C.  *Shortfall volume:* This is the volume of water which must be supplied after the assured volume has been withdrawn from the well. It is calculated by subtracting the assured volume from the peak load.

D.  *Shortfall time:* This is the time within which the well must supply the shortfall volume. It is calculated by subtracting the assured time from the peak time.

E.  *Aquifer contribution volume:* This is the volume of water the aquifer is predicted to supply to the well during the shortfall time. During the shortfall time water is assumed to be moving from the aquifer to the well at the aquifer contribution rate measured during the constant head test. Aquifer contribution volume is calculated by multiplying the aquifer contribution rate by the shortfall time.

F.  *Additional well storage volume:* This is the volume of water which must be withdrawn from well storage after the assured volume has been pumped. It is calculated by subtracting the aquifer contribution volume from the shortfall volume.

G.  *Additional well drawdown:* This is the additional drawdown required beyond the constant head drawdown. It is calculated by dividing the required additional well storage volume by the storage capacity of well casing (1.4 gallons per foot for a 6-inch well).

Equations for this sequence of calculations are:

A.  assured volume
    = constant head drawdown X storage capacity per foot of well casing

B.  assured time = assured volume/peak demand rate

C.  shortfall volume = peak load - assured volume

6

D.  shortfall time = peak time - assured time

E.  aquifer contribution volume
    = shortfall time X aquifer contribution rate

F.  additional well storage volume
    = shortfall volume - aquifer contribution volume

G.  additional well drawdown
    = additional well storage volume/storage capacity
    per foot of well casing of well

The additional well drawdown is the predicted drawdown below the constant head level at the end of peak time. It is used to calculate the required well depth as shown in *Total Well Depth and Pump Placement*, below. After deepening or redrilling, both parts of the pump test must be redone during a single testing session.

## Total Well Depth and Pump Placement

It is recommended that the pump be placed at least 10 feet below the depth to water measured at the end of a successful peak demand pump test. Also, in order to prevent siltation problems, the pump should be placed at least 10 feet above the bottom of the well. An equation for minimum pump depth is:

minimum pump depth
    = depth to water at end of successful peak demand
    pump test + 10 feet

As an additional safety precaution the pump depth can be increased beyond that suggested here, thus allowing for more drawdown than measured during testing. This will require a well deeper than the minimum shown above and may necessitate a more powerful pump.

## EXAMPLES

Example 1: Well passes both parts of test

A builder proposes a 3 bedroom, 2 bathroom house. From table 1 the pump test should last for at least 50 minutes. From table 2, the pumping rate should be at least 6 gpm.

A well was drilled, allowed to stand for 12 hours to come to its static water level, then was pumped at 8 gpm for 1 hour. After 1 hour of pumping the drawdown was 22.0 feet. In being pumped at 8 gpm for 1 hour, the well more than met the peak demand requirements and passed the peak demand test.

The constant head test was begun without pause in a second hour of pumping, also at 8 gpm. At the end of the second hour the drawdown was 22.2 feet. Because of this small change in the drawdown the well was properly considered to be at a constant head level. 8 gpm is equivalent to more than 11,000 gallons per day (table 3). A 3 bedroom house will require 600 gallons per day (table 4). The well thus passed the constant head test.

Because the well passed both parts of the pump test the production pump can be set according to the minimum safety allowance at 10 feet lower than the drawdown measured during the peak demand pump test; this is 32 feet below the static water level. Setting the pump deeper would provide an additional margin of safety. The pump should be placed at least 10 feet above the bottom of the well as a precaution against siltation.

Example 2: Well fails peak demand test

A 5 bedroom, 3 bathroom house is proposed. A 6-inch well 250 feet deep has been drilled and allowed to stand for 12 hours to come to its static water level.

From table 4, the daily household water demand will be 1000 gallons and the peak load 500 gallons. From tables 1 and 2 the peak demand pump test should last at least 55.5 minutes and the discharge rate should be at least 9 gpm.

Before pumping, the static water level was measured at 25 feet below the ground surface. At a pumping rate of 9 gpm the well failed after 33 minutes. The driller therefore proceeded directly to the constant head test. After some trial and error, the driller determined that 2 gpm was the maximum rate at which a constant head could be maintained. At the end of the constant head test the water level in the well was 230 feet below the top of the casing.

An aquifer contribution of 2 gpm is equivalent to 2,880 gallons per day (table 3) and will satisfy long-term needs. The well passed the constant head test. It failed, however, the peak demand test. The problem therefore lies in an inadequate volume of storage. The driller decided to deepen the well to provide additional storage in the borehole and calculated the additional required drawdown as outlined in *Depth Required For Adequate Storage* and below:

The constant head drawdown at a 2 gpm pumping rate is the constant head pumping level of 230 feet minus the static water level of 25 feet. This comes to 205 feet.

The assured volume (the volume of water stored in the well between the static water level and the level measured in the constant head test) is 205 feet of drawdown times 1.4 gallons per foot of casing. This comes to 287 gallons.

The assured time (the time it will take to pump out the assured volume) is 287 gallons divided by the peak demand rate of 9 gpm, or 31.9 minutes.

The shortfall time is 55.5 minutes (the peak time) minus 31.9 minutes (the assured time), or 23.6 minutes.

The shortfall volume is 500 gallons (the peak load) minus 287 gallons (the assured volume), or 213 gallons.

During the shortfall time the aquifer contribution is constant at 2 gpm. The total aquifer contribution volume is thus 2 gpm times 23.6 minutes, or 47.2 gallons.

The volume of water which must be stored in the well below the constant head level is 213 gallons (the shortfall volume) minus 47.2 gallons (the aquifer contribution volume during peak time). This comes to 165 gallons.

The additional required drawdown is calculated by dividing the shortfall volume of 165 gallons by the storage capacity of 1.4 gallons per foot of well casing. This comes to 118 feet.

The maximum predicted drawdown is 205 feet (the maximum drawdown during the constant head test) plus 118 feet (the additional drawdown for peak needs calculated above) plus 10 feet (the minimum safety allowance). This comes to 323 feet below the static water level.

The total well depth then should be at least 25 feet (the static water level) plus 323 feet (the maximum predicted drawdown below the static water level) plus 10 feet (pump placement above well bottom). This comes to a total of 368 feet.

The well must therefore be deepened to a minimum depth of 368 feet, then retested.

7

## ADDITIONAL CONSIDERATIONS

The two-part pump test is a means of evaluating the capability of a well to supply domestic needs. The interpretation of test results is mathematical and may not take into account all the physical factors that affect a particular well. Some additional considerations are listed below:

1. *Seasonal recharge variations:* Pump tests performed during times of seasonally high ground water may not accurately predict performance during times of reduced water availability. A well that passes a pumping test in the spring, during high water-table conditions, may not be able to provide an adequate supply in summer or during drought periods when the water table is lower. Tests performed between June and October are more reliable than those performed in the rest of the year in determining if a well will satisfy household water demands.

2. *Low aquifer contribution:* The aquifer contribution as defined in this report is the volume of water that flows from the aquifer to the well during pumping. Table 4 shows that a one-bedroom house requires 200 gallons per day, which is equivalent to 0.14 gallons per minute. This is an extremely low aquifer contribution value. Experience has shown that a well with an aquifer contribution of less than 0.5 gpm (720 gpd) is a marginally dependable source of water for domestic use. It is recommended that a minimum cut-off of 0.5 gpm be established for the aquifer contribution.

3. *Addition or withdrawal of water during drilling:* During drilling water may be added to the aquifer or withdrawn depending upon the drilling method used. Immediately after completion of drilling the heads in the aquifer near the well may not be at static (unstressed) levels. The pump test should be conducted after any stresses induced by the drilling process have dissipated. A 12-hour recovery period is recommended between completion of drilling and performance of the pump test.

4. *Storage capacity:* Chapter 199 of NJAC 7:10-3.85 specifies minimum hydropneumatic tank sizes. These tanks maintain pressure in household water systems. They provide some storage, but this is not their primary purpose. They never empty completely, so the available storage in these tanks is less than their total volume. The presence of tanks specified in the regulations, or larger tanks dedicated solely to water storage, may be taken into account when calculating peak demand volumes.

5. *Large households:* The assumptions that relate dwelling size to household water demand may not be applicable to large dwellings that are not fully occupied. For homes with more than 5 bedrooms or 3 bathrooms it may prove advisable to use a different method to predict peak demand time, peak load and peak demand rate.

6. *Pump discharge:* In areas of vertical fracturing and thin or permeable overburden, water discharged at the surface may quickly infiltrate to the water table. If a pump test is conducted at such a site and the water pumped from the well is discharged at the well head, the water may return to the well as artificial recharge. Because this recharge will not be present during normal operation of the well, aquifer contribution will be overestimated. Water should be discharged at a distance from the well head in order to minimize this possibility.

7. *Drawdown safety factor:* A drawdown safety factor of 10 feet was recommended above. That is, the pump is to be set 10 feet below the drawdown level measured during the peak demand pump test. This level may be increased if a well is drilled in an aquifer known for large water level fluctuations.

8. *Well diameter:* Normally, domestic wells drilled in rock aquifers have a diameter of 6 inches. This provides approximately 1.4 gallons of storage per foot of drawdown in the casing. Increasing the well diameter will increase the storage per foot of well depth. The increased storage can be accounted for by using the appropriate value for the storage capacity per foot of drawdown.

## REFERENCES CITED

Hunt, Joel, 1978, How much is enough? A minimum well formula: Water Well Journal, v. 32, no. 2, p. 53-55.

Linaweaver, F. P., Jr., Geyer, John C., and Wolff, Jerome B., 1967, A study of residential water use: A report prepared for the Technical Studies Program of the Federal Housing Administration, Department of Housing and Urban Development, at Department of Environmental Engineering Science, The Johns Hopkins University, Baltimore, Maryland.

Orndorff, John R., 1966, Domestic water use differences in individual well and public water supplies: Report III on Phase Two of the Residential Water Use Research Project, Department of Environmental Engineering Science, The Johns Hopkins University, Baltimore, Maryland.

U. S. Department of Commerce, 1981, Statistical abstract of the United States, Bureau of the Census.

U. S. Environmental Protection Agency, 1978, Management of small waste flows: EPA-600/2-78-173, Cincinnati, Ohio.

8

## TABLES

Table 1: Duration of **peak time** in minutes as a function of the numbers of bedrooms and bathrooms in a dwelling.

|  |  | Number of Bathrooms | | | | |
|---|---|---|---|---|---|---|
|  |  | 1 | 1½ | 2 | 2½ | 3 |
|  | 1 | 33.3 | 22.2 | 16.7 | 13.3 | 11.1 |
| Number | 2 | 66.7 | 44.4 | 33.3 | 26.7 | 22.2 |
| of | 3 | 100.0 | 66.7 | 50.0 | 40.0 | 33.3 |
| Bedrooms | 4 | 133.3 | 88.8 | 66.7 | 53.3 | 44.4 |
|  | 5 | 166.7 | 111.1 | 83.3 | 66.7 | 55.5 |

Table 2: **Peak demand rate** as a function of the number of bathrooms in a dwelling.

| Number of Bathrooms | Peak Demand Rate (gpm) |
|---|---|
| 1 | 3 |
| 1½ | 4.5 |
| 2 | 6 |
| 2½ | 7.5 |
| 3 | 9 |

Table 3: **Flow volumes** in gallons per minute corresponding to flow volumes in gallons per day.

| Flow Volume (gpm) | Flow Volume (gpd) |
|---|---|
| 0.01 | 14.4 |
| 0.02 | 28.8 |
| 0.05 | 72.0 |
| 0.1 | 144.0 |
| 0.2 | 288.0 |
| 0.3 | 432.0 |
| 0.4 | 576.0 |
| 0.5 | 720.0 |
| 0.6 | 864.0 |
| 0.7 | 1,008.0 |
| 0.8 | 1,152.0 |
| 0.9 | 1,296.0 |
| 1.0 | 1,440.0 |
| 2.0 | 2,880.0 |
| 5.0 | 7,200.0 |
| 10.0 | 14,400.0 |

Table 4: **Daily demand volume and peak load** as a function of the number of bedrooms in a dwelling.

| Number of Bedrooms | Daily Demand Volume (gallons) | Peak Load (gallons) |
|---|---|---|
| 1 | 200 | 100 |
| 2 | 400 | 200 |
| 3 | 600 | 300 |
| 4 | 800 | 400 |
| 5 | 1,000 | 500 |

9

# DOMESTIC WELL WORKSHEET
# FOR TWO-PART PUMP TEST

**(sheet 1 of 2)**

## PUMP TEST

─────────────────────────── Test Design ───────────────────────────

**Preliminary Well Summary**

1. Depth of well ............................... _____ feet
2. Static water level (depth to water from top of casing) ................... _____ feet
3. Number of hours between well completion and measurement of static water level ...................... _____ hours

**Dwelling Summary**

4. Number of bedrooms .................... _____
5. Number of bathrooms ................. _____

**Peak Demand Test Requirements**

6. Peak time (required minimum duration of test, from table 1) .. _____ minutes
7. Peak demand rate (required minimum discharge rate from pump during test, from table 2) ................................... _____ gpm
8. Peak load (from table 4) ............. _____ gallons

─────────────────────────── Test Measurements ───────────────────────────

**Peak Demand Test**

9. Depth to water at beginning of test (static water level) ............. _____ feet
10. Depth to pump at end of test ...... _____ feet
11. Discharge rate measured during test (use minimum observed) .... _____ gpm
12. Duration of test ........................... _____ minutes
13. Depth to water at end of test ...... _____ feet
14. Drawdown at end of peak demand test .............. line 13—line 9 = _____ feet

**Constant Head Test**

15. Constant head pumping rate ........ _____ gpm
16. Duration of pumping at constant head rate ................................. _____ minutes
17. Depth to water at end of test ...... _____ feet
18. Drawdown at end of constant head test .............. line 17—line 9 = _____ feet

─────────────────────────── Evaluation of Results ───────────────────────────

19. Peak demand test duration. If line 12 is less than line 6 then well fails peak demand test ...... _____ pass or fail

20. Peak demand pump test rate. If line 11 is less than line 7 then well fails peak demand test ...... _____ pass or fail

21. Calculate aquifer contribution (multiply line 15 by 1440 or use table 3) ..................................... _____ gpd
22. Daily home water demand (from table 4) ..................................... _____ gpd
23. Aquifer contribution rate. If line 21 is less than line 22 then well fails constant head pump test ... _____ pass or fail

10

# DOMESTIC WELL WORKSHEET
## FOR TWO-PART PUMP TEST

**(sheet 2 of 2)**

──────────── Actions Based on Test Results ────────────

| Peak demand test | Constant head test | Action | Peak demand test | Constant head test | Action |
|---|---|---|---|---|---|
| 24. pass | pass | Go to *Pump Placement and Minimum Well Depth* (lines 35-37). | | | to lines 28-34 (*Additional Drawdown for a 6-inch Well With Insufficient Storage*) |
| 25. fail | pass | The well must be developed to increase yield, deepened to increase storage or surface storage installed. If the well is deepened or developed, it must be retested. Go | 26. pass | fail | The well must be developed, deepened or redrilled at a new location to increase yield. It must then be retested. |
| | | | 27. fail | fail | |

---

## ADDITIONAL DRAWDOWN FOR A 6-INCH DIAMETER WELL WITH INSUFFICIENT STORAGE

28. Assured volume
......... line 18 X 1.4 gallons/foot = _____ gallons

29. Assured time ......... line 28/line 7 = _____ minutes

30. Shortfall volume
..................... line 8 - line 28 = _____ gallons

31. Shortfall time .... line 6 - line 29 = _____ minute

32. Aquifer contribution volume
..................... line 15 X line 31 = _____ gallons

33. Required additional storage
..................... line 30 - line 32 = _____ gallons

34. Additional drawdown needed in well
..................... line 33/1.4 gal/ft. = _____ feet

---

## TOTAL WELL DEPTH AND PUMP PLACEMENT

35. Minimum total drawdown needed
..................... line 14 + 10 feet = _____ feet

36. Depth below top of casing to place
pump ............ line 9 + line 35 = _____ feet

37. Minimum total depth of well
..................... line 36 + 10 feet = _____ feet

11

## GLOSSARY

*Aquifer contribution:* the proportion of the well flow at any given time which comes directly from the aquifer.

*Aquifer contribution rate:* the maximum rate at which water can flow from an aquifer to a well. Here assumed to equal the pumping rate measured in the constant head test.

*Aquifer contribution volume:* the total volume of water which flows from the aquifer to the well during the shortfall time.

*Assured time:* the time it will take to pump the assured volume from the well at the peak demand rate.

*Assured volume:* the volume of water in a well below the static level and above the constant head level.

*Constant head:* a stable water level attained under a constant pumping rate. For this report a rate of change of less than 0.5 feet (6 inches) per hour is taken as stable.

*Constant head drawdown:* the drawdown in a well when a constant head condition has been attained. Here measured from the static water level at the end of the constant head test.

*Constant head level:* the water level in a well at the end of the constant head test. Measured from the top of the casing.

*Constant head pumping rate:* a constant pumping rate at which a stable water level is attained. The pumping rate during the constant head test.

*Constant head test:* a pumping test in which pumping rate and drawdown are kept constant with time. For this report a rate of change of less than 0.5 feet (6 inches) per hour is taken as constant.

*Drawdown:* the decline in the water level in a well during pumping. Measured from the static water level prior to pumping.

*Hydropneumatic tank:* a tank which uses compressed air to maintain pressure in a water supply system. It is only secondarily a water storage tank.

*Peak demand rate:* the average rate of water use by a household during peak demand periods.

*Peak demand test:* a pumping test conducted to evaluate the capability of a well to supply peak demand needs of a household. The test is conducted at a rate equal to or greater than the peak demand rate for the peak time.

*Peak load:* the volume of water required by a household during each peak demand period. In this report, the peak load is assumed to be half the estimated total daily household water consumption.

*Peak time:* the length in minutes of each of two daily peak demand periods.

*Shortfall time:* the time needed to pump the shortfall volume from a well at the peak demand pumping rate.

*Shortfall volume:* the volume of water needed in addition to the assured volume to make up the peak load.

*Static level:* the water level in a well before a pumping test when all effects of drilling and previous pumping on the aquifer have dissipated and the well is in equilibrium with atmospheric pressure.

*Storage contribution:* the proportion of the well flow at any given time which comes from storage in the well.

*Well flow:* the flow rate of water from a well at a given time. It is the sum of the aquifer contribution and the well storage contribution.

*Well storage:* the volume of water stored within a well which is available for pumping.

## UNITS OF MEASUREMENT

Foot-pound-second (english) units of measurement are used in this report. These can be converted to International Standard (SI) units as follows:

| Multiply | by | to obtain |
|---|---|---|
| inches | 2.54 | centimeters |
| feet | 0.305 | meters |
| gallons | $3.79 \times 10^{3}$ | cubic meters |
| gallons/minute | $6.31 \times 10^{2}$ | liters/second |
| gallons/day | $3.79 \times 10^{3}$ | cubic meters/day |

12

# ALTERNATE METHODS TO PREDICT NITRATE-NITROGEN LEVELS AT PROPERTY BOUNDARIES

## CALCULATIONS

Purpose:  To determine the Nitrate - Nitrogen levels at the property
boundary of a particular lot.

Calculation of Dilution

Area of lot $(A')Ft^2$ = Width (ft) X Length (ft)

Infiltration rate from precipitation (IP)* = 20 in/yr
= 0.0045 feet/day

* (Average precipitation does not take into account drought conditions)

Infiltration over Area (IA) $ft^3$/day = A' X IP

Septic Discharge (SD) gallons/day = (400 gal/day/home**) X (NO[#] of homes)
= SD gallons/day ÷ 7.48 gallons/$ft^3$'
SD($ft^3$/day)

**Note:  100 gallons/persons/day X 4 persons/home = 400 gallons/home
(as prescribed by Chapter 199)

Therefore, the dilution factor (DF) is equal to the total liquid (TL = SD + IA)
entering the subsurface, divided by the septic discharge.  In equation form,
this can be described as follows:

Dilution factor (DF) = $\frac{SD + IA}{SD}$ (dimesion less)

Therefore, at a starting average Nitrate - Nitrogen concentration of 40 ppm,
the dilution would be calculated as follows:

$\frac{NO_3 - N}{DF}$ (ppm) = Diluted valve for $NO_3$ - N
at property line.

Source: N.J. Department of Environmental Protection

CALCULATIONS - ALTERNATE METHOD

Purpose:   To determine the Nitrate - Nitrogen levels at the property
boundary of a particular lot.

Calculation of Dilution

Area of lot $(A')Ft^2$= Width (ft) X Length (ft)

Infiltration rate from precipitation $(IP)$* = 13 in/yr
= 0.0030 feet/day

*(In this case precipitation takes into account drought conditions where
a drought is defined as a standard deviation from the mean precipitation)

Infiltration over Area (IA) $ft^3$/day = A' X IP

Septic Discharge (SD) gallons/day = (262.5 gal/day/home**) X (NO# of homes)
= SD gallons/day $\div$ 7.48 gallons/$ft^3$'
SD $(ft^3$/day)

**Note: 75 gallons/persons/day X 3.5 persons/home = 262.5 gallons/home

Therefore, the dilution factor (DF) is equal to the total liquid (TL = SD + IA)
entering the subsurface, divided by the septic discharge.  In equation form,
this can be described as follows:

$$\text{Dilution factor (DF)} = \frac{SD + IA}{SD} \text{ (dimension less)}$$

Therefore, at a starting average Nitrate - Nitrogen concentration of 40 ppm,
the dilution would be calculated as follows:

$$\frac{NO_3 - N}{DF} \text{ (ppm)} = \text{Diluted valve for } NO_3 - N \text{ at property line.}$$

Source: N.J. Department of Environmental Protection

# Part C

## MODEL ORDINANCE TO PROTECT CRITICAL CARBONATE BEDROCK FORMATIONS

AN ORDINANCE TO PROTECT THE WATER RESOURCES OF _____ TOWNSHIP LOCATED IN THE CARBONATE ROCK FORMATION OF THE TOWNSHIP

BE IT ORDAINED by the Township of _____, State of _____ County of _____ as follows:
Pursuant to the provisions of _____ the zoning map of _____ Township is amended to provide an overlay zone which is designated The Critical Geologic Formation Zone and shall be superimposed upon maps of those areas of the Township where carbonate rock formations are found.   Within this Zone all Environmental Impact Statements required by Section _____ of the Code of the Township as amended herein shall include the additional information required by this Ordinance.   This Critical  Geologic Formation Zone shall consist of the Critical Geologic Formation Area (C.G.F.A.) and the adjacent Critical Formation Watershed Protection Area (C.F.W.P.A.).   The Critical Geologic Formation Zone shall be shown and delineated on the Official Map of _____ Township and shall be available for inspection and will be on file, after adoption, in the office of the Township Clerk.

General Requirements in The Critical Geologic Formation Zone:
For all development proposals in the Critical Geologic Formation Area, whether residential or non-residential, a geologic investigation program to determine the potential for development shall be prepared and conducted in accordance with the requirements of Subsection _____.   The geologic investigation program shall be designed to produce information and provide recommendations for site planning, engineering design and construction techniques which shall meet or exceed the standards of the geology study provisions of the Geologic Segment of the EIS.  The Geologic Segment of the EIS may be completed and filed prior to the completion of the other portions of the EIS at the applicant's option.

Section A is amended to read as follows:
(A)  When required:  An EIS is required as part of any application for development involving new buildings or any land disturbance where Planning Board approval is required. Additionally, any proposal for development within the Critical Geologic Formation Area as shown on the official map of the Township requires an analysis in accordance with the geologic segment of the EIS.  The Planning Board may grant an exemption from the requirement of part or all of an EIS under subsection (G) below.  An EIS is also required for all public and quasi-public projects unless they are exempt from the requirements of local law or by supervening county, state or federal law.

The geologic segment of the EIS shall in addition to the information referred to herein shall include a discussion of the probable effects of the proposed development upon township water resources as related to existing geologic conditions and investigation results, a presentation of proposed engineering solutions (specifically as to design and construction aspects, including alternate solutions where appropriate), provisions for inspection and monitoring procedures during construction, and any log-term monitoring/inspections which may be recommended.
Section _____ is hereby amended to include subsection (n) as follows:
(n)  Geologic Conditions Report.  For all tracts located in the Critical Geologic Formation Area, the EIS shall contain data obtained during an appropriate site investigation.  A comprehensive site investigation program shall be conducted by the applicant to provide the Planning Board with sufficient data to define the nature of all existing geologic conditions that may limit construction and land use activities on the site. Specifically, the investigations shall yield information, which shall demonstrate that the proposed development will identify any existing geologic conditions for which appropriate engineering solutions may be necessary to minimize any adverse environmental impact caused by the proposal.
Section _____ is hereby amended to add Critical Geologic Formation Area Investigation Program:
(a)  An investigation program shall be commenced by completing Checklist #I.  Said checklist #I shall be submitted to the Planning Board Secretary and shall be reviewed by the Township Geotechnical Consultant (GTC) and a report rendered to the Planning Board

within 30 days of submission. The GTC in his report shall recommend to the Planning Board that checklist #II be prepared and submitted, or in the alternative, that portions or all of the requirements required by checklist #II be waived. If checklist #II is required by the Planning Board, the applicant shall then prepare and submit checklist #II to the Planning Board Secretary. Checklist #II shall be reviewed by the GTC and a report shall promptly be made to the Planning Board advising whether the checklists are complete. The Planning Board shall rule on the completeness of the checklist within 45 days of the date of submission. The report shall also advise the applicant as to whether any proposed testing methodology is prohibited because of the potential danger the methodology may pose to the integrity of the site or the health, safety, and welfare of the community. The geotechnical consultant may also recommend waiver of some or all of the required investigations in appropriate cases pursuant to subsection (G). At the applicant's option, both checklist I and checklist II may be submitted simultaneously in which event the Planning Board shall deem the checklists completed or incomplete within 45 days of submission.

After checklists I and II have been deemed complete by the Planning Board and the GTC has advised the Planning Board that the testing methodology poses no danger to the integrity of the site or the health, safety and welfare of the community, a permit shall be issued to the applicant authorizing the commencement of the testing procedure.

[1] Any on-site investigations and tests shall not begin until the applicant has received approval of his investigative plan and a permit has been issued. Additionally, actual notification at least 15 days in advance of the testing to commence in writing by certified mail, return receipt requested or personal service of said notice on the Township Clerk shall be given.

[2] The applicant shall arrange to have the proposed development site open for on-site inspection by the GTC or designated township inspectors at all times while the field investigation program is in progress, and testing data and results shall be made available to township officials and township inspectors on demand, but at no less frequent intervals than bi-monthly.

[3] At the completion of the field investigation a formal written report including the following shall be submitted: a description of the project, a general plan, to scale, of the entire project, showing the location of the project with respect to surface water, existing wells (within 1/2 mile) and adjacent property owners, logs of all borings, test pits and probes including evidence of incipient cavity formations, loss of circulation during drilling, voids encountered, type of drilling or excavation techniques employed, drawings of monitoring or observation wells as installed, time and dates of explorations and tests, reports of chemical analyses of on-site surface and ground water, names of individuals conducting tests if other than the P.E. referred to in the checklist, analytical methods used on soils, water samples, and rock samples, a 1" to 100' scale topographic map of the site (at a contour interval of two feet) locating all test pits, borings, wells, seismic, or electromagnetic, conductivity or other geographical surveys, an analysis of the ground water regime with rate and direction of flow; a geologic interpretation of the observed subsurface conditions, including soil and rock type, jointing (size and spacing), faulting, voids, fracturing, grain size, and sinkhole formation.

[4] The site investigation report should define the extent of geotechnical concerns at the site in relation to the planned development or land use. The proposed engineering solutions to minimize environmental impact as a result of the project, both during construction and in the foreseeable future, must be clearly detailed, together with the bases for the conclusions reached. Special consideration should be given to innovative control of surface water flows, as well as protection and replenishment of ground water.

[5] All samples taken shall be preserved and shall be available for examination by the Township upon request until final action is taken by the Planning Board on the application.

(A)    The following shall be added as Section _____.

At the applicant's option results of the investigation program required for the geologic segment of the EIS may be submitted to the Planning Board prior to the completion of other segments of the EIS, and if so submitted, shall be reviewed by the GTC and a report of his findings, conclusions, and recommendations shall be submitted to the Planning Board within 45 days of the submission of the data. The GTC shall confer with the Township Environmental Commission and request their input and non-binding recommendations within said 45 day time period. If the geotechnical consultant recommends the disapproval of the program, the recommendation shall include suggestions on alternate methodology which the GTC suggests would provide the requisite data.

(B) During his review of the geologic segment of the EIS for proposed development in the C.F.W.P.A. the GTC shall consider the data, formal reports, maps, drawings and related submission materials and shall advise the planning board whether or not the applicant has provided the Township with:

1. Sufficient design, construction and operational information to insure that the proposed development of the tract will not adversely impact on the health, safety, and welfare of the community.

2. The proposed method of development of the tract, will: minimize any deleterious effects on the quality of surface or subsurface water, and will not alter the character of surface and subsurface water flow in a manner deleterious to known conditions on-tract or off-tract;

3. Specific details insuring that design concepts and construction and operational procedures intended to protect surface and subsurface waters in critical zones will be properly implemented;

4. The submission provides specific details on inspection procedures to be followed during construction.

(4) The Planning Board shall at the request of the applicant within 45 days of the receipt of the report from the geotechnical consultant approve or disapprove the proposed geotechnical aspects of the development plan and associated construction techniques. In the event the Planning Board disapproves of the proposed development plan and associated construction procedures the Board shall state in the resolution its reasons for disapproval.

Section _____ is amended to add subsection I.

I. Geologic Segment Review. For any application requiring an EIS geologic segment submission, there shall be an application fee of $500.00 plus $100.00 for each acre of the site included within the Critical Geologic Formation Area. Additionally, there shall be posted with the Township a review escrow of $1,000.00 plus $500.00 per acre for each acre within the Critical Geologic Formation Areas, the escrow to be administered in accordance with the review escrow provisions of the Clinton Township Code. With regard to applications requiring review where the site is located all or in part of the C.F.W.P.A., the escrow fee shall be $500.00 plus $200.00 per acre.

CHECKLIST I

[  ]  Review of C.G.F.Z. map of township

[  ]  Review of geological survey maps

[  ]  Review of USDA publications and maps and the Township Natural Resource Inventory.

[  ]  Submit with this checklist a site plan map at a scale of 1:24,000 identifying proposed development site and boundaries of site that are within the C.G.F.A. and/or C.F.W.P.A. as designated on the C.G.F.Z. Map.

[  ]  Aerial photographs for the proposed site and surrounding area (at a minimum scale of 1" - 1,000', obtained during periods of little or no foliage cover).

[  ]  Summary of all known water production well logs and previous known subsurface investigations in the immediate area.

[  ]  Submission of a site map at a scale of 1" - 100' with a contour interval of two feet identifying existing surface water bodies, topogrpahy of the site, location of any existing water production wells.

[  ]  Submit any other published geologic information available to applicant which applicant deems pertinent.  Please specify_____
_____

CHECKLIST II

Proposed investigation program to be conducted in C.G.F.A. in _____ Township.

A.  General Requirements:

1.  Test borings and test pits are to be used as the primary means of identifying potential geologic hazards.   Percussion probes geophysical techniques (e.g. seismic refraction and reflection, ground penetrating radar, magnetic, gravity and conductivity) can be used to provide date between test borings and pits.

2.  Proposed exploration techniques which are not outlined in this checklist may be submitted to the GTC for review and possible inclusion in the approved investigation program.  Alterations to the planned program can be made during the progress of the field investigation by request to the GTC if so required by the nature of the encountered subsurface conditions.

B.  The intention of the site investigation program is to define the nature and limits of possible design, construction and operating concerns that could result from the existance of carbonate soil and/or rock formations underlying the proposed development site.

C.  List name and address of New Jersey licensed engineer:


List name and address of New Jersey licensed well driller:


|  | TO BE COMPLETED BY GTC | | |
|---|---|---|---|
|  | Accepted | Rejected | See Attached Comments |

1.  DIRECT TESTING PROCEDURES

[  ]  TEST BORINGS

| | Accepted | Rejected | See Attached Comments |
|---|---|---|---|
| (a) number proposed _____ | [  ] | [  ] | [  ] |
| (b) depths anticipated _____ | [  ] | [  ] | [  ] |

(Note:  if rock encountered is within 40' of ground surface, a minimum of 10' of rock is to be cored.   Rock cores shall be a minimum of 2" in diameter, to be obtained by double tube, split barrel coring device)

| | Accepted | Rejected | See Attached Comments |
|---|---|---|---|
| (c) boring techniques to be utilized: _____ _____ | [  ] | [  ] | [  ] |

(Note:   unless written approval is authorized, all test borings will be drilled using rotary wash boring procedures without use of drilling muds. Water losses in borings are to be monitored as to depth and quantities.

|  | Accepted | Rejected | See Attached Comments |
|---|---|---|---|
| (d) proposed bore hole grouting techniques:_____ _____ _____ | [ ] | [ ] | [ ] |
| (e) soil and rock sampling to be performed in accordance with ASTM Standard D 420, D 1586, D 1587, and D 2113 | [ ] | [ ] | [ ] |
| (f) logging of all test borings or test pits in accordance with the Unified Soil Classification System and in relation to the geologic origin of the constituents of the encountered materials, e.g. light yellow brown silty clay (CH), with occasional angular dolomite fragments, moderately stiff, residual soils, some stained paleo jointing. | [ ] | [ ] | [ ] |

[ ] TEST PITS

|  | Accepted | Rejected | See Attached Comments |
|---|---|---|---|
| (a) number and depth of proposed pits_____ | [ ] | [ ] | [ ] |
| (Note: to be acceptable, minimum bottom area of pits shall be 10 square feet and shall encounter rock surface over 50% of the pit area) | | | |
| (b) method of backfill to be employed_____ | [ ] | [ ] | [ ] |
| (Note: Test pit backfill shall be composed of excavated material, placed in layers and compacted to pre-excavation density, unless authorized otherwise by GTC.) | | | |

[ ] PIEZOMETERS, LYSIMETERS AND WATER TABLE DATA

|  | Accepted | Rejected | See Attached Comments |
|---|---|---|---|
| (a) number, locations and types of to be used _____ _____ | [ ] | [ ] | [ ] |

TO BE COMPLETED BY GTC

|  | Accepted | Rejected | See Attached Comments |
|---|---|---|---|
| (b) other methods to be used | [ ] | [ ] | [ ] |

_____

_____

_____

Note:    These shall be installed and
monitored in sufficient locations to
identify depth to seasonably high
water table and rate and direction of
ground water flow.

[ ] GEOCHEMICAL TESTING OF PROPERTIES OF
SOILS, ROCK AND WATER:

| A.  Methods proposed: | [ ] | [ ] | [ ] |
|---|---|---|---|
| _____ | [ ] | [ ] | [ ] |
| _____ | [ ] | [ ] | [ ] |

2.    INDIRECT TESTING PROCEDURES

[ ] Percussion Probes   [ ]   [ ]      [ ]

a.   number proposed_____

b.   depths anticipated_____

c.   measuring   techniques   to   be      [ ]      [ ]         [ ]
utilized (air loss, drilling speed
and rod drops must also be monitored)

[ ] Geophysical Studies

(a) seismic   refraction   and   reflec-      [ ]      [ ]         [ ]
tion; location and number of runs
anticipated; equipment to be used

(b) ground penetrating radar specify      [ ]      [ ]         [ ]
procedures and location of traverses

(c) magnetic, gravity or conductivity      [ ]      [ ]         [ ]
techniques--specify   procedures   and
location of surveys

[ ] Geologic Reconnaissance

Factors to be examined--soil types,      [ ]      [ ]         [ ]
rock types, vegetative changes, ob-
servable seeps, or ground water discharge, circular depressions, swales.

TO BE COMPLETED BY GTC

| | Accepted | Rejected | See Attached Comments |
|---|---|---|---|

[  ] Additional field investigation tech-   [  ]   [  ]   [  ]
niques proposed_____
_____
_____

MAPS, DRAWINGS AND OTHER DOCUMENTATION

    (a) Location of site on 1:24,000   [  ]   [  ]   [  ]
scale USGS topo map (See checklist I)

    (b) General site plan showing loca-
tion of all field testing procedures
in relation to planned development at
a scale of 1"-100'

    (c) Timetable of proposed field in-   [  ]   [  ]   [  ]
vestigation, laboratory testing, test
data receipt and final report to the
Township.

    (d) Proposed technical inspection   [  ]   [  ]   [  ]
procedures during investigation
(continuous technical supervision of
field investigations is strongly recommended)

    (e) Has an EIS without a geologic   [  ]   [  ]   [  ]
segment report already been submitted
for this site?   Date:

    (f) Submission of application fees   [  ]   [  ]   [  ]
(572-13)
Amount: _____
Date: _____
Future payments anticipated:_____
_____

    (g) Special factors or conditions   [  ]   [  ]   [  ]
applicant wishes to bring to the
attention of the GTC: _____
_____
_____

TOWNSHIP GTC REVIEW

    Approval of checklists I and II   [  ]   [  ]   [  ]

TO BE COMPLETED BY GTC

Accepted    Rejected    See Attached Comments

Date of completion of checklist
I:_____
Date of completion of checklist
II:_____
Conditions to be imposed on
approval:_____
_____
_____
_____

Date investigation to commence:
_____

[  ] Denial of Checklists I and II--Items
needed for completion_____
_____
_____
_____
_____

[  ] Waiver(s) (if any) deemed appropriate
by GTC _____
_____
_____
_____
_____

PLANNING BOARD REVIEW

[  ] Date receipt of initial submission _____

[  ] Dates checklists I and II deemed complete by GTC:_____

[  ] Permit to be issues

   Date:

[  ] Date of accepted start and completion of program (Subsection C, item 4 Timetables)
_____

[  ] Permit denied

CALIFORNIA DIVISION OF
MINES AND GEOLOGY

CDMG
NOTE **43**

# RECOMMENDED GUIDELINES
## FOR DETERMINING THE MAXIMUM CREDIBLE
## AND THE MAXIMUM PROBABLE EARTHQUAKES

The following guidelines were suggested by the Geotechnical Subcommittee of the State Building Safety Board on 3 February 1975 to assist those involved in the preparation of geologic/seismic reports as required by regulations of the California Administrative Code, Title 17, Chapter 8, Safety of Construction of Hospitals. CDMG is currently using these guidelines when reviewing geologic/seismic reports.

### Maximum credible earthquake

The maximum credible earthquake is the maximum earthquake that appears capable of occurring under the presently known tectonic framework. It is a rational and believable event that is in accord with all known geologic and seismologic facts. In determining the maximum credible earthquake, little regard is given to its probability of occurrence, except that its likelihood of occurring is great enough to be of concern. It is conceivable that the maximum credible earthquake might be approached more frequently in one geologic environment than in another.

The following should be considered when deriving the maximum credible earthquake:

(a) The seismic history of the vicinity and the geologic province;
(b) the length of the significant fault or faults which can affect the site within a radius of 100 kilometers; (See CDMG Preliminary Report 13);

(c) the type(s) of faults involved;
(d) the tectonic and/or structural history;
(e) the tectonic and/or structural pattern or regional setting (geologic framework);
(f) the time factor shall not be a parameter.

### Maximum probable earthquake
### (functional-basis earthquake)

The maximum probable earthquake is the maximum earthquake that is likely to occur during a 100-year interval. It is to be regarded as a probable occurrence, not as an assured event that will occur at a specific time.

The following should be considered when deriving the "functional-basis earthquake":

(a) The regional seismicity, considering the known past seismic activity;
(b) the fault or faults within a 100 kilometer radius that may be active within the next 100 years;
(c) the types of faults considered;
(d) the seismic recurrence factor for the area and faults (when known) within the 100 kilometer radius;
(e) the mathematic probability or statistical analysis of seismic activity associated with the faults within the 100 kilometer radius (the recurrence information should be plotted graphically);
(f) the postulated magnitude shall not be lower than the maximum that has occurred within historic time.

PYA, JES, RWS 2/75

STATE OF CALIFORNIA                    THE RESOURCES AGENCY                    DEPARTMENT OF CONSERVATION

For a list of geologic maps and reports available from the California Division of Mines and Geology, write to the California Division of Mines and Geology, P.O. Box 2980, Sacramento, CA 95812, or visit our District offices in SACRAMENTO, 2815 "O" Street, (916) 445–5716; SAN FRANCISCO, Room 2022, Ferry Building, (415) 557–0633, LOS ANGELES, Room 1065, 107 South Broadway, (213) 620–3560

CALIFORNIA DEPARTMENT OF CONSERVATION
DIVISION OF MINES AND GEOLOGY

**DMG NOTE 42**

# GUIDELINES TO GEOLOGIC/SEISMIC REPORTS

The following guidelines are taken from "Geology and earthquake hazards: Planners guide to the seismic safety element" prepared by Grading Codes Advisory Board and Building Code Committee of the Southern California Section, Association of Engineering Geologists, July, 1973. They are reprinted here courtesy of the Association of Engineering Geologists.

## I. *Introduction*

This is a suggested guide or format for the seismic section of engineering geologic reports. These reports may be prepared for projects ranging in size from a single lot to a master plan for large acreage, in scope from a single family residence to large engineered structures, and from sites located on an active fault to sites a substantial distance from the nearest known active fault. Because of this wide variation, the order, format, and scope should be flexible and tailored to the seismic and geologic conditions, and intended land use. The following suggested format is intended to be relatively complete, and not all items would be applicable to small projects or low risk sites. In addition, some items would be covered in separate reports by soil engineers, seismologists, or structural engineers.

## II. *The Investigation*

### A. *Regional Review*

A review of the seismic or earthquake history of the region should establish the relationship of the site to known faults and epicenters. This would be based primarily on review of existing maps and technical literature and would include:

1. Major earthquakes during historic time and epicenter locations and magnitudes, near the site.

2. Location of any major or regional fault traces affecting the site being investigated, and a discussion of the tectonic mechanics and other relationships of significance to the proposed construction.

### B. *Site Investigation*

A review of the geologic conditions at or near the site that might indicate recent fault or seismic activity. The degree of detail of the study should be compatible with the type of development and geologic complexity. The investigation should include the following:

1. Location and chronology of local faults and the amount and type of displacement estimated from historic records and stratigraphic relationships. Features normally related to activity such as sag ponds, alignment of springs, offset bedding, disrupted drainage systems, offset ridges, faceted spurs, dissected alluvial fans, scarps, alignment of landslides, and vegetation patterns, to name a few, should be shown on the geologic map and discussed in the report.

2. Locations and chronology of other earthquake induced features caused by lurching, settlement, liquefaction, etc. Evidence of these features should be included with the following:
    a. Map showing location relative to proposed construction.
    b. Description of the features as to length, width and depth of disturbed zone.
    c. Estimation of the amount of disturbance relative to bedrock and surficial materials.

3. Distribution, depth, thickness and nature of the various unconsolidated earth materials, including ground water, which may affect the seismic response and damage potential at the site, should be adequately described.

### C. *Methods of Site Investigation*

1. Surface investigation
    a. Geologic mapping.
    b. Study of aerial photographs.
    c. Review of local ground water data such as water level fluctuation, ground water barriers or anomalies indicating possible faults.

2. Subsurface investigation
    a. Trenching across any known active faults and suspicious zones to determine location and recency of movement, width of disturbance, physical condition of fault zone materials, type of displacement, and geometry.

Public Information Office, 1516 Ninth Street, Fourth floor, Sacramento, CA 95814, (916) 445 5716. MAIL ORDER, P.O. Box 2980, Sacramento, CA 95812. Los Angeles Office, 107 South Broadway, Room 1065, Los Angeles, CA 90012. (213) 620-3560. San Francisco Bay Area Office, 380 Civic Drive, Pleasant Hill, CA 94523-1997. (415) 671-4920.

GORDON K. VAN VLECK, Secretary
THE RESOURCES AGENCY

GEORGE DEUKMEJIAN, Governor
STATE OF CALIFORNIA

RANDALL M. WARD, Director
DEPARTMENT OF CONSERVATION

b. Exploratory borings to determine depth of unconsolidated materials and ground water, and to verify fault–plane geometry. In conjunction with the soil engineering studies, obtain samples of soil and bedrock material for laboratory testing.

c. Geophysical surveys which may indicate types of materials and their physical properties, ground water conditions, and fault displacements.

III. *Conclusions and Recommendations*

At the completion of the data accumulating phase of the study, all of the pertinent information is utilized in forming conclusions of potential hazard relative to the intended land use or development. Many of these conclusions will be revealed in conjunction with the soil engineering study.

A. *Surface Rupture Along Faults*

1. Age, type of surface displacement, and amount of reasonable anticipated future displacements of any faults within or immediately adjacent to the site.

2. Definition of any areas of high risk.

3. Recommended building restrictions or use–limitations within any designated high risk area.

B. *Secondary Ground Effects*

1. Estimated magnitude and distance of all relevant earthquakes.

2. Lurching and shallow ground rupture.

3. Liquefaction of sediments and soils.

4. Settlement of soils.

5. Potential for earthquake-induced landslide.

IV. *Presentation of Data*

Visual aids are desirable in depicting the data and may include:

A. *General data*

1. Geologic map of regional and/or local faults.

2. Map(s) of earthquake epicenters.

3. Fault strain and/or creep map.

B. *Local or site data*

1. Geologic map.

2. Geologic cross–sections illustrating displacement and/or rupture.

3. Local fault pattern and mechanics relative to existing and proposed ground surface.

4. Geophysical survey data.

5. Logs of exploratory trenches and borings.

V. *Other Essential Data*

A. *Sources of data*

1. Reference material listed in bibliography.

2. Maps and other source data referenced.

3. Compiled data, maps, plates included or referenced.

B. *Vital support data*

1. Maximum credible earthquake.

2. Maximum probable earthquake.

3. Maximum expected bedrock acceleration.

C. *Signature and license number of geologist registered in California*

5/86

# Part D

# EVALUATION CRITERIA FOR SELECTION OF SITES FOR LAND APPLICATION OF WASTEWATER

The process of site selection for land-application systems should include an initial evaluation on the basis of criteria presented in this section. The environmental setting should be described and the individual site characteristics should be analyzed. Each site should then be reevaluated in light of considerations of treatment methods, design, and expected impacts.

## C.1. GENERAL DESCRIPTION

A preliminary step in site evaluation should be a general description of the land involved. The environmental setting should be described with emphasis on:

- The location of the site

- The relationship to the overall land-use plan

- The proximity to surface water

- The number and size of available land parcels

- Location and use of any existing potable wells (I-C.2.e.6).

### C.1.a. Location

The description of site location should include both the distance and elevation difference from the treatment plant or wastewater collection area. Both will affect the feasibility and economics of the transmission of the wastewater to the site. Any significant obstructions to transmission, such as rivers, freeways, or developed residential areas, should be noted.

### C.1.b. Compatibility with Overall Land-Use Plan

Of significant importance in site selection is the compatibility of the intended use with regional land-use plans. The regional planners or the planning commission should be consulted as to the future use of potential sites.

During a visit to the site, the current use, adjacent land use, and proximity to areas developed for residential, commercial, or recreational activities can be ascertained. On the basis of a review of master plans or discussions with local planners, the proposed future use, zoning, and proposed development of the adjacent area can be determined.

### C.1.c.  Proximity to Surface Water

In many cases, the proximity of the potential site to a surface-water body may be of significance.  For overland flow systems, and systems with underdrains or pumped withdrawal, discharge of renovated water to a surface-water body may be necessary.  In such a case, the feasibility and cost of transmission may become important considerations.  The relationship of surface water to the overall hydrology of the area, and particularly to the groundwater, should be evaluated.  Water-quality aspects and site drainage are considered later in this section.

### C.1.d.  Number and Size of Available Land Parcels

The relative availability of land at potential sites, together with the probable price per acre, must be defined early in the evaluation.  The number and size of available parcels will be of significance, especially in relation to the complexity of land acquisition and control — a subject that is discussed at the end of this section.

### C.2.  DESCRIPTION OF ENVIRONMENTAL CHARACTERISTICS

The environmental characteristics of a potential site that may affect the future selection of a land-application method and the subsequent design of the treatment system include:  climate, topography, soil characteristics, geologic formations, groundwater, and receiving water.  The degree of detail required for the evaluation of any one particular characteristic is highly variable and dependent upon the size of the project and the severity of local conditions.  This discussion cannot cover all conceivable aspects, but the major environmental factors will be discussed.

### C.2.a.  Climate

Local climatic conditions will affect a large number of design decisions including: the method of land application, storage requirements, total land requirements, and loading rates.  The National Weather Service, local airports, and universities are potential sources of climatological data.  The data base should encompass a long enough period of time so that long-term averages and frequencies of extreme conditions can be established.  Each of the climatic factors is discussed in the following paragraphs.

C.2.a.1.  Precipitation — Analysis of rainfall data should be conducted with respect to both quantities and seasonal distribution.  Quantities should be expressed in terms of averages, maximums, and minimums for the period of record.  A frequency analysis should be made to determine the design annual precipitation, which will normally be the maximum precipitation values having a return period of a given number of years (the wettest year in a given number of years). The plot of precipitation against return period on probability paper, a method commonly used to display the results of the frequency analysis, is illustrated in Figure 2. Different return periods may often be used for the determination of liquid loading rates (I-E. 1. a) and the determination of storage capacity (I-E. 4.).

In cold regions, an analysis of the snow conditions with respect to depth and period of snow cover may also be required. In most cases, except for some infiltration-percolation systems, periods of snow cover will necessitate storage of the effluent for later application.

C. 2. a. 2.  Storm Intensities — An investigation of storm data for the period of record should be included in the precipitation study. A frequency analysis

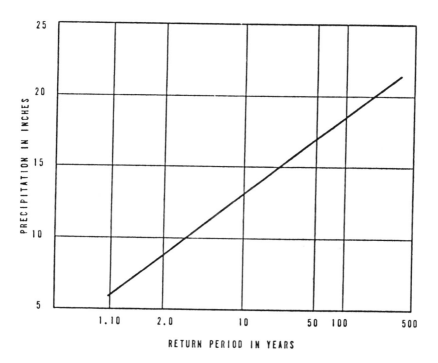

Figure 2: Typical frequency analysis for total annual precipitation

should be performed to determine the relationship between storm intensity, duration, and frequencies or return periods. The design storm event can then be analyzed for the amount of runoff it would produce and the need for any runoff control features can be determined.

C.2.a.3.  Temperature — Temperature analysis should include the range of temperatures during the various seasons. Maximum periods of freezing conditions, particularly periods in which the ground is frozen, are of special interest in determining periods of inoperation. The effects of temperature are of importance in the selection of a land-application method, the design of the loading schedule, and in the determination of storage requirements. For irrigation of annual crops, the probable early and late season frost dates need to be determined.

C.2.a.4.  Evapotranspiration — Evapotranspiration is the evaporation of water from the soil surface and vegetation plus the transpiration of water by plants. Evapotranspiration rates are dependent upon a number of factors, including humidity, temperature, and wind, and will significantly affect the water balance in almost all cases. Typical monthly totals are available in most areas from the National Weather Service, nearby reservoirs, the Agricultural Extension Service, or Agricultural Experiment Stations.

C.2.a.5.  Wind — Analysis of wind velocity and direction may be required, and should contain seasonal variations and frequency of windy conditions. Wind analysis is of importance primarily for spray application systems, where windy conditions may require large buffer zones or temporary cessation of application.

C.2.b.  Topography

The topography of the site and adjacent land is critical to the design of land-application systems. Normally, a detailed topographic map of the area will be necessary for site selection and the subsequent system design. Topographic maps are available from the U.S. Geological Survey. Information to be gained from an analysis of the topography is listed in the following discussion.

C.2.b.1.  Ground Slope — Ground slope, usually expressed as a percentage, is an important site characteristic for the determination of the land treatment method and application technique. For example, the success of an overland flow system is highly dependent upon ground slope, and irrigation by flooding normally requires slopes of less than 1 percent. Foliated hillsides with slopes of up to 40 percent have been sprayed successfully with effluent [140, 142]. Ranges of values for successful operation are given in Section D.

C.2.b.2.  Description of Adjacent Land — The topography of land adjacent to the potential site should be included in the topographic evaluation. Of primary concern are the effects of storm runoff, both from adjacent land onto the site and from the site onto adjacent lands and surface water bodies. Also of concern will be areas downslope from the site where seeps may occur as a result of increased groundwater levels.

C.2.b.3.  Erosion Potential — The erosion potential of the site and adjacent land should be predicted, and any required corrective action outlined. Both wastewater application rates and storm runoff should be considered. The typical Soil Conservation Service (SCS) evaluation of soils includes an analysis of erosion potential, which is valuable in determing the possible extent of the problem.

C.2.b.4.  Flood Potential — The site topography should be evaluated and historical data reviewed to determine the possibility of flooding on the site or adjacent areas. Sites prone to flooding, such as flood plains, may still be suitable for land application but normally only if the physical equipment is protected and off-site storage is provided.

C.2.b.5.  Extent of Clearing and Field Preparation Necessary — The extent of clearing and field preparation is largely dependent upon the selection of land-application method, the application technique, and the existing vegetation. Included in the evaluation should be:

- The extent of clearing of existing vegetation (if necessary)

- Disposition of cleared material

- Necessary replanting

- Earthwork required

Some of this information would be developed in detail in the environmental assessment.

### C.2.c.  Soil Characteristics

Soil characteristics are often the most important factors in selection of both the site and the land-application method. Definite requirements for soil characteristics exist for each of the method alternatives, with overland flow and infiltration-percolation having the strictest requirements. Information on soil characteristics can be obtained from the Soil Conservation Service, many universities, and the Agricultural Extension Service.

C.2.c.1.  Type and Description — The soil at the potential site should be described in terms of its physical and chemical characteristics. Important physical characteristics include texture and structure, which are largely influenced by the relative percentages of the mechanical, or particle-size, classes (gravel, sand, silt, and clay). Chemical characteristics which may be of importance are: pH, salinity, nutrient levels, and adsorption and fixation capabilities for various inorganic ions. The following series of tests is suggested:

- pH

- Salinity or electrical conductivity

- Organic matter

- Total exchangeable cations

- Levels of nitrogen, phosphorus, potassium, magnesium, calcium, and sodium

- Percent of the base exchange capacity occupied by sodium, potassium, magnesium, calcium, and hydrogen

Reference is suggested to the University of California manual for analysis of soils, plants, and waters [26].

C.2.c.2.  Infiltration and Percolation Potential — The potential of the soil for both infiltration and percolation is of great importance in the site selection and selection of application method.  Infiltration, the entry of water into the soil, is normally expressed as a rate in inches per hour.  The rate generally decreases with wetting time and previous moisture content of the soil; consequently, it should be determined under conditions similar to those expected during operation.  Percolation is the movement of water beneath the ground surface both vertically and horizontally, but above the water table.  It is normally, dependent upon several factors, including soil type; constraints to movement, such as lenses of clay, hardpan, or rock; and degree of soil saturation.  The limiting rate (either infiltration or percolation) must be determined and reported in inch/day (cm/day) or inch/week (cm/week).

The standard percolation test is not recommended for determination of infiltration or percolation rates.  The test results are not reproducible by different fieldmen [182] and are affected by hole width, gravel packing of holes, depth of water in holes, and the method of digging the holes.  More importantly, if subsurface lenses exist, the water in the test hole will move laterally, with the result being a fairly high percolation rate.  Designing a liquid loading rate on that basis would be disasterous because, when the entire field is loaded, the only area for flow is the few feet of depth to the lens times the field perimeter.  Instead of using the percolation test, it is suggested that several or more of the following approaches be used as a basis of determining infiltration and percolation rates:  (1) consultation with Agriculture Extension Service agents, state or local government soil scientists, or independent soil specialists; (2) engineering analysis of several soil borings and soil classifications; (3) engineering analysis of soil profiles supplied by the Soil Conservation Service (SCS); (4) consultation with county agents, agronomists, or persons having farming experience with the same, similar, or nearby soils; and (5) experience from pilot studies on parts of the field to be used.

C.2.c.3. Soil Profile — The soil profile, or relation of soil characteristics to depth, will normally be required for all site evaluations. Generally, the profile should be determined to depths of 2 to 5 feet (0. 61 to 1. 52 m) for overland flow, at least 5 feet (1. 52 m) for irrigation, and at least 10 feet (3. 05 m) for infiltration-percolation. The underlying soil layers should be evaluated principally for their renovation and percolation potentials. Lenses or constraints to flow below these levels should be located.

C.2.c.4. Evaluation by Soil Specialists — In most cases, an evaluation by soil specialists will be necessary to determine the overall suitability of the soil characteristics for the intended use. SCS representatives, soil scientists, agronomists, and Agricultural Extension Service representatives are possible sources to be consulted.

C.2.d. Geologic Formations

A basic description of the geologic conditions present and their effects should be required for all site evaluations. Infiltration-percolation sites and sites with suspected adverse geological conditions will require a relatively detailed analysis, while considerably less is required for most overland flow sites and many irrigation systems. Data on geological formations are available from the U. S. Geological Survey, state geology agencies, and occasionally from SCS or U. S. Bureau of Reclamation publications.

C.2.d.1. Type and Description — The geologic formations should be considered in terms of: the structure of the bedrock, the depth to bedrock, the lithology, degree of weathering, and the presence of any special conditions, such as glacial deposits. The presence of any discontinuities, such as sink holes, fractures or faults, which may provide short circuits to the groundwater, should be noted and thoroughly investigated. In addition, an evaluation of the potential of the area for earthquakes and their probable severity will often be of importance to the future design of the system.

C.2.d.2. Evaluation by Geologists — In many situations, an evaluation by a geologist or geohydrologist will be necessary. The geologist will be of value both in the investigation of the geologic conditions and in the evaluation of their effects. Of primary importance in the evaluation are the effects of the geology on the percolation of applied wastewater and the movement of groundwater.

C.2.e. Groundwater

An investigation of groundwater must be conducted for each site, with particular detail for potential infiltration-percolation and irrigation sites. Evaluations should be made by the engineer to determine both the effect of groundwater levels on renovation ca abilities and the effects of the applied wastewater on groundwater movement and quality with respect to the BPT requirements.

C.2.e.1. Depth to Groundwater — The depth to groundwater should be determined at each site, along with variations throughout the site, and seasonal variations. Depth to groundwater is important because it is a measure of the aeration zone in which renovation of applied wastewater takes place. Generally, the groundwater depth requirements are:

- Overland flow — sufficient depth not to interfere with plant growth

- Irrigation — at least 5 feet (1.52 m)

- Infiltration-percolation — preferably 15 feet (4.57 m) or more

Lesser depths may be acceptable where underdrains or pumped withdrawal systems are utilized.

When several layers of groundwater underlie a particular site, depths should be determined to each, unless they are separated by a continuous impervious stratum. The quality and current and planned use of each layer should also be determined.

C.2.e.2. Groundwater Flow — In most cases, the groundwater should be evaluated for direction and rate of flow and for the permeability of the aquifer. This evaluation may be unnecessary when percolation is minimal, as with an overland flow and some irrigation systems. For systems designed for high percolation rates, effects on the groundwater flow must be predicted.

Additionally, data on aquifer permeability may be evaluated, together with groundwater depth data, to predict the extent of the recharge mound. The direction of flow is important to the design of the monitoring system and should be traced to determine whether the groundwater will come to the surface, be intercepted by a surface water, or join another aquifer.

C.2.e.3. Perched Water — Perched water tables are the result of impermeable or semipermeable layers of rock, clay, or hardpan above the normal water table and may be seasonal or permanent. Perched water can cause problems for land-application systems by reducing the effective renovative depth. Sites should be investigated both for existing perched water tables and for the potential for development of new ones resulting from percolating wastewater. The effect of perched water tables should be evaluated, and the possibility of using underdrains investigated. A distinction should be made between permanent groundwater protected by impermeable strata and perched groundwater above such strata.

C.2.e.4. Quality Compared to Requirements — The quality of the groundwater is of great interest, especially in cases in which it is used for beneficial purposes or differs substantially from the expected quality of the renovated wastewater. The existing quality should be determined and compared to quality requirements for its current or intended use. The proposed requirements for BPT [3] include limitations for chemical constituents, pesticide levels, and bacteriological quality as discussed in I-B.4.

C.2.e.5.  Current and Planned Use — Both current and planned use of the ground-water should be determined, and the quality requirements for the various uses detailed.  The distance from the site to the use areas may also be of importance, because further renovation may occur during lateral movement.

C.2.e.6.  Location of Existing Wells — Much of the data required for ground-water evaluation may be determined through use of existing wells.  Wells that could be used for monitoring should be listed and their relative location described. Historical data on quality, water levels, and quantities pumped that may be available from the operation of existing wells may be of value.  Such data might include seasonal groundwater-level variations, as well as variations over a period of years.  Logs containing soil data may be available from the drillers of these wells, and this information could augment data from soil borings or geological maps.  It should be noted that much information on private wells can be obtained only with the owner's consent.  Determining ownership and locating owners can be difficult and time-consuming.

C.2.f.  Receiving Water (Other than Groundwater)

Land-application systems in which renovated water is recovered, particularly overland flow systems, may require discharge into a receiving surface water body.  Such a discharge would require a permit under the National Pollution Discharge Elimination System (NPDES).  If the receiving water is designated as effluent limited, the requirements for secondary treatment apply.  If the receiving water is designated as water-quality limited, pursuant to Section 303 of P.L. 92-500, treatment must be provided consistent with the established water-quality standards.  Included in the evaluation should be descriptions of: the type of body (lake, stream, etc.), its current use and water quality, pre-scribed water-quality standards and effluent limitations, and water-rights considerations.  Special water-quality requirements and other considerations may exist when the potential receiving water is an intermittent stream.  The current use of the water, together with its prescribed water-quality standards, will determine the degree of treatment necessary by the land-application system.

Water-rights considerations may require that certain quantities of renovated water be returned to a particular water body, particularly in the western states. In cases in which a change in method of disposal or point of discharge is contem-plated, the state agency of other cognizant authority should be contacted, and the status of all existing water rights thoroughly investigated.

C.3.  METHODS OF LAND ACQUISITION OR CONTROL

After potential sites have been selected, alternative methods of land acquisition or control should be assessed.  Alternative methods include:  (1) outright pur-chase of land with direct control, (2) appropriate lease of land with direct control, (3) purchase of land with lease back to farmer for the purpose of land application, and (4) contract with user of wastewater.  An appropriate lease would be one in which the investment of funds for construction of the land-application system would be protected and direct control of the effluent application would be retained by the municipality or district.

The selection of an acquisition and control method is highly dependent on the selected method of application. Infiltration-percolation and overland flow systems normally require a high degree of control and may often be suitable only if outright purchase of the land is possible. Because land control requirements are more flexible for irrigation systems, the leasing of land to agricultural users may be possible. Leasing of required land is often best suited to pilot studies and temporary systems.

Grant eligibility has not been considered in the discussion of these methods. For land acquisition to be eligible for a construction grant, under P. L. 92-500, the land must be an integral part of the treatment process or is to be used for ultimate disposal of residues resulting from such treatment.

Source:  United States Environmental Protection Agency (Ref. 5.11)

# GUIDELINES FOR PREPARING AN ENVIRONMENTAL ASSESSMENT FOR PROPOSED LAND APPLICATION SITES

## ENVIRONMENTAL ASSESSMENT

The impact of the project on the environment, including public health, social, and economic aspects must be assessed for each land-application alternative. Environmental assessments are required for all federally funded projects, and similar reports are required by many state and local governments. This section is not intended to replace existing guidelines (40 CFR 6) for the preparation of environmental assessments, but instead is designed to highlight some of the important considerations particular to land application.

In accordance with existing guidelines, environmental assessment will generally consist of:

- Description of the environmental setting

- Determination of components affected

- Evaluation of possible methods of mitigation of adverse effects

- Determination of unavoidable adverse effects

- Evaluation of overall and long-term effects

Environmental component interactions should be considered and measurable parameters identified if possible.

## F.1. ENVIRONMENTAL IMPACT

Environmental components that may be affected by land-application systems include: (1) soil and vegetation, (2) groundwater, (3) surface water, (4) animal and insect life, (5) air quality, and (6) local climate. Effects on the soil, vegetation, and groundwater are normally the most critical, with the effects on surface water being critical at times.

### F.1.a.  Soil and Vegetation

The effects of land application on the soil and vegetation can be either beneficial or adverse, with the overall effect most often being mixed. Effects on surrounding land and vegetation may be brought about by changes in various conditions, such as groundwater levels, drainage areas, and microclimates.

Soil conditions, including drainage characteristics and levels of chemical constituents, may be affected by land application. Infiltration and percolation capacities may decrease as a result of clogging by suspended solids, although proper management techniques including resting periods and soil surface raking may help to mitigate this condition. Rates may also increase or decrease as a result of changing chemical conditions, such as the pH and sodium content of the soil. Long-term effects on the soil chemistry, such as the buildup of certain constituents to toxic levels, may be critical in land-application systems, Effects on soil conditions should be predicted initially, and appropriate monitoring requirements should be defined. Various references, particularly Thomas and Law [167], may be helpful in predicting soil effects.

The effects on vegetation are usually beneficial for a well-operated system. Virtually all essential plant nutrients are found in wastewater and should stimulate plant growth. Toxic levels of certain constituents in the soil, which may reduce growth or render crops unsuitable for the intended use must be evaluated [27]. Excess hydraulic loadings or poor soil aeration may also be harmful to plant growth.

### F.1.b.  Groundwater

The groundwater quality and level will be affected by most land-application systems. Exceptions would be many overland flow, underdrained, and pumped withdrawal systems. Wastewater constituents that are not used by the plants, degraded by microorganisms, or fixed in the soil may leach to the groundwater. Nitrate nitrogen is the constituent of most concern; however, heavy metals, phosphorus, organics, total dissolved solids, and other elements discussed in I-B.4 may also be of significance.

Groundwater levels may be affected by land application, particulary for infiltration-percolation systems. In turn, groundwater flow may be affected with respect to both rate and direction of movement. The direction and effects of the altered groundwater flow must be predicted, and appropriate monitoring requirements defined.

### F.1.c.  Surface Water

Surface waters may be affected directly by (1) discharge from an overland flow, underdrained, or pumped withdrawal system, (2) interception of seepage from an infiltration-percolation system, or (3) undesired surface runoff from the site. Both surface water quality and rate of flow may be influenced. Changes in water quality will be regulated by federal, state, or regional standards. Effects on surface water flow should be investigated both with respect to possible increased and decreased rates of flow. Wastewater reuse systems, used to replace systems previously discharging to a surface water, will result in decreased flows with possible adverse consequences to previous downstream users, or existing fisheries.

### F.1.d.  Animal and Insect Life

Treatment by land application may result in changes in conditions, either favorably or adversely affecting certain indigenous terrestrial or aquatic species.  Beneficial effects, such as the increased nutritive value of animal forage, should be compared to possible adverse effects, such as the disruption of natural habitat, for each species of concern.  Little information exists on this subject, but Sopper [148] reports some initial findings.  The possibility of insects or rodents acting as disease vectors is discussed separately under Public Health Effects (I-F.2.b.).

### F.1.e.  Air Quality

Air quality may possibly be affected through the formation of aerosols from spray systems and through odors.  With aerosols, the primary concern is with transmission of pathogens, which will be discussed further under Public Health Effects.  Odors are caused principally by anaerobic conditions at the site or in the applied wastewater.  Correction of these conditions is the only permanent cure.

### F.1.f.  Climate

Land-application systems, particularly large irrigation or overland flow systems, may have a limited but noticeable effect on the local climate.  Air passing over a site will pick up moisture and be cooled, resulting in a localized reduction in temperature.  Original conditions are normally regained within a short distance from the site [125].

### F.2.  PUBLIC HEALTH EFFECTS

When evaluating the overall environmental impact of an alternative, special consideration should be given to those effects that relate directly to the public health.  In many cases, state health regulations and guidelines serve to protect against many of the effects.  Public health effects that should be considered include: groundwater quality, insects and rodents, runoff from site, aerosols, and contamination of crops.  Overviews of public health effects that may be helpful are contained in references [13, 130, 143, 152].

### F.2.a.  Groundwater Quality

The quality of the groundwater will be of major concern when it is to be used as a potable water supply, particularly when an infiltration-percolation system is planned.  A sufficient degree of renovation will be required to meet the BPT requirements for groundwater protection. Nitrates are the most common problem, but other constituents, including stable organics, dissolved salts, trace elements, and pathogens should be considered. Extensive monitoring and control practices must be planned.

### F.2.b.  Insects and Rodents

Because of the possibility of contamination from pathogens in the wastewater, the control of insects and rodents on a land-application site is more critical than on a conventional irrigation site. Conventional methods of control will normally be required for most pests.

Mosquitoes are a special problem because they will propagate in water standing for only a few days. Elimination of unnecessary standing water and sufficient drying periods between applications are the most effective methods of control.

### F.2.c.  Runoff from Site

Applied effluent should not be allowed to run off the site except in systems designed for surface runoff (e.g., overland flow). The extent to which runoff from storm events must be controlled depends upon the water quality objectives of the surface water and the possible effects of such runoff on water quality. Few data are available to assess storm runoff effects from land-application sites.

### F.2.d.  Aerosols

Generally, the danger of aerosols lies in their potential for the transmission of pathogens. Aerosols are microscopic droplets that conceivably could be inhaled into the throat and lungs. Aerosol travel and pathogen survival rate are dependent on several factors, including wind, temperature, humidity, vegetative screens, and other factors. Methods of reduction should be employed to ensure that transmission of aerosols is minimized, with probable travel under normal conditions being limited to an acceptable area. This area should be determined on the basis of the proximity of public access. Sorber [152] and Sepp [143] present discussions of this issue and discuss the research on the subject.

Safeguard measures that may be employed against aerosol transmission include:

- Buffer zones around the field area

- Sprinklers that spray laterally or downward with low nozzle pressure

- Rows of trees or shrubs

- Cessation of spraying or spraying only interior plots during high winds

- Combinations of the enumerated measures with adequate disinfection

### F.2.e.  Contamination of Crops

The effect of effluent irrigation on crops, with regard to safety for consumption, is a matter of some concern.  Many states have regulations dealing with the types of crops that may be irrigated with wastewater, degrees of preapplication treatment required for various crops, and purposes for which the crops may be used.  The proposed California regulations are included in Appendix E, and are offered as an example.  Individual state health departments should be consulted, since regulations vary widely from state to state. Additional information on the contamination of crops may be found in Sepp [143], Rudolfs [135], and Bernarde [13], or by contacting the FDA or other applicable agencies.

### F.3.  SOCIAL IMPACT

The overall effects of the proposed system should be evaluated in light of their impact on the sociological aspects of the community.  Included in the evaluation should be considerations of:  relocation of residents, effects on greenbelts and open space, effects on recreational activities, effects on community growth, and effects on the quality of life.

### F.3.a.  Relocation of Residents

The requirement for large quantities of land, particularly for irrigation and overland flow systems, often necessitates the purchase of land and possibly the relocation of residents.  For federally funded projects, the acquisition of land and relocation of residents must be conducted in accordance with the Uniform Relocation Assistance and Land Acquisition Policies Act of 1970. In such cases, the advantages of the proposed treatment system must be weighed against the inconvenience caused affected residents, and then compared with other alternatives.

### F.3.b.  Greenbelts and Open Spaces

Proposed treatment systems should be evaluated from an aesthetic point of view and with respect to the creation or destruction of greenbelts and open spaces.  Disruption of the local scenic character is often unnecessary and undesirable, while through proper design and planning, the beauty of the landscape can often be enhanced.  Reforestation and reclamation of disturbed areas, such as those resulting from strip mining operations, are possible beneficial effects.

### F.3.c.  Recreational Activities

The net result of the treatment system on recreational facilities should be considered.  Existing open space or parks may be disrupted; however, other recreational areas may be created or upgraded.  Irrigation of new parks or golf courses and recreational use of renovated water are possibilities for increasing the overall value of a proposed treatment system.

### F.3.d.  Community Growth

The effects of a new treatment system may stimulate or discourage the growth of a community, both in terms of economics and population.  Often, improved wastewater treatment service may allow new construction or expansion in the service area.  Such growth may consequently tax other existing community services.  The potential of the treatment system for affecting community growth should be evaluated, and the subsequent effects on other aspects of the community documented.

### F.4.  ECONOMIC IMPACT

An evaluation of the economic impact should include an analysis of all economic factors directly and indirectly affected by the treatment system.  Many factors common to conventional systems apply; however, additional factors may be applicable to various land-application systems.  Possible additional factors include:

- Change in value of the land used and adjacent lands

- Loss of tax revenues as a result of governmental purchase

- Conservation of resources and energy

- Change in quality of ground or surface waters

- Availability of an inexpensive source of water for irrigation

The effect of the treatment system on the overall local economy should then be appraised, especially with respect to financing and the availability of funds for the long-term operation and maintenance of the system.

Source:  United States Environmental Protection Agency (Ref. 5.11)

# SUGGESTED PROTOCOL FOR SAMPLING MONITORING WELLS

A.  **PLANNING**

Prior to any sampling, the objectives of the program must be established and understood.  A monitoring network that is sampled without specific objectives in mind can
generate considerable useless data.   Objectives of a ground-water sampling program
might include:

1.  Determining the potability of a private or municipal supply well.
2.  Investigating the presence or absence of contamination in a given study area.
3.  Investigating the magnitude and the vertical and horizontal extent of a known or
    suspected contaminant plume.

Additionally, the design of a ground water sampling program can be affected by the
presence, location, pumping schedule, and construction of existing monitoring wells.
In cases of suspected contaminant sources, there will usually be no monitor wells
available to sample.   In this case, private wells may be used to help define the presence or absence of contamination.   However, this should not be a standard practice.
Private supply wells should be used largely to evaluate potability and to gather preliminary data.  These wells should not be used as a substitute for properly located and
well-designed monitor wells.  See later section on sampling private (domestic) wells.

Access to monitor wells may be difficult and the wells themselves hard to locate in the
field.   Obtain information on location, access, permission, etc. before visiting the
site.   Monitor wells usually have a friction cap or screw cap, and should be locked.
Tools for removing caps and keys to unlock the wells are often necessary.

If several monitor wells must be sampled, proper designation of each is essential.  The
well number assigned by the well owner (or by the agency) should be known.  If numbers
are not assigned, a precise field description is essential so as to avoid confusing the
well sample results.  Field notes are essential.  Sketch maps also are helpful.

B.  **PREPARING THE WELL**

In order to obtain a representative ground water sample, the water that has stagnated
and/or thermally stratified in the well casing must be evacuated.  This allows ground
water from the surrounding formation to enter the well.  Evacuation of three to five
well volumes is recommended; however in wells with low recoveries this may not be practicable.   In these instances, the well may be evacuated to emptiness and allowed to
recover prior to sampling.

All newly constructed monitoring wells should be allowed to stabilize for a minimum of
two weeks prior to sampling.

Adapted From: N.J. Dept. of Environmental Protection, Field Procedures Manual, 3rd Printing, 1987.

The capacity of common casing diameters are as follows:

| Casing Diameter (ft) | Gallons/Linear (ft) |
|---|---|
| 2 - inch (0.1667) | 0.1632 |
| 4 - inch (0.3333) | 0.6528 |
| 6 - inch (0.5000) | 1.4688 |
| 8 - inch (0.6667) | 2..6112 |
| 10 - inch (0.8333) | 4.0800 |
| 12 - inch (1.0000) | 5.8752 |

The amount of water within the well casing is calculated by multiplying the linear feet of water by the appropriate volume per foot.

Example:

| | |
|---|---|
| Total depth of casing | 100 ft. |
| Depth to water | - 20 ft. |
| Depth of water column | 80 ft. |
| 2-inch casing | x0.1632 |
| Amount of water in casing | 13.06 gallons |

Alternately, use this formula to determine the gallons in any diameter pipe:

No. gallons = $5.8752 \times C^2 \times H$

Where C = casing diameter in feet
      H = height of water column in feet

Evacuation of the well can be accomplished in several ways. It is paramount to ensure that the evacuation procedure does not cause cross contamination from one well to the next. Therefore, the preferred method employs dedicated tubing and pumps. Since it is not always practical to dedicate pumps it is acceptable to decontaminate equipment between wells, if approved methods are used. Tubing should always be dedicated.

The selection of an evacuation method is usually dictated by the depth to water. If the static water level is less than 25 feet, a surface pump (centrifugal, diaphram) with new dedicated linear polyethylene (ASTM Drinking Water Grade) is the preferred method. If the water level is greater than 25 feet, however, a submersible pump should be used. The pump should be washed, rinsed, and a sample of the rinse water collected as a field blank to ensure the integrity of the sample. Hand-bailing may be utilized with a static level greater than 25 feet if no submersible pump is available. Hand-bailing is not preferred, however, unless necessary, due to the potential of aerating the sample during collection or possibly introducing contaminants during the bailing procedure.

Pumping will draw down the water level. The pump must either be lowered as the level declines or the pump set deep into the water column before evacuation begins. The distance between the pump intake and the water surface during evacuation will depend on several factors such as the presence of floating contaminants (e.g., gasoline) and the probable depth of contamination within the aquifer. In any case, the pump should not be set opposite the well screen (unconsolidated formations) in order to avoid possible clogging of the pump intake.

The disposal or discharge of floating solvents or hydrocarbons, and the discharge of highly contaminated water during evacuation requires special procedures.

After evacuation of the required volume of water from the well, sampling can begin. Sampling of the monitor well should occur as soon as possible after evacuation, preferably immediately and be consistent for each well. In most cases, the time lapse between evacuation and sampling should not exceed two hours.  Never collect a sample for volatile organics from the surface of the water column.

If the sample must be collected from the pump, reduce the discharge to a trickle when ready to sample.  This will reduce agitation and loss of volatile contaminants.

When several wells will be sampled, the <u>least</u> contaminated well should be sampled first, and the wells then sampled in ascending order of contamination.  If the contaminant levels are not known, particular attention must be paid to decontamination of the pump and hose.

C.  **SAMPLING EQUIPMENT**

The equipment utilized for ground water sampling will vary greatly, depending on the following key factors, among others:

1.  Type of well (monitoring, water supply, etc.)
2.  Depth of well
3.  Diameter of well casing
4.  Depth to water
5.  Suspected contaminants
6.  Analytes of interest
7.  Length of open hole (bedrock well)
8.  Slot size of screen and screen type

Equipment to be utilized for ground water sampling generally falls into two categories: those used to evacuate the well casing and those used to grab a discrete sample.  In some instances, the device utilized for evacuation may be the same utilized for sample withdrawal.  In most instances, however, the evacuation devices should <u>not</u> be used in actual sample collection.

Types of equipment available for monitoring well evacuation and/or sampling include the following:

1.  Bottom fill bailers (Teflon or stainless steel)
    a.  Single check valve (bottom)
    b.  Double check valve (top and bottom)
2.  Suction lift pumps/centrifugal pumps
3.  Portable submersible pumps
4.  Air lift pumps
5.  Bladder (gas squeeze) pumps
6.  Gas displacement pumps
7.  Gas piston pumps
8.  Packer pumps
9.  Continuous organics sampling system in conjunction with peristaltic pump
10. Syringe sampler
11. Bacon bomb
12. Kemmerer bottle

The sampling devices should be laboratory cleaned, preferably by the laboratory performing the analysis, utilizing recommended cleaning procedures.  The device should then be wrapped in cleaned and autoclaved aluminum foil.  The sampler should remain wrapped in this manner until immediately prior to use.

In addition to the evacuation and sampling devices, other equipment necessary for a thoroughly documented sampling event may include:

1.  Water level indicators
    a.  Steel tape and chalk
    b.  Electric tape (e.g., slope indicator, M-Scope)

2.  Sample containers, proper size and composition
3.  Preservatives, as needed
4.  Ice or ice packs
5.  Field safety instrumentation, as needed (e.g., H-NU, OVA, Photovac-TIP)
6.  Field and travel blanks
7.  Bound field notebook
8.  Sample analysis request forms
9.  Chain of custody forms
10. Chain of custody seals
11. Sample labels, indelible
12. Appropriate personal safety equipment (e.g., gloves, masks)
13. Appropriate hand tools
14. Keys to locked wells
15. Quality Assurance/Quality Control (QA/QC) "blanks".

D.  **INITIAL FIELD OBSERVATIONS AND MEASUREMENTS**

Once a well has been located and identified, the field measurements listed below should be noted in a field notebook. Be certain that the proper well is being selected. The misidentification of sampling point in the field will result in erroneous data that may affect important decisions.

a.  Physical Measurements

    i.    Presence/absence of a protective casing
    ii.   Lock and well number
    iii.  Diameter and construction material of the well casing
    iv.   Total depth of well from the top of casing (TOC) or surveyor's mark, if present
    v.    Depth from top of casing (TOC) (or mark) to water
    vi.   Calculate the linear feet of water in the well by subtracting depth-to-water from total depth of well

In addition to the physical measurements and other information that may identify the well, the following chemical information should be recorded as needed, during evacuation, and prior to sampling:

    i.    pH
    ii.   Temperature
    iii.  Specific conductance

E.  **QUALITY ASSURANCE**

Because of the considerable cost involved, extreme care must be taken to insure that the ground water data is accurate. Laboratory work, no matter how sophisticated, is only as good as the quality of the sample supplied to the analyst. Field procedures are relatively primitive when compared to the high technology of modern analytical chemistry. There are many areas to which specific attention must be paid; some of these relate to well drilling procedures and well construction materials. Issues that are discussed below include cleanliness and the construction materials used for sampling equipment.

a.  Field measurement and sampling equipment that will enter the well must be cleaned prior to its entry, using approved methodologies. Whenever possible, sampling equipment should be laboratory cleaned and wrapped and dedicated to a specific well for the day's sampling. This shall apply to all bailers to be utilized to collect samples. Pumps and equipment not amendable to laboratory cleaning must be field cleaned using approved methodologies.

b. The materials involved in ground water sampling are critical to the collection of valid monitoring information. Reliance on inexpensive materials (e.g., PVC) may lead to the collection of unreliable data, particularly when volatile, pH sensitive, or valence reduced chemical constituents are being evaluated. Sampling methods which minimize agitation, air content, gas exchange, and depressurization are preferable. The construction materials of the sampler which contact the water are as critical as the laboratory containers. Recommended materials for bailers, pump parts, tubing, samplers and associated apparatus in decreasing order of preference are: Teflon, stainless steel 316, stainless steel 304, polypropylene, polyethylene, linear polyethylene, Viton, conventional polyethylene, PVC.

Studies have been completed and further work is underway relative to these materials and their suitability in trace organic sampling. Bailers constructed of Teflon and/or stainless steel are recommended for use in critical projects. Additionally, any materials contacting the water to be sampled (i.e., material used to lower bailer) should be constructed of Teflon or stainless steel. This is especially true when high concentrations of contaminants which will degrade PVC are present.

Tubing utilized in well evacuation may consist of materials other than Teflon, but may not be utilized for sample collection and must be dedicated for use in only one well.

## F. SAMPLING PRIVATE (DOMESTIC) WELLS

An important step when sampling a domestic well is to obtain as much information as possible from the home owner, such as well depth, screen setting, diameter of casing, any taste, odor or aesthetic problems, when was the well drilled and by whom. This information should be used with caution unless it can be verified with drilling logs, etc.

An operating domestic well must be pumped to waste prior to sample collection. Samples taken immediately will be water within the plumbing and not directly from the aquifer. It is therefore essential to evacuate the plumbing and water storage tank. House storage tanks vary in capacity but 50 gallons is not unusual. A minimum of 15 minutes before collection is a good rule-of-thumb; longer is desirable. Listen for the pump or the electric circuit to the pump to come on, indicating that the plumbing is being evacuated. Opening all faucets, flushing the toilet, etc., will use water and shorten the waiting time.

Inquire of the well owner regarding any treatment equipment installed on his system. Softening, iron removal, turbidity removal, disinfection, pH adjustment and other equipment is often used; these will give misleading analyses, depending on the parameters of interest. Home carbon filters for the removal of organics also are increasingly popular. Basement and outside faucets will often avoid such treated water.

A brief inspection of the system should be performed to locate the well, pump, storage tanks and any treatment devices.

NOTE: If a sample must be taken following a treatment unit, the type, size and purpose of the unit should be noted on sample sheets and the field notes.

Samples should be taken as close to the pumping well as possible and prior to any storage tanks or treatment systems. Therefore, basement and outside faucets are usually the best sample points.

Home faucets, particularly kitchen faucets, usually have a screen installed on the discharge. This must be removed prior to sampling for bacteria, or for volatile organics, since the screen tends to agitate the discharge and some organics may be lost. When sampling for bacteria, do not take a sample from a swivel faucet, since the joint may harbor a significant bacterial population.

NOTE:    Home owner's plumbing systems should not be tampered with in any way, except for removal of the faucet screen (with permission of the home owner).

For long-term monitoring projects which utilize domestic wells a specific tap or faucet should be designated as the target sample access point for accurate reproducibility in future samples.

# TYPICAL WASTEWATER FLOWS FROM VARIOUS SOURCES

### Typical Wastewater Flows from Institutional Sources

| Source | Unit | Wastewater Flow | |
| --- | --- | --- | --- |
| | | Range | Typical |
| | | gpd/unit | |
| Hospital, Medical | Bed | 132  - 251 | 172 |
| | Employee | 5.3 -  15.9 | 10.6 |
| Hospital, Mental | Bed | 79.3 - 172 | 106 |
| | Employee | 5.3 -  15.9 | 10.6 |
| Prison | Inmate | 79.3 - 159 | 119 |
| | Employee | 5.3 -  15.9 | 10.6 |
| Rest Home | Resident | 52.8 - 119 | 92.5 |
| | Employee | 5.3 -  15.9 | 10.6 |
| School, Day:<br>  With Cafeteria, Gym,<br>    Showers | Student | 15.9 -  30.4 | 21.1 |
| With Cafeteria Only | Student | 10.6 -  21.1 | 15.9 |
| Without Cafeteria, Gym,<br>    Showers | Student | 5.3 -  17.2 | 10.6 |
| School, Boarding | Student | 52.8 - 106 | 74.0 |

Source:  Ref. 5.9

## Typical Wastewater Flows from Recreational Sources

| Source | Unit | Wastewater Flow Range | Typical |
|--------|------|-------|---------|
| | | gpd/unit | |
| Apartment, Resort | Person | 52.8 - 74 | 58.1 |
| Cabin, Resort | Person | 34.3 - 50.2 | 42.3 |
| Cafeteria | Customer | 1.1 - 2.6 | 1.6 |
| | Employee | 7.9 - 13.2 | 10.6 |
| Campground (developed) | Person | 21.1 - 39.6 | 31.7 |
| Cocktail Lounge | Seat | 13.2 - 26.4 | 19.8 |
| Coffee Shop | Customer | 4.0 - 7.9 | 5.3 |
| | Employee | 7.9 - 13.2 | 10.6 |
| Country Club | Member Present | 66.0 - 132 | 106 |
| | Employee | 10.6 - 15.9 | 13.2 |
| Day Camp (no meals) | Person | 10.6 - 15.9 | 13.2 |
| Dining Hall | Meal Served | 4.0 - 13.2 | 7.9 |
| Dormitory, Bunkhouse | Person | 19.8 - 46.2 | 39.6 |
| Hotel, resort | Person | 39.6 - 63.4 | 52.8 |
| Laundromat | Machine | 476 - 687 | 581 |
| Store Resort | Customer | 1.3 - 5.3 | 2.6 |
| | Employee | 7.9 - 13.2 | 10.6 |
| Swimming Pool | Customer | 5.3 - 13.2 | 10.6 |
| | Employee | 7.9 - 13.2 | 10.6 |
| Theater | Seat | 2.6 - 4.0 | 2.6 |
| Visitor Center | Visitor | 4.0 - 7.9 | 5.3 |

Source: Ref. 5.9

## Typical Wastewater Flows from Commercial Sources

| Source | Unit | Wastewater Flow Range | Typical |
|--------|------|-------|---------|
| | | gpd/unit | |
| Airport | Passenger | 2.1 - 4.0 | 2.6 |
| Automobile Service Station | Vehicle Served | 7.9 - 13.2 | 10.6 |
| | Employee | 9.2 - 15.8 | 13.2 |
| Bar | Customer | 1.3 - 5.3 | 2.1 |
| | Employee | 10.6 - 15.8 | 13.2 |
| Hotel | Guest | 39.6 - 58.0 | 50.1 |
| | Employee | 7.9 - 13.2 | 10.6 |
| Industrial Building (excluding industry and cafeteria) | Employee | 7.9 - 17.2 | 14.5 |
| Laundry (self-service) | Machine | 475 - 686 | 580 |
| | Wash | 47.5 - 52.8 | 50.1 |
| Motel | Person | 23.8 - 39.6 | 31.7 |
| Motel with Kitchen | Person | 50.2 - 58.1 | 52.8 |
| Office | Employee | 7.9 - 17.2 | 14.5 |
| Restaurant | Meal | 2.1 - 4.0 | 2.6 |
| Rooming House | Resident | 23.8 - 50.1 | 39.6 |
| Store, Department | Toilet room | 423 - 634 | 528 |
| | Employee | 7.9 - 13.2 | 10.6 |
| Shopping Center | Parking Space | 0.5 - 2.1 | 1.1 |
| | Employee | 7.9 - 13.2 | 10.6 |

Source: Ref. 5.9

## POLLUTANT CONCENTRATIONS AND CONTRIBUTIONS OF
## RESIDENTIAL WASTEWATER

### Pollutant Contributions of Major Residential Wastewater Fractions[a]
### (gm/cap/day)

| Parameter | Garbage Disposal | Toilet | Basins, Sinks, Appliances | Approximate Total |
|-----------|------------------|--------|---------------------------|-------------------|
| BOD$_5$ | 18.0<br>10.9 - 30.9 | 16.7<br>6.9 - 23.6 | 28.5<br>24.5 - 38.8 | 63.2 |
| Suspended Solids | 26.5<br>15.8 - 43.6 | 27.0<br>12.5 - 36.5 | 17.2<br>10.8 - 22.6 | 70.7 |
| Nitrogen | 0.6<br>0.2 - 0.9 | 8.7<br>4.1 - 16.8 | 1.9<br>1.1 - 2.0 | 11.2 |
| Phosphorus | 0.1<br>0.1 - 0.1 | 1.2<br>0.6 - 1.6 | 2.8<br>2.2 - 3.4 | 4.0 |

[a] Means and ranges of results reported

Source: Ref. 5.9

### Pollutant Concentrations of Major Residential Wastewater Fractions[a] (mg/l)

| Parameter | Garbage Disposal | Toilet | Basins, Sinks, Appliances | Combined Wastewater |
|-----------|------------------|--------|---------------------------|---------------------|
| BOD$_5$ | 2380 | 280 | 260 | 360 |
| Suspended Solids | 3500 | 450 | 160 | 400 |
| Nitrogen | 79 | 140 | 17 | 63 |
| Phosphorus | 13 | 20 | 26 | 23 |

[a]Based on the average results presented in the table above and the following wastewater flows: Garbage disposal—2 gpcd (8 lpcd); toilet—16 gpcd (61 lpcd); basins, sinks and appliances—29 gpcd (110 lpcd); total—47 gpcd (178 lpcd).

## Indicator Organism and Pathogen Concentrations in Domestic Septage

| Parameter | Typical Range counts/100 ml |
|---|---|
| Total Coliform | $10^7 - 10^9$ |
| Fecal Coliform | $10^6 - 10^8$ |
| Fecal Streptococci | $10^6 - 10^7$ |
| Ps. aeruginosa | $10^1 - 10^3$ |
| Salmonella sp. | $<1 - 10^2$ |
| Parasites<br>Toxacara, Ascaris lumbricoides, Trichuris trichiura, Trichuris vulpis | Present |

Source:  Ref. 5.9

## Characteristics of Typical Residential Wastewater[a]

| Parameter | Mass Loading gm/cap/day | Concentration mg/l |
|---|---|---|
| Total Solids | 115 - 170 | 680 - 1000 |
| Volatile Solids | 65 - 85 | 380 - 500 |
| Suspended Solids | 35 - 50 | 200 - 290 |
| Volatile Suspended Solids | 25 - 40 | 150 - 240 |
| $BOD_5$ | 35 - 50 | 200 - 290 |
| Chemical Oxygen Demand | 115 - 125 | 680 - 730 |
| Total Nitrogen | 6 - 17 | 35 - 100 |
| Ammonia | 1 - 3 | 6 - 18 |
| Nitrites and Nitrates | <1 | <1 |
| Total Phosphorus | 3 - 5 | 18 - 29 |
| Phosphate | 1 - 4 | 6 - 24 |
| Total Coliforms[b] | - | $10^{10}$ - $10^{12}$ |
| Fecal Coliforms[b] | - | $10^8$ - $10^{10}$ |

[a] For typical residential dwellings equipped with standard water-using fixtures and appliances (excluding garbage disposals) generating approximately 45 gpcd (170 lpcd).

[b] Concentrations presented in organisms per liter.

Source: Ref. 5.9

## RESIDENTIAL WATER USE BY ACTIVITY[a]

| Activity | Gal/use | Uses/cap/day | gpcd[b] |
|----------|---------|--------------|---------|
| Toilet Flush | 4.3<br>4.0 - 5.0 | 3.5<br>2.3 - 4.1 | 16.2<br>9.2 - 20.0 |
| Bathing | 24.5<br>21.4 - 27.2 | 0.43<br>0.32 - 0.50 | 9.2<br>6.3 - 12.5 |
| Clotheswashing | 37.4<br>33.5 - 40.0 | 0.29<br>0.25 - 0.31 | 10.0<br>7.4 - 11.6 |
| Dishwashing | 8.8<br>7.0 - 12.5 | 0.35<br>0.15 - 0.50 | 3.2<br>1.1 - 4.9 |
| Garbage Grinding | 2.0<br>2.0 - 2.1 | 0.58<br>0.4 - 0.75 | 1.2<br>0.8 - 1.5 |
| Miscellaneous | - | - | 6.6<br>5.7 - 8.0 |
| Total | - | - | 45.6<br>41.4 - 52.0 |

[a] Mean and ranges of results reported

[b] gpcd may not equal gal/use multiplied by uses/cap/day due to difference in the number of study averages used to compute the mean and ranges shown.

Source: Ref. 5.9

## PERCOLATION TEST PROCEDURES AND OTHER ALTERNATIVES

Percolation Test

(a)  The following equipment is required for the percolation test:

1.    A soil auger, post-hole digger or other means of preparing a test hole as prescribed in (b) below;

2.    A knife or trowel for removing smeared or compacted surfaces from the walls of the test hole;

3.    Fine (from two to 10 millimeter in diameter) gravel (optional);

4.    A water supply (50 gallons is generally adequate);

5.    A straight board (to serve as fixed reference point for water level measurements); (Note:  A mechanical float gauge is preferred for more precise measurements)

6.    A clock and a ruler (12 inches or longer, engineering scale);

7.    An automatic siphon or float valve (optional); and

8.    A hole liner consisting of a 14 inch section of slotted pipe or well screen, or a 14 inch length of one-quarter inch hardware cloth or other similar material rolled into a tube (optional).  The hole liner shall be no smaller than two inches in dimater less than the test hole.

(b)  Percolation tests shall not be conducted in frozen ground or in holes which have been allowed to remain open to the atmosphere for periods greater than three days.  The following procedure shall be used in preparation of the test hole.

1.    Step One:  Excavate a test hole having horizontal dimensions of eight to 12 inches at a depth such that the lower six inches of the test hole are contained entirely within the soil horizon or layer of fill material being tested.  In order to facilitate access to the lower portion of the hole, the test hole may be excavated from the bottom of a shallow pit provided that the vertical axis of the test hole is a minimum of 14 inches measured from the bottom of the pit to the bottom of the test hole.

2.    Step Two:  In soil textures other than sands or loamy sands, remove smeared or compacted soil from the sides and bottom of the test hole by inserting the tip of a knife or trowel into the soil surface and gently prying upward and outward.  Remove the loose soil from the test hole.

3.    Step Three:  At this point, a one-half inch layer of fine gravel may be placed in the bottom of the hole to protect the soil surface from disturbance or siltation when water is added to the hole.  If additional protection is desired, a hole liner as described in (a)8 above may be placed in the hole and the space between the liner and the sides of the hole may be filled with fine gravel.

4.    Step Four:  Place and secure a straight board horizontally across the top of the test hole to serve as a fixed point for depth of water measurements to be made at appointed time intervals throughout the test.

(c)  All soils, except for sandy textured soils which meet the requirements of (d) below, shall be pre-soaked using the following procedure.  Any soil which exhibits cracks or fissures between soil aggregates shall be pre-soaked regardless of the texture.  Pre-soak as follows:

1.    Fill the test hole with water and maintain a minimum depth of 12 inches for a period of four hours by refilling as necessary or by means of an automatic siphon or float valve.

2.    At the end of four hours, cease adding water to the hole and allow the hole to drain for a period of from 16 to 24 hours.

(d)  In sandy textured soils, including sands, loamy sands and sandy loams, where a rapid percolation rate is anticipated, fill the test hole to a depth of 12 inches and allow to drain completely.  Refill the hole to a depth of 12 inches and record the time required for the hole to drain completely.  If this time is less than 60 minutes, the test procedure may begin as prescribed in (e) below without further pre-soaking.  If water remains in the test hole after 60 minutes, the hole must be pre-soaked as prescribed in (c) above before proceeding with the test.

(e)  Immediately following the pre-soak procedure (no more than 28 hours after the start of the pre-soak procedure), the percolation rate shall be determined using the following procedure:

1.    Step One:  If water remains in the test hole after the completion of the pre-soak period, the test shall be terminated and the percolation rate shall be reported as greater than 60 minutes per inch.  If no water remains in the test hole, fill to a depth of

seven inches.  At a five to 30 minute time interval, depending upon the rate of fall, record the drop in water level to the nearest one-tenth of an inch.  Refill the hole at the end of each time interval and repeat this procedure using the same time interval until a constant rate of fall is attained.  A constant rate of fall is attained when the difference between the highest and lowest of three consecutive measurements is no greater than two-tenths of an inch.

    2.    Step Two:  Immediately after the completion of Step One, refill the test hole to a depth of seven inches and record the time required for exactly six inches of water to seep away.  This time divided by six will be the percolation rate in minutes per inch.

    (f)    The results of the percolation test shall be interpreted as follows:

    1.    When the purpose of the test is to determine the design permeability at the level of infiltration, the slowest percolation rate determined within the proposed disposal field shall be used for design purposes.  If any of the measured percolation rates are slower than 60 minutes per inch or faster than three minutes per inch the application shall not be approved.  A percolation rate may be the result of a single percolation test or the average of several replicate tests, as allowed in N.J.A.C. 7:9A-6.1(e)2.

    2.    When the result of the test(s) is an average percolation rate slower than 60 minutes per inch, the horizon or subtratum in question shall be considered hydraulically restrictive.

    3.    When the result of the test(s) is an average percolation rate faster than three minutes per inch, the horizon or substratum in question shall be considered excessively coarse.

    4.    When a seepage pit is proposed, the design percolation rate shall be calculated by adding the products of the percolation rate and the thickness of each individual horizon tested and dividing the result by the total thickness of all the horizons tested.  Any horizon with a percolation rate slower than 40 minutes per inch shall be excluded from this computation.

**Pit-Bailing Test**

(a) The following equipment is required for performing a pit-bailing test (See Figure 8):
1.   A back-hoe;
2.   Wooden or metal stakes, string and a hanging level;
3.   A steel measuring tape;
4.   A pump (optional);
5.   A stop-watch; and
6.   A perforated pipe, with a three inch diameter or greater.

(b) The following procedure shall be used for preparation of the test pit:
1.   Step One: Excavate a test pit extending into but not below the soil horizon or layer to be tested. The bottom of the pit should be a minimum of 1.5 feet below the observed water level and a minimum of six feet below the proposed level of infiltration. The bottom of the pit should be relatively flat and level. The shape of the pit within the depth interval tested should be approximately square or round. A rectangular or elliptical pit may be used provided that, within the depth interval tested, the length of the long dimension is no more than twice thhee length of the short dimension. The excavation made for a soil profile pit as prescribed in N.J.A.C. 7:9A-5.2 may be used provided that all the above requirements are met.
2.   Step Two: Allow the water level to rise in the pit for a minimum of two hours and until the sides have stabilized. If large volumes of soil have slumped into the pit, this soil must be removed before proceeding with the test. If the sides of the pit continue to slump and cannot be stabilized, the test shall be abandoned. If water is observed seeping into the pit from horizons above the zone of saturation in which the test is being conducted, adequate means shall be taken to intercept and divert this water away from the test pit, otherwise the pit-bailing test shall not be used. If, during the excavation of the pit, the water level in the pit rises suddenly after a hydraulically restrictive horizon is penetrated, and continues to rise above the bottom of the hydraulically restrictive horizon, the pit-bailing test shall not be used.

(c) The following procedure shall be used for performance of the pit-bailing test and the calculation of test results:
1.   Step One: Establish a fixed reference point for depth to water level measurements which will not be disturbed during removal of water from the pit or which can be temporarily removed and later re-positioned in exactly the same place. One way to establish a removable reference mark is as follows:
      i.   Drive stakes firmly into the ground on opposite sides of the test pit, several feet beyond the edge, where they will not be disturbed.
      ii.  Next, stretch a string with hanging level from stake to stake, over the pit, and adjust the string to make it level.
      iii. Finally, secure the string to the stakes and mark or notch the positions on the stakes where the string is attached so that the string may be removed temporarily and later repositioned exactly in its place.
2.   Step Two: Measure the distance from the reference level to the bottom of the pit and to the observed water level.
3.   Step Three: Lower the water in the pit by at least one foot, by pumping or bailing. If the back-hoe bucket is used to remove water from the pit, it may be necessary to remove the reference level marker prior to bailing and re-position it in its original position prior to beginning step four.
4.   Step Four: Choose a time interval, based upon the observed rate of water level rise. At the end of each time interval, measure and record the information indicated in (c)4i through iii below and repeat these measurements until the water level in the pit has risen a total of one foot or more.
      i.   Time, in minutes (the time interval, in minutes, between measurements should be chosen to allow the water level to rise by several inches);
      ii.  Depth of water level below the reference string at the end of each time interval, to the nearest eighth of an inch or one-hundredth of a foot; and
      iii. Area of water surface, in square feet. Measure appropriate dimensions of the water surface, depending on the shape of the pit, to permit calculation of the area of the water surface at the time of each water level depth measurement. Entering a soil pit

excavated below the water table can be extremely dangerous and should be avoided unless the pit is relatively shallow and the sides of the pit have been stepped and sloped at prescribed in N.J.A.C. 7:9A-5.2(e)3 to eliminate the likelihood of sudden and severe cave-in of the pit.  The distance between two opposite edges of the water surface can be measured accurately, without entering the pit, as follows.  Place a board on the ground, perpendicular to the side of the pit and extending out over the edge.  Using a plumb-bob, position this board so that its end is directly over the edge of the water surface in the pit, below.  Position a second board, in the same manner, on the opposite side of the pit. Measure the distance between the ends of the boards to determine the length of the water surface below.

5.    Step Five:    Determine whether an adequately consistent set of data has been obtained in accordance with (e)5i and ii below.

i.    Calculate the permeability for each time interval using the following equation:
$$K_a = (h_{rise}/t) \times [A_{av}/2.27 \ (H^2 - h^2) \times 60 \ min/hr$$
where:

$K_a$ = permeability, in inches per hour;

$h_{rise}$ = difference in depth to water level at the beginning and end of the time interval, in inches;

t = length of time interval, minutes;

$A_{av}$ = average of water surface area at the beginning of time interval (end of previous time interval) and at the end of the time interval, in square feet;

H = difference between depth to assumed static water level and actual or assumed depth to impermeable stratum, in feet (Depth to impermeatble stratum, if unknown, is assumed to be one and one-half times the depth of the pit.); and

h = difference between average depth of water levels at beginning and end of time interval and actual or assumed depth to the impermeable stratum, in feet.

ii.    If the calculated values of $K_a$ for successive time intervals show either an increasing or a decreasing trend, repeat Steps Three and Four until consecutive values of $K_a$ are approximately equal.

6.    Step Six:    Remove as much water as possible from the pit.  Continue excavating the pit until an impermeable stratum is encountered or as deep as possible considering the limitations of the excavating equipment used and the nature of the soil conditions encountered.  Where no impermeable stratum is encountered, the impermeable stratum shall be assumed to be at the bottom of the excavation.

7.    Step Seven:    Immediately place upright in the pit a piezometer consisting of a perforated pipe, three inches in diameter or larger, of sufficient length to extend from the bottom of the pit to above the ground surface.  Cap and hold the piezometer vertical while backfilling the pit.  When the pit is almost completely filled and the piezometer is secure in its position, record the height of the piezometer above the reference level. Finish backfilling the pit and mound soil around the piezometer and over the pit to divert surface water.

8.    Step Eight:    Return 24 hours or more later and record the depth from the top of the piezometer to the static water level.  Correct this depth by subtracting from it the height of the piezometer above the reference level measured in Step Seven.

9.    Step Nine:    Re-calculate the permeability, K, using the following formula:
$$K = (h_{rise}/t) \times [A_{av}/2.27 \ (H^2-h^2)] \times 60 \ min/hr$$
where:

K = permeability, inches per hour;

The values of $h_{rise}$, t, and $A_{av}$ are the values recorded for these parameters in the last time interval of Step Four of this subsection;

H = difference between depth to actual corrected static water level and actual or assumed depth to impermeable stratum, recorded in Steps Six and Eight, in feet; and

h = difference between the average depth of water levels at the beginning and end of the last time interval recorded in Step Four and the actual or assumed depth to the impermeable stratum recorded in Step Six, in feet.

(d)  When the permeability calculated in Step Eight of (c) above is slower than 0.2 inch per hour, the horizon(s) being tested shall be considered a hydraulically restrictive horizon and shall not be considered an acceptable zone of wastewater disposal.

(e)  *[When piezometers are installed for the purpose of conducting this test, the piezometers shall be removed or filled with cement grout after completion of the test

except in those cases where the piezometers will be utilized for monitoring ground water
levels or for ground water sampling as required by the administrative authority or by the
Department.  Piezometers used for monitoring ground water levels over extended periods of
time, or for ground water sampling in connection with water quality monitoring, may be considered to be monitoring wells requiring installation by a licensed well driller and a
permit issued by the Department pursuant to State law (N.J.S.A. 58:4-1 et seq.).   The
applicant shall contact the Department for a determination of whether or not a permit is
required.]*  *When piezometers are used for conducting this test, they shall be installed
and removed in accordance with the Department's procedures pursuant to N.J.S.A. 58:4A-4.1
et seq.*

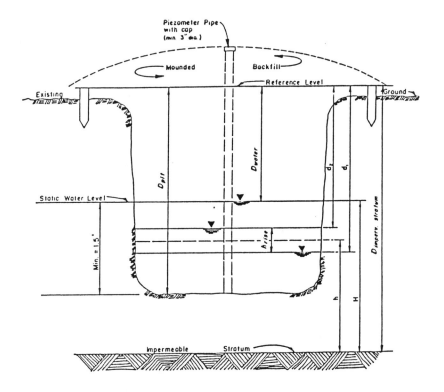

$$K \, (\text{in./hr.}) = \left[ (^{\Delta} rise)/t \right] \; \times \; \left[ A_{ww}/2.27 \, (H^2 - h^2) \right] \; \times \; 60 \, \text{min./hr.}$$

Figure 8: Pit-bailing Test.

Soil Permeability Class Rating

(a)  Determination of permeability by the soil permeability class rating technique is based upon a hydrometer analysis performed as prescribed in (f) below, and a sieve analysis performed as prescribed in (g) below, together with evaluation of soil morphological properties by the use of a soil log pit or soil boring(s).  As an alternate to the hydrometer analysis procedure prescribed in (f) below, the hydrometer analysis procedure given in ASTM STANDARD D 422, published by the American Society for Testing and Materials, may be used to determine the percent by weight of sand and the percent by weight of clay in the sample.

(b)  The following equipment is required:

1.  A two-millimeter sieve, with an eight inch or larger diameter frame;

2.  A set of two sieves, with five inch or larger diameter frames, with covers and pans.  The sieves shall meet the following specifications:

i.  The first sieve shall be 0.25 millimeter, 60-mesh, Bureau of Standards, phosphor bronze wire cloth; and

ii.  The second sieve shall be 0.047 millimeter, 300-mesh, Bureau of Standards, phospor bronze wire cloth (0.0015 wire);

3.  A wooden rolling pan or mortar with rubber-tipped pestle;

4.  An oven;

5.  A scale (0.1 gram accuracy);

6.  Distilled water;

7.  A sodium hexametaphosphate solution of 50 grams of the salt dissolved in one liter of distilled water;

8.  The electric mixer (see Section 2.1.1 of ASTM Standard D 422) or mechanical shaker;

9.  A 1000 milliliter graduated cylinder with rubber stopper;

10.  A soil hydrometer calibrated to read in grams per liter at 68 degrees Fahrenheit (ASTM #152H);

11.  A thermometer;

12.  A clock with second hand; and

13.  A sieve shaker

(c)  A loose sample of soil, 200 grams or more, shall be collected from the soil horizon or substratum to be tested.

(d)  The soil sample shall be prepared as follows:

1.  Pass the soil sample to be tested, which has been allowed to air dry, through a two millimeter sieve to remove coarse fragments.  Use moderate pressure with a wooden rolling pin or mortar with rubber-tipped pestle to break soil aggregates (but not soft rock fragments) which are larger than two millimeters.

2.  Weigh both the material retained and the material which passes through the sieve.  This method shall not be used where the weight of coarse fragments retained on the sieve exceeds 75 percent of the total sample weight.

3.  Discard the coarse fragments.

(e)  Dispersion of the soil sample shall be accomplished using a motor-mixed or a reciprocating shaker as prescribed below.  This procedure shall be followed for each replicate sample tested.

1.  Step One:  Place 40 grams of air dry soil which has been passed through a two millimeter sieve into a mixing cup or one liter shaker bottle together with 100 milliliters of sodium hexametaphosphate solution and 400 milliliters of distilled water.  Weigh out an additional 40 gram sample for determination of oven dry weight.  Re-weigh the latter sample after keeping it in an oven at 105 degrees Centigrade for 24 hours.  (Only one sample is required for determination of oven-dry weight regardless of the number of replicate samples used for the hydrometer analysis.)

2.  Step Two:  If a motor mixer is used, allow the soil to soak in the cup for 10 minutes, place the cup on the mixer and mix the sample for five minutes.  Next, transfer the suspension completely to the cylinder.  Rinse the mixing cup with distilled water and pour the rinse water into the cylinder so that none of the suspension is left in the mixing cup.  Bring the volume of the suspension in the cylinder up to the 1000 milliliter mark with distilled water.  Allow the suspension to reach room temperature.

3.  Alternate Step Two:  If a reciprocating shaker is used in lieu of the mixer, shake the sample for 12 hours, at a rate of approximately 120 strokes per minute, and

transfer to the cylinder rinsing the shaking bottles with distilled water.  Bring the volume of the suspension in the cylinder to the 1000 milliliter mark with distilled water. Allow the suspension to reach room temperature.

(f)  The following procedure shall be used for the hydrometer analysis:

1.  Step One:  Calibrate the hydrometer as follows:  Add 100 milliliters of sodium hexametaphosphate solution to a 1000 milliliter cylinder and fill to the 1000 milliliter mark with distilled water.  Place the stopper in the cylinder and shake vigorously in a back and forth motion.  Place the cylinder on the table and lower the hydrometer into the solution.  Determine the scale reading at the upper edge of the meniscus surrounding the hydrometer stem.  This is the hydrometer calibration, Rc.  Record the temperature in degrees Fahrenheit ($^{\circ}$F).

2.  Step Two:  Place a stopper in the cylinder containing the dispersed soil sample, shake the cylinder using a back and forth motion (avoid causing circular currents in the cylinder) and place the cylinder on the table.  Record the time immediately.  After 20 seconds carefully lower the hydrometer into the cylinder and, after exactly 40 seconds, read the hydrometer.  Repeat this step until two successive readings are obtained which agree within 0.5 gram per liter.

3.  Step Three:  Determine the temperature of the suspension and correct the hydrometer reading as follows:

i.  Subtract the reading obtained in Step One, Rc, from the hydrometer reading.

ii.  For each degree Fahrenheit above 68 add 0.2 gram to the reading or for each degree Fahrenheit below 68 subtract 0.2 gram.

4.  Step Four:  Remove the hydrometer, stopper the cylinder, and shake the hydrometer as in Step Two.  Remove the stopper and immediately place the cylinder on a table where it will not be disturbed.  Take a hydrometer reading after exactly two hours and correct the hydrometer reading as in Step Three.

5.  Step Five:  Using test data reporting forms provided          , record, the following data:

i.  Oven dry weight of soil, Wt (from Step One of (e) above);

ii.  Hydrometer calibration, Rc (Step One);

iii. Hydrometer reading at 40 seconds, R1 (Step Two);

iv.  Temperature of suspension (Step Three);

v.  Corrected hydrometer reading, R1' (Step Three);

vi.  Hydrometer reading at two hours, R2' (Step Four); and

vii. Corrected hydrometer reading, R2' (Step Four);

6.  Step Six:  Calculate the percent of sand and percent of clay as follows:

i.  Percent of sand = (Wt. - R')/Wt. x 100

ii.  Percent of clay = R2'/Wt. x 100

NOTE:  The hydrometer analysis may not be carried out in a room where the temperature varies more than two degrees during the time required to perform the test.

(g)  A sieve analysis shall be performed as prescribed below for each replicate sample used in the hydrometer analysis except when the content of sand determined as prescribed in Step Six of (f) above is less than 25 percent.

1.  Step One:  After the completion of Step Four in (f) above, pour the suspension from the sedimentation cylinder into a 0.047 millimeter sieve and wash the fine material through the sieve using running water.

2.  Step Two:  Dry the sieve and its contents in an oven.  Cool the sieve and transfer the sand to a pre-weighed evaporating dish (or similar heat resistant vessel) carefully, using a soft brush.

3.  Step Three:  Place the dish and its contents in an oven at 105 degrees Centigrade, for two hours, to dry.  Cool the dish and its contents and weigh to the nearest 0.01 gram.  Determine the weight of the sand by subtracting the weight of the dish.

4.  Step Four:  Assemble a stack of sieves as specified in (a)2 above, consisting of the pan, the 0.047 millimeter sieve and the 0.25 millimeter sieve, from bottom to top, respectively.  Inspect sieves carefully before using to make sure that they are clean and undamaged.  Transfer the sand from the evaporating dish to the top sieve using a soft brush to complete the transfer.

5.  Step Five:  Put the cover on the top sieve, firmly fasten the sieves to the sieve shaker and shake for three minutes.  Disassemble the stack of sieves, transfer the

contents of each sieve to a weighing dish separately.  Weigh the contents of each sieve to
the nearest 0.01 gram.  Record the following data:

    i.    Total weight of sand fraction from Step Three;

    ii.    Weight of sand passing the 0.25 millimeter sieve (retained in the 0.047 milli-
meter sieve);

    iii. Percent fine plus very fine sand:  Divide weight of sand passing 0.25 millimeter
sieve by total weight of sand fraction and multiply this value by 100.

    (h)  The following procedure shall be used to determine the soil permeability class:

    1.    Step One:  Using the soil permeability/textural triangle (Figure 6) determine
the soil permeability class of the soil horizon being tested, based upon the average
percentage of sand and the average percentage of clay in the replicate samples tested as
prescribed in (f) above.

    2.    Step Two:  If the average percentage of fine plus very fine sand in the repli-
cate samples tested, determined as prescribed in Step Five of section (g) above, is 50
percent or greater, adjust the permeability class determined in Step One of this subsection
to the next slowest class.

    3.    Step Three:  If the soil horizon being tested is found to have a massive or
platy structure or a hard, very hard, firm, very firm or extremely firm consistence, determined as prescribed in N.J.A.C. 7:9A-5.3, adjust the permeability class determined in Step
One of this subsection to the next slowest class.

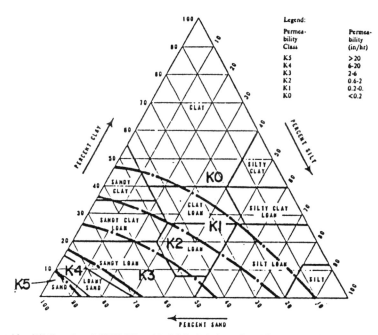

Adapted from N.N. Hantzsche et al. (1982) Soil Textural Analysis for Onsite Sewage Disposal Evaluation, Proc. 3rd Nat. Symposium
on Individual and Small Community Sewage Treatment, Am. Soc. Agric. Eng., St. Joseph, Michigan

Figure 6: Soil Permeability/Textural Triangle.

**Piezometer Test**

(a)   The following equipment is required for the piezometer test:

1.   A screw type soil auger, minimum of one inch in diameter, with extensions;

2.   A piezometer tube consisting of a metal pipe beveled on the outside lower edge, with an inside diameter about one-sixteenth of an inch larger than the diameter of the soil auger;

3.   A maul or hammer to drive pipe into the ground;

4.   A pump with tubing, to evacuate water from piezometer tube;

5.   A stop watch;

6.   A means for accurately measuring the water level within the piezometer tube as a function of time, which may consist of one of the following:

i.   A light-weight rod with measuring scale mounted on a cylindrical float with a diameter one-quarter inch or more smaller than the inside diameter of the piezometer tube;

ii.   An electric probe consisting of a thin wire embedded in and protruding from the tapered end of a wooden rod, graduated in inches, and connected in series to a limiting resistor, a millimeter and a 33-volt hearing-aid battery, the opposite terminal of which is connected to the piezometer tube; or

iii.  For depths greater than six feet, an electric sounder or the "wetted tape" method should be used.

(b)   The following procedure shall be used for the piezometer test:

1.   Step One:   Remove any sod, vegetation or leaf litter from the ground surface where the test hole will be excavated.  The test hole may be excavated from the existing ground surface or from the bottom of a large excavation or soil profile pit.

2.   Install the piezometer in accordance with Step Two A and Two B outlined in (b)2i and ii below or Alternate Step Two outlined in (b)2iii below.

i.   Step Two A:   Using the soil auger, drill test hole down to a depth of six inches.  Remove the auger and drive the piezometer tube into the hole to a depth of five inches.  Re-insert the soil auger through the piezometer tube and into the test hole and drill down six inches further.  Remove the soil auger, drive the piezometer tube six inches deeper, re-insert the auger and drill six inches deeper, repeating this procedure until the test hole reaches the top of the soil horizon or zone within a soil horizon to be tested.

ii.   Step Two B:   Using the soil auger, extend the test hole exactly four inches below the bottom of the piezometer tube (see Figure 9).  In coarse-textured soils lacking cohesion, where the unlined cavity at the bottom of the test hole may be unstable, use a piezometer tube with closely spaced perforations in the lower four inches of its length and drive the tube down to the bottom of the test hole.

iii.  Alternate Step Two:   Power equipment may be used in lieu of the hand auger to drill the test hole and install the piezometer casing provided that the casing fits tightly into the hole or the installation is sealed with bentonite so that leakage does not occur around the outside of the casing and provided that a suitable unlined cavity is provided at the bottom of the bore hole as required in Step Two B above.

3.   Step Three:   Allow the lower portion of the test hole to fill with ground water and pump the water out one or more times to minimize the effect of soil puddling and to flush the soil pores in the unlined portion of the test hole.

4.   Step Four:   Allow the water level to rise within the piezometer until the water level becomes relatively stable.  Note the approximate rate of rise and record the static water level using the top of the piezometer tube as a reference point.

5.   Step Five:   Pump most of the water out of the piezometer tube.  Record the time and the depth of the water level below the top of the tube.  After an appropriate interval of time, record the new depth of the water level.  Choose the length of the time interval based upon the rate of rise observed in Step Four so that the difference in water levels at the beginning and end of the time interval will be large enough to permit an accurate measurement, but do not allow the water level to rise to within eight inches of the static level determined in Step Four.

6.   Step Six:   Repeat Step Five of this subsection, lowering the water level to approximately the same depth and using the same time interval, until consistent results are obtained.

7. Step Seven: Allow the water level in the piezometer tube to rise and, a minimum of 24 hours later, record the depth of the water table for use in the calculation of permeability.

(c) The permeability of the soil horizon tested shall be determined as follows:

1. Step One: Determine the value of the A-parameter from Figure 10 based upon D, the diameter of the soil auger (or drill bit).

2. Step Two: Calculate the permeability, K, in inches per hour, using the following formula:

$$K = 60 \text{ min/hr} \times (3.14R^2)At \times \ln (d_1 - D_{stat}/d_2 - D_{stat}) \text{ where:}$$

K = the permeability of the soil horizon tested, in inches per hour;

R = the inside radius of the piezometer tube, in inches;

ln = the natural logarithm;

$D_{stat}$ = the depth of the static water level below the top of the piezometer tube determined in Step Seven, in inches;

$d_1$ = depth of the water level below the top of the piezometer tube at the beginning of the last time interval, in inches;

$d_2$ = depth of the water level below the top of the piezometer tube at the end of the last time interval, in inches;

t = length of time interval, in minutes; and

A = value determined in Step One above, in inches.

(d) When the permeability calculated in (c)2 above is less than 0.2 inch per hour, the horizon or substratum in question shall be considered hydraulically restrictive and shall not be considered an acceptable zone of wastewater disposal.

(e) *[When piezometers are installed for the purpose of conducting this test, the piezometers shall be removed or filled with cement grout after the completion of the test except in those cases where the piezometers will be utilized for monitoring ground water levels or for ground water sampling as required by the administrative authority or by the Department. Piezometers used for monitoring ground water levels over extended periods of time, or for ground water sampling in connection with water quality monitoring, may be considered to be monitoring wells requiring installation by a licensed well driller and a permit issued by the Department pursuant to state law (N.J.S.A. 58:4-1 et seq.). The applicant shall contact the Department for a determination of whether or not a permit is required.]* *When piezometers are used for conducting this test, they shall be installed and removed in accordance with the Department's procedures pursuant to N.J.S.A. 58:4A-4.1 et seq.*

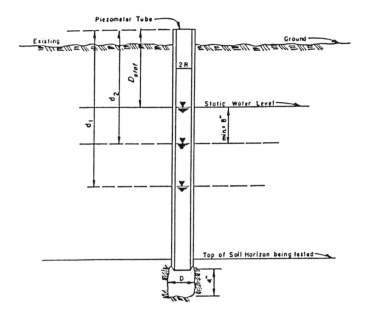

$$K = 60 \text{ min./hr.} \times (3.14 \, R^2 / At) \times ln \left[ (d_1 - D_{stat}) / (d_2 - D_{stat}) \right]$$

Figure 9: Piezometer Test.

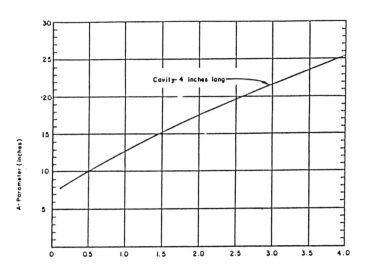

Figure 10: "A" Parameter for Piezometer Test.

**Basin Flooding Test**

(a)    The following equipment is required for basin flooding test:

1.    Excavating equipment capable of producing a test basin as prescribed in (b) below;

2.    A water supply (minimum of 375 gallons per basin filling); and

3.    A means for accurately measuring the water level within the basin as required in (c) below.

(b)    A test basin meeting the following requirements shall be excavated within or immediately adjacent to the proposed disposal field.

1.    The bottom of the test basin shall be at a depth between six and eight feet below the bottom of the proposed level of infiltration.

2.    The bottom area of the basin shall be a minimum of 50 square feet.

3.    A soil profile pit excavated as prescribed in N.J.A.C. 7:9A-5.2 may be utilized for this test provided that the requirements of (b)1 and 2 above are satisfied.

4.    The bottom of the basin should be made as level as possible so that high areas of rock do not project above the water level when the basin is flooded as prescribed in (c) below.

5.    If ground water is observed within the test basin, the basin flooding test shall not be used.

(c)    The following test procedure shall be used for the basin flooding test:

1.    Step One:  Fill the test basin with exactly 12 inches of water and record the time.  Allow the basin to drain completely.  If the time required for the basin to drain completely is greater than 24 hours, the test shall be terminated and the limiting zone in question shall be considered to be a massive rock substratum.

2.    Step Two:  If the basin drains completely within 24 hours after the first flooding, immediately refill the basin to a depth of 12 inches and record the time.  If the basin drains completely within 24 hours of the second filling, the limiting zone in question shall be considered to be fractured rock substratum.  If water remains in the basin after 24 hours the limiting zone in question shall be considered to be a massive rock substratum.

(d)    Due to the potential safety hazards which are posed by the excavation of a large test basin such as that required for this test, adequate safety measures shall be taken including the use of stepped and sloped sidewalls to permit safe access to the test basin during the test procedure as well as the use of warning signs or a fence to limit access to the basin by the public during periods when the basin is left unattended, or both.

(e)    The basin flooding test shall not be conducted in rock strata which have been blasted with explosives.

Tube Permeameter Test

(a)  The following equipment is required for the tube permeameter test:

1.  A thin-walled (one millimeter or less in thickness) metal tube, from one and one-half to three inches in diameter, six inches in length, beveled on the lower outside edge;

2.  A wooden block with dimensions broader than the diameter of the tube in (a)1 above and a hammer, to drive the tube into the soil;

3.  A small trowel;

4.  A knife (to trim core);

5.  Muslin or similar open-textured cloth and a rubberband;

6.  A soaking basin of adequate size and depth to soak cores as prescribed in (c) below;

7.  Fine gravel (from two to 10 millimeters in diameter);

8.  A test basin of adequate length (generally 10 inches or greater) and width (generally four inches or greater) to accomodate one or more replicate samples at a time. The depth of the basin should be adequate to allow placement of the sample on a layer of gravel while keeping the bottom of the core several inches below the rim of the basin, as prescribed in (d) below (See Figure).

9.  A stopper which fits water-tight into the top of the sample tube and which is fitted with a glass standpipe from three to five inches long and from 0.25 to 0.75 inches in diameter. The standpipe should have a scale for measuring changes in water level over time as required in (d) below;

10.  A small laboratory wash bottle for refilling standpipe;

11.  A clock or watch with second hand;

12.  A ruler (engineering scale is best);

13.  One gallon of water per test. The water should be allowed to stand in an open container until clear of dissolved air. Boiling may be used to remove air provided that the water is allowed to cool down to room temperature before use; and

14.  A two millimeter sieve.

(b)  When the tube permeameter test is used, undisturbed samples shall be collected as prescribed in (d) below. When the texture of the soil to be tested is a sand or loamy sand and lack of soil cohesion or the presence of large amounts of coarse fragments, roots or worm channels prevent the taking of undisturbed samples, disturbed samples shall be taken as presribed in (e) below. When the texture of the soil is other than a sand or loamy sand and undisturbed samples cannot be taken, the tube permeameter test shall not be used.

(c)  When the tube permeameter test is used, a minimum of two replicate samples shall be taken and the procedures outlined in this section shall be followed for each replicate sample to be tested. It is recommended that more than two replicate samples be taken to avoid the necessity of re-sampling in the event that samples are damaged in transport or the results of one or more replicate tests must be rejected due to extreme variability of results, as required in (i) below. Replicate samples shall be taken from within the same soil horizon at the same location within the area of the proposed disposal field.

(d)  The following procedure shall be used to collect each replicate sample:

1.  Step One:  Expose an undisturbed horizontal surface within and a minimum of three inches above the bottom of the soil horizon or layer to be tested.

2.  Step Two:  Position the sampling tube on the soil surface at the point chosen for sampling. Care should be taken to avoid large gravel or stones, large roots, worm holes or any discontinuity which might influence results. If the soil is excessively dry it may be moistened, but not saturated, provided that the force of falling water is not allowed to act directly upon the soil surface.

3.  Step Three:  Hold the wooden block on the top of the sampling tube and drive the tube into the soil a distance of from two to four inches *(but not entirely through the horizon)* using light even blows with the hammer. Care should be taken to hit the block squarely in the center and to drive the tube straight down into the soil. Do not attempt to straighten the tube by pushing or by hitting the tube on the side with the hammer.

4.  Step Four:  When the tube has been driven to the desired depth, carefully remove the soil around the outside of the tube, insert a trowel into the soil below the tube and, exerting pressure from below, lift the sampling tube out of the soil.

5.    Step Five:  Trim the bottom of the soil core flush with the sampling tube using a knife and taking care not to smear the soil surface.  Carefully invert the sampling tube and tap the side lightly with the handle of the knife or similar implement to remove any loose soil which may be resting on the top of the soil core and to verify that an undisturbed sample has been obtained.  Omit this step in the case of sandy-textured non-cohesive soils with single grain structure.  Check the top and bottom surfaces of the core sample and discard any sample which has worm holes or large cracks caused by handling.

6.    Step Six:  After the core has been checked for worm holes or signs of disturbance, stretch a piece of muslin cloth over the bottom of the tube and secure with a strong rubberband.

(e)  The following procedure shall be used for the collection of disturbed samples for the tube permeameter test:

1.    Step One:  Collect an adequate volume of the soil or fill material to be tested. Spread the soil on a clean surface and allow to dry in the air until dry to the touch.  As oven may be used to accelerate drying provided that the soil is allowed to cool down to room temperature before testing.

2.    Step Two:  Pass the soil through a two millimeter sieve to remove gravel and stones.

3.    Step Three:  Stretch a piece of muslin cloth over the bottom of the sampling tubes and place the tubes on a flat surface.  Slowly pour the soil into each sampling tube while gently tapping the side of the tube with a hard instrument.  Fill the tubes to a depth of three to four inches.  Check the bulk density of the sample by dividing the weight of the sample (weight of sample tube containing sample minus the weight of the empty sample tube) by the volume of the sample (length of sample multiplied by 3.14 $r^2$, where r is the internal radius of the sample tube).  The minimum acceptable bulk density for disturbed samples is 1.2 grams per cubic centimeter.

(f)  The following procedure shall be used for pre-soaking undisturbed or disturbed core samples for the tube permeameter test:

1.    Step One:  Place the soil core in the pre-soak basin and fill the basin with water to a point just below the top of the soil core.  Never fill the basin to a level which is higher than the top of the soil core.  Never use water directly from the tap to soak cores.  Use only de-aired water as prescribed in (a)13 above.  Allow the sample to soak until the top surface of the core is saturated with water.  This may require only a few minutes of soaking for sandy textured soils or several days for clay textured soils. Failure to soak the sample for sufficient time may result in greatly reduced permeability measurements due to entrapped air.

2.    Step Two:  When the sample has soaked for sufficient time, place a one inch layer of fine gravel (from two to 10 millimeters in diameter) on top of the soil core in the sampling tube.  Slowly fill the tube with de-aired water taking care not to disturb the surface of the core.  A small spatula or similar implement may be used to break the fall of the water as it is poured into the tube.

3.    Step Three:  Immediately transfer the soil core to the test basin in which a layer of gravel has been placed and gently press the soil core into the gravel so that it stands vertically with its base positioned at the desired depth below the rim of the test basin.

(g)  The following procedure shall be used to conduct the tube permeameter test:

1.    Step One:  When the soil core has been positioned at the desired height within the test basin, fill the test basin to overflowing with de-aired water.  (Note:   The hydraulic head used in the test depends upon the height of the top of the sample tube or standpipe above the rim of the test basin as shown in Figure 5.  In general, a higher hydraulic head should be used for heavy textured soils to expedite the test and a lower head should be used for sandy textured soils to prevent an excessively fast flow rate).

2.    Step Two:  Fill the tube to overflowing with de-aired water and record the time, in minutes, required for the water level in the tube to drop a standard distance such as one-half inch, one inch, or two inches.  Repeat the difference between the highest and lowest of three successive readings is less than five percent.  When the readings are less than 20 minutes in length the time should be reported to the nearest second.

3.    Alternate Step Two:  When the rate of fall observed in "Step Two" ((g)2 above) is slow, the flow rate may be increased by use of a standpipe as shown in Figure 5.  Carefully insert the standpipe into the top of the sample tube and fill with de-aired water.

The apparatus should be checked for leaks where the standpipe fits into the sample tube. Silicon jelly, petroleum jelly or a similar material may be used to prevent leakage. Measure the rate of fall of the water level in the standpipe as in Step Two.

(h)   The permeability of each replicate sample tested shall be calculated using the following formula:

1.   $K$ (in/hr) = 60 min/hr x $L$(in)/$T$(min) x $r^2/R^2$ x ln ($H_1/H_2$)

Where:

$K$ is the permeability of the soil sample;

$L$ is the length of the soil core, in inches;

$T$ is the time, in minutes, required for the water level to drop from $H_1$ to $H_2$ during the final test interval;

$r$ is the radius of the standpipe, in centimeters or inches;

$R$ is the radius of the soil core, in the same units as "$r$";

ln is the natural logarithm

$H_1$ is the height of the water level above the rim of the test basin at the beginning of each test interval, in inches; and

$H_2$ is the height of the water level above the rim of the test basin at the end of each test interval, in inches.

Note:   When the standpipe is not used, the term $r^2/R^2$ is omitted from the equation.

(i)   Variability of test results shall be evaluated as follows:

1.   Soil permeability classes are defined as follows:

| Measured Permeability | Soil Permeability Class |
|---|---|
| Greater than 20 inches per hour ("in/hr") | K5 |
| 6-20 in/hr | K4 |
| 2-6 in/hr | K3 |
| 0.6-2 in/hr | K2 |
| 0.2-0.6 in/hr | K1 |
| Less than 0.2 in/hr | K0 |

2.   The variability of soil permeability test results shall be considered acceptable only where the results of all replicate tests fall within one soil permeability class or two adjacent permeability classes.

3.   Where the results of replicate tests differ by more than one soil permeability class, the samples shall be examined for the following defects:

i.   Cracks, worm channels, large root channels or poor soil tube contact within the sample yielding the highest permeability value(s);

ii.   Large pieces of gravel, roots or unsaturated soil within the interior of the sample yielding the slowest permeability value(s); or

iii.   Smearing or compaction of the upper or lower surface of the sample yielding the lowest permeability value(s).

4.   If any of the defects described in (i)3 above are found, the defective core(s) shall be discarded and the test repeated using a new replicate sample for each defective replicate sample.

(j)   When the test results have been obtainedd with an acceptable range of variability as defined in (i) above, the results shall be interpreted as follows:

1.   When the purpose of the test is to determine the design permeability at the level of infiltration, the slowest of the test replicate results shall be used for design purposes.

2.   When the purpose of the test is to identify a hydraulically restrictive horizon or substratum above the water table, the horizon or substratum in question shall be considered hydraulically restrictive if the average permeability of the replicate sample tested falls within soil permeability class K0 as defined in (i)1 above.

3.   When the purpose of the test is to identify an excessively coarse horizon or substratum above the water table, the horizon or substratum in question shall be considered excessively coarse if the average permeability of the replicate samples tested falls within permeability class K5 as defined in (i)1 above.

(k)   Where results of replicate tests exceed the limits of variability allowed in (i)2 above, the results shall be interpreted as follows:

1.   When the purpose of the test is to determine the design permeability at the depth of infiltration, the slowest of the test replicate results shall be used for design purposes.

2.   When the purpose of the test is to identify a hydraulically restrictive horizon or substratum above the water table, the horizon or substratum in question shall be considered hydraulically restrictive if the slowest permeability of the replicate samples tested falls within soil permeability class K0 as defined in (i)1 above.

3.   When the purpose of the test is to identify an excessively coarse horizon or substratum above the water table, the horizon or substratum in question shall be considered excessively coarse if the fastest permeability of the replicate samples tested falls within permeability class K5 as defined in (i)1 above.

Source:   New Jersey Department of Environmental Protection

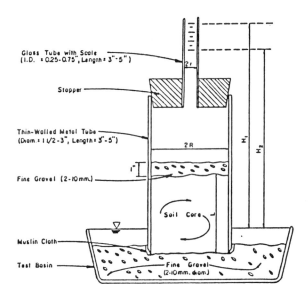

$$K \,(\text{in./hr.}) = 60 \,(\text{min./hr.}) \times r^2/R^2 \times L\,(\text{in.})/T\,(\text{min.}) \times \ln\,(\,H_1/H_2\,)$$

Figure 5: Tube Permeameter (with standpipe).

# Part E

# FARMLAND PRESERVATION STATUTES

**AGRICULTURE—DOMESTIC ANIMALS**                                    **4:1B–15**

**4:1B–15.  Severability**

If any clause, sentence, paragraph, section or part of this act shall be adjudged by any court of competent jurisdiction to be invalid, such judgment shall not affect, impair, or invalidate the remainder thereof, but shall be confined in its operation to the clause, sentence, paragraph, section or party thereof directly involved in the controversy in which such judgment shall have been rendered.

L.1976, c. 50, § 15, eff. July 22, 1976.

**Library References**
   Statutes ⊂=64(2).
   C.J.S. Statutes § 96 et seq.

## CHAPTER 1C.  AGRICULTURAL DEVELOPMENT AND FARMLAND PRESERVATION

**Last additions in text indicated by <u>underline</u>; deletions by ~~strikeouts~~**

Source:  Reprinted with permission from New Jersey Statutes Annotated, Title 4 and Title 54, Copyright © 1989 by West Publishing Co. Inc., St. Paul, MN.

## 4:1C-1                    AGRICULTURE—DOMESTIC ANIMALS

Section

### 4:1C-1.   Short title

This act shall be known and may be cited as the "Right to Farm Act."

L.1983, c. 31, § 1, eff. Jan. 26, 1983.

### Senate Natural Resources and Agriculture Committee Statement
### Senate, No. 854—L.1983, c. 31

The principal purpose of this bill is to promote, to the greatest extent practicable and feasible, the continuation of agriculture in the State of New Jersey while recognizing the potential conflicts among all lawful activities in the State. To this end, the bill provides for the establishment of the State Agriculture Development Committee.

The purpose of the committee is to aid in the coordination of State policies which affect the agricultural industry in a manner which will mitigate unnecessary constraints on essential farming practices by recommending to appropriate State departments a program of agriculture management practices which, if consistent with relevant federal and State law, and nonthreatening to the public health and safety, would afford the farmer protection against municipal regulations and private nuisance suits.

The bill further establishes a 120-day period of negotiation during which time the committee and State instrumentalities regulating agriculture in a manner inconsistent with recommended agriculture management practices would seek to negotiate a mutually agreeable solution to the conflict.

---

#### Historical Note

Title of Act:
An Act concerning agriculture and supplementing Title 4 of the Revised Statutes. L.1983, c. 31.

Library References
Agriculture ⬦1, 2.
C.J.S. Agriculture §§ 2 to 8 et seq., 27 to 29, 37, 38, 51, 65, 67, 96, 100, 102, 134, 162, 175.

---

zoning ordinance prohibiting commercial logging operations, so as to enable taxpayer, who had contracted to sell lumber, to qualify for farmland assessment based on showing of anticipated gross sales within a reasonable time following the tax year and, in any event, any legislative preemption is limited to commercial farms. L & Z Realty Co., Inc. v. Borough of Ringwood, 6 N.J.Tax 450 (1984).

#### Notes of Decisions

1.   In general
Right to Farm Act (§ 4:1C-1 et seq.) is not retroactive and, hence, Act did not override local

### 4:1C-2.   Legislative findings

The Legislature finds and declares that:

a.   The retention of agricultural activities would serve the best interest of all citizens of this State by insuring the numerous social, economic and environmental benefits which accrue from one of the largest industries in the Garden State;

b.   Several factors have combined to create a situation wherein the regulations of various State agencies and the ordinances of individual municipalities may unnecessarily constrain essential farm practices;

c.   It is necessary to establish a systematic and continuing effort to examine the effect of governmental regulation on the agricultural industry;

**Last additions in text indicated by underline; deletions by ~~strikeouts~~**

AGRICULTURE—DOMESTIC ANIMALS                    **4:1C-4**

d.  All State departments and agencies thereof should encourage the maintenance of agricultural production and a positive agricultural business climate;

e.  It is the express intention of this act to establish as the policy of this State the protection of commercial farm operations from nuisance action, where recognized methods and techniques of agricultural production are applied, while, at the same time, acknowledging the need to provide a proper balance among the varied and sometimes conflicting interests of all lawful activities in New Jersey.

L.1983, c. 31, § 2, eff. Jan. 26, 1983.

**Historical Note**

Statement: Committee statement to Senate, No. 854—L.1983, c. 31, see § 4:1C-1.

### 4:1C-3.  Definitions

As used in this act:

a.  "Commercial farm" means any place producing agricultural or horticultural products worth $2,500.00 or more annually;

b.  "Committee" means the State Agriculture Development Committee established pursuant to section 4 of this act.[1]

L.1983, c. 31, § 3, eff. Jan. 26, 1983.

[1] Section 4:1C-4.

**Historical Note**

Statement: Committee statement to Senate, No. 854—L.1983, c. 31, see § 4:1C-1.

**Library References**

Words and Phrases (Perm. Ed.)

### 4:1C-4.  State agriculture development committee; establishment; membership; terms; vacancies; compensation; meetings; minutes; staff

a.  In order that the State's regulatory action with respect to agricultural activities may be undertaken with a more complete understanding of the needs and difficulties of agriculture, there is established in the Executive Branch of the State Government a public body corporate and politic, with corporate succession, to be known as the State Agriculture Development Committee.  For the purpose of complying with the provisions of Article V, Section IV, paragraph 1 of the New Jersey Constitution, the committee is allocated within the Department of Agriculture, but, notwithstanding that allocation, the committee shall be independent of any supervision or control by the State Board of Agriculture, by the department or by the secretary or any officer or employee thereof, except as otherwise expressly provided in this act.  The committee shall constitute an instrumentality of the State, exercising public and essential governmental functions, and the exercise by the committee of the powers conferred by this or any other act shall be held to be an essential governmental function of the State.

b.  The committee shall consist of 11 members, five of whom shall be the Secretary of Agriculture, who shall serve as chairman, the Commissioner of Environmental Protection, the Commissioner of Community Affairs, the State Treasurer and the Dean of Cook College, Rutgers University, or their designees, who shall serve ex officio, and six citizens of the State, to be appointed by the Governor with the advice and consent of the Senate, four of whom shall be actively engaged in farming, the majority of whom shall own a portion of the land that they farm, and two of whom shall represent the general public.  With respect to the members actively engaged in farming, the State Board of Agriculture shall recommend to the Governor a list of potential candidates and their alternates to be considered for each appointment.

c.  Of the six members first to be appointed, two shall be appointed for terms of 2 years, two for terms of 3 years and two for terms of 4 years.  Thereafter, all appointments shall be made for terms of 4 years.  Each of these members shall hold office for the term of the appointment and until a successor shall have been appointed and qualified.  A member shall be eligible for reappointment for no more than two consecutive terms.  Any vacancy in the membership occurring other than

**Last additions in text indicated by underline; deletions by strikeouts**

**4:1C–4**                                    AGRICULTURE—DOMESTIC ANIMALS

by expiration of term shall be filled in the same manner as the original appointment but for the unexpired term only.

d.  Members of the committee shall receive no compensation but the appointed members may, subject to the limits of funds appropriated or otherwise made available for these purposes, be reimbursed for expenses actually incurred in attending meetings of the committee and in performance of their duties as members thereof.

e.  The committee shall meet at the call of the chairman as soon as may be practicable following appointment of its members and shall establish procedures for the conduct of regular and special meetings, including procedures for the notification of departments of State regulating the activities of commercial agriculture, provided that all meetings are conducted in accordance with the provisions of the "Open Public Meetings Act," P.L. 1975, c. 231 (C. 10:4–6 et seq.).

f.  A true copy of the minutes of every meeting of the committee shall be prepared and forthwith delivered to the Governor. No action taken at such meeting by the commission shall have force or effect until 15 days, exclusive of Saturdays, Sundays and public holidays, after such copy of the minutes shall have been so delivered. If, in said 15-day period, the Governor returns such copy of the minutes with a veto of any action taken by the commission at such meeting, such action shall be null and void and of no force and effect.

g.  The department shall provide any personnel that may be required as staff for the committee.

L.1983, c. 31, § 4, eff. Jan. 26, 1983.

**Historical Note**

Statement: Committee statement to Senate, No. 854—L.1983, c. 31, see § 4:1C–1.

C.J.S. Agriculture §§ 8 et seq.; 51, 67, 96, 100, 134, 175.

**Library References**

Agriculture ⋘2.

### 4:1C–5.  Powers of committee

The committee may:

a.  Adopt bylaws for the regulation of its affairs and the conduct of its business;

b.  Adopt and use a seal and alter the same at its pleasure;

c.  Sue and be sued;

d.  Apply for, receive, and accept from any federal, State, or other public or private source, grants or loans for, or in aid of, the committee's authorized purposes;

e.  Enter into any agreement or contract, execute any instrument, and perform any act or thing necessary, convenient, or desirable for the purposes of the committee or to carry out any power expressly given in this act;

f.  Adopt, pursuant to the "Administrative Procedure Act," P.L. 1968, c. 410 (C. 52:14B–1 et seq.), rules and regulations necessary to implement the provisions of this act;

g.  Request assistance and avail itself of the services of the employees of any State, county or municipal department, board, commission or agency as may be made available for these purposes.

L.1983, c. 32, § 5, eff. Jan. 26, 1983, operative Jan. 26, 1983.

**Administrative Code References**

Acquisition of development easements, preservation of agricultural lands, see N.J.A.C. 2:76–6.1 et seq.

Agricultural management practices, see N.J.A.C. 2:76–2.1 et seq.

Creation of farmland preservation programs, see N.J.A.C. 2:76–3.1 et seq., 2:76–4.1 et seq.

Soil and water conservation project cost-sharing, rules, see N.J.A.C. 2:76–5.1 et seq.

### 4:1C–6.  Duties of committee

The committee shall:

**Last additions in text indicated by underline; deletions by strikeouts**

## AGRICULTURE—DOMESTIC ANIMALS                4:1C–7

a. Consider any matter relating to the improvement of farm management practices;

b. Review and evaluate the proposed rules, regulations and guidelines of any State agency in terms of feasibility, effect and conformance with the intentions and provisions of this act;

c. Study, develop and recommend to the appropriate State departments and agencies thereof a program of agricultural management practices which shall include, but not necessarily be limited to, air and water quality control, noise control, pesticide control, fertilizer application, integrated pest management, and labor practices;

d. Upon a finding of conflict between the regulatory practices of any State instrumentality and the agricultural management practices recommended by the committee, commence a period of negotiation not to exceed 120 days with the State instrumentality in an effort to reach a resolution of the conflict, during which period the State instrumentality shall inform the committee of the reasons for accepting, conditionally accepting or rejecting the committee's recommendations and submit a schedule for implementing all or a portion of the committee's recommendations.

e. Within 1 year of the effective date of this act and at least annually thereafter, recommend to the Governor, the Legislature and the appropriate State departments and agencies thereof any actions which should be taken that recognize the need to provide a proper balance among the varied and sometimes conflicting interests of all lawful activities in the State, minimize unnecessary constraints on essential agricultural activities, and are consistent with the promotion of the public health, safety and welfare.

L.1983, c. 31, § 5, eff. Jan. 26, 1983.

**Historical Note**

Statement: Committee statement to Senate, No. 854—L.1983, c. 31, see § 4:1C–1.

C.J.S. Agriculture §§ 8 et seq., 51, 67, 96, 100, 134, 175.

**Library References**

Agriculture ⬤=2.

### 4:1C–7. Additional duties of committee

The committee shall:

a. Establish guidelines and adopt criteria for identification of agricultural lands suitable for inclusion in agricultural development areas and farmland preservation programs to be developed and adopted by a board applying for moneys from the fund;

b. Certify to the secretary that the board has approved the agricultural development area and the farmland preservation program within the area where matching grants from the fund shall be expended;

c. Review State programs and plans and any other public or private action which would adversely affect the continuation of agriculture as a viable use of the land in agricultural development areas and recommend any administrative action, executive orders or legislative remedies which may be appropriate to lessen these adverse effects;

d. Study, develop and recommend to the departments and agencies of State government a program of recommended agricultural management practices appropriate to agricultural development areas, municipally approved programs (provided that these practices shall not be more restrictive than for those areas not included within municipally approved programs) and other farmland preservation programs, which program shall include but not necessarily be limited to: air and water quality control; noise control; pesticide control; fertilizer application; soil and water management practices; integrated pest management; and labor practices;

e. Review and approve, conditionally approve or disapprove all applications for funds pursuant to the provisions of this act; and

**Last additions in text indicated by ~underline~; deletions by ~~strikeouts~~**

**4:1C-7**                                    AGRICULTURE—DOMESTIC ANIMALS

f. Generally act as an advocate for and promote the interests of productive agriculture and farmland retention within the administrative processes of State government.

L.1983, c. 32, § 6, eff. Jan. 26, 1983, operative Jan. 26, 1983.

**Administrative Code References**

Creation of farmland preservation programs, see N.J.A.C. 2:76-3.1 et seq., 2:76-4.1 et seq.

**C.J.S.** Agriculture §§ 8 et seq., 51, 67, 96, 100, 134, 175.

**Library References**

Agriculture ⟐2.

**4:1C-8.  Appropriated moneys; use by secretary**

The secretary shall use the sum of money appropriated by section 31 of this act, and any other sums as may be appropriated from time to time for like purposes, to assist the committee in administering the provisions of this act to make grants to assist boards or any other local units as authorized herein, to acquire development easements, to purchase fee simple absolute titles to farmland for resale with agricultural deed restrictions for farmland preservation purposes, and to make grants to landowners to fund soil and water conservation projects, on land devoted to farmland preservation programs within duly ~~adopted~~ certified agricultural development areas.

With respect to moneys to be utilized to make grants for soil and water conservation projects, the secretary shall not approve any grant unless it shall be for a project which is also part of a farm conservation plan approved by the local soil conservation district.

L.1983, c. 32, § 4, eff. Jan. 26, 1983, operative Jan. 26, 1983.  Amended by L.1988, c. 4, § 2, eff. March 9, 1988.

**Historical Note**

Statement: Committee statement to Senate, No. 1974—L.1988, c. 4, see § 4:1C-13.

**C.J.S.** Agriculture §§ 8 et seq., 51, 67, 96, 100, 134, 175.

**Library References**

Agriculture ⟐2.

**4:1C-9.  Commercial farm owners and operators; permissible activities**

The owner or operator of a commercial farm which meets the eligibility criteria for differential property taxation pursuant to the "Farmland Assessment Act of 1964," P.L. 1964, c. 48 (C. 54:4-23.1 et seq.) and the operation of which conforms to agricultural management practices recommended by the committee and all relevant federal or State statutes or rules and regulations adopted pursuant thereto and which does not pose a direct threat to public health and safety may:

a. Produce agricultural and horticultural crops, trees and forest products, livestock, and poultry and other commodities as described in the Standard Industrial Classification for agriculture, forestry, fishing and trapping;

b. Process and package the agricultural output of the commercial farm;

c. Provide for the wholesale and retail marketing of the agricultural output of the commercial farm, and related products that contribute to farm income, including the construction of building and parking areas in conformance with municipal standards;

d. Replenish soil nutrients;

e. Control pests, predators and diseases of plants and animals;

f. Clear woodlands using open burning and other techniques, install and maintain vegetative and terrain alterations and other physical facilities for water and soil conservation and surface water control in wetland areas; and

g. Conduct on-site disposal of organic agricultural wastes.

L.1983, c. 31, § 6, eff. Jan. 26, 1983.

**Last additions in text indicated by underline; deletions by ~~strikeouts~~**

## AGRICULTURE—DOMESTIC ANIMALS

## 4:1C–12

Historical Note   –
Statement: Committee statement to Senate, No.
854—L.1983, c. 31, see § 4:1C–1.

Library References
Taxation ☞348.1(3).
C.J.S. Taxation § 411.

**4:1C–10.   Commercial agricultural operation, activity or structure; presumption**

In all relevant actions filed subsequent to the effective date of this act, there shall exist a rebuttable presumption that no commercial agricultural operation, activity or structure which conforms to agricultural management practices recommended by the committee, and all relevant federal or State statutes or rules and regulations adopted pursuant thereto and which does not pose a direct threat to public health and safety, shall constitute a public or private nuisance, nor shall any such operation, activity or structure be deemed to otherwise invade or interfere with the use and enjoyment of any other land or property.

L.1983, c. 31, § 7, eff. Jan. 26, 1983.

Historical Note
Statement: Committee statement to Senate, No.
854—L.1983, c. 31, see § 4:1C–1.

Administrative Code References

Development easements for preservation of agricultural lands, acquisition and deed restrictions, see N.J.A.C. 2:76–6.15.

Library References
Nuisance ☞33, 49(1), 84.
C.J.S. Nuisances §§ 101, 121 et seq., 127, 149.

**4:1C–11.   Short title**

This act shall be known and may be cited as the "Agriculture Retention and Development Act."

L.1983, c. 32, § 1, eff. Jan. 26, 1983, operative Jan. 26, 1983.

Historical Note
Section 32 of L.1983, c. 32, approved Jan. 26, 1983, provides:

"This act shall take effect immediately, but shall remain inoperative until the "Right to Farm Act," P.L. 198[3], c. [31] (C. [4:1C–1 et seq.]) (now pending before the Legislature as Senate Bill No. 854 Scs) is enacted and becomes effective [L.1983, c. 31, eff. Jan. 26, 1983]."

Title of Act:

An Act concerning agricultural development and farmland preservation, providing for the establishment of county agriculture development boards, providing for the establishment of voluntary farmland preservation programs, authorizing the purchase of development easements and the funding of soil and water conservation projects on agricultural land, and making an appropriation. L.1983, c. 32.

Administrative Code References

Creation of farmland preservation programs, see N.J.A.C. 2:76–3.1 et seq.
Environmental and health impact statement requirements, see N.J.A.C. 7:26–2.9.
Sanitary landfill, environmental performance standard, see N.J.A.C. 7:26–2A.6.

Library References
Comments.

Agricultural preservation, see N.J.P. vol. 35, Pane, § 556.

WESTLAW Electronic Research
See WESTLAW Electronic Research Guide following the Preface.

**4:1C–12.   Legislative findings and declarations**

The Legislature finds and declares that:

a.   The strengthening of the agricultural industry and the preservation of farmland are important to the present and future economy of the State and the welfare of the citizens of the State, and that the Legislature and the people have demonstrated recognition of this fact through their approval of the "Farmland Preservation Bond Act of 1981," P.L. 1981, c. 276;

b.   All State departments and agencies thereof should encourage the maintenance of agricultural production and a positive agricultural business climate;

c.   It is necessary to authorize the establishment of State and county organizations to coordinate the development of farmland preservation programs within identified areas where agriculture will be presumed the first priority use of the land

**Last additions in text indicated by underline; deletions by ~~strikeouts~~**

**4:1C-12**                                    AGRICULTURE—DOMESTIC ANIMALS

and where certain financial, administrative and regulatory benefits will be made available to those landowners who choose to participate, all as hereinafter provided.

L.1983, c. 32, § 2, eff. Jan. 26, 1983, operative Jan. 26, 1983.

### 4:1C-13.  Definitions

As used in this act:

a.  "Agricultural development areas" means areas identified by a county agricultural development board pursuant to the provisions of section 11 of this act [1] and certified by the State Agriculture Development Committee;

b.  "Agricultural use" means the use of land for common farmsite activities, including but not limited to: production, harvesting, storage, grading, packaging, processing and the wholesale and retail marketing of crops, plants, animals and other related commodities and the use and application of techniques and methods of soil preparation and management, fertilization, weed, disease and pest control, disposal of farm waste, irrigation, drainage and water management, and grazing;

c.  "Board" means a county agriculture development board established pursuant to section 7 [2] or a subregional agricultural retention board established pursuant to section 10 of this act;[3]

d.  "Committee" means the State Agriculture Development Committee established pursuant to section 4 of the "Right to Farm Act," P.L.1983, c. 31 (C. 4:1C-4);

e.  "Cost," as used with respect to cost of <u>fee simple absolute title,</u> development easements or soil and water conservation projects, includes, in addition to the usual connotations thereof, interest or discount on bonds; cost of issuance of bonds; the cost of inspection, appraisal, legal, financial, and other professional services, estimates and advice; and the cost of organizational, administrative and other work and services, including salaries, supplies, equipment and materials necessary to administer this act;

f.  "Development easement" means an interest in land, less than fee simple absolute title thereto, which enables the owner to develop the land for any nonagricultural purpose as determined by the provisions of this act and any relevant rules or regulations promulgated pursuant hereto;

g.  "Development project" means any proposed construction or capital improvement for nonagricultural purposes;

h.  "Farmland preservation program" or "municipally approved farmland preservation program" (hereinafter referred to as municipally approved program) means any voluntary program, the duration of which is at least 8 years, authorized by law enacted subsequent to the effective date of the "Farmland Preservation Bond Act of 1981," P.L.1981, c. 276, which has as its principal purpose the long-term preservation of significant masses of reasonably contiguous agricultural land within agricultural development areas adopted pursuant to this act and the maintenance and support of increased agricultural production as the first priority use of that land.  Any municipally approved program shall be established pursuant to section 14 of this act; [4]

i.  "Fund" means the "Farmland Preservation Fund" created pursuant to the "Farmland Preservation Bond Act of 1981," P.L.1981, c. 276;

j.  "Governing body" means, in the case of a county, ~~the board of chosen freeholders~~ <u>the governing body of the county,</u> and in the case of a municipality, the commission, council, board or body, by whatever name it may be known, having charge of the finances of the municipality;

k.  "Secretary" means the Secretary of Agriculture;

l.  "Soil and water conservation project" means any project designed for the control and prevention of soil erosion and sediment damages, the control of pollution on agricultural lands, the impoundment, storage and management of water for agricultural purposes, or the improved management of land and soils to achieve maximum agricultural productivity;

m.  "Soil conservation district" means a governmental subdivision of this State organized in accordance with the provisions of R.S. 4:24-1 et seq.;

**Last additions in text indicated by <u>underline</u>; deletions by ~~strikeouts~~**

## AGRICULTURE—DOMESTIC ANIMALS                4:1C-14

n. <u>"Agricultural deed restrictions for farmland preservation purposes" means a statement containing the conditions of the conveyance and the terms of the restrictions set forth in P.L.1983, c. 32 and as additionally determined by the committee on the use and the development of the land which shall be recorded with the deed in the same manner as originally recorded.</u>

L.1983, c. 32, § 3, eff. Jan. 26, 1983, operative Jan. 26, 1983. Amended by L.1988, c. 4, § 1, eff. March 9, 1988.

1 Section 4:1C-18.
2 Section 4:1C-14.
8 Section 4:1C-17.
4 Section 4:1C-21.

### Senate Natural Resources and Agriculture Committee Statement
### Senate, No. 1974—L.1988, c. 4

The Senate Natural Resources and Agriculture Committee favorably reported Senate Bill No. 1974.

This bill implements the amendments to the State farmland preservation program authorized by the voters in the general election held in November 1987, pursuant to P.L.1987, c. 240. The bill increases the maximum proportion of the purchase costs of farmland development easements on which moneys from the "Farmland Preservation Bond Act of 1981" (P.L.1981, c. 276) may be spent from 50% to 80% generally, and 100% under emergency conditions. The bill also provides for a mechanism for the purchase, by the State Agriculture Development Committee, of fee simple absolute titles to farmland for resale with agricultural deed restrictions for farmland preservation purposes.

The bill provides that the State Agriculture Development Committee is responsible for the maintenance of the property while it is in their possession. In addition, the bill provides for State payments to municipalities while the State is in possession of the property so the municipalities do not suffer any loss of tax revenue. Finally, the bill clarifies provisions related to the donation, by an individual, of all or a portion of the value of a development easement.

---

Library References

Zoning and Land Planning ☞68, 279.
C.J.S. Zoning and Land Planning §§ 25, 28, 129.

Words and Phrases (Perm. Ed.)

### 4:1C-14. County agriculture development board; membership terms; vacancies; compensation; chairman; existing public bodies

a. The governing body of any county may, by resolution duly adopted, establish a public body under the name and style of "The County Agriculture Development Board," with all or any significant part of the name of the county inserted. Every board shall consist of three non-voting members as follows: a representative of the county planning board; a representative of the local soil conservation district; and the county agent of the New Jersey Cooperative Extension Service whose jurisdiction encompasses the boundaries of the county; and seven voting members who shall be residents of the county, four of whom shall be actively engaged in farming, the majority of whom shall own a portion of the land they farm, and three of whom shall represent the general public, appointed by the board of chosen freeholders, or, in the counties operating under the county executive plan or county supervisor plan pursuant to the provisions of the "Optional County Charter Law," P.L. 1972, c. 154 (C. 40:41A-1 et seq.), by the county executive, or the county supervisor, as the case may be, with the advice and consent of the board of chosen freeholders. With respect to the members actively engaged in farming, the county board of agriculture shall recommend to the board of chosen freeholders, the county executive or the

**Last additions in text indicated by <u>underline</u>; deletions by ~~strikeouts~~**

17

county supervisor, as appropriate, a list of potential candidates and their alternates to be considered for each appointment.

b.   Of the seven members first to be appointed, three shall be appointed for terms of 2 years, two for terms of 3 years, and two for terms of 4 years.  Thereafter, all appointments shall be made for terms of 4 years.  Each of these members shall hold office for the term of the appointment and until a successor shall have been appointed and qualified.  A member shall be eligible for reappointment for no more than two consecutive terms.  Any vacancy in the membership occurring other than by expiration of term shall be filled in the same manner as the original appointment but for the unexpired term only.

c.   The board of chosen freeholders, county executive or county supervisor, as appropriate, may appoint such other advisory members to the board as they may deem appropriate.

d.   Members of the board shall receive no compensation but the appointive members may, subject to the limits of funds appropriated or otherwise made available for these purposes, be reimbursed for expenses actually incurred in attending meetings of the board and in performance of their duties as members thereof.

e.   The board shall meet as soon as may be practicable following the appointment of its members and shall elect a chairman from among its members and establish procedures for the conduct of regular and special meetings, provided that all meetings are conducted in accordance with the provisions of the "Open Public Meetings Act," P.L. 1975, c. 231 (C. 10:4-6 et seq.).  The chairman shall serve for a term of 1 year and may be reelected.

f.   The chairman shall appoint three members actively engaged in farming to serve with the representatives of the general public for the purpose of mediating disputes pursuant to the provisions of section 19 of this act.[1]

g.   Notwithstanding the provisions of subsections a. and b. of this section, any public body established by the governing body of any county prior to May 3, 1982 which was established to carry out functions substantially similar to the functions of boards pursuant to this act and which proposes to apply for grants pursuant hereto may carry out the functions authorized herein, provided that within 5 years following the effective date of this act those boards established prior to May 3, 1982 shall reorganize so that the board reflects no more than a simple majority of members actively engaged in farming or equal representation of the general public and those actively engaged in farming.

L.1983, c. 32, § 7, eff. Jan. 26, 1983, operative Jan. 26, 1983.

[1] Section 4:1C-26.

**Cross References**
Low-interest farm loan program to preserve farmland, see '0:23-12.1 et  q.

C.J.S. Agriculture §§ 8 et seq., 51, 67, 96, 100, 134, 175.

**Library References**
Agriculture ⚖2.

### 4:1C-15.  Duties

Every board shall:

a.   Develop and adopt, after public hearings, agricultural retention and development programs, which shall have as their principal purpose the long-term encouragement of the agricultural business climate and the preservation of agricultural land in the county;

b.   Establish the minimum acreage of significant masses of reasonably contiguous land required for the creation of a municipally approved program or other farmland preservation programs;

c.   Establish minimum standards for the inclusion of land in a municipally approved program or other farmland preservation programs;

d.   Review and approve, conditionally approve or disapprove petitions for the formation of a municipally approved program or other farmland preservation programs, and monitor the operations thereof;

**Last additions in text indicated by underline; deletions by strikeouts**

## AGRICULTURE—DOMESTIC ANIMALS                    4:1C–17

e. Review and approve, conditionally approve or disapprove, prior to any applications to the committee, any request for financial assistance authorized by this act;

f. Monitor and make appropriate recommendations to the committee and to county and municipal governing bodies and boards with respect to resolutions, ordinances, regulations and development approvals which would threaten the continued viability of agricultural activities and farmland preservation programs within agricultural development areas;

g. At the request of a municipality, require that any person proposing any nonagricultural development in an agricultural development area prepare and submit a statement as to the potential impact the proposed development would have on agricultural activities in the area.

L.1983, c. 32, § 8, eff. Jan. 26, 1983, operative Jan. 26, 1983.

**Administrative Code References**

Agricultural development areas, regulation, see
N.J.A.C. 2:76–1.1 et seq.

### 4:1C–16.  Powers

Every board may:

a. Develop an educational and informational program concerning farmland preservation techniques and recommended agricultural management practices to advise and assist municipalities, farmers and the general public with respect to the implementation of these techniques;

b. Provide assistance to farm operators concerning permit applications and information regarding the regulatory practices of State government agencies.

L.1983, c. 32, § 9, eff. Jan. 26, 1983, operative Jan. 26, 1983.

**Administrative Code References**

Agricultural development areas, regulation, see
N.J.A.C. 2:76–1.1 et seq.
Agricultural management practices, see N.J.A.
C. 2:76–2.1 et seq.

**Library References**

Agriculture ⟨=2.
C.J.S. Agriculture §§ 8 et seq., 51, 67, 96, 100, 134, 175.

### 4:1C–17.  Subregional agricultural retention board; membership; dissolution

a. If any board of chosen freeholders has not created a board within 1 year of the effective date of this act, the governing body of any municipality located within that county may, singly or jointly by parallel ordinance with other contiguous municipalities within the county, establish a subregional agricultural retention board, which shall have the same responsibilities as a county board, except that its jurisdiction shall not exceed the boundaries of the municipality or municipalities establishing the board. Every subregional agricultural retention board may receive State moneys from the fund pursuant to the provisions of this act.

b. The members of a subregional agricultural retention board shall be appointed in the same manner as a county board, except that the planning board representative shall be from the municipal planning board and the appointive members shall be residents of the municipality. If two or more municipalities jointly create a subregional board, the number of members thereof shall be multiplied by the number of municipalities involved.

c. If the governing body of the county creates a board subsequent to the establishment of a subregional agricultural retention board, the subregional body shall, within 90 days of the date of the creation of the board, be dissolved but may remain advisory to the board. The board shall honor any contractual commitments of the subregional agricultural retention board.

L.1983, c. 32, § 10, eff. Jan. 26, 1983, operative Jan. 26, 1983.

**Library References**
Agriculture ⟨=2.

C.J.S. Agriculture §§ 8 et seq., 51, 67, 96, 100, 134, 175.

**Last additions in text indicated by underline; deletions by ~~strikeouts~~**

**4:1C–18. Agricultural development area; recommendation and approval**

The board may, after public hearing, identify and recommend an area as an agricultural development area, which recommendation shall be forwarded to the county planning board. The board shall document where agriculture shall be the preferred, but not necessarily the exclusive, use of land if that area:

a. Encompasses productive agricultural lands which are currently in production or have a strong potential for future production in agriculture and in which agriculture is a permitted use under the current municipal zoning ordinance or in which agriculture is permitted as a nonconforming use;

b. Is reasonably free of suburban and conflicting commercial development;

c. Comprises not greater than 90% of the agricultural land mass of the county;

d. Incorporates any other characteristics deemed appropriate by the board.

Approval of the agricultural development area by the board shall be in no way construed to authorize exclusive agricultural zoning or any zoning which would have the practical effect of exclusive agricultural zoning, nor shall the adoption be used by any tax official to alter the value of the land identified pursuant hereto or the assessment of taxes thereon.

L.1983, c. 32, § 11, eff. Jan. 26, 1983, operative Jan. 26, 1983.

Library References
Agriculture ⬅2.
Zoning and Land Planning ⬅68, 279.
C.J.S. Agriculture §§ 8 et seq., 51, 67, 96, 100, 134, 175.

C.J.S. Zoning and Land Planning §§ 25, 55, 129.

**4:1C–19. Land acquisition or construction in agriculture development area; notice of intent; review; hearing**

a. Any public body or public utility which intends to exercise the power of eminent domain, pursuant to the provisions of the "Eminent Domain Act of 1971," P.L. 1971, c. 361 (C. 20:3–1 et seq.), for the acquisition of land included in an agricultural development area, or which intends to advance a grant, loan, interest subsidy or other funds within an agricultural development area for the construction of dwellings, commercial or industrial facilities, transportation facilities, or water or sewer facilities to serve nonfarm structures, shall file a notice of intent with the board and the committee, the provisions of any other law, rule or regulation to the contrary notwithstanding, 30 days prior to the initiation of this action. This notice shall contain a statement of the reasons for the acquisition and an evaluation of alternatives which would not include action in the agricultural development area.

b. Within 30 days of the receipt of this notice of intent, the board and the committee shall review the proposed action to determine its effect upon the preservation and enhancement of agriculture in the agricultural development area, the municipally approved program, and upon overall State agricultural preservation and development policies. If the board or the committee finds that the proposed action would cause unreasonably adverse effects on the agricultural development area, or State agricultural preservation and development policies, the board or the committee may direct that no action be taken thereon for 60 days, during which time a public hearing shall be held by the board or the committee in the agricultural development area and a written report containing the recommendations of the board or the committee concerning the proposed acquisition or development project shall be made public. Notice of the hearing shall be afforded in accordance with the provisions of the "Open Public Meetings Act," P.L. 1975, c. 231 (C. 10:4–6 et seq.).

c. The secretary may, upon finding that the provisions of this section have been violated, request the Attorney General to bring an action to enjoin the acquisition or development project.

L.1983, c. 32, § 12, eff. Jan. 26, 1983, operative Jan. 26, 1983.

Library References
Eminent Domain ⬅45, 180, 181, 273.

C.J.S. Eminent Domain §§ 65 et seq., 242, 243, 244, 401.

Last additions in text indicated by <u>underline</u>; deletions by ~~strikeouts~~

AGRICULTURE—DOMESTIC ANIMALS                    **4:1C–21**

#### 4:1C–20.   Petition for farmland preservation program;  approval;  agreement between board and landowner

a.  Any one or more owners of land which qualifies for differential property tax assessment pursuant to the "Farmland Assessment Act of 1964," P.L. 1964, c. 48 (C. 54:4–23.1 et seq.), and which is included in an agricultural development area, may petition the board for the creation of a farmland preservation program.  The petition shall include a map of the boundaries of the proposed farmland preservation program and any other information deemed appropriate by the board.

b.  Approval of the petition by the board and creation of the farmland preservation program shall be signified by an agreement between the board and the landowner to retain the land in agricultural production for a minimum period of 8 years.  The agreement shall constitute a restrictive covenant and shall be filed and recorded with the county clerk in the same manner as a deed.

L.1983, c. 32, § 13, eff. Jan. 26, 1983, operative Jan. 26, 1983.

**Library References**
  Taxation ⊕348.1(3).
  C.J.S. Taxation § 411.

#### 4:1C–21.   Petition for municipally approved program;  content;  review

a.  Any one or more owners of land which qualifies for differential property tax assessment pursuant to the "Farmland Assessment Act of 1964," P.L. 1964, c. 48 (C. 54:4–23.1 et seq.), and which is included in an agricultural development area may petition the board for the creation of a municipally approved program comprising that land;  provided that the owner or owners own at least the minimum acreage established by the board.  The petition shall include a map of the boundaries of the municipally approved program and any other information deemed appropriate by the board.

b.  Upon receipt thereof, the board shall review this petition for conformance with minimum eligibility criteria as established by the committee and the board.  If the board finds that the criteria have been met, it shall immediately forward a copy of the petition to the county planning board, the governing body of any municipality wherein the proposed municipally approved program is located, and to the planning board of each affected municipality.

c.  Within 60 days of receipt of the petition, the municipal planning board shall review and report to the municipal governing body the potential effect of the proposed municipally approved program upon the planning policies and objectives of the municipality.

d.  The municipal governing body shall, after public hearing and within 120 days of receipt of the report, recommend to the board, by ordinance duly adopted, that the municipally approved program boundaries be approved, conditionally approved with proposed geographical modifications, or disapproved.

e.  Upon receipt of a recommendation by the governing body to approve the petition, the board shall forward the petition for the creation of the municipally approved program and the municipal ordinance approving the municipally approved program to the county planning board.  This action shall constitute creation of a municipally approved program.

f.  Upon receipt of a recommendation by the governing body to conditionally approve the petition with proposed geographical modifications, the board shall review the recommendation for conformance with minimum eligibility criteria.  If the board finds that the criteria have been met and that the proposed modifications encourage agriculture retention and development to the greatest practicable extent, the petition shall be forwarded and adopted pursuant to subsection e. of this section.

g.  Upon receipt of a recommendation by the governing body to disapprove the petition, the board shall take no further action and the proposed municipally approved program shall not be adopted.

h.  If the governing body proposes modifications to the petition which exclude any land from being included within a municipally approved program, the owner thereof

**Last additions in text indicated by underline;  deletions by strikeouts**

**4:1C–21**                                    AGRICULTURE—DOMESTIC ANIMALS

may request that the board mediate on behalf of the landowner with the municipal governing body prior to acting on the recommendation thereof. The landowner may request mediation by the committee with respect to any action taken by the board.

i.   If any municipal governing body fails to act on a petition to create a municipally approved program within the time prescribed in subsection d. of this section, the board or the landowner may appeal to the committee to intervene, and the committee may approve or disapprove a petition for the creation of a municipally approved program pursuant to the provisions of this section.

j.   The board shall advise owners of any land contiguous to the proposed municipally approved program that a petition has been received, solicit opinions concerning inclusion of this land and, if the board deems appropriate, encourage the inclusion of the land in the municipally approved program.

Any landowner not included in the municipally approved program as initially created may, within 2 years following the creation date, request inclusion, and upon review by the board and municipal governing body, and a finding that this inclusion is warranted, become part of the municipally approved program; provided that the landowner enters into an agreement pursuant to section 17 of this act [1] for the remaining duration of the municipally approved program.

L.1983, c. 32, § 14, eff. Jan. 26, 1983, operative Jan. 26, 1983.

[1] Section 4:1C–24.

**Library References**
Taxation ⇒348.1(3).
C.J.S. Taxation § 411.

### 4:1C–22.  Documentation of municipally approved program

The creation of a municipally approved program shall be documented in the following manner:

a.   The petition in its final form shall be filed and recorded, in the same manner as a deed, with the county clerk and shall be filed with the municipal clerk;

b.   The petition, the municipal ordinance of adoption and the county resolution or ordinance of adoption, as the case may be, shall be filed with the committee; and

c.   The petition in its final form shall be filed with the municipal tax assessor for the purpose of qualifying for the exemption from property taxation on new farm structures and improvements within the municipally approved program, as authorized and provided in the Constitution.

The documentation of the creation of the municipally approved program as prescribed herein shall in no way be construed to constitute or in any other way authorize exclusive agricultural zoning.

L.1983, c. 32, § 15, eff. Jan. 26, 1983, operative Jan. 26, 1983.

**Library References**
Taxation ⇒348.1(3).
C.J.S. Taxation § 411.

### 4:1C–23.  Zoning of land in program

Notwithstanding the provisions of P.L. 1975, c. 291 (C. 40:55D–1 et seq.) or any other law, rule or regulation to the contrary, no municipality shall alter its zoning ordinance as it pertains to land included within a municipally approved program in any way so as to provide for exclusive agricultural zoning or zoning which has the practical effect of exclusive agricultural zoning for a period of 11 years from the date of the creation of the municipally approved program, unless all landowners within that municipally approved program who entered into an agreement pursuant to the provisions of section 17 of this act [1] agree to that alteration by express written consent at the end of the minimum period required by section 17 of this act. [1]

L.1983, c. 32, § 16, eff. Jan. 26, 1983, operative Jan. 26, 1983.

[1] Section 4:1C–24.

**Last additions in text indicated by underline; deletions by strikeouts**

## AGRICULTURE—DOMESTIC ANIMALS                    4:1C-25

Library References
  Zoning and Land Planning ⟜68.
  C.J.S. Zoning and Land Planning §§ 25, 55.

**4:1C-24.  Agreement to retain land in agricultural production; soil and water conservation project grants; sale of development easement**

a. Landowners within a municipally approved program or other farmland preservation program shall enter into an agreement with the board, and the municipal governing body, if appropriate, to retain the land in agricultural production for a minimum period of 8 years. The agreement shall constitute a restrictive covenant and shall be filed with the municipal tax assessor and recorded with the county clerk in the same manner as a deed.

b. The landowner shall be eligible to apply to the local soil conservation district and the board for a grant for a soil and water conservation project approved by the State Soil Conservation Committee and to the board to sell a development easement on the land, subject to the provisions of this act.

c. The landowner or farm operator as an agent for the landowner may apply to the local soil conservation district and the board for a grant for a soil and water conservation project approved by the State Soil Conservation Committee on land included within a municipally approved program or other farmland preservation program and restricted by an agreement entered into pursuant to subsection a. of this section.

d. Approval by the local soil conservation district and the board for grants for soil and water conservation projects shall be contingent upon a written agreement by the person who would receive funds that the project shall be maintained for a specified period of not less than 3 years, and shall be a component of a farm conservation plan approved by the local soil conservation district.

e. If the landowner applying for funds for a soil and water conservation project pursuant to this section provides 50% of those funds without assistance from the county, the local soil conservation district shall review, approve, conditionally approve or disapprove the application. The committee shall certify that the land on which the soil and water conservation project is to be conducted is part of a municipally approved program or other farmland preservation program and restricted by an agreement entered into pursuant to the provisions of this section.

L.1983, c. 32, § .17, eff. Jan. 26, 1983, operative Jan. 26, 1983.

Cross References
  Price of development easement, see § 4:1C-31.

Administrative Code References
  Soil and water conservation project cost sharing, see N.J.A.C. 2:90-2.1 et seq.

Water conservation project cost sharing, procedural rules, see N.J.A.C. 2:90-3.1 et seq.

Library References
  Taxation ⟜348.1(3).
  C.J.S. Taxation § 411.

**4:1C-25.  Eminent domain; funding for construction of facilities to serve nonfarm structures**

The provisions of any law to the contrary notwithstanding, no public body shall exercise the power of eminent domain for the acquisition of land in a municipally approved program, nor shall any public body advance a grant, loan, interest subsidy or other funds within a municipally approved program for the construction of dwellings, commercial facilities, transportation facilities, or water or sewer facilities to serve nonfarm structures unless the Governor declares that the action is necessary for the public health, safety and welfare and that there is no immediately apparent feasible alternative. If the Governor so declares, the provisions of section 12 of this act[1] shall apply.

L.1983, c. 32, § 18, eff. Jan. 26, 1983, operative Jan. 26, 1983.

[1] Section 4:1C-19.

**Last additions in text indicated by <u>underline</u>; deletions by ~~strikeouts~~**

## 4:1C-25

**Library References**
Eminent Domain ⊜45.
C.J.S. Eminent Domain § 65 et seq.

### 4:1C-26.  Actions; presumption; complaint

a. In all relevant actions filed subsequent to the effective date of this act, there shall exist an irrebuttable presumption that no agricultural operation, activity or structure which is conducted or located within a municipally approved program and which conforms to agricultural management practices approved by the committee, and all relevant federal or State statutes or rules and regulations adopted pursuant thereto and which does not pose a direct threat to public health and safety shall constitute a public or private nuisance, nor shall any such operation, activity or structure be deemed to otherwise invade or interfere with the use and enjoyment of any other land or property.

b. In the event that any person wishes to file a complaint to modify or enjoin an agricultural operation or activity under the belief that the operation or activity violates the provisions of subsection a. of this section, that person shall, 30 days prior to instituting any action in a court of competent jurisdiction, petition the board to act as an informal mediator.

c. The board shall, in the course of its regular or special meetings but within 30 days of receipt of the petition, seek to facilitate the resolution of any dispute.  No statement or expression of opinion made in the course of a meeting concerning the dispute shall be deemed admissible in any subsequent judicial proceeding thereon.

L.1983, c. 32, § 19, eff. Jan. 26, 1983, operative Jan. 26, 1983.

**Library References**
Nuisance ⊜33, 49(1), 84.
C.J.S. Nuisances §§ 101, 121 et seq., 127, 149.

### 4:1C-27.  Emergency restrictions on use of water and energy

The provisions of any law, rule, regulation or ordinance to the contrary notwithstanding, agricultural activities on land in a municipally approved program shall be exempt from any emergency restrictions instituted on the use of water and energy supplies unless the Governor declares that the public safety and welfare require otherwise.

L.1983, c. 32, § 20, eff. Jan. 26, 1983, operative Jan. 26, 1983.

**Library References**
Waters and Water Courses ⊜201, 249.
C.J.S. Waters §§ 277 et seq., 352 et seq.

### 4:1C-28.  Farm structure design

a. The provisions of any law, rule, regulation or ordinance to the contrary notwithstanding, any criteria developed by a land grant college or a recognized organization of agricultural engineers and approved by the committee for farm structure design shall be the acceptable minimum construction standard for a farm structure located in a municipally approved program or other farmland preservation program.

b. The use by a farm owner or operator of a farm structure design approved pursuant to subsection a. of this section shall, the provisions of any law, rule, regulation or ordinance to the contrary notwithstanding, be exempt from any requirement concerning the seal of approval or fee of an architect or professional engineer.

L.1983, c. 32, § 21, eff. Jan. 26, 1983, operative Jan. 26, 1983.

### 4:1C-29.  Length of program; termination; inclusion of additional landowners

a. The municipally approved program shall remain in effect for a minimum of 8 years, provided that a review of the practicability and feasibility of its continuation

**Last additions in text indicated by <u>underline</u>; deletions by ~~strikeouts~~**

AGRICULTURE—DOMESTIC ANIMALS                    **4:1C–31**

shall be conducted by the board and the municipal governing bodies within the year immediately preceding the termination date of the municipally approved program.

b. If subsequent to notification by the board, none of the parties to the agreement entered into pursuant to section 17 of this act[1] notify the board within this 1 year period that they wish to terminate the municipally approved program, the municipally approved program shall continue in effect for another 8-year period and may continue for succeeding 8-year periods, provided that no notice of termination is received by the board during subsequent periods of review.

c. Termination of the municipally approved program at the end of any 8-year period shall occur following the receipt by the board of any notice of termination. The municipal tax assessor shall be notified by the board if the municipally approved program is terminated.

d. Nothing in this section shall be construed to preclude the reformation of a municipally approved program, as initially created pursuant to the provisions of this act.

e. Any landowner not included in a municipally approved program may request inclusion at any time during the review conducted pursuant to subsection a. of this section. If the board and the municipal governing body find that this inclusion would promote agricultural production, the inclusion shall be approved.

L.1983, c. 32, § 22, eff. Jan. 26, 1983, operative Jan. 26, 1983.

[1] Section 4:1C–24.

### 4:1C–30. Withdrawal of land; taxation

a. Withdrawal of land from the municipally approved program or other farmland preservation program prior to its termination date may occur in the case of death or incapacitating illness of the owner or other serious hardship or bankruptcy, following a public hearing conducted pursuant to the "Open Public Meetings Act," P.L. 1975, c. 231 (C. 10:4–6 et seq.) and approval by the board and in the case of a municipally approved program, the municipal governing body, at a regular or special meeting thereof. The approval shall be documented by the filing with the county clerk and county planning board, by the board and municipal governing body, of a resolution or ordinance, as appropriate, therefor. The local tax assessor shall also be notified by the board of this withdrawal.

b. Following approval to withdraw from the municipally approved program, the affected landowner shall pay to the municipality, with interest at the rate imposed by the municipality for nonpayment of taxes pursuant to R.S. 54:4–67, any taxes not paid as a result of qualifying for the property tax exemption for new farm structures or improvements in the municipally approved program, as authorized and provided in the Constitution, and shall repay, on a pro rata basis as determined by the local soil conservation district, to the board or the committee, or both, as the case may be, any remaining funds from grants for soil and water conservation projects provided pursuant to the provisions of this act, except in the case of bankruptcy, death or incapacitating illness of the owner, where no such payback of taxes or grants shall be required.

L.1983, c. 32, § 23, eff. Jan. 26, 1983, operative Jan. 26, 1983.

**Administrative Code References**

Deed restrictions on lands included in farmland preservation programs, withdrawal of land, see N.J.A.C. 2:76–3.5, 2:76–4.5.
Farmland preservation programs, withdrawal of lands, see N.J.A.C. 2:76–3.11, 2:76–4.10.

**Library References**

Agriculture ⬅2.
C.J.S. Agriculture §§ 8 et seq., 51, 67, 96, 100, 134, 175.

### 4:1C–31. Offer to sell developmental easement; price; evaluation of suitability of land; appraisal

a. Any landowner applying to the board to sell a development easement pursuant to section 17 of this act[1] shall offer to sell the development easement at a price which, in the opinion of the landowner, represents a fair value of the development potential of the land for nonagricultural purposes, as determined in accordance with the provisions of this act.

**Last additions in text indicated by <u>underline</u>; deletions by ~~strikeouts~~**

## 4:1C–31                    AGRICULTURE—DOMESTIC ANIMALS

b.  Any offer shall be reviewed and evaluated by the board and the committee in order to determine the suitability of the land for development easement purchase. Decisions regarding suitability shall be based on the following criteria:

(1) Priority consideration shall be given, in any one county, to offers with higher numerical values obtained by applying the following formula:

$$\frac{\text{nonagricultural development value} \quad - \quad \text{agricultural value} \quad - \quad \text{landowner's asking price}}{\text{nonagricultural development value} \quad - \quad \text{agricultural value}}$$

(2) The degree to which the purchase would encourage the survivability of the municipally approved program in productive agriculture; and

(3) The degree of imminence of change of the land from productive agriculture to nonagricultural use.

The board and the committee shall reject any offer for the sale of development easements which is unsuitable according to the above criteria and which has not been approved by the board and the municipality.

c.  Two independent appraisals paid for by the board shall be conducted for each parcel of land so offered and deemed suitable. The appraisals shall be conducted by independent, professional appraisers selected by the board and the committee from among members of recognized organizations of real estate appraisers. The appraisals shall determine the current overall value of the parcel for nonagricultural purposes, as well as the current market value of the parcel for agricultural purposes. The difference between the two values shall represent an appraisal of the value of the development easement. If a development easement is purchased using moneys appropriated from the fund, the State shall provide 50% no more than 80%, except 100% under emergency conditions specified by the committee pursuant to rules or regulations, of the cost of the appraisals conducted pursuant to this section.

d.  Upon receiving the results of the appraisals, the board and the committee shall compare the appraised value and the landowner's offer and, pursuant to the suitability criteria established in subsection b. of this section:

(1) Approve the application to sell the development easement and rank the application in accordance with the criteria established in subsection b. of this section; or

(2) Disapprove the application, stating the reasons therefor.

e.  Upon approval by the committee and the board, the secretary is authorized to provide the board, within the limits of funds appropriated therefor, an amount equal to 50% no more than 80%, except 100% under emergency conditions specified by the committee pursuant to rules or regulations, of the purchase price of the development easement, as determined pursuant to the provisions of this section. The board shall match that amount provide its required share and accept the landowner's offer to sell the development easement. The acceptance shall cite the specific terms, contingencies and conditions of the purchase.

f.  The landowner shall accept or reject the offer within 30 days of receipt thereof. Any offer not accepted within that time shall be deemed rejected.

g.  Any landowner whose application to sell a development easement has been rejected for any reason other than insufficient funds may not reapply to sell a development easement on the same land within 2 years of the original application.

h.  No development easement shall be purchased at a price greater than the appraised value determined pursuant to subsection c. of this section.

i.  The appraisals conducted pursuant to this section or the fair market value of land restricted to agricultural use shall not be used to increase the assessment and taxation of agricultural land pursuant to the "Farmland Assessment Act of 1964," P.L.1964, c. 48 (C. 54:4–23.1 et seq.).

L.1983, c. 32, § 24, eff. Jan. 26, 1983, operative Jan. 26, 1983.  Amended by L.1988, c. 4, § 3, eff. March 9, 1988.

1 Section 4:1C–24.

**Last additions in text indicated by underline; deletions by strikeouts**

## AGRICULTURE—DOMESTIC ANIMALS                4:1C-31.1

**Historical Note**
Statement: Committee statement to Senate, No.
1974—L.1988, c. 4, see § 4:1C-13.

**Library References**
Easements ⊄2.
C.J.S. Easements § 9.

**4:1C-31.1.  Farmland within certified agricultural development area; sale by landowner; acquisition by committee; resale; payment of taxes by state**

a.  Any landowner of farmland within an agricultural development area certified by the committee may apply to the committee to sell the fee simple absolute title at a price which, in the opinion of the landowner, represents a fair market value of the property.

b.  The committee shall evaluate the offer to determine the suitability of the land for purchase.  Decisions regarding suitability shall be based on the eligibility criteria for the purchase of development easements listed in subsection b. of section 24 of P.L.1983, c. 32 (C. 4:1C-31) and the criteria adopted by the committee and the board of that county.  The committee shall also evaluate the offer taking into account the amount of the asking price, the asking price relative to other offers, the location of the parcel relative to areas targeted within the county by the board and among the counties, and any other criteria as the committee has adopted pursuant to rule or regulation.  The committee may negotiate reimbursement with the county and include the anticipated reimbursement as part of the evaluation of an offer.

c.  The committee shall rank the offers according to the criteria to determine which, if any, should be appraised.  The committee shall reject any offer for the purchase of fee simple absolute title determined unsuitable according to any criterion in this subsection or adopted pursuant to this subsection, or may defer decisions on offers with a low ranking.  The committee shall state, in writing, its reasons for rejecting an offer.

d.  Appraisals of the parcel shall be conducted to determine the fair market value according to procedures adopted by regulation by the committee.

e.  The committee shall notify the landowner of the fair market value and negotiate for the purchase of the title in fee simple absolute.

f.  Any land acquired by the committee pursuant to the provisions of this amendatory and supplementary act shall be held of record in the name of the State and shall be offered for resale by the State, notwithstanding any other law, rule or regulation to the contrary, within a reasonable time of its acquisition with agricultural deed restrictions for farmland preservation purposes as determined by the committee pursuant to the provisions of this act.

 g.  The committee shall be responsible for the operation and maintenance of lands acquired and shall take all reasonable steps to maintain the value of the land and its improvements.

h.  To the end that municipalities may not suffer loss of taxes by reason of acquisition and ownership by the State of New Jersey of property under the provisions of this act, the State shall pay annually on October 1 to each municipality in which property is so acquired and has not been resold a sum of money equal to the tax last assessed and last paid by the taxpayer upon this land and the improvement thereon for the taxable year immediately prior to the time of its acquisition.  In the event that land acquired by the State pursuant to this act had been assessed at an agricultural and horticultural use valuation in accordance with provisions of the "Farmland Assessment Act of 1964," P.L.1964, c. 48 (C. 54:4-23.1 et seq.), at the time of its acquisition by the State, no rollback tax pursuant to section 8 of P.L.1964, c. 48 (C. 54:4-23.8) shall be imposed as to this land nor shall this rollback tax be applicable in determining the annual payments to be made by the State to the municipality in which this land is located.

All sums of money received by the respective municipalities as compensation for loss of tax revenue pursuant to this section shall be applied to the same purposes as is the tax revenue from the assessment and collection of taxes on real property of these municipalities, and to accomplish this end the sums shall be apportioned in the

**Last additions in text indicated by <u>underline</u>; deletions by ~~strikeouts~~**

## 4:1C–31.1

AGRICULTURE—DOMESTIC ANIMALS

same manner as the general tax rate of the municipality for the tax year preceding the year of receipt.

L.1988, c. 4, § 5, eff. March 9, 1988.

**Historical Note**

Statement: Committee statement to Senate, No. 1974—L.1988, c. 4, see § 4:1C–13.

**Title of Act:**

An Act concerning farmland preservation and amending and supplementing P.L. 1983, c. 32. L.1988, c. 4.

### 4:1C–31.2.  Rules and regulations

The committee shall adopt rules and regulations necessary to carry out the purposes of this amendatory and supplementary act according to the "Administrative Procedure Act," P.L.1968, c. 410 (C. 52:14B–1 et seq.).

L.1988, c. 4, § 6, eff. March 9, 1988.

**Historical Note**

Statement: Committee statement to Senate, No. 1974—L.1988, c. 4, see § 4:1C–13.

### 4:1C–32.  Conveyance of easement following purchase;  conditions and restrictions;  payment

a. No development easement purchased pursuant to the provisions of this act shall be sold, given, transferred or otherwise conveyed in any manner.

b. Upon the purchase of the development easement by the board, the landowner shall cause a statement containing the conditions of the conveyance and the terms of the restrictions on the use and development of the land to be attached to and recorded with the deed of the land, in the same manner as the deed was originally recorded. These restrictions and conditions shall state that any development for nonagricultural purposes is expressly prohibited, shall run with the land and shall be binding upon the landowner and every successor in interest thereto.

c. At the time of settlement of the purchase of a development easement, the landowner and the board may agree upon and establish a schedule of payment which provides that the landowner may receive consideration for the easement in a lump sum, or in installments over a period of up to 10 years from the date of settlement, provided that:

(1) If a schedule of installments is agreed upon, the State Comptroller shall retain in the fund an amount of money sufficient to pay the landowner pursuant to the schedule;

(2) The landowner shall receive annually, interest on any unpaid balance remaining after the date of settlement. The interest shall accrue at a rate established in the installment contract.

L.1983, c. 32, § 25, eff. Jan. 26, 1983, operative Jan. 26, 1983.

**Library References**

Easements ⟜24.
C.J.S. Easements § 46.

### 4:1C–33.  Enforcement of conditions or restrictions

The committee or the board is authorized to institute, in the name of the State, any proceedings intended to enforce the conditions or restrictions on the use and development of land on which a development easement has been purchased pursuant to this act.

L.1983, c. 32, § 26, eff. Jan. 26, 1983, operative Jan. 26, 1983.

**Library References**

Easements ⟜61(3, 7).
C.J.S. Easements §§ 103, 104, 109.

**Last additions in text indicated by underline; deletions by ~~strikeouts~~**

## AGRICULTURE—DOMESTIC ANIMALS                    4:1C–37

### 4:1C–34.  Persons acquiring developmental easement; sale to board

Any person or organization acquiring a development easement, by purchase, gift or otherwise, may apply to sell that development easement to the board, provided that the land on which the development easement was acquired shall be subject to the conditions and provisions of this act and that the board and the committee make a determination to purchase the development easement in the manner prescribed in section 24 of this act.[1]

L.1983, c. 32, § 27, eff. Jan. 26, 1983, operative Jan. 26, 1983.

[1] Section 4:1C-31.

**Library References**
Easements ⚖24.
C.J.S. Easements § 46.

### 4:1C–35.  Donation of development easement to board

If a person wishes to donate all or a portion of the value of the development easement to the board, the value of the donation shall be appraised pursuant to the provisions of section 24 of this act. For the purpose of qualifying This requirement shall apply only if the board is requesting State funds.  In order to qualify for State funds, pursuant to the provisions of this act, the county may shall make up the difference between its required share of the total appraised value of the easement and the appraised value of the donation and 50% of the total appraised value of the easement.  In the event the value of the donation exceeds the required county share, the amount in excess shall be deducted from the State share.

L.1983, c. 32, § 28, eff. Jan. 26, 1983, operative Jan. 26, 1983.  Amended by L.1988, c. 4, § 4, eff. March 9, 1988.

[1] Section 4:1C-31.

| Historical Note | Library References |
|---|---|
| Statement: Committee statement to Senate, No. 1974—L.1988, c. 4, see § 4:1C–13. | Easements ⚖24. C.J.S. Easements § 46. |

### 4:1C–36.  Pinelands area

Nothing herein contained shall be construed to prohibit the creation of a municipally approved program or other farmland preservation program, the purchase of development easements, or the extension of any other benefit herein provided on land, and to owners thereof, in the Pinelands area, as defined pursuant to section 3 of P.L. 1979, c. 111 (C. 13:18A–3).

L.1983, c. 32, § 29, eff. Jan. 26, 1983, operative Jan. 26, 1983.

**Library References**
Easements ⚖2.
C.J.S. Easements § 9.

### 4:1C–37.  Joint legislative oversight committee; duties

The Senate Natural Resources and Agriculture Committee and the Assembly Agriculture and Environment Committee are designated as the Joint Legislative Oversight Committee on Agricultural Retention and Development.  The duties and responsibilities of the joint oversight committee shall be as follows:

a. To monitor the operation of the committee and its efforts to retain farmland in productive agricultural use and to recommend to the committee any rule, regulation, guideline, or revision thereto which it deems necessary to effectuate the purposes and provisions of this act;

b. To review and evaluate the implementation of development easement purchases on agricultural land;

c. To review and evaluate all relevant existing and proposed statutes, rules, regulations and ordinances, so as to determine their individual effect upon the conduct of agricultural activities in this State; and

**Last additions in text indicated by <u>underline</u>; deletions by ~~strikeouts~~**

## 4:1C–37                    AGRICULTURE—DOMESTIC ANIMALS

d. To recommend to the Legislature any legislation which it deems necessary in order to effectuate the purposes of this act.

L.1983, c. 32, § 30, eff. Jan. 26, 1983, operative Jan. 26, 1983.

### CHAPTER 2A.  BREEDING AND RAISING OF ANIMALS

#### 4:2A–1.  Definitions

**Administrative Code References**

. Regulation of raising or breeding of nutria, see N.J.A.C. 2:2–6.1 et seq.

Requirements for importation of nutria, see N.J.A.C. 2:3–7.1.

### CHAPTER 3.  POULTRY AND EGGS

#### 4:3–1.  Annual educational program and exhibits

**Administrative Code References**

New Jersey–United States poultry and turkey improvement plans, see N.J.A.C. 2:2–10.3.

#### 4:3–11.12.  Establishment of standards, grades and size-weight classes; rules and regulations

**Administrative Code References**

New Jersey standards for quality of individual shell eggs, see N.J.A.C. 2:71–1.23 et seq.

#### 4:3–12.  Permit for sale of baby chicks

**Administrative Code References**

Interstate health certificate for poultry, see N.J. A.C. 2:3–1.1 et seq.

### CHAPTER 4.  FEEDING STUFFS FOR LIVE STOCK AND POULTRY

#### 4:4–20.1.  Short title

**Historical Note**

**Title of Act:**
An Act to regulate the manufacture and distribution of commercial feeds in the State of New Jersey and repealing R.S. 4:4–1 through R.S. 4:4–20.  L.1970, c. 338.

**Administrative Code References**

Commercial feeding stuffs, see N.J.A.C. 2:68–1.1.

#### 4:4–20.9.  Inspection fees and reports

a.  An inspection fee at the rate of $0.15 per ton shall be paid on commercial feeds distributed in this State by the person who distributes the commercial feed to the consumer subject to the following:

(1) No fee shall be paid on a commercial feed if the payment has been made by a previous distributor.

(2) No fee shall be paid on customer formula feeds if the inspection fee is paid on the commercial feeds which are used as ingredients therein.

(3) No fee shall be paid on commercial feeds which are used as ingredients for the manufacture of commercial feeds which are subject to the inspection fee.  If the fee has already been paid, credit shall be given for such payment.

(4) In the case of a person who manufacturers or distributes commercial feed in the State, a minimum annual fee of $25.00 shall be paid.

**Last additions in text indicated by underline; deletions by strikeouts**

## FARMLAND TAX ASSESSMENT STATUTES

**54:4–23a**                                    TAXATION OF PROPERTY

### Library References

Taxation ⇐65.
C.J.S. Taxation § 72.

### Notes of Decisions

**1. Validity**

Statute prohibiting newly constructed single-family dwelling from being added to real property tax assessment list until certificate of occupancy has been issued and dwelling has been actually occupied is not unconstitutional "special" tax exemption legislation. New Jersey State League of Municipalities v. Kimmelman, 204 N.J.Super. 323, 498 A.2d 1266 (A.D. 1985) certification granted 102 N.J. 384, 508 A.2d 246.

This section prohibiting newly constructed single-family dwelling from being added to real property tax assess-

ment list until certificate of occupancy has been issued and dwelling has been actually occupied did not violate principle condemning tax exemption statutes based upon personal status of owner where, even though this section was sponsored by builders and developers, this section served public interest in achieving overall objective of coping with critical housing shortage. New Jersey State League of Municipalities v. Kimmelman, 197 N.J.Super. 89, 484 A.2d 59 (L.1984) reversed on other grounds 204 N.J.Super. 323, 498 A.2d 1266, certification granted 102 N.J. 384, 508 A.2d 246.

## 54:4–23.1.  Short title

This act shall be known and referred to by its short title, the "Farmland Assessment Act of 1964."

L.1964, c. 48, § 1.

### Historical Note

**Title of Act:**

An Act concerning the valuation, assessment and taxation of land actively devoted to agricultural or horticultural

uses; defining such uses; providing for penalties and tax lien; supplementing Title 54 of the Revised Statutes; and making an appropriation. L.1964, c. 48.

### Cross References

Value of land actively devoted to agricultural or horticultural use, see Const. Art. 8, § 1, par. 1.

### Administrative Code References

Farmland Assessment Act, see N.J.A.C. 18:15–1.1 et seq.
Municipally approved farmland preservation programs, requisite of qualification for farmland assessment, see N.J.A.C. 2:76–4.3.
Planned real estate development full disclosure regulations, see N.J.A.C. 5:26–1.1 et seq.

478

## ASSESSMENT OF REAL ESTATE

**54:4–23.1**

### Law Review Commentaries

Agricultural preservation in the New Jersey courts. Lewis Goldshore (1980) 105 N.J.L.J. 475.

Environmental Law for 1979 and beyond. Lewis Goldshore (Fall 1979) No. 89 N.J. Lawyer 41.

Future of farmland and preservation. (1981) 12 Rutgers L.J. 713.

Preservation of farmland. Lewis Goldshore (1977) 9 Rutgers-Camden L.J. 21.

### Notes of Decisions

### 1. Validity

Farmland Assessment Act, by providing that forestland actively and exclusively devoted to agricultural use of commercial production of forest products can be granted farmland assessment, did not exceed scope and purpose of enabling constitutional amendment providing that land which is determined to be actively devoted to agricultural or horticultural use and has been for at least two successive years immediately preceding tax year is entitled to be assessed only at its value for such use. Urban Farms, Inc. v. Wayne Tp., Passaic County, 159 N.J.Super. -61, 386 A.2d 1357 (A.D.1978).

### 2. Construction and application

Taxation of farm property at favorable rates as provided in Farmland Assessment Act, being in nature of exemption from taxation in that it favors certain taxpayers at expense of remaining taxpayers in taxing district, should be construed to exist only when such legislative intention is clear and unmistakable. Department of Environmental Protection v. Franklin Tp., 3 N.J. Tax 105, 181 N.J.Super. 309, 437 A.2d 353 (Tax 1981) affirmed 5 N.J.Tax 476.

### 3. Purpose

New Jersey favors preservation of farmland and open spaces over that of development for residential or commercial uses or even over uses which maximize municipal tax revenues. Mindel v.

Township Council of Franklin Tp., 167 N.J.Super. 461, 400 A.2d 1244 (L.1979).

Primary-goal of Farmland Assessment Act was to save family farm and provide farmers with some economic relief by permitting farmlands to be taxed at a lower assessment. Andover Tp. v. Kymer, 140 N.J.Super. 399, 356 A.2d 418 (A.D.1976).

Farmland Assessment Act of 1964 (§ 54:4–23.1 et seq.), is not intended to treat in any way or to confer any special, different, or new tax status upon municipally-owned watersheds. City of East Orange v. Livingston Tp., 102 N.J. Super. 512, 246 A.2d 178 (L.1968) affirmed 54 N.J. 96, 253 A.2d 546.

Purposes of Farmland Assessment Act were not based upon humanitarian policies to ameliorate injustice to victims, but, rather, its purpose was to grant to farmers tax incentives to maintain land in agricultural or horticultural use, thereby -restraining overdevelopment and retaining "open spaces." Galloway Tp. v. Petkevis, 2 N.J. Tax 85 (1980).

### 4. Local regulations

Farmland Assessment Act does not preclude municipalities from prohibiting, restricting or conditioning use of property for commercial farming or timbering operations, where appropriate in light of considerations legislatively mandated by the Municipal Land Use Law (§ 40:55D–1 et seq.). Borough of Kinnelon v. South Gate Associates, 172 N.J. Super. 216, 411 A.2d 724 (A.D.1980) certification denied 85 N.J. 94, 425 A.2d 260.

Where benefits from continued use of land for farming were enormous, where adverse effect on adjoining farmers from farming land was minimal, and where evidence demonstrated that only

## 54:4–23.1
**Note 4**

### TAXATION OF PROPERTY

respect in which farming of land was offensive was that it was less economically lucrative to township than other uses, strict application of township zoning ordinance to land so as to prohibit farming on ground that farming was inconsistent with township's master plan was unreasonable and arbitrary. Mindel v. Township Council of Franklin Tp., 167 N.J.Super. 461, 400 A.2d 1244 (L.1979).

Neither the Farmland Assessment Act nor its underlying constitutional authorization preempt local zoning regulations from prohibiting, conditioning or restraining the use of property for commercial timbering operations and, hence, where zoning ordinance in effect at end of pretax year prohibited commercial logging operations, the landowner, seeking farmland assessment for woodland, could not satisfy statutory requirements as regards anticipated yearly gross sales within a reasonable period of time following the tax year. L & Z Realty Co., Inc. v. Borough of Ringwood, 6 N.J.Tax 450 (1984).

Land use in violation of local zoning requirements is ineligible for farmland assessment. L & Z Realty Co., Inc. v. Borough of Ringwood, 6 N.J.Tax 450 (1984).

**5.  Freeze Act**

Fact that portion of 1977 tax year judgment directed application of farm-land assessment did not defeat taxpayer's rights under Freeze Act (§ 54:2–43 [repealed; see, now, 54:51A–8]), so long as assessments could be apportioned and no part of farmland assessment of base year judgment was carried over to Freeze Act judgment years. Belmont v. Wayne Tp., 5 N.J. Tax 110 (1983).

**6.  Review**

Prior proceedings of county tax board were not binding in subsequent proceedings before Tax Court dealing with assessments for later years in absence of evidence that same issue of farmland assessment was litigated as claims for farmland assessment in subsequent years generated new cause of action. Bass River Tp. (Burlington County) v. Hogwallow Inc., 182 N.J.Super. 584, 442 A.2d 1055 (A.D.1982).

Nothing in the language of the constitutional amendment, the history of the constitutional referendum, or the history of the legislation passed pursuant to the mandate of the constitutional amendment reflects any intention to except farmland classification and assessment cases from the statutory appeal process. Bunker Hill Cranberry Co., Inc. v. Jackson Tp., Ocean County, 144 N.J.Super. 230, 365 A.2d 204 (A.D.1976).

## 54:4–23.2.   Value of land actively devoted to agricultural or horticultural use

For general property tax purposes, the value of land, not less than 5 acres in area, which is actively devoted to agricultural or horticultural use and which has been so devoted for at least the 2 successive years immediately preceding the tax year in issue, shall, on application of the owner, and approval thereof as hereinafter provided, be that value which such land has for agricultural or horticultural use.

L.1964, c. 48, § 2.

**Constitutional Provisions**

Constitutional provisions, see Const.
Art. 8, § 1, par. 1.

**Administrative Code References**

Value of land qualifying for farmland assessment, see N.J.A.C. 18:15–4.1 et seq.

480

ASSESSMENT OF REAL ESTATE                    54:4–23.3

**Library References**

Taxation ☞348.1(3).
C.J.S. Taxation § 411.


## 54:4–23.3.   Land deemed in agricultural use

Land shall be deemed to be in agricultural use when devoted to the production for sale of plants and animals useful to man, including but not limited to: forages and sod crops; grains and feed crops; dairy animals and dairy products; poultry and poultry products; livestock, including beef cattle, sheep, swine, horses, ponies, mules or goats, including the breeding and grazing of any or all of such animals; bees and apiary products; fur animals; trees and forest products; or when devoted to and meeting the requirements and qualifications for payments or other compensation pursuant to a soil conservation program under an agreement with an agency of the Federal Government.

L.1964, c. 48, § 3.

**Library References**

Taxation ☞348.1(3).
C.J.S. Taxation § 411.


**Notes of Decisions**

Bees  6
Construction and application  1
Contiguous land  4
Evidence  8
Marginal use  3
Multiple use  2
Water reserve  5
Woodlands  7

———

**1. Construction and application**

Municipally-owned lands utilized as public water supply from which city derived income by sales of hay, timber and cordwood was not "actively devoted" to "agricultural use" within meaning of § 54:4–23.2 where principal use of land was a watershed, and land was taxable as land used for purpose and for protection of public water supply and not as farmland. City of East Orange v. Livingston Tp., 102 N.J.Super. 512, 246 A.2d 178 (L.1968) affirmed 54 N.J. 96, 253 A.2d 546.

Mere haphazard use of land resulting in sufficient income to meet income re-

quirements of Farmland Assessment Act of 1964 does not necessarily qualify land for farmland assessment. Brunetti v. Lacey Tp., 6 N.J.Tax 565 (1984).

Neither the Farmland Assessment Act nor its underlying constitutional authorization preempt local zoning regulations from prohibiting, conditioning or restraining the use of property for commercial timbering operations and, hence, where zoning ordinance in effect at end of pretax year prohibited commercial logging operations, the landowner, seeking farmland assessment for woodland, could not satisfy statutory requirements as regards anticipated yearly gross sales within a reasonable period of time following the tax year. L & Z Realty Co., Inc. v. Borough of Ringwood, 6 N.J.Tax 450 (1984).

Where taxpayer did nothing on land between planting of trees in 1975 and mowing of part of one lot more than five years later, taxpayer did not qualify for farmland assessment, as his land was not actively devoted to agricultural or

**54:4–23.3**
Note 1

horticultural activity. Princeton Research Lands, Inc. v. Upper Freehold Tp., 4 N.J.Tax 402 (1982).

Where farmer on which had formerly been small family farm did not, in 1978, plant corn and did not harvest corn he had planted in 1977, where his fields had been taken over by weeds in 1978 and could no longer be characterized as hayfields, and apples on trees were stunted and wormy from lack of pruning and spraying, there was insufficient farming activity on the property in 1978 for property to have qualified for farmland assessment. Jackson Tp. v. Paolin, 3 N.J. Tax 39, 181 N.J.Super. 293, 437 A.2d 344 (Tax 1981).

Where horse farm was located in three taxing districts, the entire farm had previously been accorded farmland assessment treatment, and three taxing district parcels were acquired by taxpayer in one transaction, each unit of land did not have to satisfy requirements of farmland assessment statutes in order to qualify for farmland assessment treatment. Borough of Califon v. Stonegate Properties, Inc., 2 N.J. Tax 153 (1981).

Where two tracts of land for which farmland assessment was sought were neither contiguous nor appurtenant to a tract entitled to a farmland assessment, the two tracts were not reasonably required for the purpose of maintaining farmland tract in agricultural use within meaning of Farmland Assessment Act, and thus such tracts were not entitled to farmland assessment. Bass River Tp. v. Hogwallow, Inc., 1 N.J.Tax 612 (1980).

**2.  Multiple use**

Where principal use of 210-acre tract was for production and sale of hay and tract met statutory requirement as to gross sales, tract was entitled to lower farmland tax assessment, despite fact that taxpayer might be holding land for eventual resale or for speculation. Andover Tp. v. Kymer, 140 N.J.Super. 399, 356 A.2d 418 (A.D.1976).

Although multiple uses of a parcel of land, some agricultural and some nonagricultural do not automatically disqualify it for farmland assessment, agricultural use must predominate. Green Pond Corp. v. Rockaway Tp., 2 N.J. Tax 273

(1981) affirmed in part, reversed in part on other grounds 4 N.J.Tax 534.

Where land was partially used for production and sale of horses, but primary use was breeding and training of horses belonging to other individuals and training of riders, service operations which were nonqualifying uses, land was not qualified for farmland assessment. Bloomingdale Indus. Park v. Borough of Bloomingdale, 1 N.J. Tax 145 (1980).

**3.  Marginal use**

Woodland and other acreage having marginal value for agricultural use may be given preferential tax treatment pursuant to Farmland Assessment Act when it is appurtenant to and reasonably required for purpose of maintaining land actually devoted to agricultural use, particularly where it has been part of farm for a number of years; thus, woodland which is not producing sufficient income from agricultural or horticultural use may in fact qualify for farmland assessment. Urban Farms, Inc. v. Wayne Tp., Passaic County, 159 N.J.Super. 61, 386 A.2d 1357 (A.D.1978).

**4.  Contiguous land**

Contiguously owned, separately acquired woodland tracts, which had never functionally been part of farm, which had not been integrated with it in any documentary manner and on which there had been undertaken independent commercial operations, not for benefit of farm, but as completely separate business activity, did not qualify for farmland assessment on theory that tracts were both legally and functionally part of qualifying farm. Wiesenfeld v. South Brunswick Tp., 166 N.J.Super. 90, 398 A.2d 1342 (A.D.1979).

There was sufficient credible evidence in the record as a whole to support the trial judge's finding that the woodlands and reservoirs contiguous to the petitioner's cranberry bogs were reasonably required for the purpose of maintaining the land in agricultural use within the meaning of the regulations promulgated by the director of the division of taxation to effectuate the purposes of the Farmland Assessment Act and, as such, were within the comprehension of lands actively devoted to agricultural use. Bunker Hill Cranberry Co., Inc. v. Jack-

## ASSESSMENT OF REAL ESTATE

**54:4-23.3**

son Tp., Ocean County, 144 N.J.Super. 230, 365 A.2d 204 (A.D.1976).

Where, based upon testimony of field man employed by county assessor's office and defendant wife and agricultural agent, one-to-one and one-quarter additional acres of land surrounding chicken coops and shed were used for farm purposes and for no other purpose, in addition to 4.05 acres used for fields and actual farm structures, and there was a preponderance of evidence that woodland and swampy area of lot were a part of the farm, entire farm, except one acre under and actually used in connection with farmhouse, was devoted to agricultural use and entitled to farmland assessment treatment. Jackson Tp. v. Hamburger, 2 N.J. Tax 430 (1981).

Evidence concerning value and temperament of horses on horse farm, fact that many were owned by third parties, and the degree of care necessary for the safety of both public and the horses was sufficient to establish that tract of land which was contiguous to balance of horse farm was necessary as a buffer area to the proper operation of the horse farm, and thus such buffer area qualified for farmland assessment under farmland assessment statute; however, other tract which was separated from balance of the farm by a roadway was not required for use as a buffer zone for operation of the farm, and thus could not qualify for farmland assessment. Borough of Califon v. Stonegate Properties, Inc., 2 N.J. Tax 153 (1981).

### 5. Water reserve

City's water reserve property located in other municipalities was not assessable as farmland under Farmland Assessment Act of 1964, and § 54:4-3.3 dealing with local taxation of water reserve land did not apply. City of East Orange v. Livingston Tp., 54 N.J. 96, 253 A.2d 546 (1969).

Application of city board of water commissioners with soil conservation districts requesting assistance in matter of possible training programs in soil conservation did not entitled city to have water reserve assessed as farmland on theory that land met requirements for payments pursuant to soil conservation program undertaken with Federal Government. City of East Orange v.

Livingston Tp., 102 N.J.Super. 512, 246 A.2d 178 (L.1968) affirmed 54 N.J. 96, 253 A.2d 546.

### 6. Bees

Five-acre tract upon which beehives were located was not "devoted to" honey production so as to qualify for a reduced property tax assessment for lands actively devoted to agricultural or horticultural use, notwithstanding that the hives produced more than enough honey, sold at both wholesale and retail, to satisfy income requirement of Farmland Assessment Act, where bees from hives on that property gathered their nectar to a greater extent from other property. Barrett v. Borough of Frenchtown, 6 N.J.Tax 558 (1984).

### 7. Woodlands

Taxpayer's cutting of firewood from dead trees and cedar trees for holiday decoration, from which he received no income, was a minimal agricultural activity which was consistent with maintaining a woodland pursuant to classification set forth by state farmland evaluation advisory committee, and was not consistent with maintaining a timber operation or tree farm, and thus, classification of his land should have been as "woodland" rather than as "cropland harvested" for farmland assessment purposes. Pinoon v. Bernards Tp., 1 N.J.Tax 351 (1980).

### 8. Evidence

Forestland owner's admitted intended use for land currently devoted to woodland management for production of commercial lumber, that is, development for industrial and residential use, was not to be considered in determining whether land was to be granted farmland assessment pursuant to Farmland Assessment Act. Urban Farms, Inc. v. Wayne Tp., Passaic County, 159 N.J.Super. 61, 386 A.2d 1357 (A.D.1978).

Corporate taxpayers did not establish that five acres of woodlands were devoted to agricultural use in 1973 and 1974, the two years preceding tax year 1975, so as to qualify for farmland assessment of woodlands for 1975 where only agricultural use made of woodlands resulted from sale of firewood and sale of timber, there was nothing offered as to where

**TAXATION OF PROPERTY**

from among 555 acres of woodland the firewood was harvested, and, although there was testimony that all timbering activity took place in one particular area of woodland, such area could not be identified as to location or size. Green Pond Corp. v. Rockaway Tp., 2 N.J. Tax 273 (1981) affirmed in part, reversed in part on other grounds 4 N.J.Tax 534.

## 54:4–23.4.  Land deemed in horticultural use

Land shall be deemed to be in horticultural use when devoted to the production for sale of fruits of all kinds, including grapes, nuts and berries; vegetables; nursery, floral, ornamental and greenhouse products; or when devoted to and meeting the requirements and qualifications for payments or other compensation pursuant to a soil conservation program under an agreement with an agency of the Federal Government.

L.1964, c. 48, § 4.

**Library References**

Taxation ☞348.1(3).
C.J.S. Taxation § 411.

## 54:4–23.5.  Land deemed actively devoted to agricultural or horticultural use

Land, five acres in area, shall be deemed to be actively devoted to agricultural or horticultural use when the gross sales of agricultural or horticultural products produced thereon together with any payments received under a soil conservation program have averaged at least $500.00 per year during the 2-year period immediately preceding the tax year in issue, or there is clear evidence of anticipated yearly gross sales and such payments amounting to at least $500.00 within a reasonable period of time.

In addition, where the land is more than five acres in area, it shall be deemed to be actively devoted to agricultural or horticultural use when the gross sales of agricultural or horticultural products produced on the area above five acres together with any payments received under a soil conservation program have averaged at least $5.00 per acre per year during the 2-year period immediately preceding the tax year in issue, or there is clear evidence of anticipated yearly gross sales and such payments amounting to an average of at least $5.00 per year within a reasonable period of time; except in the case of woodland and wetland, where the minimum requirement shall be an average of $0.50 per acre on the area above five acres.

Land previously qualified as actively devoted to agricultural or horticultural use under the act; but failing to meet the additional

ASSESSMENT OF REAL ESTATE                    **54:4–23.5**

requirement on acreage above five acres shall not be subject to the roll-back tax because of such disqualification, but shall be treated as land for which an annual application has not been submitted. L.1964, c. 48, § 5. Amended by L.1973, c. 99, § 1, eff. May 2, 1974.

### Historical Note

The 1973 amendment inserted, at the beginning of the first paragraph, ", five acres in area," and added the second and third paragraphs.

Section 2 of L.1973, c. 99, approved May 2, 1973, provides:

"This act shall take effect immediately except that the tax year 1973 shall be the first tax to which the amendatory provisions for this act shall apply."

### Library References

Taxation ⚮348.1(3).
C.J.S. Taxation § 411.

### Notes of Decisions

Agricultural or horticultural use   2
Burden of proof   9
Construction and application   1
Date of assessment   6
Evidence, sales   5
Intent   7
Local regulations   8
Payments, sales   4
Review   10
Sales   3

---

**1. Construction and application**

Where 210-acre tract was in fact principally used for agriculture and statutory requirements as to area and gross sales were met, entire tract was entitled to tax assessment as farmland, despite fact that only about 100 acres of tract were actually being farmed and that portions of tract were wooded, swampy, or rocky. Andover Tp. v. Kymer, 140 N.J. Super. 399, 356 A.2d 418 (A.D.1976).

**2. Agricultural or horticultural use**

Where logging on property lasted only 12 days until stopped by borough, and where forestry operation during years in question was not a permitted use of the property under terms of borough's zoning ordinance, property was not actively devoted to agricultural or horticultural use during entire period of two calendar years in question so as to be eligible for

farmland assessment for successive year under Farmland Assessment Act of 1964. Clearview Estates, Inc. v. Borough of Mountain Lakes, 188 N.J.Super. 99, 456 A.2d 111 (A.D.1982).

Agriculture or horticultural use of land otherwise eligible for farmland assessment under Farmland Assessment Act must be lawful, that is, a permitted use. Clearview Estates, Inc. v. Borough of Mountain Lakes, 188 N.J.Super. 99, 456 A.2d 111 (A.D.1982).

When land is devoted to multiple uses, agriculture or horticulture must be dominant use in order for tract to qualify for taxation under Farmland Assessment Act. Department of Environmental Protection v. Franklin Tp., 3 N.J. Tax 105, 181 N.J.Super. 309, 437 A.2d 353 (Tax 1981) affirmed 5 N.J.Tax 476.

**3. Sales**

Where fact that gross sales in excess of $500 reported on plaintiffs' income tax returns for the years 1975, 1976 and 1977, in corroboration with claimed gross sales in excess of $500 was of insufficient weight to satisfy plaintiffs' burden of proving at least $500 in gross income for two previous years, and less than five acres of subject property was devoted to the nursery business, plaintiffs were not entitled to have certain property assessed as farmland for tax

**54:4–23.5**
**Note 3**

purposes. Checchio v. Scotch Plains Tp., 2 N.J. Tax 450 (1981).

This provision of Farmland Assessment Act governing income requirements for farmland assessment must be interpreted in light of §§ 54:4–23.3 and 23.4 stating that land shall be deemed to be in agricultural or horticultural use when devoted to production of plants, animals, fruits and like products. Gottdiener v. Roxbury Tp., 2 N.J. Tax 206 (1981).

Although income derived from sale of firewood taken from 49.6 acres of 60-acre tract satisfied minimum statutory requirements for farmland assessment of entire tract, activities conducted on 49.6 acres in 1973 and 1974 were not of such nature as to qualify entire tract for favored treatment under farmland assessment as woodlands in 1975. Spiotta Bros. v. Mine Hill Tp., 1 N.J.Tax 42 (1980).

**4. —— Payments, sales**

In determining whether taxpayer corporations satisfied income requirements of statute for two years preceding each of tax years 1975, 1977 and 1978, for which farmland assessment of woodlands was sought, although taxpayers introduced form of United States Department of Agriculture dated October 16, 1975, authorizing payment to them of $200 for a timber stand improvement program and farmland assessment report prepared for taxpayers described such payments as a "soil conservation payment" which could be used to satisfy income requirements of this section, where record did not reveal when, if ever, payment was actually made and Department of Agricultural form was signed on February 12, 1977, by general manager of taxpayer corporations applying for payment, it could only be assumed that such payment, if it was ever made, was made some time after February 12, 1977, and therefore had to be ignored. Green Pond Corp. v. Rockaway Tp., 2 N.J. Tax 273 (1981) affirmed in part, reversed in part on other grounds 4 N.J.Tax 534.

Although corporate taxpayers proved firewood sales for woodland property for which farmland assessment was sought by invoices indicating billings only, not receipt of payment, taxpayers nevertheless satisfied statutory income requirements to qualify for farmland assessment in tax year 1975, even if gross sales requirement of this section required actual receipt of payment in two preassessment years, where, if firewood payments were received in years of sale, then a yearly average was made out for two years preceding tax year 1975 satisfying this section, if all firewood payments were received in 1974 same average was established, and if insufficient payments were made prior to 1975, remaining amounts due plus amounts anticipated from lumber company would satisfy statute under "reasonably anticipated income" test. Id.

**5. —— Evidence, sales**

While contracts authorizing timber buyer to cut wood for a fixed amount of money satisfied income requirements of Farmland Assessment Act of 1964 for qualifying land for farmland assessment, where the number of acres from which wood was cut or location of such acres could not be determined and there was no evidence of anticipated yearly gross sales amounting to at least $500 within a reasonable time, land did not qualify for farmland assessment as land devoted to production of trees and forest products for sale. Brunetti v. Lacey Tp., 6 N.J.Tax 565 (1984).

Professional forester's estimates of timber accruals for two-year period prior to tax year for which farmland assessment was sought were mere expression of increases in value and did not constitute "gross sales" for purpose of satisfying statutory income requirements. L & Z Realty Co., Inc. v. Borough of Ringwood, 6 N.J.Tax 450 (1984).

Corporate taxpayers had not satisfied income requirements of this section by showing likelihood of sales of forest products from their woodlands within a reasonable period of time following tax year so as to qualify woodlands for farmland assessment, where, although evidence of stumpage value was introduced demonstrating potential for yearly income from sale of forest products in excess of requisite amount, taxpayers did not offer sufficient evidence of anticipated sales themselves nor testimony regarding age of trees or their readiness for harvest so as to satisfy reasonable time prerequisite. Green Pond Corp. v. Rockaway Tp., 2 N.J.Tax 273 (1981) affirmed in part, reversed in part on other grounds 4 N.J.Tax 534.

Purported proceeds from sale of cattle would not be considered as "farm income" for 1973 within purview of Farmland Assessment Act where taxpayers failed to present evidence establishing how long cattle were on the property and whether they were products of property except for sketchy references to fact that they were introduced in later part of 1972 and attempts to establish that they were consigned for sale from April to October 1973, and thus failed to prove that cattle were produced on land in question. Gottdiener v. Roxbury Tp., 2 N.J.Tax 206 (1981).

In action for farmland assessment of property, deposit slips and copies of cancelled checks in support of two sales of hay, together with assessors' admissions that hay was growing on farm during time in question was sufficient proof of farm income within meaning of this section governing income requirements for farmland assessment purposes. Gottdiener v. Roxbury Tp., 2 N.J. Tax 206 (1981).

**6.  Date of assessment**

Taxpayers' claim for farmland assessment of property for 1974 would fail by reason of failure to actively devote premises to agricultural-horticultural use in 1972, even if they satisfied all requirements in 1973. Gottdiener v. Roxbury Tp., 2 N.J. Tax 206 (1981).

**7.  Intent**

Claim that taxpayer is holding property for speculation is not to be considered in determining qualification for farmland assessment status. Spiotta Bros. v. Mine Hill Tp., 1 N.J. Tax 42 (1980).

**8.  Local regulations**

Neither the Farmland Assessment Act nor its underlying constitutional authorization preempt local zoning regulations from prohibiting, conditioning or restraining the use of property for commercial timbering operations and, hence, where zoning ordinance in effect at end of pretax year prohibited commercial logging operations, the landowner, seeking farmland assessment for woodland, could not satisfy statutory requirements as regards anticipated yearly gross sales within a reasonable period of time following the tax year. L & Z Realty Co., Inc. v. Borough of Ringwood, 6 N.J.Tax 450 (1984).

**9.  Burden of proof**

In seeking to qualify land for farm land assessment under Farm Land Assessment Act of 1964, landowner has burden of proving that income from property resulted from "products produced" on land in issue. Brunetti v. Lacey Tp., 6 N.J.Tax 565 (1984).

To establish right to assessment of land on farmland basis, taxpayer had burden of establishing ownership, area of at least five acres devoted to agricultural or horticultural use at least two years prior to tax year, gross sales of products from land totaling at least $500 per year and fact of application for the farmland assessment. Kugler v. Wall Tp., 1 N.J. Tax 10 (1980).

**10.  Review**

A review of a determination of the assessor as to whether lands are actively devoted to agricultural of horticultural uses so as to qualify for a farmland assessment is not limited to a proceeding in lieu of prerogative writs in the Superior Court and may be obtained by a proceeding in the State Division of Tax Appeals. Bunker Hill Cranberry Co., Inc., v. Jackson Tp., Ocean County, 144 N.J. Super. 230, 365 A.2d 204 (A.D.1976).

## 54:4–23.6.  Qualifications for valuation, assessment and taxation as land actively devoted to agricultural or horticultural use

Land which is actively devoted to agricultural or horticultural use shall be eligible for valuation, assessment and taxation as herein provided when it meets the following qualifications:

**54:4–23.6**                    TAXATION OF PROPERTY

(a) It has been so devoted for at least the 2 successive years immediately preceding the tax year for which valuation under this act is requested;

(b) The area of such land is not less than 5 acres when measured in accordance with the provisions of section 11 hereof; and

(c) Application by the owner of such land for valuation hereunder is submitted on or before August 1 of the year immediately preceding the tax-year to the assessor of the taxing district in which such land is situated on the form prescribed by the Director of the Division of Taxation.

L.1964, c. 48, § 6.   Amended by L.1970, c. 243, § 1.

### Historical Note

The 1970 amendment, in subsec. (c), substituted "August 1" for "October 1".

Section 4 of L.1970, c. 243, approved Oct. 28, 1970, provided:

"This act shall take effect January 1, 1971."

### Administrative Code References

Proof to support application for farmland assessment, see N.J.A.C. 18:15–3.1 et seq.

### Library References

Taxation ⊜348.1(3).
C.J.S. Taxation § 411.

### Notes of Decisions

Application for agricultural assessment 4
Area 2
Burden of proof 6
Construction and application 1
Duration of use 3
Multiple parcels of land; application for agricultural assessment 5
Review 7

#### 1. Construction and application

This section authorizing farmland assessment of woodland property on condition that five acres of woodlands have been devoted to agricultural use for two-year period preceding tax year, should be strictly construed against the taxpayer, since it provides benefits to the taxpayer at the expense of all the remaining taxpayers in the taxing district.

Green Pond Corp. v. Rockaway Tp., 2 N.J. Tax 273 (1981) affirmed in part, reversed in part on other grounds 4 N.J.Tax 534.

Where horse farm was located in three taxing districts, the entire farm had previously been accorded farmland assessment treatment, and three taxing district parcels were acquired by taxpayer in one transaction, each unit of land did not have to satisfy requirements of farmland assessment statutes in order to qualify for farmland assessment treatment. Borough of Califon v. Stonegate Properties, Inc., 2 N.J. Tax 153 (1981).

#### 2. Area

Where woodlands were not reasonably required for maintaining cultivated area in horticultural use, taxpayers could not

488

## ASSESSMENT OF REAL ESTATE

aggregate woodlands acreage with acreage under cultivation so as to qualify for farmland assessment. Mason v. Wyckoff Tp., 1 N.J.Tax 433 (1980).

### 3. Duration of use

Where logging on property lasted only 12 days until stopped by borough, and where forestry operation during years in question was not a permitted use of the property under terms of borough's zoning ordinance, property was not actively devoted to agricultural or horticultural use during entire period of two calendar years in question so as to be eligible for farmland assessment for successive year under Farmland Assessment Act of 1964. Clearview Estates, Inc. v. Borough of Mountain Lakes, 188 N.J.Super. 99, 456 A.2d 111 (A.D.1982).

Statutory requirement for woodlands to qualify for farmland assessment that five acres of woodland be devoted to agricultural use in two years preceding tax year in question could not be satisfied by showing that taxpayers' woodlands had been under forestry management program for one year and eight months preceding tax year 1977; taxpayers had to demonstrate the devotion of five acres to agricultural use for two-year period commencing January 1, 1975, and such a showing had not been made. Green Pond Corp. v. Rockaway Tp., 2 N.J. Tax 273 (1981) affirmed in part, reversed in part on other grounds 4 N.J.Tax 534.

### 4. Application for agricultural assessment

Since applications for agricultural assessments were due by August 1 of each pretax year, taxpayer's failure to so file rendered land ineligible on August 1 of each such pretax year and proper time for utilizing added assessments procedure was during each of related tax years. Cherry Hill Indus. Properties v. Voorhees Tp., 186 N.J.Super. 307, 452 A.2d 673 (A.D.1982) affirmed as modified on other grounds 91 N.J. 526, 453 A.2d 850.

Property made ineligible for agricultural assessments by failure of taxpayer to file timely application did not become omitted property until proper time for utilizing added assessments procedure, during each of related tax years,

had run. Cherry Hill Indus. Properties v. Voorhees Tp., 186 N.J.Super. 307, 452 A.2d 673 (A.D.1982) affirmed as modified on other grounds 91 N.J. 526, 453 A.2d 850.

Failure to timely file a farmland application disqualifies land from receiving a farmland assessment. Belmont v. Wayne Tp., 3 N.J. Tax 382 (1981).

Statutes requiring filing of farmland assessment application by August 1 of pretax year are not ambiguous and reflect clear intention that filing date contained therein be complied with, and word "shall" as used therein is construed as to mean "must" and filing deadline of August 1 of pretax year is mandatory and may not be tolled. Galloway Tp. v. Petkevis, 2 N.J. Tax 85 (1980).

Provision of Farmland Assessment Act which provides that application for valuation of farmland must be "submitted" to tax assessor on or before August 1 means that application must be "filed" with assessor by that date; mere mailing of application to assessor prior to August 1 does not satisfy requirements of Act with respect to timeliness. Hashomer Hatzair, Inc. v. East Windsor Tp., 1 N.J. Tax 115, 176 N.J.Super. 250, 422 A.2d 808 (Tax 1980).

### 5. —— Multiple parcels of land, application for agricultural assessment

Where separate, noncontiguous parcels of land in agricultural or horticultural use, in single ownership, are located in same taxing district, a separate application for farmland assessment must be made with respect to each parcel. Jackson Tp. v. Paolin, 3 N.J. Tax 39, 181 N.J.Super. 293, 437 A.2d 344 (Tax 1981).

Where no application for farmland assessment was made with respect to lot 70 for any year, no farmland assessment could be granted for that lot, notwithstanding application for lot which was proximate in location to lot 70. Id.

### 6. Burden of proof

In action concerning difference between full assessment and assessment paid on property which had allegedly improperly received beneficial farmland

TAXATION OF PROPERTY

assessments, burden of proof was on taxpayer to demonstrate that he had properly filed and was entitled to farmland assessment. Cherry Hill Indus. Properties v. Voorhees Tp., 186 N.J.Super. 307, 452 A.2d 673 (A.D.1982) affirmed as modified on other grounds 91 N.J. 526, 453 A.2d 850.

7. Review

Refusal of the Division of Tax Appeals to credit unsubstantiated testimony of taxpayer's predecessor in title as to his gross sales from produce derived from property did not require reversal of determination that property was not being actively devoted to agricultural or horticultural uses so as to qualify for a farmland valuation. Franklin Estates,

Inc. v. Edison Tp., 142 N.J.Super. 179, 361 A.2d 53 (A.D.1976) affirmed 73 N.J. 462, 375 A.2d 658.

Res judicata was not applicable to bind Tax Court to follow county board of taxation's 1973 judgment granting farmland assessments to tracts of land, in that findings of fact and conclusions of law arrived at by a county board of taxation are not binding upon Tax Court, and a judgment concluding that land is entitled to benefits of Farmland Assessment Act for any prior year cannot bar a municipality from questioning its qualification in future years. Bass River Tp. v. Hogwallow, Inc., 1 N.J. Tax 612 (1980).

## 54:4–23.7.  Considerations of assessor in valuing land

The assessor in valuing land which qualifies as land actively devoted to agricultural or horticultural use under the tests prescribed by this act, and as to which the owner thereof has made timely application for valuation, assessment and taxation hereunder for the tax year in issue, shall consider only those indicia of value which such land has for agricultural or horticultural use.  In addition to use of his personal knowledge, judgment and experience as to the value of land in agricultural or horticultural use, he shall, in arriving at the value of such land, consider available evidence of agricultural and horticultural capability derived from the soil survey data at Rutgers, The State University, the National Co-operative Soil Survey, and the recommendations of value of such land as made by any county or State-wide committee which may be established to assist the assessor.

L.1964, c. 48, § 7.

### Cross References

Eligibility, determination for each tax year, application, added assessment, see § 54:4–23.13.
Roll back taxes, determination of amounts, see § 54:4–23.8.

### Administrative Code References

Indicia of value to be used, see N.J.A.C. 18:15–4.2.

### Library References

Taxation ⟺348(2), 348.1(3).
C.J.S. Taxation § 411.

ASSESSMENT OF REAL ESTATE                    54:4–23.8

### Notes of Decisions

Advisory committee recommendations
2
Review 3
Valuation 1

_____

### 1. Valuation

Action of Division of Tax Appeals in classifying certain forestland as "cropland harvested" and therefore valuing it at $720 per acre, the highest suggested valuation for farmland in the county, in which cropland was valued at from $60 to $720 per acre depending on soil rating and woodland from $21 to $33, was error, in that only applicable suggested valuations submitted to division were those for woodland, and no other evidence or method of valuation was before division to support determination. Urban Farms, Inc. v. Wayne Tp., Passaic County, 159 N.J.Super. 61, 386 A.2d 1357 (A.D.1978).

### 2. Advisory committee recommendations

Where trial judge adopted assessment rate recommended by farmland evaluation advisory committee in its report for farmland in the county wherein lands were located, but report did nothing more than establish ranges of suggested values for each of several classifications of farmlands in various areas of state, values which were merely designed as guidelines for assessor, and did not provide an accurate means of attributing a specific value to a specific piece of property, trial judge's adoption of lowest recommended value for farmlands in county, without more, in fixing value of petitioner's farmlands was totally arbitrary. Banker Hill Cranberry Co., Inc. v. Jackson Tp., Ocean County, 144 N.J.Super. 230, 365 A.2d 204 (A.D.1976) certification denied 73 N.J. 59, 372 A.2d 324.

Recommendations of state farmland evaluation advisory committee, which assessor is required to consider in determining value for farmland, are not mandated but are only guidelines which are to be considered along with assessor's personal knowledge, judgment and experience. Pinson v. Bernards Tp., 1 N.J. Tax 351 (1980).

### 3. Review

Owner of farmland was required to have timely paid taxes due for year in order to be entitled to pursue appeal challenging determination of county board of taxation as to whether the farmland was, for valuation purposes, "cropland harvested" or "woodland." West Cap Associates, Inc. v. West Milford Tp., 4 N.J.Tax 364 (1982).

## 54:4–23.8.  Roll-back taxes; determination of amounts

When land which is in agricultural or horticultural use and is being valued, assessed and taxed under the provisions of this act, is applied to a use other than agricultural or horticultural, it shall be subject to additional taxes, hereinafter referred to as roll-back taxes, in an amount equal to the difference, if any, between the taxes paid or payable on the basis of the valuation and the assessment authorized hereunder and the taxes that would have been paid or payable had the land been valued, assessed and taxed as other land in the taxing district, in the current tax year (the year of change in use) and in such of the 2 tax years immediately preceding, in which the land was valued, assessed and taxed as herein provided.

If in the tax year in which a change in use of the land occurs, the land was not valued, assessed and taxed under this act, then such land shall be subject to roll-back taxes for such of the 2 tax years,

54:4–23.8                          TAXATION OF PROPERTY

immediately preceding, in which the land was valued, assessed and taxed hereunder.

In determining the amounts of the roll-back taxes chargeable on land which has undergone a change in use, the assessor shall for each of the roll-back tax years involved, ascertain:

(a) The full and fair value of such land under the valuation standard applicable to other land in the taxing district;

(b) The amount of the land assessment for the particular tax year by multiplying such full and fair value by the county percentage level, as determined by the county board of taxation in accordance with section 3 of P.L.1960, chapter 51 (C. 54:4–2.27);

(c) The amount of the additional assessment on the land for the particular tax year by deducting the amount of the actual assessment on the land for that year from the amount of the land assessment determined under (b) hereof;  and

(d) The amount of the roll-back tax for that tax year by multiplying the amount of the additional assessment determined under (c) hereof by the general property tax rate of the taxing district applicable for that tax year.

L.1964, c. 48, § 8.   Amended by L.1970, c. 243, § 2.

### Historical Note

Effective date, see Historical Note under § 54:4–23.6.

The 1970 amendment substituted, in subsec. (b), "county percentage level" for "average real property assessment ratio of the taxing district" and "in accordance with section 3 of P.L.1960, chapter 51 (C. 54:4–2.27)" for "for the purposes of the county equalization table for such year, pursuant to sections 54:3–17 to 19 of the Revised Statutes".

### Cross References

Notice of possible rollback taxes, see §§ 54:5–12, 54:5–13.

### Administrative Code References

Roll-back taxes, see N.J.A.C. 18:15–7.1 et seq.

### Library References

Taxation ⟜348.1(3), 362.
C.J.S. Taxation §§ 411, 508.

### Notes of Decisions

Amount of assessment  6
Burden of proof  15

Change in use  3
Condemnation or eminent domain  8

ASSESSMENT OF REAL ESTATE                    **54:4–23.8**

### 1.  Construction and application

Even if a rollback tax could fairly be characterized as a penalty, amount of penalty is limited to amount of property tax taxing district would have received if property had not been assessed as qualified farmland for relevant years under the Farmland Assessment Act; thus, rollback tax assessments must comply with Constitution's tax clause (Const. Art. 8, § 1, par. 1). V.B.R. Associates v. Bernards Tp., 6 N.J.Tax 241 (1984).

Rollback taxes are not triggered until the land is no longer devoted to an agricultural or horticultural use even though the land does not qualify in a subsequent year for a farmland assessment because it has failed to meet all of the statutory requirements to classify it as actively devoted. Belmont v. Wayne Tp., 3 N.J. Tax 382 (1981).

Rollback taxes are technically not omitted or added assessments; instead, they are additional taxes and, in adopting a method to be followed by the assessor in making such an assessment the legislature specifically directed that the procedure to be followed is that set forth in the original method for assessment of omitted property; the alternate method is not to be used for rollback taxes. Atlantic City Development Corp. v. Hamilton Tp., 3 N.J. Tax 363 (1981).

### 2.  Purpose

Rollback taxes were designed to protect a municipality from pure speculation by returning to its coffers some of the tax benefits the owner received while the land qualified under the Farmland Assessment Act. Gardiner v.

State, 196 N.J.Super. 529, 483 A.2d 442 (L.1984).

### 3.  Change in use

Failure of department of environmental protection to submit timely farmland assessment applications on property it had acquired for public water supply purposes and for public recreation and conservation purposes did not constitute change in use of property such that rollback provisions of the act became applicable where imposition of rollback taxes hinged upon change in use and not filing of application form and where filing of application went to issue of eligibility for farmland assessment. Department of Environmental Protection v. Franklin Tp., 3 N.J. Tax 105, 181 N.J.Super. 309, 437 A.2d 353 (Tax 1981) affirmed 5 N.J.Tax 476.

Although prior to conveyance of parcel to sewage authority, it was part of, appurtenant to, or reasonably required for purpose of maintaining portion of 112-acre tract which was actively devoted to agricultural use, where this relationship was severed when subject parcel was split up and conveyed to third party for purposes wholly unrelated to agriculture or horticulture, i.e., recreation, and where, by very fact of its conveyance, it was apparent that land was no longer needed to support remaining, unsold portion of farm, "change in use" occurred at time of transfer which beckoned to a rollback assessment. Hinck v. Wall Tp., 3 N.J. Tax 96 (1981).

Farmer's failure to devote his property actively to agriculture within intent of Farmland Assessment Act in year 1978 was not an "application of the property to a use other than agriculture" and was not a "change in use" of the property within intent of Farmland Assessment Act so as to trigger imposition of rollback taxes upon property. Jackson Tp. v. Paolin, 3 N.J. Tax 39, 181 N.J.Super. 293, 437 A.2d 344 (Tax 1981).

### 4.  —— Date of assessment, change in use

Since, under Farmland Assessment Act, roll back taxes are assessed in the year of change in use from agricultural to other purposes, rather than when change in possession or title occurs, fact that filing of declaration, which gave

## 54:4–23.8
**Note 4**

Turnpike Authority, a right to immediate possession of the land, occurred in 1971 and 1973 did not mean that roll back assessments arising upon change in use of land from agricultural purposes should have been imposed in 1971 and 1973 rather than in 1972 and 1974, when the changes in use occurred. New Jersey Turnpike Authority v. Washington Tp., 137 N.J.Super. 543, 350 A.2d 69 (A.D.1975) affirmed 73 N.J. 180, 373 A.2d 652.

Following change in use of subject land in tax year 1981, township could not collect rollback taxes under the Farmland Assessment Act for each of tax years 1976 and 1977, where assessments placed on subject land for each of tax years 1979 and 1980 were not farmland assessments but, instead, were regular assessments. Eagle Plaza Associates v. Voorhees Tp., 6 N.J.Tax 582 (1984).

Proper rollback assessment for year of change in use is based upon valuation and assessment of property as of assessing date, not date of change in use. Plushanski v. Union Tp., 1 N.J. Tax 520, 176 N.J.Super. 626, 424 A.2d 473 (Tax 1980).

### 5. Full and fair value

Division of Tax Appeals, in purporting to fix "roll-back" taxes, erred in failing to apply the statutory mandate requiring the assessor to ascertain the "full and fair value of such land under the valuation standard applicable to other land in the taxing district." Schere v. Freehold Tp., 150 N.J.Super. 404, 375 A.2d 1218 (A.D.1977).

Homesite approach to valuation is accepted valuation approach to determined fair market value of undeveloped land and may properly be used by assessor as part of valuation process, including determination of rollback taxes. Plushanski v. Union Tp., 1 N.J. Tax 520, 176 N.J.Super. 626, 424 A.2d 473 (Tax 1980).

Value of lot subject to rollback tax could not be determined solely by direct proportional allocation of value of entire tract upon which lot was located, but consideration had to be given to value of portion of tract containing lot subject to rollback tax. Plushanski v. Union Tp., 1 N.J. Tax 520, 176 N.J.Super. 626, 424 A.2d 473 (Tax 1980).

### 6. Amount of assessment

While the Division of Tax Appeals, which purported to fix "roll-back" taxes for the tax years 1972 and 1973 on property sold by petitioners to copetitioner, asserted that it was stipulated that other vacant lands in the area were assessed at $3,000 an acre, nowhere in the record did it appear that counsel for the municipality so stipulated, and even if it were stipulated that the assessment for "roll-back" purposes was to be based on the "assessment" of other properties, the stipulation was irrelevant, for this section provides for a clear, four-step process to calculate the amount of "roll-back" taxes due, with one step in the process mandating a determination of the "full and fair value" of the subject property, which calculation has nothing to do with the "assessment" of other properties. Schere v. Freehold Tp., 150 N.J.Super. 404, 375 A.2d 1218 (A.D. 1977).

Under the Farmland Assessment Act of 1964, rollback tax assessments must be within the "common level range" of assessment, i.e., that range which is plus or minus 15 percent of the average ratio for that district. N.J.S.A. V.B.R. Associates v. Bernards Tp., 6 N.J.Tax 241 (1984).

### 7. Other governmental units

Section 27:23–12 which specifically grants Turnpike authority tax exemption for turnpike projects or for property acquired under the provisions of the Turnpike Authority Act did not preclude assessment against Turnpike Authority, which had purchased certain land which had been used as farm land and which was to be used for turnpike purposes, from being assessed for the roll back taxes resulting from change in the use of the land from agricultural purposes. New Jersey Turnpike Authority v. Washington Tp., 137 N.J.Super. 543, 350 A.2d 69 (A.D.1975) affirmed 73 N.J. 180, 373 A.2d 652.

Property whose use was changed from farmland to other use because of acquisition by state department of transportation was subject to roll back assessments for the year of change of use and the two tax years immediately preceding. State v. Washington Tp., 73 N.J. 182, 373 A.2d 652 (1977).

## ASSESSMENT OF REAL ESTATE

Rollback provisions of Farmland Assessment Act applied to land acquired by department of environmental protection for public water supply purposes and for public recreation and conservation purposes, even though such purposes were consistent with purposes of the act, where such uses were not agricultural or horticultural as defined by the act. Department of Environmental Protection v. Franklin Tp., 3 N.J. Tax 105, 181 N.J.Super. 309, 437 A.2d 353 (Tax 1981) affirmed 5 N.J.Tax 476.

### 8.  Condemnation or eminent domain

Under Farmland Assessment Act, where no change of use was made by the seller but the land was in effect taken out of circulation by the condemnation action of the state, the latter must bear all rollback taxes, sellers being completely free of any such obligation. Gardiner v. State, 196 N.J.Super. 529, 483 A.2d 442 (L.1984).

### 9.  Sale of land

Where there was no collateral agreement between the sellers and buyers for the sellers to pay roll-back taxes, where the sellers had maintained the subject land as a farm qualifying under the Farmland Assessment Act up to the date of settlement, and where the buyers in fact changed the use of the property to one other than farming subsequent to the date of settlement, the buyers thereby incurred liability for the payment of the roll-back taxes assessed upon the change of use of the land. Paz v. DeSimone, 139 N.J.Super. 102, 352 A.2d 609 (Ch.1976).

### 10.  Discriminatory valuation

In action by taxpayer for review of valuation and assessment method employed by assessor in imposing rollback taxes, there was insufficient proof of discrimination and therefore no ratio of assessment could be applied to actual assessment of property for years in question. Plushanski v. Union Tp., 1 N.J. Tax 520 (1980).

### 11.  Title insurance

Title insurance policy would be reformed to reflect an exception for liability under the Farmland Assessment Act, as the testimony of the title company's clerk and the revised testimony of the attorney who represented the buyers at settlement supported the proposition that the exception for liability under the Act was to remain on the policy of title insurance to be issued, and as it was also shown that the clerk through honest mistake did not include the exception on the marked-up settlement report given the typist who typed up the issued policy. Paz v. DeSimone, 139 N.J.Super. 102, 352 A.2d 609 (Ch.1976).

### 12.  Tax liens

Tax search certificate issued by township was insufficient to put county, as purchaser of property within township, on notice as to presence of rollback taxes pursuant to Farmland Assessment Act, even though such taxes were lien on subject property, where certificate contained caveat which would vitiate legislative scheme which was intended to be liberally interpreted in favor of principle of security in land title. Mahwah Tp. v. Bergen County, 2 N.J. Tax 479 (1981).

### 13.  Waiver

If taxing district neglects to state definitively whether or not subject property has been assessed under Farmland Assessment Act in last three years, it waives right to pursue rollback tax liability under Farmland Assessment Act against purchaser taxpayer. Mahwah Tp. v. Bergen County, 2 N.J. Tax 479 (1981).

### 14.  Estoppel

Fact that municipalities which imposed roll back assessment upon Turnpike Authority when it acquired land which had been used for agricultural purposes did not participate in the condemnation proceedings in which the Authority acquired the land did not estop them from later asserting the claims for roll back taxes. New Jersey Turnpike Authority v. Washington, Tp., 137 N.J.Super. 543, 350 A.2d 69 (A.D.1975) affirmed 73 N.J. 180, 373 A.2d 652.

### 15.  Burden of proof

Burden of proof to establish a change in use of lands so as to subject it to rollback tax is on the party claiming the change. Belmont v. Wayne Tp., 3 N.J. Tax 382 (1981).

**54:4—23.9**                    TAXATION OF PROPERTY

## 54:4—23.9.   Procedure for assessment, collection, payment, etc., of roll-back taxes

The assessment, collection, apportionment and payment over of the roll-back taxes imposed by section 8,[1] the attachment of the lien for such taxes, and the right of a taxing district, owner or other interested party to review any judgment of the county board of taxation affecting such roll-back taxes, shall be governed by the procedures provided for the assessment and taxation of omitted property under chapter 413 of the laws of 1947.[2]  Such procedures shall apply to each tax year for which roll-back taxes may be imposed, notwithstanding the limitation prescribed in section 1 of said chapter respecting the periods for which omitted property assessments may be imposed.

L.1964, c. 48, § 9.

[1] Section 54:4—23.8.

[2] Section 54:4—63.12 et seq.

### Cross References

Assessment of omitted property, see § 54:4—63.12 et seq.

### Library References

Taxation ⊂327 et seq., 362, 515 et seq., 544 et seq.

C.J.S. Taxation §§ 393, 508, 607, 618, 630, 640, 641, 1082, 1088.

### Notes of Decisions

Estoppel   1
Review   2

———

**1.  Estoppel**

Equitable estoppel did not excuse taxpayers' filing of complaint to review farmland assessment rollback judgments more than eight months after service of such judgments upon taxpayers, where taxing district did nothing within appeal period to prevent taxpayers from appealing, there was no misrepresentation or concealment of any fact by tax-ing district, and tax bills, which allegedly contained erroneous information, were not received by taxpayers until after their time for appeal had expired. Gale Builders, Inc. v. Hunterdon County Bd. of Taxation, 8 N.J.Tax 16 (1985).

**2.  Review**

Farmland assessment rollback judgment entered by county board of taxation must be appealed to Tax Court within 45 days of service of the judgment. Gale Builders, Inc. v. Hunterdon County Bd. of Taxation, 8 N.J.Tax 16 (1985).

## 54:4—23.10.   Determination of true value of land for purposes of state school aid and determining apportionment valuation

The Director of the Division of Taxation in equalizing the value of land assessed and taxed under this act for the purposes of State

## ASSESSMENT OF REAL ESTATE    54:4–23.11

school aid, and each county board of taxation in equalizing such land for the purposes of determining the "apportionment valuation" under section 54:4–49 of the Revised Statutes, shall determine the true value of such land on the basis of its agricultural or horticultural use. The director shall promulgate rules and regulations to effectuate the purposes of this section.

L.1964, c. 48, § 10.

### Administrative Code References

Apportionment valuation of land, classification, assessment ratio, see N.J.A.C. 18:15–9.1.

### Library References

Taxation ⟨⇐299.
C.J.S. Taxation § 354.

## 54:4–23.11.  Area of land included

In determining the total area of land actively devoted to agricultural or horticultural use there shall be included the area of all land under barns, sheds, silos, cribs, greenhouses and like structures, lakes, dams, ponds, streams, irrigation ditches and like facilities, but land under and such additional land as may be actually used in connection with the farmhouse shall be excluded in determining such total area.

L.1964, c. 48, § 11.

### Library References

Taxation ⟨⇐348.1(3).
C.J.S. Taxation § 411.

### Notes of Decisions

**1. Construction and application**

Particularly where such land has been part of the farm for a number of years, woodland, wet areas and other acreage having marginal value for agricultural or horticultural use may be given special farmland tax treatment, so long as such acreage is part of, appurtenant to, or reasonably required for the purpose of maintaining land actually devoted to farm use. Andover Tp. v. Kymer, 140 N.J.Super. 399, 356 A.2d 418 (A.D.1976).

Where farmhouse lot of one-quarter of an acre could not be extended without encroaching upon land under cultivation, even though property was located in residential-agricultural zone of township requiring minimum lot size of approximately one and one-quarter acres, under Farmland Assessment Act of 1964 the one-quarter acre in actual use as farmhouse lot and not one and one-quarter acres was taxable at residential rate rather than at lower farmland rate. Terhune v. Franklin Tp., 107 N.J.Super. 218, 258 A.2d 18 (A.D.1969).

Ravine area created by brook which was approximately 43 feet below level of buildings could not be aggregated with

TAXATION OF PROPERTY

acreage under cultivation so as to comply with minimum acreage requirement for farmland assessment. Mason v. Wyckoff Tp., 1 N.J. Tax 433 (1980).

Where woodlands were not reasonably required for maintaining cultivated area in horticultural use, taxpayers could not aggregate woodlands acreage with acreage under cultivation so as to qualify for farmland assessment. Mason v. Wyckoff Tp., 1 N.J. Tax 433 (1980).

Additional woodland, uncultivated, unused and unneeded for primary function of production of shrubs, could not be added to bring owner within statutory five-acre requirement for assessment as farmland. Kugler v. Wall Tp., 1 N.J. Tax 10 (1980).

## 54:4–23.12.  Valuation, assessment and taxation of structures

All structures, which are located on land in agricultural or horticultural use and the farmhouse and the land on which the farmhouse is located, together with the additional land used in connection therewith, shall be valued, assessed and taxed by the same standards, methods and procedures as other taxable structures and other land in the taxing district; provided, however, that the term "structures" shall not include temporary demountable plastic covered framework made up of portable parts with no permanent understructures or related apparatus, commonly known as seed starting plastic greenhouses.

L.1964, c. 48, § 12.  Amended by L.1979, c. 70, § 1, eff. April 10, 1979.

### Historical Note

The 1979 amendment added the provision excluding "seed starting plastic greenhouses" from the definition of "structures".

### Administrative Code References

Structures, see N.J.A.C. 18:15–4.5.

### Library References

Taxation ⬅348(6), 348.1(3).
C.J.S. Taxation § 411.

### Notes of Decisions

**1.  Construction and application**

Taxpayer's log cabin and one-half acre lot were not actively devoted to agricultural or horticultural use as required to qualify for preferential tax treatment of Farmland Assessment Act, where cabin was used primarily for shelter and rest while taxpayer was on property on weekends and during vacations, and used occasionally as overnight accommodation unrelated to taxpayer's tree-farming operation, and cabin was never used primarily as storage shed or what could be considered a farm building. Warselle Land Corp. v. Tewksbury Tp., 3 N.J. Tax 565 (1981).

### 54:4–23.13.   Determination of eligibility for each tax year; application; added assessment

Eligibility of land for valuation, assessment and taxation under this act shall be determined for each tax year separately.  Application shall be submitted by the owner to the assessor of the taxing district in which such land is situated on or before August 1 of the year immediately preceding the tax year for which such value n, assessment and taxation are sought.  If the application is file y delivery through the mails or a commercial courier or messenger service, compliance with the time limit for filing shall be established if there is satisfactory evidence that it was committed for delivery to the United States Postal Service or to the courier or messenger service within the time allowed for filing.  An application once filed with the assessor for the ensuing tax year may not be withdrawn by the applicant after August 1 of the pretax year.

If a change in use of the land occurs between August 1 and December 31 of the pretax year, either the assessor or the county board of taxation shall deny or nullify such application and, after examination and inquiry, shall determine the full and fair value of said land under the valuation standard applicable to other land in the taxing district and shall assess the same, according to such value.  If, notwithstanding such change of use, the land is valued, assessed and taxed under the provisions of this act in the ensuing year, the assessor shall enter an assessment, as an added assessment against such land, in the "Added Assessment List" for the particular year involved in the manner prescribed in chapter 397 of the laws of 1941.[1]  The amount of the added assessment shall be in an amount equal to the difference, if any, between the assessment imposed under this act and the assessment which would have been imposed had the land been valued and assessed as other land in the taxing district.  The enforcement and collection of additional taxes resulting from any additional assessment so imposed shall be as provided by said chapter.  The additional assessment imposed under this section shall not affect the roll-back taxes, if any, under section 8 of this act.[2]

L.1964, c. 48, § 13.  Amended by L.1972, c. 146, § 1, eff. Sept. 7, 1972; L.1982, c. 72, § 1, eff. July 19, 1982.

[1] Section 54:4–63.1 et seq.

[2] Section 54:4–23.8.

**54:4–23.13**                                    TAXATION OF PROPERTY

Assembly Agriculture and Environment Committee Statement
Assembly, No. 1126—L.1982, c. 72

• • • • • •

The bill requires "satisfactory evidence" of committal for delivery [of the application] within the time allowed, such as a postmark, postal certification or registration, or an appropriate receipt from a commercial service.

### Historical Note

The 1972 amendment substituted "August 1" for "October 1" throughout the section.

The 1982 amendment inserted the third sentence in the first paragraph.

Section 2 of L.1982, c. 72, approved July 19, 1982, provides:

"This act shall take effect immediately and shall apply to applications filed for the 1982 tax year and thereafter."

### Cross References

Added assessment of real estate, see § 54:4–63.1 et seq.

### Library References

Taxation ⟷348.1(1), (3).
C.J.S. Taxation § 411.

### Notes of Decisions

Change in valuation 4
Construction and application 1
Denial or nullification of application 3
Purpose 2

_____

**1. Construction and application**

Procedure for assessing omitted property was not inapplicable merely because property had not been omitted but was alleged to have been assessed at less than full value when it was assessed as farmland for years at issue. Cherry Hill Indus. Properties v. Voorhees Tp., 186 N.J.Super. 307, 452 A.2d 673 (A.D. 1982) affirmed as modified on other grounds 91 N.J. 526, 453 A.2d 850.

Statutes requiring filing of farmland assessment application by August 1 of pretax year are not ambiguous and reflect clear intention that filing date contained therein be complied with, and word "shall" as used therein is construed as to mean "must" and filing deadline of August 1 of pretax year is mandatory and may not be tolled. Galloway Tp. v. Petkevis, 2 N.J. Tax 85 (1980).

**2. Purpose**

The goals of the Freeze Act (§ 54:2–43), to avoid harassment of taxpayers and to relieve taxpayers of expense of filing petition for two succeeding years, are equally as applicable to valuation aspect of farmland assessment cases as to cases involving valuation of other types of property. Sirota v. Howell Tp., 1 N.J. Tax 280 (1980).

**3. Denial or nullification of application**

Since applications for agricultural assessments were due by August 1 of each pretax year, taxpayer's failure to so file rendered land ineligible on August 1 of each such pretax year and proper time for utilizing added assessments procedure was during each of related tax years. Cherry Hill Indus. Properties v. Voorhees Tp., 186 N.J.Super. 307, 452 A.2d 673 (A.D.1982) affirmed as modified on other grounds 91 N.J. 526, 453 A.2d 850.

Where taxpayer should have been aware of disallowance of its application for farmland assessment in 1975, challenge to that disallowance could not be considered in connection with taxpayer's timely petition concerning disallowance

ASSESSMENT OF REAL ESTATE                    **54:4–23.13b**

of claim for 1976. Princeton Research Lands, Inc. v. Upper Freehold Tp., 4 N.J.Tax 402 (1982).

**4.  Change in valuation**

If any facts set forth in farmland assessment application, or otherwise, indicate that farmland property has changed in value, exceptions to the Freeze Act (§ 54:2–43 [repealed; see, now, § 54:51A–8]) will permit assessor to take action to change assessment of farmland property in any succeeding year. Sirota v. Howell Tp., 1 N.J.Tax 280 (1980).

## 54:4–23.13a.  Revaluation program; inclusion in assessments; applications for lands devoted to agricultural or horticultural use

In any municipality in which a program of revaluation of all property in the municipality has been or shall be undertaken and completed in time to be reflected in the assessments for the next succeeding tax year but not in sufficient time to permit taxpayers to make applications prior to August 1 of the pretax year for the valuation, assessment and taxation of their lands for the ensuing tax year on the basis of being actively devoted to agricultural or horticultural use, any such application which has been or shall be filed with the assessor after August 1, and prior to December 31 of the pretax year, shall be deemed to have been timely made for the tax year next succeeding completion of the revaluation program, notwithstanding any provision to the contrary of the act to which this act is a supplement or of any other law, and the taxes of any applicant whose lands qualify for valuation, assessment and taxation as lands actively devoted to agricultural or horticultural use shall be adjusted accordingly for the tax year commencing January 1 next succeeding completion of the revaluation program and credited or debited, as the case may be, against any taxes due or to become due on such lands.

L.1968, c. 455, § 1, eff. Feb. 21, 1969.  Amended by L.1972, c. 146, § 2, eff. Sept. 7, 1972.

### Historical Note

The 1972 amendment substituted "August 1" for "October 1" throughout the section.

Title of Act:

A Supplement to the "Farmland Assessment Act of 1964," approved May 11, 1964 (P.L.1964, c. 48).  L.1968, c. 455.

## 54:4–23.13b.  Notice of disallowance of claim

Where an application for valuation hereunder has been filed by the owner of land within the time provided herein, the assessor of the taxing district in which such land is situated shall, on or before November 1 of the pretax year, forward to such owner a notice of disallowance by regular mail when a claim has been disallowed. The assessor shall set forth in reason or reasons therefor together

**54:4–23.13b**                    TAXATION OF PROPERTY

with a statement notifying the landowner of his right to appeal such determination to the county board of taxation on or before August 15 of the tax year.

L.1970, c. 237, § 1.

### Historical Note

Section 2 of L.1970, c. 237, approved Oct. 28, 1970, provided:

"This act shall take effect immediately and shall be applicable with respect to applications for the tax year 1971 and thereafter."

Title of Act:

A Supplement to the "Farmland Assessment Act of 1964", approved May 11, 1964 (P.L.1964, c. 48). L.1970, c. 237.

### Administrative Code References

Farmland assessment, notice of disallowance of claim, see N.J.A.C. 18:15–3.6.

### Notes of Decisions

Appeal  4
Construction and application  1
Late notice  2
Reasons for disallowance  3

**1. Construction and application**

Word "shall," within this section providing that a notice of a rejection of a claim for farmland status "shall" be forwarded to the taxpayer on or before November 1 of the pretax year, is by no means conclusive of this section's purpose, which is to declare compliance therewith a mandatory condition precedent to disallowance of a claim for farmland valuation, but is merely a presumption, which can be readily overcome within the context of the purpose of this section in which it appears.  Franklin Estates, Inc. v. Edison Tp., 142 N.J.Super. 179, 361 A.2d 53 (A.D.1976) affirmed 73 N.J. 462, 375 A.2d 658.

**2. Late notice**

Late notice of disallowance of claim for farmland valuation did not result in an automatic grant of such status in absence of evidence that delay had prejudiced taxpayer. Franklin Estates, Inc. v. Edison Tp., 142 N.J.Super. 179, 361

A.2d 53 (A.D.1976) affirmed 73 N.J. 462, 375 A.2d 658.

**3. Reasons for disallowance**

Disparity between reasons given for rejection of claim for farmland evaluation in notice and reasons found by Division of Tax Appeals to sustain rejection did not require that farmland status be granted where no prejudice was established in that taxpayer was offered full opportunity to establish compliance with mandatory statutory criteria as to amount of land being farmed and income derived therefrom, and failure to meet required burden of proof was not even claimed attributable to disparity of reasons given for rejection of claim. Franklin Estates, Inc. v. Edison Tp., 142 N.J.Super. 179, 361 A.2d 53 (A.D.1976) affirmed 73 N.J. 462, 375 A.2d 658.

**4. Appeal**

Section 54:2–43 (repealed; see, now, § 54:51A–8) did not apply to judgments granting farmland assessments; therefore, for tax years immediately following judgment tax year, freeze statute did not relieve owner from filing annual application nor does it excuse him from filing appeal from denial of farmland assessment. Belmont v. Wayne Tp., 5 N.J.Tax 110 (1983).

### 54:4–23.14.  Form of application; certification

Application for valuation, assessment and taxation of land in agricultural or horticultural use under this act shall be on a form prescribed by the Director of the Division of Taxation, and provided for the use of claimants by the governing bodies of the respective taxing districts.  The form of application shall provide for the reporting of information pertinent to the provisions of Article VIII, Section 1, paragraph 1(b) of the Constitution, as amended, and this act.  A certification by the landowner that the facts set forth in the application are true may be prescribed by the director to be in lieu of a sworn statement to that effect.  Statements so certified shall be considered as if made under oath and subject to the same penalties as provided by law for perjury.

L.1964, c. 48, § 14.

#### Cross References

Perjury and other falsification in official matters, see § 2C:28–1 et seq.

#### Administrative Code References

Application for farmland assessment, see N.J.A.C. 18:15–2.1 et seq.

#### Library References

Taxation ☞348.1(3).
C.J.S. Taxation § 411.

#### Notes of Decisions

**1.  Gross income**

Taxpayers' application for farmland assessment was not defective on ground that they failed or refused to substantiate required gross income; while such nonaction may be a basis for denial of application, it cannot be considered jurisdictional deficiency which would preclude review on the merits. Checchio v. Scotch Plains Tp., 2 N.J. Tax 450 (1981).

### 54:4–23.15.  Continuance of valuation, assessment and taxation under act

Continuance of valuation, assessment and taxation under this act shall depend upon continuance of the land in agricultural or horticultural use and compliance with the other requirements of this act and not upon continuance in the same owner of title to the land. Liability to the roll-back tax shall attach when a change in use of the land occurs but not when a change in ownership of the title takes place if the new owner continues the land in agricultural or horticultural use, under the conditions prescribed in this act.

L.1964, c. 48, § 15.

**54:4–23.15**                    TAXATION OF PROPERTY

**Administrative Code References**

Continuance of farmland assessment, see N.J.A.C. 18:15–10.1.

**Library References**

Taxation ⟺348.1(3).
C.J.S. Taxation § 411.

**Notes of Decisions**

**1.  In general**

Taxpayers, who requested rescission of subdivision approval on property which had been subject to farmland assessment prior to obtaining subdivision approval, were entitled to reinstatement of farmland assessment solely by virtue of their request for rescission of subdivision approval, even though their request was ultimately denied, where property was maintained throughout for farmland use. C.O.W. Const. Co. v. East Brunswick Tp., 2 N.J. Tax 556 (1981).

Res judicata was not applicable to bind Tax Court to follow county board of taxation's 1973 judgment granting farmland assessments to tracts of land, in that findings of fact and conclusions of law arrived at by a county board of taxation are not binding upon Tax Court, and a judgment concluding that land is entitled to benefits of Farmland Assessment Act for any prior year cannot bar a municipality from questioning its qualification in future years. Bass River Tp. v. Hogwallow, Inc., 1 N.J. Tax 612 (1980).

**54:4–23.15a.  Mailing of form to claim continuance of valuation, assessment and taxation; notice of filing requirement**

On or before July 1 the assessor shall mail to each taxpayer whose land has been valued, assessed, and taxed for the then current tax year pursuant to the "Farmland Assessment Act of 1964" [1] a copy of the form prescribed to claim a continuance of valuation, assessment and taxation under such act for the succeeding tax year together with a notice that the completed form is required to be filed with the assessor on or before August 1.

The failure of any taxpayer to receive a form for claiming continuance of a farmland assessment shall not relieve him of the requirement to claim and establish his right thereto as required by law.

L.1971, c. 400, § 1, eff. Jan. 10, 1972.

[1] Sections 54:4–23.1 to 54:4–23.23.

**Historical Note**

**Title of Act:**
A Supplement to the "Farmland Assessment Act of 1964," approved May 11, 1964 (P.L.1964, c. 48, C. 54:4–23.1 et seq.). L.1971, c. 400.

ASSESSMENT OF REAL ESTATE                    54:4-23.18

**Notes of Decisions**

**1. Filing**

Statutes (including § 54:4-23.13) requiring filing of farmland assessment application by August 1 of pretax year are not ambiguous and reflect clear intention that filing date contained therein be complied with, and word "shall" as used therein is construed as to mean "must" and filing deadline of August 1 of pretax year is mandatory and may not be tolled. Galloway Tp. v. Petkevis, 2 N.J. Tax 85 (1980).

## 54:4-23.16.  Separation or split off of part of land

Separation or split off of a part of the land which is being valued, assessed and taxed under this act, either by conveyance or other action of the owner of such land, for a use other than agricultural or horticultural, shall subject the land so separated to liability for the roll-back taxes applicable thereto, but shall not impair the right of the remaining land to continuance of valuation, assessment and taxation hereunder, provided it meets the 5-acre minimum requirement and such other conditions of this act as may be applicable.

L.1964, c. 48, § 16.

**Administrative Code References**

Liability for roll-back taxes on land split-off, see N.J.A.C. 18:15-11.1.

## 54:4-23.17.  Repealed by L.1970, c. 243, § 3

**Historical Note**

Effective date of repeal, see Historical Note under § 54:4-23.6 note.

The repealed section was added by L.1964, c. 48, § 17, and related to imposition of roll-back taxes on land taken by eminent domain.

## 54:4-23.18.  Location of contiguous land in more than one taxing district

Where contiguous land in agricultural or horticultural use in one ownership is located in more than one taxing district, compliance

54:4–23.18                    TAXATION OF PROPERTY

with the 5-acre minimum area requirement shall be determined on the basis of the total area of such land and not the area which is located in the particular taxing district.

L.1964, c. 48, § 18.

### 54:4–23.19.  Tax list and duplicate; factual details

The factual details to be shown on the assessor's tax list and duplicate with respect to land which is being valued, assessed and taxed under this act shall be the same as those set forth by the assessor with respect to other taxable property in the taxing district.

L.1964, c. 48, § 19.

#### Administrative Code References

Recordation of taxable value, see N.J.A.C. 18:15–5.3.

#### Library References

Taxation ⟨⇒411 et seq.
C.J.S. Taxation § 456.

#### Notes of Decisions

1. In general

Legislature mandated that taxing districts must pay particular attention to fact that certain property has been assessed according to Farmland Assessment Act; consequently taxing districts must state definitively whether or not the subject property has been assessed under Farmland Assessment Act in last three years. Mahwah Tp. v. Bergen County, 2 N.J. Tax 479 (1981).

### 54:4–23.20.  State farmland evaluation advisory committee

There is hereby created a State Farmland Evaluation Advisory Committee, the members of which shall be the Director of the Division of Taxation; the Dean of the College of Agriculture, Rutgers, The State University; and the Secretary of Agriculture. The committee shall meet from time to time on the call of the Secretary of Agriculture and annually determine and publish a range of values for each of the several classifications of land in agricultural and horticultural use in the various areas of the State. The primary objective of the committee shall be the determination of the ranges in fair value of such land based upon its productive capabilities when devoted to agricultural or horticultural uses. In making these annual determinations of value, the committee shall consider available evidence of agricultural or horticultural capability derived from the soil survey at Rutgers, The State University,

506

## ASSESSMENT OF REAL ESTATE                54:4–23.21

the National Co-operative Soil Survey, and such other evidence of value of land devoted exclusively to agricultural or horticultural uses as it may in its judgment deem pertinent. On or before October 1 of each year, the committee shall make these ranges of fair value available to the assessing authority in each of the taxing districts in which land in agricultural and horticultural use is located.

L.1964, c. 48, § 20.

### Administrative Code References

State farmland evaluation committee, see N.J.A.C. 18:15–14.1 et seq.

### Library References

Taxation ☞313.
C.J.S. Taxation § 375.

### Notes of Decisions

Adoption of determination    1
Use of determination    2

**1. Adoption of determination**

Where trial judge adopted both assessment rate recommended by farmland evaluation advisory committee in its report for farmland in the county wherein lands were located, but report did nothing more than establish ranges of suggested values for each of several classifications of farmlands in various areas of state, values which were merely designed as guidelines for assessor, and did not provide an accurate means of attributing a specific value to a specific piece of property, trial judge's adoption of lowest recommended value for farmlands in county, without more, in fixing value of petitioner's farmlands was totally arbitrary. Bunker Hill Cranberry Co., Inc. v. Jackson Tp., Ocean County, 144 N.J.Super. 230, 365 A.2d 204 (A.D. 1976) certification denied 73 N.J. 59, 372 A.2d 324.

**2. Use of determination**

Action of Division of Tax Appeals in classifying certain forestland as "cropland harvested" and therefore valuing it at $720 per acre, the highest suggested valuation for farmland in the county, in which cropland was valued at from $60 to $720 per acre depending on soil rating and woodland from $21 to $33, was error, in that only applicable suggested valuations submitted to division were those for woodland, and no other evidence or method of valuation was before division to support determination. Urban Farms, Inc. v. Wayne Tp., Passaic County, 159 N.J.Super. 61, 386 A.2d 1357 (A.D.1978).

Recommendations of state farmland evaluation advisory committee, which assessor is required to consider in determining value for farmland, are not mandated but are only guidelines which are to be considered along with assessor's personal knowledge, judgment and experience. Pinson v. Bernards Tp., 1 N.J. Tax 351 (1980).

## 54:4–23.21.  Rules and regulations; forms

The director is empowered to promulgate such rules and regulations and to prescribe such forms as he shall deem necessary to effectuate the purposes of this act.

L.1964, c. 48, § 21.

## FARMLAND TRESPASS STATUTES

C 521-1                                                    R. S. 4:17-2 et al.

P. L. 1983, CHAPTER 521, *approved January 17, 1984*

1982 Assembly No. 2141 (*Second Official Copy Reprint*)

AN ACT concerning trespass on agricultural or horticultural lands, amending R. S. 4:17-2, and repealing R. S. 4:17-1.

1  BE IT ENACTED *by the Senate and General Assembly of the State*
2  *of New Jersey:*

1  1. R. S. 4:17-2 is amended to read as follows:

2  4:17-2. [A person who shall trespass upon the cultivated lands
3  of another within this State after being forbidden so to do by the
4  owner, occupant, lessee or licensee thereof, or after public notice
5  given as provided in this section on the part of the owner, occupant,
6  lessee or licensee of the lands forbidding such trespass, shall be
7  subject to a fine of not more than twenty dollars ($20.00) and
8  costs, and the county district courts and the municipal courts shall
9  have jurisdiction to try such offenders and pronounce sentence as
10 provided in this article.
11 The notice shall bear the name of the owner, occupant, lessee or
12 licensee of the lands, and be posted conspicuously on the corners
13 of the property or along the roads or highways abutting the prop-
14 erty, on all boundary lines bordering on adjoining lands and at each
15 point where a stream enters upon or leaves the property.] *Any*
16 *person who trespasses upon the agricultural or horticultural lands*
17 *of another***[*, regardless of whether public notice is posted.*]** *is*
18 *liable to a penalty of not less than $100.00, to be collected in a civil*
19 *action by a summary proceeding under "the penalty enforcement*

EXPLANATION—Matter enclosed in bold-faced brackets [thus] in the above bill
  is not enacted and is intended to be omitted in the law.
    Matter printed in italics *thus* is new matter.
Matter enclosed in asterisks or stars has been adopted as follows:
    *—Assembly committee amendments adopted February 14, 1983.
    **—Senate committee amendments adopted June 23, 1983.

20  law" (N. J. S. 2A:58–1 et seq.). The Superior Court and county
21  district court shall have jurisdiction to enforce the "penalty enforce-
22  ment law." If the violation is of a continuing nature, each day dur-
23  ing which it continues constitutes an additional, separate and dis-
24  tinct offense. As used in this act, "agricultural or horticultural
25  lands" means lands devoted to the production for sale of plants and
26  animals useful to man, encompassing plowed or tilled fields, stand-
27  ing crops or their residues, **cranberry** bogs **and appurtenant
28  dams, dikes, canals, ditches and pump houses, including impound-
29  ments, man-made reservoirs and the adjacent shorelines thereto**,
30  orchards, **nurseries** and lands with a maintained fence for the
31  purpose of restraining domestic livestock. **[*This section shall
32  not apply to a person in possession of a valid hunting, fishing or
33  trapping license engaged in an activity authorized by one of these
34  licenses.*]** **"Agricultural or horticultural lands" shall also in-
35  clude lands in agricultural use, as defined in section 3 of P. L. 1983,
36  c. 32 (C. 4:1C–13) where public notice prohibiting trespass is given
37  by actual communication to the actor, conspicuous posting, or fenc-
38  ing or other enclosure manifestly designed to exclude intruders.
39  Nothing in this act shall relieve owners of agricultural or horti-
40  cultural lands from the obligation to provide conspicuous posting
41  prohibiting trespass on the waters or banks along or around any
42  waters listed for stocking with fish in the current fish code adopted
43  pursuant to section 32 of P. L. 1948, c. 448 (C. 13:1B–30) before a
44  trespass violation may be found.**

1    2. R. S. 4:17–1 is repealed.

1    3. This act shall take effect immediately.

# Part F

# MODEL ORDINANCE TO CONTROL FEEDING OF CANADIAN GEESE
# AND OTHER MIGRATORY WATERFOWL

Introduced by _____

First Reading _____   Second Reading _____

Ordinance No. _____   Council Bill No. _____

### AN ORDINANCE

enacting a new definition to Section 5.031 and enacting new Sections
5.053, 5.054, 5.055, 5.056 and 5.057, relating to prohibiting the
feeding of migratory waterfowl or creating any condition causing a
congestion of said waterfowl which results in an accumulation of
waterfowl droppings or damage to plants or property or which results
in a threat or nuisance to the public health, safety or welfare; and
fixing the time when this ordinance shall become effective.

BE IT ORDAINED BY THE COUNCIL OF THE CITY OF COLUMBIA, MISSOURI, AS FOLLOWS:

SECTION 1.   That Chaper 5 of the Revised Ordinance of the City of Columbia, Missouri,
1964, is hereby amended by enacting a new definition in Section 5.031 and enacting new
Sections 5.053, 5.054, 5.055, 5.056 and 5.057, relating to prohibiting the feeding of
migratory waterfowl or creating any condition causing a congestion of said waterfowl which
results in an accumulation of waterfowl droppings or damage to plants or property or which
results in a threat or nuisance to the public health, safety or welfare, reading in words
and figures as follows:

CHAPTER 5.   ANIMALS AND FOWL

. . .

SECTION 5.031  Definitions

. . .

MIGRATORY WATERFOWL shall include those species of birds, not otherwise defined as a
domesticated animal in this chapter, commonly known as swans, geese, brants, river and sea
ducks, and any other waterfowl falling under the jurisdiction of the Missouri Conservation
Commission or otherwise defined by the Commission as migratory waterfowl.

. . .

SECTION 5.053  Feeding Waterfowl

It shall be unlawful for any person to:

A.   Feed any migratory waterfowl.

B.   Create any condition, or allow any condition to exist, which results in a congregation
     or congestion of migratory waterfowl which:

   1.   Results in an accumulation of waterfowl feces or droppings, or

   2.   Results in damage to flora, fauna, or private or public property, or

   3.   Results in a threat or nuisance to the public health, safety or welfare, or

   4.   Results in a threat to the health, safety or welfare of said migratory waterfowl.

SECTION 5.054   Health Director's Duty to Abate a Waterfowl Nuisance

Whenever the Health Director becomes aware that migratory waterfowl are being fed in violation of city ordinance, or that a congregation or congestion of migratory waterfowl as
prohibited by ordinance exists, he shall:

A.  After notifying and consulting with the appropriate Missouri Conservation Commission
    official, declare that a nuisance exists, and

B.  Shall give notice of the nuisance to the owner or occupant of the premises upon which
    the nuisance exists or the feeding is occurring.  Notice shall be deemed sufficient if
    it shall have been served either personally, or by posting such notice upon the prop-
    erty in question, or by mailing such notice through the United States mail; mailed
    notice to be deemed served upon proof that such notice was deposited in the United
    States mail, first class postage prepaid and addressed, in the case of an individual,
    to his business address, residence address or such address as the records of the
    Assessor shall reveal to be his address for tax purposes; or in the case of a corpora-
    tion, to the registered office of agent of such corporation as the records of the
    Secretary of State of the State of Missouri shall reveal.  Such notice shall order the
    nuisance to be immediately abated and/or that such feeding shall immediately stop.

SECTION 5.055   Director's Power to Remove a Congestion

A.  The Health Director shall have all powers necessary, upon application by a citizen, or
    at his own discretion, to disperse or remove any congregation or congestion of migra-
    tory waterfowl by any lawful means.

B.  Before taking any steps to remove or disperse any congestion or congregation of migra-
    tory waterfowl, the Director shall meet and confer with the appropriate Conservation
    Commission official and with the advice and consent of that official determine the
    appropriate method and means of removing or dispersing the migratory waterfowl
    considering first the health, safety and welfare of the citizens of Columbia and,
    second, the health, safety and welfare of the migratory waterfowl.

C.  Whenever the Director, after consultation with the Department of Conservation, deter-
    mines that waterfowl shall be dispersed by scare tactics or noise tactics, he may dele-
    gate implementation of the plan developed by his consultations to any responsible citi-
    zen or citizens in the affected neighborhood.  This paragraph shall not be construed to
    authorize any private citizen to discharge any firearm or to use any device which
    launches a projectile of any type.

D.  The Director shall apply for and secure all permits necessary by law prior to taking
    any action.

SECTION 5.056   Avian Nuisances

Whenever the Health Director becomes aware that birds, not otherwise protected, regulated
or controlled by Federal or State law, rule or regulation, are congregating or flocking
together in such numbers as to cause a health hazard or are interfering with the peace,
quiet, comfort or repose of the inhabitants of the City of Columbia, he may take whatever
steps are necessary to remove or disperse the congregation or flock.

SECTION 5.057   Health Director May Delegate Certain Powers

In order to remove or disperse congregations or flocks of birds constituting a nuisance as
defined by ordinance, the Health Director may delegate the implementation of his dispersal
plan to any responsible citizen or citizens in the affected neighborhood by permit
specifying the name of the permittee, date of issuance, date of expiration, and details of
the plan so delegated.

This ordinance shall not be construed to empower the Health Director to authorize any private citizen to discharge any firearm or to use any device which launches a projectile of any type.

SECTION 2.   That this ordinance shall be in full force and effect from and after its passage.

Source:  U.S. Fish and Wildlife Service

# Index